Elementary Technical Mathematics

TENTH EDITION

Dale Ewen
Parkland Community College

C. Robert Nelson
Champaign Centennial High School

 BROOKS/COLE
CENGAGE Learning™

Australia • Brazil • Japan • Korea • Mexico • Singapore • Spain • United Kingdom • United States

BROOKS/COLE
CENGAGE Learning

Elementary Technical Mathematics,
   Tenth Edition
Dale Ewen / C. Robert Nelson

Math Editor: Marc Bove
Assistant Editor: Stefanie Beeck
Editorial Assistant: Kyle O'Loughlin
Media Editor: Heleny Wong
Marketing Manager: Ashley Pickering
Marketing Assistant: Angela Kim
Marketing Communications Manager:
   Mary Anne Payumo
Content Project Manager: Jerilyn Emori
Creative Director: Rob Hugel
Art Director: Vernon Boes
Print Buyer: Linda Hsu
Rights Acquisitions Account Manager, Text:
   Bob Kauser
Rights Acquisitions Account Manager,
   Image: Don Schlotman
Production Service: Lynn Steines,
   S4Carlisle Publishing Services
Text and Cover Designer: Roy Neuhaus
Photo Researcher: Jennifer Lim, Bill Smith
   Group
Copy Editor: Lorretta Palagi
Cover Image: Clockwise, from top: David
   Joel/Getty Images, Mark Richards/
   PhotoEdit, Monty Rakusen/Getty
   Images, Jupiterimages/Getty Images,
   David Young-Wolff/PhotoEdit,
   Thinkstock Images/Getty Images,
   Stockbyte/Getty Images
Compositor: S4Carlisle Publishing Services

For product information and technology assistance, contact us at
Cengage Learning Customer & Sales Support, 1-800-354-9706.

For permission to use material from this text or product, submit all requests online at **www.cengage.com/permissions.**
Further permissions questions can be e-mailed to
permissionrequest@cengage.com.

Library of Congress Control Number: 2009923966

Student Edition:
ISBN-13: 978-1-4390-4689-0
ISBN-10: 1-4390-4689-1

Brooks/Cole
10 Davis Drive
Belmont, CA 94002-3098
USA

Cengage Learning is a leading provider of customized learning solutions with office locations around the globe, including Singapore, the United Kingdom, Australia, Mexico, Brazil, and Japan. Locate your local office at **www.cengage.com/global.**

Cengage Learning products are represented in Canada by Nelson Education, Ltd.

To learn more about Wadsworth, visit **www.cengage.com/brookscole**

Purchase any of our products at your local college store or at our preferred online store **www.ichapters.com.**

Printed in the United States of America
1 2 3 4 5 6 7 13 12 11 10 09

Contents

16   Binary and Hexadecimal Numbers   565

Appendixes

List of Applications

Preface

Elementary Technical Mathematics, Tenth Edition, is intended for technical, trade, allied health, or Tech Prep programs. This book was written for students who plan to learn a technical skill, but who have minimal background in mathematics or need considerable review. To become proficient in most technical programs, students must learn basic mathematical skills. To that end, Chapters 1 through 4 cover basic arithmetic operations, fractions, decimals, percent, the metric system, and numbers as measurements. Chapters 5 through 11 present essential algebra needed in technical and trade programs. The essentials of geometry—relationships and formulas for the most common two- and three-dimensional figures—are given in detail in Chapter 12. Chapters 13 and 14 present a short but intensive study of trigonometry that includes right-triangle trigonometry as well as oblique triangles and graphing. The concepts of statistics that are most important to technical fields are discussed in Chapter 15. An introduction to binary and hexadecimal numbers is found in Chapter 16 for those who requested this material.

We have written this text to match the reading level of most technical students. Visual images engage these readers and stimulate the problem-solving process. We emphasize that these skills are essential for success in technical courses.

The following important text features have been retained from previous editions:

- We use a large number of applications from a wide variety of technical areas, including auto/diesel mechanics, industrial and construction trades, electronics, agriculture, allied health, CAD/drafting, HVAC, manufacturing, welding, aviation, and natural resources.

- Chapter 1 reviews basic concepts in such a way that individuals, groups of students, or the entire class can easily study only those sections they need to review.

- A comprehensive introduction to basic algebra is presented for those students who need it as a prerequisite to more advanced algebra courses. However, the book has been written to allow the omission of selected sections or chapters without loss of continuity, to meet the needs of specific students.

- More than 6340 exercises assist student learning of skills and concepts.

- More than 720 detailed, well-illustrated examples, many with step-by-step comments, support student understanding of skills and concepts.

- A chapter summary with a glossary of basic terms, a chapter review, and a chapter test appear at the end of each chapter as aids for students in preparing for quizzes and exams. Each chapter test is designed to be completed by an average student in no more than approximately 50 minutes.

Chapter 3   Review

Give the metric prefix for each value:

1. 0.001   **2.** 1000

Give the SI abbreviation for each prefix:

3. mega   **4.** micro

Write the SI abbreviation for each quantity:

5. 42 millilitres   **6.** 8.3 nanoseconds

Write the SI unit for each abbreviation:

7. 18 km   **8.** 350 mA   **9.** 50 μs

Which is larger?

10. 1 L or 1 mL   **11.** 1 kW or 1 MW
12. 1 km² or 1 ha   **13.** 1 m³ or 1 L

Fill in each blank:

14. 650 m = _____ km   **15.** 750 mL = _____ L
16. 6.1 kg = _____ g   **17.** 4.2 A = _____ μA
18. 18 MW = _____ W   **19.** 25 μs = _____ ns

28. 180 lb = _____ kg
29. 126 ft = _____ m
30. 360 cm = _____ in.
31. 275 in² = _____ cm²
32. 18 yd² = _____ ft²
33. 5 m³ = _____ ft³
34. 15.0 acres = _____ ha

Choose the most reasonable quantity:

35. Jorge and Maria drive **a.** 1600 cm, **b.** 470 m, **c.** 12 km, or **d.** 2400 mm to college each day.
36. Chuck's mass is **a.** 80 kg, **b.** 175 kg, **c.** 14 μg, or **d.** 160 Mg.
37. A car's gas tank holds **a.** 18 L, **b.** 15 kL, **c.** 240 mL, or **d.** 60 L of gasoline.
38. Jamilla, being of average height, is **a.** 5.5 m, **b.** 325 mm, **c.** 55 cm, or **d.** 165 cm tall.
39. A car's average gas consumption is **a.** 320 km/L, **b.** 15 km/L, **c.** 35 km/L, or **d.** 0.75 km/L.

Chapter 3   Test

1. Give the metric prefix for 1000.
2. Give the metric prefix for 0.01.
3. Which is larger, 200 mg or 1 g?
4. Write the SI unit for the abbreviation 240 μL.
5. Write the abbreviation for 30 hectograms.
6. Which is longer, 1 km or 25 cm?

Fill in each blank:

7. 4.25 km = _____ m   **8.** 7.28 mm = _____ μm
9. 72 m = _____ mm   **10.** 256 hm = _____ cm
11. 12 dg = _____ mg   **12.** 16.2 g = _____ mg
13. 7.236 metric tons = _____ kg
14. 310 g = _____ cg   **15.** 72 hg = _____ mg
16. 1.52 dL = _____ L   **17.** 175 L = _____ m³
18. 2.7 m³ = _____ cm³   **19.** 400 ha = _____ km²
20. 0.2 L = _____ mL
21. What is the basic SI unit of time?

Fill in each blank:

23. 280 W = _____ kW
24. 13.9 mA = _____ A
25. 720 ps = _____ ns
26. What is the basic SI unit for temperature?
27. What is the freezing temperature of water on the Celsius scale?

Fill in each blank, rounding each result to three significant digits when necessary:

28. 25°C = _____ °F   **29.** 28°F = _____ °C
30. 98.6°F = _____ °C   **31.** 100 km = _____ mi
32. 200 cm = _____ in.   **33.** 1.8 ft³ = _____ in³
34. 37.8 ha = _____ acres   **35.** 80.2 kg = _____ lb

- The text design and second color help to make the text more accessible, highlight important concepts, and enhance the art presentation.
- A reference of useful, frequently referenced information—such as metric system prefixes, U.S. weights and measures, metric and U.S. conversion, and formulas from geometry—is printed on the inside covers.

- The use of a scientific calculator has been integrated in an easy-to-use format with calculator flowcharts and displays throughout the text to reflect its nearly universal use in technical classes and on the job. The instructor should inform the students when *not* to use a calculator.

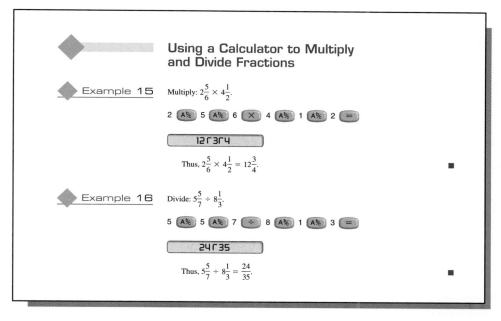

- Cumulative reviews are provided at the end of every even-numbered chapter to help students review for comprehensive exams.

- Studies show that current students will experience several career changes during their working lives. The chapter-opening pages illustrate various career paths for students to consider as their careers, technology, and the workplace evolve. The basic information provided in the chapter openers about a technical career is explored in further detail on the Brooks/Cole book companion website at www.cengage.com/mathematics/ewen.

Mathematics at Work

The nation's construction industry depends on a technical and competent workforce. This workforce includes, but is not limited to, carpenters who cut, fit, and assemble wood and other materials in construction projects; plumbers, pipefitters, and steamfitters who install, maintain, and repair many different types of pipe systems that carry water, steam, air, and other liquids; painters who apply paint, stain, varnish, and other finishes to buildings and other structures; electricians who install, maintain, and repair electrical wiring, equipment, and fixtures; bricklayers and stonemasons who build walls and other structures with bricks, blocks, stones, and other masonry materials; and structural and reinforcing metal workers who use materials made from iron, steel, and other materials to construct highways, bridges, buildings, and towers.

Construction trade workers often learn their own trade through apprenticeship programs administered by local joint union–management committees or through community college or trade school programs, some of which are offered in partnership with the local joint union–management committees. For more information, go to the website listed below.

Construction Trades
Carpenter framing a building.

- Special application exercises in the areas of auto/diesel mechanics, industrial and construction trades, electronics, agriculture, allied health, CAD/drafting, HVAC, manufacturing, welding, aviation, and natural resources have been submitted by faculty in these technical areas and are marked with related icons.

15. Approximately how many studs are needed for the exterior walls in the building shown in Illustration 1? (See Example 4.)

ILLUSTRATION 1

16. A pipe 24 ft long is cut into four pieces: the first 4 ft long, the second 5 ft long, and the third 7 ft long. What is the length of the remaining piece? (Assume no waste from cutting.)

17. A welder needs to weld together pipes of lengths 10 ft, 15 ft, and 14 ft. What is the total length of the new pipe?

18. A welder ordered a 125-ft³ cylinder of argon gas, a shielding gas for TIG welding. After a few days, only 78 ft³ remained. How much argon was used?

Divide (use the remainder form with r):

27. $4\overline{)7236}$ **28.** $5\overline{)308,736}$

29. $4668 \div 12$ **30.** $15,648 \div 36$

31. $67,560 \div 80$ **32.** $\dfrac{188,000}{120}$

33. A car uses gas at the rate of 31 miles per gallon (mi/gal or mpg) and has a 16-gallon tank. How far can it travel on one tank of gas?

34. A car uses gas at a rate of 12 kilometres per litre (km/L) and has a 65-litre tank. How far can it travel on one tank of gas?

35. A four-cylinder engine has a total displacement of 1300 cm³. Find the displacement of each piston.

36. A car travels 1274 mi and uses 49 gal of gasoline. Find its mileage in miles per gallon.

37. A car travels 2340 km and uses 180 L of gasoline. Find its gas consumption in kilometres per litre.

38. To replace some damaged ductwork, 20 linear feet of 8-in. × 16-in. duct is needed. The cost is $13 per 4 linear feet. What is the cost of replacement?

- Group activity projects are included at the end of each chapter.

 Chapter 3 Group Activities

1. Mathematics is used in a variety of places. One location where mathematics is used frequently is in the medical profession. In small groups, brainstorm about the places in a hospital where you think math is used. Think of the different departments and the different professions in the hospital such as radiology, general surgery, etc. After you have thought about this, divide and go to a hospital to check your theory of where and how math is used. Get permission from the proper authorities to ask the employees how they use math. One example is pediatricians who use math in prescribing medication to children. They must be careful to get the weight of a child and use this information to prescribe the proper dosage. The prescription notifies the pharmacist of the amount of medication to give the patient. Make a report on your findings of how math is used in the medical field and make special note of the conversions that doctors and nurses must use. Plan a similar activity for another workplace/profession.

2. Do the following:
   a. Write how old you are to the day. Convert this to days. Convert this to hours and then to minutes.
   b. Write how tall you are. Convert this to feet, to yards, to inches, to metres, and to centimetres.
   c. Write how much you weigh. Convert this to kilograms and to grams.

   Do a little research and see what gravity is on earth and how your weight is determined by gravity. Further research what gravity is on the moon and how your weight would differ on the moon compared to on earth. ($W = mg$)

- An instructor's edition that includes all the answers to exercises is available.

Significant changes in the tenth edition include the following:

- The following topics were added by special requests of users:
- **New** category of natural resources application exercises that includes forestry, soil management, wildlife management, parks, recycling, and related areas
- **New** Section 1.16, Applications Involving Percent: Personal Finance
- **New** Section 15.9, Standard Deviation for Grouped Data, and other changes/ updates in Chapter 15
- **New** Appendix B Exponential Equations
- Signed number drill exercises have been added to assist students to learn addition, subtraction, and multiplication of signed numbers
- More than 330 new exercises have been added.
- Chapter objectives have been added.

Useful ancillaries available to qualified adopters of this text include the following:

- **Instructor's Edition** The Instructor's Edition features an appendix containing the answers to all problems in the book. (1-4390-4724-3)
- **PowerLecture™ CD-ROM with ExamView®** This CD-ROM provides dynamic media tools for teaching. Create, deliver, and customize tests (both print and online) in minutes with ExamView® Computerized Testing Featuring Algorithmic Equations. Easily build solution sets for homework or exams using Solution Builder's online solutions manual. Microsoft® PowerPoint® lecture slides, figures from the book, and a Test Bank, in electronic format, are also included. (1-4390-4752-9)
- **Solutions Builder** Easily build solution sets for homework or exams using Solution Builder's online solutions manual. (1-4390-4753-7)

- **WebAssign** WebAssign, the most widely used homework system in higher education, allows you to assign, collect, grade, and record homework assignments via the Web. Through a partnership between WebAssign and Cengage Learning Brooks/Cole, this proven homework system has been enhanced to include links to textbook sections, video examples, and problem-specific tutorials. (0-538-73899-5)

We are grateful for the courtesy of the L. S. Starrett Company in allowing us to use photographs of their instruments in Chapter 4. The authors also thank the many faculty members who used earlier editions and who offered suggestions. In particular, we thank William G. Camp, Professor, Cornell University and Professor Emeritus, Virginia Tech and Martin Alderman, Cornell University PhysTEC Teacher in Residence for writing natural resources applications problems and the following reviewers: Amir F. Arabi, Central Virginia Community College; Cynthia Broughton, Arizona Western College; Nancy Jo Buchli, Southeast Community College–Milford; James Carpenter, College of the Mainland; Amy Curry, College of Lake County; Royetta S. Ealba, Henry Ford Community College; Jonathan Greer, Grand Rapids Community College; Mehran Hassanpour, South Texas Community College; Paul McCombs, Rock Valley College; Gray McCracken, Shelton State Community College; Lorie McFee, North Buncomble High School; Carol McVey, Florence-Darlington Technical College; Lara Michaels, Green River Community College; Linda Nokes, Southwestern Michigan College; Arthur M. Peck, Lane Community College; Catherine Pellish, Front Range Community College; Gary Rattray, Central Maine Community Collge; Fran Seigle, Lakes Region Community College; Richard Watikins, Tidewater Community College; Emily E. White, Enka High School; and Carol L. Williams, Des Moines Area Community College.

Anyone wishing to correspond regarding suggestions or questions should write Dale Ewen through the publisher.

For all their help, we thank our editor, Marc Bove; assistant editor, Stefanie Beeck; and the staff of Cengage Learning Brooks/Cole. We also greatly appreciate the diligent, personal, and professional efforts of Lynn Steines, S4Carlisle Publishing Services, in coordinating production; Lorretta Palagi for a great job copy editing; and Curtis Nunn for checking the answers.

Dale Ewen
C. Robert Nelson

1 Basic Concepts

Mathematics at Work

Automotive service technicians inspect, maintain, and repair automobiles, light trucks, and vans. In the past, these workers were called mechanics. The increasing sophistication of automotive technology now requires workers to be able to use computerized shop equipment and work with electronic components in addition to the traditional hand tools. When a mechanical or electronic problem occurs, the technician uses a diagnostic approach to repair the problem based on information from the owner and the information obtained from the service equipment and computerized databases and service manuals.

The National Automotive Technicians Education Foundation (NATEF), an affiliate of the National Institute for Automotive Service Excellence (ASE), certifies automotive service technician, collision repair and refinish technician, engine specialist, and medium/heavy truck technician training programs offered by community colleges, postsecondary trade schools, technical institutes, and high schools. Although voluntary, NATEF certification signifies that the program meets uniform standards for instructional facilities, equipment, staff credentials, and curriculum. Various automobile manufacturers and their participating dealers also sponsor two-year associate degree programs at postsecondary schools across the United States. For more information, go to the website listed below.

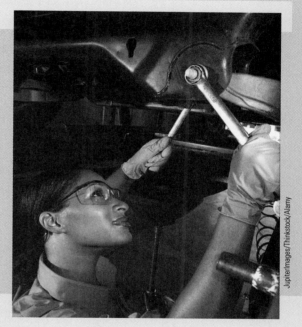

Automotive Service Technician
Automotive service technician working on an automobile.

> **Objectives**
>
> ■ Add, subtract, multiply, and divide whole numbers.
>
> ■ Add, subtract, multiply, and divide whole numbers with a scientific calculator.
>
> ■ Apply the rules for order of operations.
>
> ■ Find the area and volume of geometric figures.
>
> ■ Evaluate formulas.
>
> ■ Find the prime factorization of whole numbers.
>
> ■ Add, subtract, multiply, and divide fractions.
>
> ■ Add, subtract, multiply, and divide fractions with a scientific calculator.
>
> ■ Use conversion factors to change from one unit to another within the U.S. system of weights and measures.
>
> ■ Add, subtract, multiply, and divide decimal fractions.
>
> ■ Add, subtract, multiply, and divide decimal fractions with a scientific calculator.
>
> ■ Round numbers to a particular place value.
>
> ■ Apply the percent concept; change a percent to a decimal, a decimal to a percent, a fraction to a percent, and a percent to a fraction.
>
> ■ Solve application problems involving the addition, subtraction, multiplication, and division of whole numbers, fractions, and decimal fractions and percents.
>
> ■ Find powers and roots of numbers using a scientific calculator.
>
> ■ Solve personal finance problems involving percent.

Unit 1A

REVIEW OF OPERATIONS WITH WHOLE NUMBERS

1.1   Review of Basic Operations

The **positive integers** are the numbers 1, 2, 3, 4, 5, 6, and so on. They can also be written as +1, +2, +3, and so on, but usually the *positive* (+) sign is omitted. The **whole numbers** are the numbers 0, 1, 2, 3, 4, 5, 6, and so on. That is, the whole numbers consist of the positive integers and zero.

The value of any digit in a number is determined by its place in the particular number. Each place represents a certain power of ten. By powers of ten, we mean the following:

$$10^0 = 1$$
$$10^1 = 10$$

$10^2 = 10 \times 10 = 100$ (the second power of 10)
$10^3 = 10 \times 10 \times 10 = 1000$ (the third power of 10)
$10^4 = 10 \times 10 \times 10 \times 10 = 10{,}000$ (the fourth power of 10) and so on.

Note: A small superscript number (such as the 2 in 10^2) is called an *exponent*.

The number 2354 means 2 thousands plus 3 hundreds plus 5 tens plus 4 ones.

In the number 236,895,174, each digit has been multiplied by some power of 10, as shown below.

	(ten millions)		(hundred thousands)		(thousands)		(tens)	
	10^7		10^5		10^3		10^1	
	|		|		|		|	
2	3	6,	8	9	5,	1	7	4
	|		|		|		|	
10^8		10^6		10^4		10^2		10^0
(hundred millions)		(millions)		(ten thousands)		(hundreds)		(units)

The "+" (plus) symbol is the sign for addition, as in the expression $5 + 7$. The result of adding the numbers (in this case, 12) is called the **sum**. Integers are added in columns with the digits representing like powers of ten in the same vertical line. (*Vertical* means up and down.)

◆ **Example 1**

Add: $238 + 15 + 9 + 3564$.

```
  238
   15
    9
 3564
 3826
```
■

Subtraction is the inverse operation of addition. Therefore, subtraction can be thought of in terms of addition. The "−" (minus) sign is the symbol for subtraction. The quantity $5 - 3$ can be thought of as "what number added to 3 gives 5?" The result of subtraction is called the **difference**.

To check a subtraction, add the difference to the second number. If the sum is equal to the first number, the subtraction has been done correctly.

◆ **Example 2**

Subtract: $2843 - 1928$.

Subtract:

```
  2843     first number
 −1928     second number
   915     difference
```

Check:

```
  1928     second number
 +915      difference
  2843     This sum equals the first number, so
           915 is the correct difference.
```
■

Next, let's study some applications. To communicate about problems in electricity, technicians have developed a "language" of their own. It is a picture language that uses symbols and diagrams. The symbols used most often are listed in Table 2 of Appendix A. The circuit diagram is the most common and useful way to show a circuit. Note how each component (part) of the picture (Figure 1.1a) is represented by its symbol in the circuit diagram (Figure 1.1b) in the same relative position.

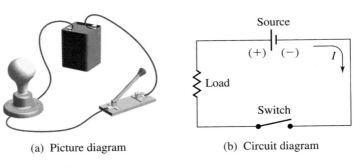

(a) Picture diagram   (b) Circuit diagram

FIGURE 1.1
Components in a circuit

FIGURE 1.2

The light bulb may be represented as a resistance. Then the circuit diagram in Figure 1.1b would appear as in Figure 1.2, where

represents the resistor

represents the switch

represents the source. The short line represents the negative terminal of a battery, and the long line represents the positive terminal. The current flows from negative to positive.

There are two basic types of electrical circuits: series and parallel. An electrical circuit with only one path for the current, I, to flow is called a *series* circuit (Figure 1.3a). An electrical circuit with more than one path for the current to flow is called a *parallel* circuit (Figure 1.3b). A circuit breaker or fuse in a house is wired in series with its outlets. The outlets themselves are wired in parallel.

(a) Series circuit   (b) Parallel circuits

FIGURE 1.3
Two basic types of electrical circuits

◆ **Example 3**

In a series circuit, the total resistance equals the sum of all the resistances in the circuit. Find the total resistance in the series circuit in Figure 1.4. Resistance is measured in ohms, Ω.

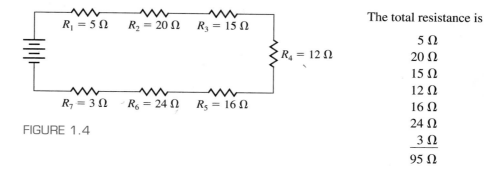

FIGURE 1.4

The total resistance is

5 Ω
20 Ω
15 Ω
12 Ω
16 Ω
24 Ω
<u>3 Ω</u>
95 Ω

■

◆ **Example 4**

Studs are upright wooden or metal pieces in the walls of a building, to which siding, insulation panels, drywall, or decorative paneling are attached. (A wall portion with seven studs is shown in Figure 1.5.) Studs are normally placed 16 in. on center and are placed double at all internal and external corners of a building. The number of studs needed in a wall can be estimated by finding the number of linear feet (ft) of the wall. How many studs are needed for the exterior walls of the building in Figure 1.6?

FIGURE 1.5

FIGURE 1.6

The outside perimeter of the building is the sum of the lengths of the sides of the building:

48 ft
15 ft
15 ft
9 ft
32 ft
8 ft
6 ft
6 ft
5 ft
<u>10 ft</u>
154 ft

Therefore, approximately 154 studs are needed in the outside wall.

■

Repeated addition of the same number can be shortened by multiplication. The "×" (times) and the "·" (raised dot) are used to indicate multiplication. When adding the lengths of five pipes, each 7 ft long, we have 7 ft + 7 ft + 7 ft + 7 ft + 7 ft = 35 ft of pipe. In multiplication, this would be 5 × 7 ft = 35 ft. The 5 and 7 are called *factors*. The result of multiplying numbers (in this case, 35) is called the **product**. Computing areas, volumes, forces, and distances requires skills in multiplication.

Example 5   Multiply: 358 × 18.

$$
\begin{array}{r}
358 \\
\times\ 18 \\
\hline
2864 \\
358 \\
\hline
6444
\end{array}
$$
■

Division is the inverse operation of multiplication. The following symbols are used to show division: $15 \div 5$, $5\overline{)15}$, 15/5, and $\frac{15}{5}$. The quantity $15 \div 5$ can also be thought of as "what number times 5 gives 15?" The answer to this question is 3, which is 15 divided by 5. The result of dividing numbers (in this case, 3) is called the **quotient**. The number to be divided, 15, is called the *dividend*. The number you divide by, 5, is called the *divisor*.

Example 6   Divide: 84 ÷ 6.

$$
\begin{array}{r}
14 \quad \leftarrow \text{quotient} \\
6\overline{)84} \quad \leftarrow \text{dividend} \\
6 \\
\hline
24 \\
24 \\
\hline
0 \quad \leftarrow \text{remainder}
\end{array}
$$

divisor ⟶

■

Example 7   Divide: 115 ÷ 7.

$$
\begin{array}{r}
16 \quad \leftarrow \text{quotient} \\
7\overline{)115} \quad \leftarrow \text{dividend} \\
7 \\
\hline
45 \\
42 \\
\hline
3 \quad \leftarrow \text{remainder}
\end{array}
$$

divisor ⟶

■

The *remainder* (when not 0) is usually written in one of two ways: with an "r" preceding it or with the remainder written over the divisor as a fraction, as shown in Example 8. (Fractions are discussed in Unit 1B.)

Example 8   Divide: 534 ÷ 24.

$$
\begin{array}{r}
22 \text{ r } 6 \quad \text{or} \quad 22\tfrac{6}{24} \\
24\overline{)534} \\
48 \\
\hline
54 \\
48 \\
\hline
6
\end{array}
$$
This quotient may be written 22 r 6 or $22\tfrac{6}{24}$.

■

Example 9

Ohm's law states that in a simple electrical circuit, the current I (measured in amps, A) equals the voltage E (measured in volts, V) divided by the resistance R (measured in ohms, Ω). Find the current in the circuit of Figure 1.7.

The current $I = \dfrac{E}{R} = \dfrac{110}{22} = 5$ A.

FIGURE 1.7

 Example 10

An 8-row corn planter costs $50,400. It has a 10-year life and a salvage value of $5000. What is the annual depreciation? (Use the straight-line depreciation method.)

The straight-line depreciation method means that the difference between the cost and the salvage value is divided evenly over the life of the item. In this case, the difference between the cost and the salvage value is

$50,400	cost
−5,000	salvage
$45,400	difference

This difference divided by 10, the life of the item, is $4540. This is the annual depreciation. ■

Using a Scientific Calculator

Use of a scientific calculator is integrated throughout this text. To demonstrate how to use a common scientific calculator, we show which keys to use and the order in which they are pushed. We have chosen to illustrate the most common types of algebraic logic calculators. Yours may differ. If so, consult your manual.

Note: We will always assume that your calculator is cleared before you begin any calculation.

Use a calculator to add, subtract, multiply, and divide as shown in the following examples.

 Example 11

Add: 9463
      125
        9
       80

9463 [+] 125 [+] 9 [+] 80 [=]

$$\boxed{9677}$$

The sum is 9677. ■

Example 1**2**   Subtract: 3500
                           1628

3500 $\boxed{-}$ 1628 $\boxed{=}$

1872

The result is 1872.

 Example 1**3**   Multiply: 125×68.

125 $\boxed{\times}$ 68 $\boxed{=}$

8500

The product is 8500.

 Example 1**4**   Divide: $8700 \div 15$.

8700 $\boxed{\div}$ 15 $\boxed{=}$

580

The quotient is 580.

Note: Your instructor will indicate which exercises should be completed using a calculator.

Exercises 1.1 _____

Add:

1. $832 + 9 + 56 + 2358$

2. $324 + 973 + 66 + 9430$

3. 384
       291
       147
       632

4.   78
        107
         45
        217
          9
        123

5. $197 + 1072 + 10{,}877 + 15{,}532 + 768{,}098$

6. $160{,}000 + 19{,}000 + 4{,}160{,}000 + 506{,}000$

Subtract and check:

7. 7561
      2397

8. 4000
        702

9. $98{,}405 - 72{,}397$

10. $417{,}286 - 287{,}156$

11. 4000
        1180

12. 60,000
         9,876

Find the total resistance in each series circuit:

13.

14.

15. Approximately how many studs are needed for the exterior walls in the building shown in Illustration 1? (See Example 4.)

ILLUSTRATION 1

16. A pipe 24 ft long is cut into four pieces: the first 4 ft long, the second 5 ft long, and the third 7 ft long. What is the length of the remaining piece? (Assume no waste from cutting.)

17. A welder needs to weld together pipes of lengths 10 ft, 15 ft, and 14 ft. What is the total length of the new pipe?

18. A welder ordered a 125-ft³ cylinder of argon gas, a shielding gas for TIG welding. After a few days, only 78 ft³ remained. How much argon was used?

19. Total the following input and output (I-O) entries in cubic centimetres (cm³)* for a patient.
Input: 300 cm³, 550 cm³, 150 cm³, 75 cm³, 150 cm³, 450 cm³, 250 cm³
Output: 325 cm³, 150 cm³, 525 cm³, 250 cm³, 175 cm³

20. A student pilot must complete 40 h of total flight time as required for her private pilot certificate. She had already entered 31 h of flight time in her logbook. Monday she logged another 2 h, then Wednesday she logged another 3 h, and Friday she logged yet another 2 h. If she can fly 3 h more on Saturday, will she have enough total time as required for the certificate?

Multiply:

21. 567
      48

22. 8374
      203

23. 71,263 × 255

24. 1520 × 320

25. 6800 × 5200

26. 30,010 × 4080

---

*Although cm³ is the "official" metric abbreviation and will be used throughout this book, some readers may be more familiar with the abbreviation "cc," which is still used in some medical and allied health areas.

Divide (use the remainder form with r):

27. 4⟌7236

28. 5⟌308,736

29. 4668 ÷ 12

30. 15,648 ÷ 36

31. 67,560 ÷ 80

32. $\dfrac{188,000}{120}$

33. A car uses gas at the rate of 31 miles per gallon (mi/gal or mpg) and has a 16-gallon tank. How far can it travel on one tank of gas?

34. A car uses gas at a rate of 12 kilometres per litre (km/L) and has a 65-litre tank. How far can it travel on one tank of gas?

35. A four-cylinder engine has a total displacement of 1300 cm³. Find the displacement of each piston.

36. A car travels 1274 mi and uses 49 gal of gasoline. Find its mileage in miles per gallon.

37. A car travels 2340 km and uses 180 L of gasoline. Find its gas consumption in kilometres per litre.

38. To replace some damaged ductwork, 20 linear feet of 8-in. × 16-in. duct is needed. The cost is $13 per 4 linear feet. What is the cost of replacement?

39. The bill for a new transmission was received. The total cost for labor was $402. If the car was serviced for 6 h, find the cost of labor per hour.

40. The cost for a set of four Pirelli P4000 Super-touring tires of size 215/70ZR15 is $508. What is the price for each tire?

41. A small Cessna aircraft has enough fuel to fly for 4 h. If the aircraft cruises at a ground speed of 125 miles per hour (mi/h or mph), how many miles can the aircraft fly in the 4 h?

42. A small plane takes off and climbs at a rate of 500 ft/min. If the plane levels off after 15 min, how high is the plane?

43. Inventory shows the following lengths of 3-inch steel pipe:

5 pieces 18 ft long
42 pieces 15 ft long
158 pieces 12 ft long
105 pieces 10 ft long
79 pieces 8 ft long
87 pieces 6 ft long

What is the total linear feet of pipe in inventory?

44. An order of lumber contains 36 boards 12 ft long, 28 boards 10 ft long, 36 boards 8 ft long, and 12 boards 16 ft long. How many boards are contained in the order? How many linear feet of lumber are contained in the order?

45. Two draftpersons operating the same computer plotter work 8 hours each, on a day and night shift basis. One produces 80 drawings per hour; the other produces 120 drawings per hour. What is the difference in their outputs after 30 days?

46. A shipment contains a total of 5232 linear feet of steel pipe. Each piece of pipe is 12 ft long. How many pieces should be expected?

47. How should a window 75 in. wide be placed so that it is centered on a wall 17 ft 5 in. wide?

48. A farmer expects a yield of 165 bushels per acre (bu/acre) from 260 acres of corn. If the corn is stored, how many bushels of storage are needed?

49. A farmer harvests 6864 bushels (bu) of soybeans from 156 acres. What is his yield per acre?

50. A railroad freight car can hold 2035 bu of corn. How many freight cars are needed to haul the expected 12,000,000 bu from a local grain elevator?

51. On a given day, eight steers weighed 856 lb, 754 lb, 1044 lb, 928 lb, 888 lb, 734 lb, 953 lb, and 891 lb. **a.** What is the average weight? **b.** In 36 days, 4320 lb of feed is consumed. What is the average feed consumption per day per steer?

52. What is the weight (in tons) of a stack of hay bales 6 bales wide, 110 bales long, and 15 bales high? The average weight of each bale is 80 lb. (1 ton = 2000 lb.)

53. From a 34-acre field, 92,480 lb of oats are harvested. Find the yield in bushels per acre. (1 bu of oats weighs 32 lb.)

54. A standard bale of cotton weighs approximately 500 lb. How many bales are contained in 15 tons of cotton?

55. A tractor costs $175,000. It has a 10-year life and a salvage value of $3000. What is the annual depreciation? (Use the straight-line depreciation method. See Example 10.)

56. How much pesticide powder would you put in a 400-gal spray tank if 10 gal of spray, containing 2 lb of pesticide, are applied per acre?

Using Ohm's law, find the current I in amps (A) in each electrical circuit (see Example 9):

57.
58.

Ohm's law, in another form, states that in a simple circuit the voltage E (measured in volts, V) equals the current I (measured in amps, A) times the resistance R (measured in ohms, Ω). Find the voltage E measured in volts (V) in each electrical circuit:

59.
60.

61. A hospital dietitian determines that each patient needs 4 ounces (oz) of orange juice. How many ounces of orange juice must be prepared for 220 patients?

62. During 24 hours, a patient is given three phenobarbital tablets of 60 mg each. How many milligrams of phenobarbital does the patient receive altogether?

63. To give 800 mg of quinine sulfate from 200-mg tablets, how many tablets would you use?

64. A nurse used two 4-grain potassium permanganate tablets in the preparation of a medication. How much potassium permanganate did she use?

65. A sun room addition to a home has a wall 14 ft 6 in. long measured from inside wall to inside wall. Four windows are to be equally spaced from each other in this wall. The windows are 2 ft 6 in. wide including the inside window molding. What is the space between the wall and windows shown in Illustration 2?

ILLUSTRATION 2

66. A solid concrete block wall is being built around a rectangular storage building 12 ft 8 in. by 17 ft 4 in. using 16-in.-long by 8-in.-high by 4-in.-thick concrete block. How many blocks will be needed to build the 8-ft-high wall around the building as shown in Illustration 3? (Ignore the mortar joints.)

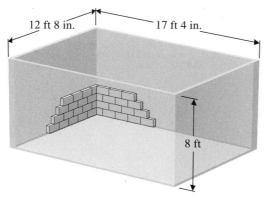

12 ft 8 in.   17 ft 4 in.

8 ft

ILLUSTRATION 3

67. A sheet of plywood 8 ft long is painted with three equally spaced stripes to mark off a hazardous area as shown in Illustration 4. If each stripe is 10 in. wide, what is the space between the end of the plywood and the first stripe?

8 ft

10 in.   1 ft 2 in.

ILLUSTRATION 4

68. In a small machine shop, eight 5-gallon drums of oil are on hand. If 2 gallons are used each day and the owner wants a 30-day supply on hand, how many drums should be ordered?

69. Using a process called "cruising timber," foresters can estimate the amount of lumber in board feet in trees before they are cut down. In a stand of 1000 trees, a forester selects a representative sample of 100 trees and estimates that the sample contains 8540 board feet of lumber. If the entire stand containing 2500 trees is harvested, how many board feet would the landowner expect to harvest?

70. In tilapia aquaculture production, a feed conversion ratio of 2 lb of high-protein pelleted feed per pound of weight gain, after death losses, is not unusual. At that rate of feed conversion, if fish food costs $520 per ton (2000 lb), what would be the feed cost per pound of live fish produced?

1.2 Order of Operations

The expression 5^3 means to use 5 as a factor 3 times. We say that 5^3 is the third *power* of 5, where 5 is called the *base* and 3 is called the *exponent*. Here, 5^3 means $5 \times 5 \times 5 = 125$. The expression 2^4 means that 2 is used as a factor 4 times; that is, $2^4 = 2 \times 2 \times 2 \times 2 = 16$. Here, 2^4 is the fourth power of 2.

Just as we use periods, commas, and other punctuation marks to help make sentences more readable, we use **grouping symbols** in mathematics, such as *parentheses* "( )" and *brackets* "[ ]," to help clarify the meaning of mathematical expressions. Parentheses not only give an expression a particular meaning, they also specify the order to be followed in evaluating and simplifying expressions.

What is the value of $8 - 3 \cdot 2$? Is it 10? Is it 2? Or is it some other number? It is very important that each mathematical expression have only one value. For this to happen, we all must not only perform the exact *same operations* in a given mathematical expression or problem but also perform them in exactly the *same order*. The following order of operations is followed by all.

Order of Operations

1. Always do the operations within parentheses or other grouping symbols first.

2. Then evaluate each power, if any. Examples:

$$4 \times 3^2 = 4 \times (3 \times 3) = 4 \times 9 = 36$$

$$5^2 \times 6 = (5 \times 5) \times 6 = 25 \times 6 = 150$$

$$\frac{5^3}{6^2} = \frac{5 \times 5 \times 5}{6 \times 6} = \frac{125}{36}$$

3. Next, perform multiplications and divisions in the order in which they appear as you read from left to right. For example,

$$60 \times 5 \div 4 \div 3 \times 2$$
$$= \quad 300 \div 4 \div 3 \times 2$$
$$= \quad 75 \div 3 \times 2$$
$$= \quad 25 \times 2$$
$$= \quad 50$$

4. Finally, perform additions and subtractions in the order in which they appear as you read from left to right.

Note: If two parentheses or a number and a parenthesis occur next to one another without any sign between them, multiplication is indicated.

By using the above procedure, we find that $8 - 3 \cdot 2 = 8 - 6 = 2$.

Example 1

Evaluate: $2 + 5(7 + 6)$.

$$= 2 + 5(13) \qquad \text{Add within parentheses.}$$
$$= 2 + \;\; 65 \qquad \text{Multiply.}$$
$$= \;\; 67 \qquad \text{Add.}$$

Note: A number next to parentheses indicates multiplication. In Example 1, $5(13)$ means 5×13. Adjacent parentheses also indicate multiplication: $(5)(13)$ also means 5×13.   ■

Example 2

Evaluate: $(9 + 4) \times 16 + 8$.

$$= 13 \quad \times 16 + 8 \qquad \text{Add within parentheses.}$$
$$= \qquad 208 \;\; + 8 \qquad \text{Multiply.}$$
$$= \qquad\qquad 216 \qquad \text{Add.}$$ ■

Example 3

Evaluate: $(6 + 1) \times 3 + (4 + 5)$.

$$= 7 \qquad \times 3 + \quad 9 \qquad \text{Add within parentheses.}$$
$$= \qquad 21 \;\; + \quad 9 \qquad \text{Multiply.}$$
$$= \qquad\qquad 30 \qquad \text{Add.}$$ ■

Example 4   Evaluate: $4(16 - 4) + \dfrac{14}{7} - 8$.

$= 4(\quad 12\quad) + \dfrac{14}{7} - 8$ Subtract within parentheses.

$=\qquad 48\ +\ 2\ -\ 8$ Multiply and divide.

$=\qquad\qquad\qquad 42$ Add and subtract.   ■

 Example 5   Evaluate: $7 + (6 - 2)^2$.

$= 7 +\quad 4^2$ Subtract within parentheses.

$= 7 +\quad 16$ Evaluate the power.

$=\quad 23$ Add.   ■

 Example 6   Evaluate: $25 - 3 \cdot 2^3$.

$= 25 - 3 \cdot 8$ Evaluate the power.

$= 25 -\quad 24$ Multiply.

$=\quad 1$ Subtract.   ■

If pairs of parentheses are nested (parentheses within parentheses, or within brackets), work from the innermost pair of parentheses to the outermost pair. That is, remove the innermost parentheses first, remove the next innermost parentheses second, and so on.

Example 7   Evaluate: $6 \times 2 + 3[7 + 4(8 - 6)]$.

$= 6 \times 2 + 3[7 + 4(\ 2\ )]$ Subtract within parentheses.

$= 6 \times 2 + 3[7 + 8\qquad]$ Multiply.

$= 6 \times 2 + 3[\ 15\qquad]$ Add within brackets.

$=\quad 12\ + 45$ Multiply.

$=\qquad 57$ Add.   ■

Exercises 1.2

Evaluate each expression:

1. $8 - 3(4 - 2)$

2. $(8 + 6)4 + 8$

3. $(8 + 6) - (7 - 3)$

4. $4 \times (2 \times 6) + (6 + 2) \div 4$

5. $2(9 + 5) - 6 \times (13 + 2) \div 9$

6. $5(8 \times 9) + (13 + 7) \div 4$

7. $27 + 13 \times (7 - 3)(12 + 6) \div 9$

8. $123 - 3(8 + 9) + 17$

9. $16 + 4(7 + 8) - 3$

10. $(18 + 17)(12 + 9) - (7 \times 16)(4 + 2)$

11. $9 - 2(17 - 15) + 18$

12. $(9 + 7)5 + 13$

13. $(39 - 18) - (23 - 18)$

14. $5(3 \times 7) + (8 + 4) \div 3$

15. $3(8 + 6) - 7(13 + 3) \div 14$

16. $6(4 \times 5) + (15 + 9) \div 6$

17. $42 + 12(9 - 3)(12 + 13) \div 30$

18. $228 - 4 \times (7 + 6) - 8(6 - 2)$

19. $38 + 9 \times (8 + 4) - 3(5 - 2)$

20. $(19 + 8)(4 + 3) \div 21 + (8 \times 15) \div (4 \times 3)$

21. $27 - 2 \times (18 - 9) - 3 + 8(43 - 15)$

22. $6 \times 8 \div 2 \times 9 \div 12 + 6$

23. $12 \times 9 \div 18 \times 64 \div 8 + 7$

24. $18 \div 6 \times 24 \div 4 \div 6$

25. $7 + 6(3 + 2) - 7 - 5(4 + 2)$

26. $5 + 3(7 \times 7) - 6 - 2(4 + 7)$

27. $3 + 17(2 \times 2) - 67$

28. $8 - 3(9 - 2) \div 21 - 7$

29. $28 - 4(2 \times 3) + 4 - (16 \times 8) \div (4 \times 4)$

30. $6 + 4(9 + 6) + 8 - 2(7 + 3) - (3 \times 12) \div 9$

31. $24/(6 - 2) + 4 \times 3 - 15/3$

32. $(36 - 6)/(5 + 10) + (16 - 1)/3$

33. $3 \times 15 \div 9 + (13 - 5)/2 \times 4 - 2$

34. $28/2 \times 7 - (6 + 10)/(6 - 2)$

35. $10 + 4^2$

36. $4 + 2 \cdot 3^2$

37. $\dfrac{20 + (2 \cdot 3)^2}{7 \cdot 2^3}$

38. $\dfrac{(20 - 2 \cdot 5)^2}{3^3 - 2}$

39. $6[3 + 2(2 + 5)]$

40. $5((4 + 6) + 2(5 - 2))$

41. $5 \times 2 + 3[2(5 - 3) + 4(4 + 2) - 3]$

42. $3(10 + 2(1 + 3(2 + 6(4 - 2))))$

1.3 Area and Volume

To measure the length of an object, you must first select a suitable standard unit of length. To measure short lengths, choose a unit such as centimetres or millimetres in the metric system, or inches in the U.S. or, as it is still sometimes called, the English system. For long distances, choose metres or kilometres in the metric system, or yards or miles in the U.S. system.

Area

The **area** of a plane geometric figure is the number of square units of measure it contains. To measure the surface area of an object, first select a standard unit of area suitable to the object to be measured. Standard units of area are based on the square and are called square units. For example, a square inch (in^2) is the amount of surface area within a square that measures one inch on a side. A square centimetre (cm^2) is the amount of surface area within a square that is 1 cm on a side. (See Figure 1.8.)

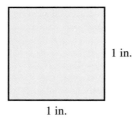

1 in.

1 in.

1 square inch (in^2)

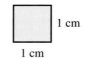

1 cm

1 cm

1 square centimetre (cm^2)

FIGURE 1.8
Square units

Example 1

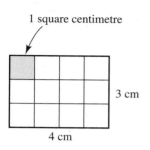

1 square centimetre

3 cm

4 cm

FIGURE 1.9

What is the area of a rectangle measuring 4 cm by 3 cm?

Each square in Figure 1.9 represents 1 cm². By simply counting the number of squares, you find that the area of the rectangle is 12 cm².

You can also find the area by multiplying the length times the width:

$$\text{Area} = \underset{\text{(length)}}{l} \times \underset{\text{(width)}}{w}$$
$$= 4\,\text{cm} \times 3\,\text{cm} = 12\,\text{cm}^2 \qquad \text{Note: cm} \times \text{cm} = \text{cm}^2$$

■

Example 2

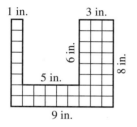

1 in.   3 in.

6 in.

8 in.

5 in.

9 in.

FIGURE 1.10

What is the area of the metal plate represented in Figure 1.10?

Each square represents 1 square inch. By simply counting the number of squares, we find that the area of the metal plate is 42 in².

Another way to find the area of the figure is to find the areas of two rectangles and then find their difference, as in Figure 1.11.

=   8 in.   −   6 in.

9 in.   5 in.

FIGURE 1.11

Area of outer rectangle: 9 in. × 8 in. = 72 in²
Area of inner rectangle: 5 in. × 6 in. = 30 in²
Area of metal plate: = 42 in²   Subtract.   ■

Volume

The **volume** of a solid geometric figure is the number of cubic units of measure it contains. In area measurement, the standard units are based on the square and called square units. For volume measurement, the standard units are based on the cube and called cubic units. For example, a cubic inch (in³) is the amount of space contained in a cube that measures 1 in. on each edge. A cubic centimetre (cm³) is the amount of space contained in a cube that measures 1 cm on each edge. A cubic foot (ft³) is the amount of space contained in a cube that measures 1 ft (or 12 in.) on each edge. (See Figure 1.12.)

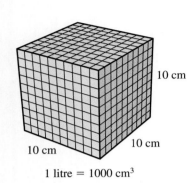

1 litre = 1000 cm³

FIGURE 1.13
Litre

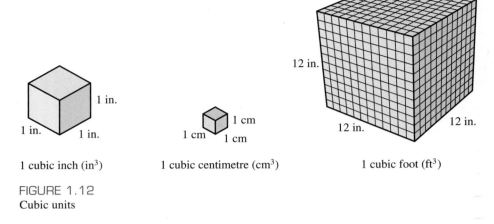

1 cubic inch (in³)   1 cubic centimetre (cm³)   1 cubic foot (ft³)

FIGURE 1.12
Cubic units

Figure 1.13 shows that the cubic decimetre (litre) is made up of 10 layers, each containing 100 cm³, for a total of 1000 cm³.

 Example 3

FIGURE 1.14

Find the volume of a rectangular box 8 cm long, 4 cm wide, and 6 cm high.

Suppose you placed one-centimetre cubes in the box, as in Figure 1.14. On the bottom layer, there would be 8 × 4, or 32, one-cm cubes. In all, there are six such layers, or 6 × 32 = 192 one-cm cubes. Therefore, the volume is 192 cm³.

You can also find the volume of a rectangular solid by multiplying the length times the width times the height:

$$V = l \times w \times h$$
$$= 8 \text{ cm} \times 4 \text{ cm} \times 6 \text{ cm}$$
$$= 192 \text{ cm}^3 \qquad \text{Note: } \text{cm} \times \text{cm} \times \text{cm} = \text{cm}^3 \qquad ■$$

Example 4

How many cubic inches are in one cubic foot?

The bottom layer of Figure 1.15 contains 12 × 12, or 144, one-inch cubes. There are 12 such layers, or 12 × 144 = 1728 one-inch cubes. Therefore, 1 ft³ = 1728 in³.   ■

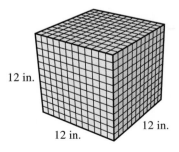

FIGURE 1.15
Cubic foot

Exercises 1.3

1. How many square yards (yd^2) are contained in a rectangle 12 yd long and 8 yd wide?

2. How many square metres (m^2) are contained in a rectangle 12 m long and 8 m wide?

3. At a small airport, Runway 11-29 is 4100 ft long and 75 ft wide. What is the area of the runway?

4. A small rectangular military operating zone has dimensions 12 mi by 22 mi. What is its area?

5. A 2009 Honda Accord LX measures 191 in. by 73 in. Find the area it occupies.

6. Five pieces of sheet metal have been cut to form a container. The five pieces are of sizes 27 by 15, 15 by 18, 27 by 18, 27 by 18, and 15 by 18 (all in inches). What is the total area of all five pieces?

In the following exercises, assume that all corners are square and that like measurements are not repeated because the figures are assumed to have consistent lengths. All three of the following mean that the length of a side is 3 cm:

3 cm   3 cm   3 cm

Find the area of each figure:

7.

8.

9.

10.

11.   12.
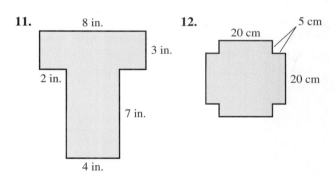

13. How many tiles 4 in. on a side should be used to cover a portion of a wall 48 in. long by 36 in. high? (See Illustration 1.)

ILLUSTRATION 1

14. How many ceiling tiles 2 ft by 4 ft are needed to tile a ceiling that is 24 ft by 26 ft? (Be careful how you arrange the tiles.)

15. How many gallons of paint should be purchased to paint 20 motel rooms as shown in Illustration 2? (Do not paint the floor.) One gallon is needed to paint 400 square feet (ft^2).

ILLUSTRATION 2

16. How many pieces of 4-ft by 8-ft drywall are needed for the 20 motel rooms in Exercise 15? All four walls and the ceiling in each room are to be drywalled. Assume that the drywall cut out for windows and doors cannot be salvaged and used again.

17. The replacement cost for construction of houses is $110/$\text{ft}^2$. Determine how much house insurance should be carried on each of the one-story houses in Illustration 3.

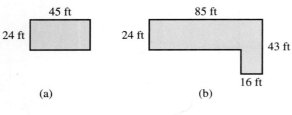

(a)   (b)

ILLUSTRATION 3

18. The replacement cost for construction of the building in Illustration 4 is $90/$\text{ft}^2$. Determine how much insurance should be carried for full replacement.

ILLUSTRATION 4

Find the volume of each rectangular solid:

19.

20.

21.

22.

23.

24.

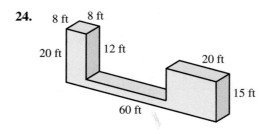

25. Find the volume of a rectangular box 10 cm by 12 cm by 5 cm.

26. A mountain cabin has a single room 20 ft by 10 ft by 8 ft high. What is the total volume of air in the room that will be circulated through the central ventilating fan?

27. Common house duct is 8-in. by 20-in. rectangular metal duct. If the length of a piece of duct is 72 in., what is its volume?

28. A furnace filter measures 16 in. by 20 in. by 1 in. What is its volume?

29. A large rectangular tank is to be made of sheet metal as follows: 3 ft by 5 ft for the top and the base, two sides consisting of 2 ft by 3 ft, and two sides consisting of 2 ft by 5 ft. Find the volume of this container.

30. Suppose an oil pan has the rectangular dimensions 14 in. by 16 in. by 4 in. Find its volume.

31. Find the weight of a cement floor that is 15 ft by 12 ft by 2 ft if 1 ft³ of cement weighs 193 lb.

32. A trailer 5 ft by 6 ft by 5 ft is filled with coal. Given that 1 ft³ of coal weighs 40 lb and 1 ton = 2000 lb, how many tons of coal are in the trailer?

33. A rectangular tank is 8 ft long by 5 ft wide by 6 ft high. Water weighs approximately 62 lb/ft³. Find the weight of water if the tank is full.

34. A rectangular tank is 9 ft by 6 ft by 4 ft. Gasoline weighs approximately 42 lb/ft³. Find the weight of gasoline if the tank is full.

35. A building is 100 ft long, 50 ft wide, and 10 ft high. Estimate the cost of heating it at the rate of $55 per 1000 ft³.

36. A single-story shopping center is being designed to be 483 ft long by 90 ft deep. Two stores have been pre-leased. One occupies 160 linear feet and the other will occupy 210 linear feet. The owner is trying to decide how to divide the remaining space, knowing that the smallest possible space should be 4000 ft². How many stores can occupy the remaining area as shown in Illustration 5?

ILLUSTRATION 5

37. A trophy company needs a shipping box for a trophy 15 in. high with an 8-in.-square base. The box company is drawing the die to cut the cardboard for this box. How large a sheet of cardboard is needed to make one box that allows 1 in. for packing and 1 in. for a glue edge as shown in Illustration 6?

ILLUSTRATION 6

38. Styrofoam "peanuts" will be used to pack the trophy in the box in Illustration 6 to prevent the trophy from being broken during shipment. Ignoring the box wall thickness, how many cubic inches of peanuts will be used for each package if the volume of the trophy is 450 in³?

39. A standard cord of wood measures 4 ft by 4 ft by 8 ft. What is the volume in cubic feet of a cord of wood?

40. A municipal wastewater treatment plant has a settling tank that is 125 ft long and 24 ft wide with an effective depth of 12 ft. What is the surface area of the liquid in the tank and what is the volume of sewerage that the settling tank will hold?

1.4 Formulas

A **formula** is a statement of a rule using letters to represent the relationship of certain quantities. In physics, one of the basic rules states that *work* equals *force* times *distance*. If a person (Figure 1.16) lifts a 200-lb weight a distance of 3 ft, we say the work done is 200 lb × 3 ft = 600 foot-pounds (ft-lb). The work, W, equals the force, f, times the distance, d, or $W = f \times d$.

200 lb

3 ft

FIGURE 1.16

A person pushes against a car weighing 2700 lb but does not move it. The work done is 2700 lb × 0 ft = 0 ft-lb. An automotive technician (Figure 1.17) moves a diesel engine weighing 1100 lb from the floor to a workbench 4 ft high. The work done in moving the engine is 1100 lb × 4 ft = 4400 ft-lb.

1100 lb

4 ft

FIGURE 1.17

To summarize, if you know the amount of force and the distance the force is applied, the work can be found by simply multiplying the force and distance. The formula $W = f \times d$ is often written $W = f \cdot d$, or simply $W = fd$. Whenever there is no symbol between a number and a letter or between two letters, it is assumed that the operation to be performed is multiplication.

Example 1

If $W = fd$, $f = 10$, and $d = 16$, find W.

$W = fd$
$W = (10)(16)$
$W = 160$ Multiply.   ■

Example 2

If $I = \dfrac{E}{R}$, $E = 450$, and $R = 15$, find I.

$I = \dfrac{E}{R}$

$I = \dfrac{450}{15}$

$I = 30$ Divide.   ■

Example 3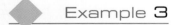

If $P = I^2R$, $I = 3$, and $R = 600$, find P.

$P = I^2R$
$P = (3)^2(600)$
$P = (9)(600)$ Evaluate the power.
$P = 5400$ Multiply.   ■

There are many other formulas used in science and technology. Some examples are given here:

a. $d = vt$ **c.** $f = ma$ **e.** $I = \dfrac{E}{R}$

b. $W = IEt$ **d.** $P = IE$ **f.** $P = \dfrac{V^2}{R}$

Formulas from Geometry

The area of a triangle is given by the formula $A = \frac{1}{2}bh$, where b is the length of the base and h, the height, is the length of the altitude to the base (Figure 1.18). (An altitude of a triangle is a line from a vertex perpendicular to the opposite side.)

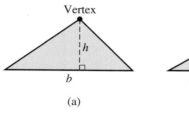

(a)   (b)

Triangle

FIGURE 1.18

 Example 4

Find the area of a triangle whose base is 18 in. and whose height is 10 in.

$$A = \frac{1}{2}bh$$

$$A = \frac{1}{2}(18 \text{ in.})(10 \text{ in.})$$

$$= 90 \text{ in}^2 \qquad \text{Note: (in.)(in.)} = \text{in}^2 \qquad ■$$

Parallelogram

FIGURE 1.19

The area of a *parallelogram* (a four-sided figure whose opposite sides are parallel) is given by the formula $A = bh$, where b is the length of the base and h is the perpendicular distance between the base and its opposite side (Figure 1.19).

 Example 5

Find the area of a parallelogram with base 24 cm and height 10 cm.

$$A = bh$$

$$A = (24 \text{ cm})(10 \text{ cm})$$

$$= 240 \text{ cm}^2 \qquad \text{Note: (cm)(cm)} = \text{cm}^2 \qquad ■$$

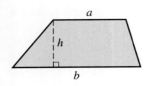

Trapezoid

FIGURE 1.20

The area of a *trapezoid* (a four-sided figure with one pair of parallel sides) is given by the formula $A = \left(\frac{a + b}{2}\right)h$, where a and b are the lengths of the parallel sides (called *bases*), and h is the perpendicular distance between the bases (Figure 1.20).

Example 6   Find the area of the trapezoid in Figure 1.21.

$a = 10$ in.

$h = 7$ in.

$b = 18$ in.

FIGURE 1.21

$$A = \left(\frac{a + b}{2}\right)h$$

$$A = \left(\frac{10 \text{ in.} + 18 \text{ in.}}{2}\right)(7 \text{ in.})$$

$$= \left(\frac{28 \text{ in.}}{2}\right)(7 \text{ in.}) \qquad \text{Add within parentheses.}$$

$$= (14 \text{ in.})(7 \text{ in.}) \qquad \text{Divide.}$$

$$= 98 \text{ in}^2 \qquad \text{Multiply.} \qquad ■$$

Exercises 1.4

Use the formula $W = fd$, where f represents a force and d represents the distance that the force is applied. Find the work done, W:

1. $f = 30, d = 20$
2. $f = 17, d = 9$
3. $f = 1125, d = 10$
4. $f = 203, d = 27$
5. $f = 176, d = 326$
6. $f = 2400, d = 120$

*From formulas **a** to **f** on page 21, choose one that contains all the given letters. Then use the formula to find the unknown letter:*

7. If $m = 1600$ and $a = 24$, find f.
8. If $V = 120$ and $R = 24$, find P.
9. If $E = 120$ and $R = 15$, find I.
10. If $v = 372$ and $t = 18$, find d.
11. If $I = 29$ and $E = 173$, find P.
12. If $I = 11$, $E = 95$, and $t = 46$, find W.

Find the area of each triangle:

13. $b = 10$ in., $h = 8$ in.
14. $b = 36$ cm, $h = 20$ cm
15. $b = 54$ ft, $h = 30$ ft
16. $b = 188$ m, $h = 220$ m

Find the area of each rectangle:

17. $b = 8$ m, $h = 7$ m
18. $b = 24$ in., $h = 15$ in.
19. $b = 36$ ft, $h = 18$ ft
20. $b = 250$ cm, $h = 120$ cm

Find the area of each trapezoid:

21. $a = 7$ ft, $b = 9$ ft, $h = 4$ ft
22. $a = 30$ in., $b = 50$ in., $h = 24$ in.
23. $a = 96$ cm, $b = 24$ cm, $h = 30$ cm
24. $a = 450$ m, $b = 750$ m, $h = 250$ m
25. The volume of a rectangular solid is given by $V = lwh$, where l is the length, w is the width, and h is the height of the solid. Find V if $l = 25$ cm, $w = 15$ cm, and $h = 12$ cm.
26. Find the volume of a box with dimensions $l = 48$ in., $w = 24$ in., and $h = 96$ in.
27. Given $v = v_0 + gt$, $v_0 = 12$, $g = 32$, and $t = 5$, find v.
28. Given $Q = CV$, $C = 12$, and $V = 2500$, find Q.
29. Given $I = \dfrac{E}{Z}$, $E = 240$, and $Z = 15$, find I.
30. Given $P = I^2R$, $I = 4$, and $R = 2000$, find P.

31. The formula for piston displacement is $P = cd^2SN$, where c is a constant, d is the cylinder bore, S is the stroke, and N is the number of cylinders. For $c = 0.7854$, $d = 3$, $S = 4$, and $N = 4$, find the piston displacement.

32. The length of a cylinder is given by $l = \dfrac{V}{cd^2}$, where c is the constant 0.785, d is the diameter of the cylinder, and V is the volume of the cylinder. Find l if $V = 47$ in^3 and $d = 2.98$ in.

33. A homeowner has a lawn with the dimensions shown in Illustration 1. Fertilizer recommendations for lawns are given in pounds per thousand square feet (tsf). What is the area of the lawn in tsf?

ILLUSTRATION 1

1.5 Prime Factorization

Divisibility

A number is **divisible** by a second number if, when you divide the first number by the second number, you get a zero remainder. That is, the second number *divides* the first number.

Example 1   12 is divisible by 3, because 3 divides 12.   ■

Example 2   124 is not divisible by 7, because 7 does not divide 124. Check with a calculator.   ■

There are many ways of classifying the positive integers. They can be classified as even or odd, as divisible by 3 or not divisible by 3, as larger than 10 or smaller than 10, and so on. One of the most important classifications involves the concept of a **prime number**: an integer greater than 1 that has no divisors except itself and 1. The first ten prime numbers are 2, 3, 5, 7, 11, 13, 17, 19, 23, and 29.

An integer is **even** if it is divisible by 2; that is, if you divide it by 2, you get a zero remainder. An integer is **odd** if it is not divisible by 2.

In multiplying two or more positive integers, the positive integers are called the *factors* of the product. Thus, 2 and 5 are factors of 10, since $2 \times 5 = 10$. The numbers 2, 3, and 5 are factors of 30, since $2 \times 3 \times 5 = 30$. Those prime numbers whose product equals the given integer are called **prime factors**. The process of finding the prime factors of a positive integer is called **prime factorization**. The prime factorization of a given number is the product of its prime factors. That is, each of the factors is prime, and their product equals the given number. One of the most useful applications of prime factorization is in finding the least common denominator (LCD) when adding and subtracting fractions. This application is found in Section 1.7.

 Example 3

Factor 28 into prime factors.

a. $28 = 4 \cdot 7$ **b.** $28 = 7 \cdot 4$ **c.** $28 = 2 \cdot 14$
       $= 2 \cdot 2 \cdot 7$ $= 7 \cdot 2 \cdot 2$ $= 2 \cdot 7 \cdot 2$

    In each case, you have three prime factors of 28; one factor is 7, the other two are 2's. The factors may be written in any order, but we usually list all the factors in order from smallest to largest. It would not be correct in the examples above to leave $7 \cdot 4$, $4 \cdot 7$, or $2 \cdot 14$ as factors of 28, since 4 and 14 are not prime numbers.   ■

    Short division, a condensed form of long division, is a helpful way to find prime factors. Find a prime number that divides the given number. Divide, using short division. Then find a second prime number that divides the result. Divide, using short division. Keep repeating this process of stacking the quotients and divisors (as shown below) until the final quotient is also prime. The prime factors will be the product of the divisors and the final quotient of the repeated short divisions.

 Example 4

Find the prime factorization of 45.

$3\lfloor\underline{45}$ Divide by 3.
$3\lfloor\underline{15}$ Divide by 3.
   5

The prime factorization of 45 is $3 \cdot 3 \cdot 5$.   ■

 Example 5

Find the prime factorization of 60.

$2\lfloor\underline{60}$ Divide by 2.
$2\lfloor\underline{30}$ Divide by 2.
$3\lfloor\underline{15}$ Divide by 3.
   5

The prime factorization of 60 is $2 \cdot 2 \cdot 3 \cdot 5$.   ■

 Example 6

Find the prime factorization of 17.

17 has no factors except for itself and 1. Thus, 17 is a prime number. When asked for factors of a prime number, write "prime" as your answer.   ■

Divisibility Tests

To eliminate some of the guesswork involved in finding prime factors, divisibility tests can be used. Such tests determine whether or not a particular positive integer divides another integer without carrying out the division. Divisibility tests and prime factorization are used to reduce fractions to lowest terms and to find the lowest common denominator. (See Unit 1B.)

    The following divisibility tests for certain positive integers are most helpful.

Divisibility by 2

If a number ends with an even digit, then the number is divisible by 2.

Note: Zero is even.

Example 7

Is 4258 divisible by 2?

Yes; since 8, the last digit of the number, is even, 4258 is divisible by 2. ■

Note: Check each example with a calculator.

Example 8

Is 215,517 divisible by 2?

Since 7 (the last digit) is odd, 215,517 is not divisible by 2. ■

Divisibility by 3

If the sum of the digits of a number is divisible by 3, then the number itself is divisible by 3.

Example 9

Is 531 divisible by 3?

The sum of the digits $5 + 3 + 1 = 9$. Since 9 is divisible by 3, then 531 is divisible by 3. ■

Example 10

Is 551 divisible by 3?

The sum of the digits $5 + 5 + 1 = 11$. Since 11 is not divisible by 3, then 551 is not divisible by 3. ■

Divisibility by 5

If a number has 0 or 5 as its last digit, then the number is divisible by 5.

Example 11

Is 2372 divisible by 5?

The last digit of 2372 is neither 0 nor 5, so 2372 is not divisible by 5. ■

Example 12

Is 3210 divisible by 5?

The last digit of 3210 is 0, so 3210 is divisible by 5. ■

Example 13

Find the prime factorization of 204.

    2|204   Last digit is even.
    2|102   Last digit is even.
    3| 51   Sum of digits is divisible by 3.
       17

The prime factorization of 204 is $2 \cdot 2 \cdot 3 \cdot 17$. ■

◆ Example 14   Find the prime factorization of 630.

$$\begin{array}{r|l} 2 & 630 \\ 3 & 315 \\ 3 & 105 \\ 5 & \ 35 \\ \hline & 7 \end{array}$$ Last digit is even.
Sum of digits is divisible by 3.
Sum of digits is divisible by 3.
Last digit is 5.

The prime factorization of 630 is $2 \cdot 3 \cdot 3 \cdot 5 \cdot 7$.   ■

Note: As a general rule of thumb:

1. Keep dividing by 2 until the quotient is not even.
2. Keep dividing by 3 until the quotient's sum of digits is not divisible by 3.
3. Keep dividing by 5 until the quotient does not end in 0 or 5.

That is, if you divide out all the factors of 2, 3, and 5, the remaining factors, if any, will be much smaller and easier to work with, and perhaps prime.

Note: Some people prefer to use the divisibility tests for 2 and 5 before using the divisibility test for 3.

Exercises 1.5

Which numbers are divisible **a.** *by 3 and* **b.** *by 4?*

1. 15   **2.** 28   **3.** 96
4. 172   **5.** 78   **6.** 675

Classify each number as prime or not prime:

7. 53   **8.** 57   **9.** 93
10. 121   **11.** 16   **12.** 123
13. 39   **14.** 87

Test for divisibility by 2:

15. 458   **16.** 12,746   **17.** 315,817
18. 877,778   **19.** 1367   **20.** 1205

Test for divisibility by 3 and check your results with a calculator:

21. 387   **22.** 1254   **23.** 453,128
24. 178,213   **25.** 218,745   **26.** 15,690

Test for divisibility by 5 and check your results with a calculator:

27. 70   **28.** 145   **29.** 366
30. 56,665   **31.** 63,227   **32.** 14,601

Test the divisibility of each first number by the second number:

33. 56; 2   **34.** 42; 3
35. 218; 3   **36.** 375; 5
37. 528; 5   **38.** 2184; 3
39. 198; 3   **40.** 2236; 3
41. 1,820,670; 2   **42.** 2,817,638; 2
43. 7,215,720; 5   **44.** 5,275,343; 3

Find the prime factorization of each number (use divisibility tests where helpful):

45. 20   **46.** 18   **47.** 66
48. 30   **49.** 36   **50.** 25
51. 27   **52.** 59   **53.** 51
54. 56   **55.** 42   **56.** 63
57. 120   **58.** 72   **59.** 171
60. 360   **61.** 105   **62.** 78
63. 252   **64.** 444

Unit 1A Review

1. Add: $33 + 104 + 75 + 29$
2. Subtract: $2301 - 506$
3. Multiply: 3709×731
4. Divide: $9300 \div 15$
5. Josh has the following lengths of 3-inch plastic pipe:
   - 3 pieces 12 ft long
   - 8 pieces 8 ft long
   - 9 pieces 10 ft long
   - 12 pieces 6 ft long

   Find the total length of pipe on hand.
6. If one bushel of corn weighs 56 lb, how many bushels are contained in 14,224 lb of corn?

Evaluate each expression:

7. $6 + 2(5 \times 4 - 2)$
8. $3^2 + 12 \div 3 - 2 \times 3$
9. $12 + 2[3(8 - 2) - 2(3 + 1)]$
10. In Illustration 1, find the area.

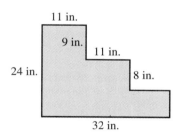

11 in.

9 in.

11 in.

24 in.

8 in.

32 in.

ILLUSTRATION 1

11. In Illustration 2, find the volume.

6 ft

8 ft   15 ft

ILLUSTRATION 2

12. If $d = vt$, $v = 45$, and $t = 4$, find d.
13. If $I = \dfrac{E}{R}$, $E = 120$, and $R = 12$, find I.
14. If $A = \dfrac{1}{2}bh$, $b = 40$, and $h = 15$, find A.

Classify each number as prime or not prime:

15. 51   16. 47

Test for divisibility of each first number by the second number:

17. 195; 3   18. 821; 5

Find the prime factorization of each number:

19. 40   20. 135

Unit 1B

REVIEW OF OPERATIONS WITH FRACTIONS

1.6 Introduction to Fractions

The U.S. system of measurement, which is derived from and sometimes called the English system, is a system whose units are often expressed as common fractions and mixed numbers. The metric system of measurement is a system whose units are expressed as decimal fractions and powers of ten. As we continue to convert from the U.S. system to the metric system, more computations will be done with decimals, which are easier—especially with

FIGURE 1.22

FIGURE 1.23

a calculator. Fewer computations will be done with fractions, which are more difficult. During the transition period, we will need to feel comfortable with both systems. The metric system is developed in Chapter 3.

A **common fraction** may be defined as the ratio or quotient of two integers (say, a and b) in the form $\frac{a}{b}$ (where $b \neq 0$). Examples are $\frac{1}{2}, \frac{7}{11}, \frac{3}{8}$, and $\frac{37}{22}$. The integer below the line is called the *denominator*. It gives the denomination (size) of equal parts into which the fraction unit is divided. The integer above the line is called the *numerator*. It numerates (counts) the number of times the denominator is used. Look at one inch on a ruler, as shown in Figures 1.22 and 1.23.

$\frac{1}{4}$ in. means 1 of 4 equal parts of an inch.

$\frac{2}{4}$ in. means 2 of 4 equal parts of an inch.

$\frac{3}{4}$ in. means 3 of 4 equal parts of an inch.

$\frac{5}{16}$ in. means 5 of 16 equal parts of an inch.

$\frac{12}{16}$ in. means 12 of 16 equal parts of an inch.

$\frac{15}{16}$ in. means 15 of 16 equal parts of an inch.

Two fractions $\frac{a}{b}$ and $\frac{c}{d}$ are *equal* or *equivalent* if $ad = bc$. That is, $\frac{a}{b} = \frac{c}{d}$ if $ad = bc$ ($b \neq 0$ and $d \neq 0$). For example, $\frac{2}{4}$ and $\frac{8}{16}$ are names for the same fraction, because $(2)(16) = (4)(8)$. There are many other names for this same fraction, such as $\frac{1}{2}, \frac{9}{18}, \frac{10}{20}, \frac{5}{10}, \frac{3}{6}$, and so on.

$\frac{2}{4} = \frac{1}{2}$, because $(2)(2) = (4)(1)$ $\frac{2}{4} = \frac{5}{10}$, because $(2)(10) = (4)(5)$

Figure 1.24 may help you visualize the relative sizes of fractions. Note that $\frac{1}{2} = \frac{2}{4} = \frac{4}{8} = \frac{8}{16}, \frac{7}{8}$ is greater than $\frac{3}{4}$, and $\frac{3}{16}$ is less than $\frac{1}{4}$.

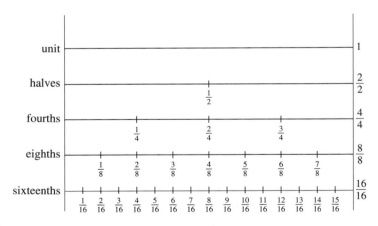

FIGURE 1.24
Relative sizes of fractions

We have two rules for finding equal (or *equivalent*) fractions.

Equal or Equivalent Fractions

1. The numerator and denominator of any fraction may be *multiplied* by the same number (except zero) without changing the value of the given fraction, thus producing an equivalent fraction. For example, $\frac{4}{9} = \frac{4 \cdot 5}{9 \cdot 5} = \frac{20}{45}$.

2. The numerator and denominator of any fraction may be *divided* by the same number (except zero) without changing the value of the given fraction. For example, $\frac{6}{10} = \frac{6 \div 2}{10 \div 2} = \frac{3}{5}$.

We use these rules for equivalent fractions (a) to reduce a fraction to lowest terms and (b) to change a fraction to higher terms when adding and subtracting fractions with different denominators.

To *simplify* a fraction means to find an equivalent fraction whose numerator and denominator are *relatively prime*—that is, a fraction whose numerator and denominator have no common divisor. This is also called **reducing a fraction to lowest terms**.

To reduce a fraction to lowest terms, write the prime factorization of the numerator and the denominator. Then divide (cancel) numerator and denominator by any pair of common factors. You may find it helpful to use the divisibility tests in Section 1.5 to write the prime factorizations.

◆———— **Example 1**

Simplify: $\dfrac{35}{50}$.

$\dfrac{35}{50} = \dfrac{\cancel{5} \cdot 7}{2 \cdot \cancel{5} \cdot 5} = \dfrac{7}{10}$ Note the use of the divisibility test for 5. A last digit of 0 or 5 indicates the number is divisible by 5.   ■

◆———— **Example 2**

Simplify: $\dfrac{63}{99}$.

$\dfrac{63}{99} = \dfrac{\cancel{3} \cdot \cancel{3} \cdot 7}{\cancel{3} \cdot \cancel{3} \cdot 11} = \dfrac{7}{11}$ Note the use of the divisibility test for 3 twice.   ■

◆———— **Example 3**

Simplify: $\dfrac{84}{300}$.

$\dfrac{84}{300} = \dfrac{\cancel{2} \cdot \cancel{2} \cdot \cancel{3} \cdot 7}{\cancel{2} \cdot \cancel{2} \cdot \cancel{3} \cdot 5 \cdot 5} = \dfrac{7}{25}$ ■

Simplifying Special Fractions

1. Any number (except zero) divided by itself is equal to 1. For example,

$$\frac{3}{3} = 1; \frac{5}{5} = 1; \frac{173}{173} = 1$$

2. Any number divided by 1 is equal to itself. For example,

$$\frac{5}{1} = 5; \frac{9}{1} = 9; \frac{25}{1} = 25$$

continued

3. Zero divided by any number (except zero) is equal to zero. For example,

$$\frac{0}{6} = 0; \quad \frac{0}{13} = 0; \quad \frac{0}{8} = 0$$

4. Any number *divided by zero* is not meaningful and is called *undefined*. For example, $\frac{4}{0}$ is undefined.

A **proper fraction** is a fraction whose numerator is less than its denominator. Examples of proper fractions are $\frac{2}{3}, \frac{5}{14}$, and $\frac{3}{8}$. An **improper fraction** is a fraction whose numerator is greater than or equal to its denominator. Examples of improper fractions are $\frac{7}{5}, \frac{11}{11}$, and $\frac{9}{4}$.

A **mixed number** is an integer plus a proper fraction. Examples of mixed numbers are $1\frac{3}{4}\left(1 + \frac{3}{4}\right), 14\frac{1}{9}$, and $5\frac{2}{15}$.

Changing an Improper Fraction to a Mixed Number

To change an improper fraction to a mixed number, divide the numerator by the denominator. The quotient is the whole-number part. The remainder over the divisor is the proper fraction part of the mixed number.

Example 4   Change $\dfrac{17}{3}$ to a mixed number.

$$\frac{17}{3} = 17 \div 3 = 3\overline{)17} \atop 5\,r\,2 \quad = 5\frac{2}{3}$$

Example 5   Change $\dfrac{78}{7}$ to a mixed number.

$$\frac{78}{7} = 78 \div 7 = 7\overline{)78} \atop 11\,r\,1 \quad = 11\frac{1}{7}$$

If the improper fraction is not reduced to lowest terms, you may find it easier to reduce it before changing it to a mixed number. Of course you may reduce the proper fraction after the division if you prefer.

Example 6   Change $\dfrac{324}{48}$ to a mixed number and simplify.

Method 1: Reduce the improper fraction to lowest terms first.
$$\frac{324}{48} = \frac{\cancel{2}\cdot\cancel{2}\cdot 3\cdot 3\cdot 3\cdot\cancel{3}}{\cancel{2}\cdot\cancel{2}\cdot 2\cdot 2\cdot\cancel{3}} = \frac{27}{4} = 4\overline{)27} \atop 6\,r\,3 = 6\frac{3}{4}$$

Method 2: Change the improper fraction to a mixed number first.
$$\frac{324}{48} = 48\overline{)324}^{\,6\,r\,36} \atop \underline{288} \atop 36 = 6\frac{36}{48} = 6\frac{\cancel{2}\cdot\cancel{2}\cdot 3\cdot\cancel{3}}{\cancel{2}\cdot\cancel{2}\cdot 2\cdot 2\cdot\cancel{3}} = 6\frac{3}{4}$$

One way to change a mixed number to an improper fraction is to multiply the integer by the denominator of the fraction and then add the numerator of the fraction. Then place this sum over the original denominator.

◆ Example 7 Change $2\frac{1}{3}$ to an improper fraction.

$$2\frac{1}{3} = \frac{(2 \times 3) + 1}{3} = \frac{7}{3}$$ ■

◆ Example 8 Change $5\frac{3}{8}$ to an improper fraction.

$$5\frac{3}{8} = \frac{(5 \times 8) + 3}{8} = \frac{43}{8}$$ ■

◆ Example 9 Change $10\frac{5}{9}$ to an improper fraction.

$$10\frac{5}{9} = \frac{(10 \times 9) + 5}{9} = \frac{95}{9}$$ ■

A number containing an integer and an improper fraction may be simplified as follows.

◆ Example 10 Change $3\frac{8}{5}$ **a.** to an improper fraction and then **b.** to a mixed number in simplest form.

a. $3\frac{8}{5} = \frac{(3 \times 5) + 8}{5} = \frac{23}{5}$

b. $\frac{23}{5} = 23 \div 5 = 5\overline{)23} = 4\frac{3}{5}$
$\phantom{\frac{23}{5} = 23 \div 5 = 5\overline{)23}}4\text{ r }3$ ■

A calculator with a fraction key may be used to simplify fractions as follows. The fraction key often looks similar to $\boxed{A\frac{b}{c}}$.

◆ Example 11 Reduce $\frac{108}{144}$ to lowest terms.

108 $\boxed{A\frac{b}{c}}$ 144 $\boxed{=}$

$$\boxed{3\ulcorner4}$$

Thus, $\frac{108}{144} = \frac{3}{4}$ in lowest terms. ■

A calculator with a fraction key may be used to change an improper fraction to a mixed number, as follows.

Example 12

Change $\dfrac{25}{6}$ to a mixed number.

25 $\boxed{A\%}$ 6 $\boxed{=}$

$$\boxed{4\Gamma1\Gamma6}$$

Thus, $\dfrac{25}{6} = 4\dfrac{1}{6}$. ■

Note: The symbol Γ separates the whole number, the numerator, and the denominator in the display.

A calculator with a fraction key may be used to change a mixed number to an improper fraction as follows. The improper fraction key often looks similar to $\boxed{d\!/_e}$.

Note: Most scientific calculators are programmed so that several keys will perform more than one function. These calculators have what is called a *second function key*. To access this function, press the second function key first.

 Example 13

Change $6\dfrac{3}{5}$ to an improper fraction.

6 $\boxed{A\%}$ 3 $\boxed{A\%}$ 5 $\boxed{d\!/_e}$ $\boxed{=}$

$$\boxed{33\Gamma5}$$

Thus, $6\dfrac{3}{5} = \dfrac{33}{5}$. ■

Exercises 1.6 _____

Simplify:

1. $\dfrac{12}{28}$ 2. $\dfrac{9}{12}$ 3. $\dfrac{36}{42}$ 16. $\dfrac{6}{8}$ 17. $\dfrac{14}{16}$ 18. $\dfrac{7}{28}$

4. $\dfrac{12}{18}$ 5. $\dfrac{9}{48}$ 6. $\dfrac{8}{10}$ 19. $\dfrac{27}{36}$ 20. $\dfrac{15}{18}$ 21. $\dfrac{12}{16}$

7. $\dfrac{13}{39}$ 8. $\dfrac{24}{36}$ 9. $\dfrac{48}{60}$ 22. $\dfrac{9}{18}$ 23. $\dfrac{20}{25}$ 24. $\dfrac{12}{36}$

10. $\dfrac{72}{96}$ 11. $\dfrac{9}{9}$ 12. $\dfrac{15}{1}$ 25. $\dfrac{12}{40}$ 26. $\dfrac{54}{72}$ 27. $\dfrac{112}{128}$

13. $\dfrac{0}{8}$ 14. $\dfrac{6}{6}$ 15. $\dfrac{9}{0}$ 28. $\dfrac{330}{360}$ 29. $\dfrac{112}{144}$ 30. $\dfrac{525}{1155}$

Change each fraction to a mixed number in simplest form:

31. $\dfrac{78}{5}$ **32.** $\dfrac{11}{4}$ **33.** $\dfrac{28}{3}$

34. $\dfrac{21}{3}$ **35.** $\dfrac{45}{36}$ **36.** $\dfrac{67}{16}$

37. $\dfrac{57}{6}$ **38.** $\dfrac{84}{9}$ **39.** $5\dfrac{15}{12}$

40. $2\dfrac{70}{16}$

Change each mixed number to an improper fraction:

41. $3\dfrac{5}{6}$ **42.** $6\dfrac{3}{4}$ **43.** $2\dfrac{1}{8}$

44. $5\dfrac{2}{3}$ **45.** $1\dfrac{7}{16}$ **46.** $4\dfrac{1}{2}$

47. $6\dfrac{7}{8}$ **48.** $8\dfrac{1}{5}$ **49.** $10\dfrac{3}{5}$

50. $12\dfrac{5}{6}$

1.7 Addition and Subtraction of Fractions

Technicians must be able to compute fractions accurately, because mistakes on the job can be quite costly. Also, many shop drawing dimensions contain fractions.

Adding Fractions

$$\frac{a}{c} + \frac{b}{c} = \frac{a + b}{c} \quad (c \neq 0)$$

That is, to add two or more fractions with the same denominator, first add their numerators. Then place this sum over the common denominator and simplify.

 Example 1 Add: $\dfrac{1}{8} + \dfrac{3}{8}$.

$$\frac{1}{8} + \frac{3}{8} = \frac{1 + 3}{8} = \frac{4}{8} = \frac{1}{2} \qquad \text{Add the numerators and simplify.} \qquad ■$$

Example 2 Add: $\dfrac{2}{16} + \dfrac{5}{16}$.

$$\frac{2}{16} + \frac{5}{16} = \frac{2 + 5}{16} = \frac{7}{16} \qquad \text{Add the numerators.} \qquad ■$$

 Example 3 Add: $\dfrac{2}{31} + \dfrac{7}{31} + \dfrac{15}{31}$.

$$\frac{2}{31} + \frac{7}{31} + \frac{15}{31} = \frac{2 + 7 + 15}{31} = \frac{24}{31} \qquad \text{Add the numerators.} \qquad ■$$

To add fractions with different denominators, we first need to find a common denominator. When reducing fractions to lowest terms, we *divide* both numerator and denominator by the same nonzero number, which does not change the value of the fraction. Similarly, we can *multiply* both numerator and denominator by the same nonzero number without changing the value of the fraction. It is customary to find the **least common denominator (LCD)** for fractions with unlike denominators. The LCD is the smallest positive integer that has all the denominators as divisors. Then, multiply both numerator and denominator of each fraction by a number that makes the denominator of the given fraction the same as the LCD.

> To find the least common denominator (LCD) of a set of fractions:
>
> **1.** Factor each denominator into its prime factors.
> **2.** Write each prime factor the number of times it appears *most* in any *one* denominator in Step 1. The LCD is the product of these prime factors.

Example 4   Find the LCD of the following fractions: $\dfrac{1}{6}, \dfrac{1}{8},$ and $\dfrac{1}{18}$.

Step 1   Factor each denominator into prime factors. (Prime factorization may be reviewed in Section 1.5.)

$$6 = 2 \cdot 3$$
$$8 = 2 \cdot 2 \cdot 2$$
$$18 = 2 \cdot 3 \cdot 3$$

Step 2   Write each prime factor the number of times it appears *most* in any *one* denominator in Step 1. The LCD is the product of these prime factors.

Here, 2 appears once as a factor of 6, three times as a factor of 8, and once as a factor of 18. So 2 appears at most *three* times in any one denominator. Therefore, you have $2 \cdot 2 \cdot 2$ as factors of the LCD. The factor 3 appears at most twice in any one denominator, so you have $3 \cdot 3$ as factors of the LCD. Now 2 and 3 are the only factors of the three given denominators. The LCD for $\frac{1}{6}, \frac{1}{8},$ and $\frac{1}{18}$ must be $2 \cdot 2 \cdot 2 \cdot 3 \cdot 3 = 72$. Note that 72 does have divisors 6, 8, and 18. This procedure is shown in Table 1.1.

Table 1.1

| | | Number of times the prime factor appears | |
| | | in given denominator | *most* in any one denominator |
Prime factor	Denominator		
2	$6 = 2 \cdot 3$	once	
	$8 = 2 \cdot 2 \cdot 2$	three times	three times
	$18 = 2 \cdot 3 \cdot 3$	once	
3	$6 = 2 \cdot 3$	once	
	$8 = 2 \cdot 2 \cdot 2$	none	
	$18 = 2 \cdot 3 \cdot 3$	twice	twice

From the table, we see that the LCD contains the factor 2 three times and the factor 3 two times. Thus, LCD $= 2 \cdot 2 \cdot 2 \cdot 3 \cdot 3 = 72$.   ■

Example 5   Find the LCD of $\frac{3}{4}$, $\frac{3}{8}$, and $\frac{3}{16}$.

$$4 = 2 \cdot 2$$
$$8 = 2 \cdot 2 \cdot 2$$
$$16 = 2 \cdot 2 \cdot 2 \cdot 2$$

The factor 2 appears at most *four* times in any one denominator, so the LCD is $2 \cdot 2 \cdot 2 \cdot 2 = 16$. Note that 16 does have divisors 4, 8, and 16. ■

 Example 6   Find the LCD of $\frac{2}{5}$, $\frac{4}{15}$, and $\frac{3}{20}$.

$$5 = 5$$
$$15 = 3 \cdot 5$$
$$20 = 2 \cdot 2 \cdot 5$$

The LCD is $2 \cdot 2 \cdot 3 \cdot 5 = 60$. ■

Of course, if you can find the LCD by inspection, you need not go through the method shown in the examples.

 Example 7   Find the LCD of $\frac{3}{8}$ and $\frac{5}{16}$.

By inspection, the LCD is 16, because 16 is the smallest number that has divisors 8 and 16. ■

After finding the LCD of the fractions you wish to add, change each of the original fractions to a fraction of equal value, with the LCD as its denominator.

 Example 8   Add: $\frac{3}{8} + \frac{5}{16}$.

First, find the LCD of $\frac{3}{8}$ and $\frac{5}{16}$. The LCD is 16. Now change $\frac{3}{8}$ to a fraction of equal value with a denominator of 16.

Write: $\frac{3}{8} = \frac{?}{16}$.   Think: $8 \times ? = 16$.

Since $8 \times 2 = 16$, we multiply both the numerator and the denominator by 2. The numerator is 6, and the denominator is 16.

$$\frac{3}{8} \times \frac{2}{2} = \frac{6}{16}$$

Now, using the rule for adding fractions, we have

$$\frac{3}{8} + \frac{5}{16} = \frac{6}{16} + \frac{5}{16} = \frac{6+5}{16} = \frac{11}{16}$$ ■

Now try adding some fractions for which the LCD is more difficult to find.

Example 9   Add: $\dfrac{1}{4} + \dfrac{1}{6} + \dfrac{1}{16} + \dfrac{7}{12}$.

First, find the LCD.

$$4 = 2 \cdot 2$$
$$6 = 2 \cdot 3$$
$$16 = 2 \cdot 2 \cdot 2 \cdot 2$$
$$12 = 2 \cdot 2 \cdot 3$$

Note that 2 is used as a factor at most four times in any one denominator and 3 as a factor at most once. Thus, the LCD = $2 \cdot 2 \cdot 2 \cdot 2 \cdot 3 = 48$.

Second, change each fraction to an equivalent fraction with 48 as its denominator.

$$\frac{1}{4} = \frac{?}{48} \qquad \frac{1 \times 12}{4 \times 12} = \frac{12}{48}$$

$$\frac{1}{6} = \frac{?}{48} \qquad \frac{1 \times 8}{6 \times 8} = \frac{8}{48}$$

$$\frac{1}{16} = \frac{?}{48} \qquad \frac{1 \times 3}{16 \times 3} = \frac{3}{48}$$

$$\frac{7}{12} = \frac{?}{48} \qquad \frac{7 \times 4}{12 \times 4} = \frac{28}{48}$$

$$\frac{1}{4} + \frac{1}{6} + \frac{1}{16} + \frac{7}{12} = \frac{12}{48} + \frac{8}{48} + \frac{3}{48} + \frac{28}{48}$$

$$= \frac{12 + 8 + 3 + 28}{48}$$

$$= \frac{51}{48}$$

Simplifying, we have

$$\frac{51}{48} = 1\frac{3}{48} = 1\frac{\cancel{3} \cdot 1}{\cancel{3} \cdot 16} = 1\frac{1}{16} \qquad ■$$

Subtracting Fractions

$$\frac{a}{c} - \frac{b}{c} = \frac{a - b}{c} \qquad (c \neq 0)$$

To subtract two or more fractions with a common denominator, subtract their mumerators and place the difference over the common denominator and simplify.

Example 10   Subtract: $\dfrac{3}{5} - \dfrac{2}{5}$.

$$\frac{3}{5} - \frac{2}{5} = \frac{3 - 2}{5} = \frac{1}{5} \qquad \text{Subtract the numerators.} \qquad ■$$

◆ Example 11   Subtract: $\dfrac{5}{8} - \dfrac{3}{8}$.

$$\frac{5}{8} - \frac{3}{8} = \frac{5-3}{8} = \frac{2}{8} = \frac{\cancel{2}\cdot 1}{\cancel{2}\cdot 4} = \frac{1}{4}$$ Subtract the numerators and simplify.   ■

To subtract two fractions that have different denominators, first find the LCD. Then express each fraction as an equivalent fraction using the LCD, and subtract the numerators.

◆ Example 12   Subtract: $\dfrac{5}{6} - \dfrac{4}{15}$.

$$\frac{5}{6} - \frac{4}{15} = \frac{25}{30} - \frac{8}{30}$$ First change the fractions to the LCD, 30.

$$= \frac{25-8}{30} = \frac{17}{30}$$ Subtract the numerators.   ■

Adding Mixed Numbers

To add mixed numbers, find the LCD of the fractions. Add the fractions, then add the whole numbers. Finally, add these two results and simplify.

◆ Example 13   Add: $2\dfrac{1}{2}$ and $3\dfrac{3}{5}$.

$$2\frac{1}{2} = 2\frac{5}{10}$$ First change the proper fractions to the LCD, 10.

$$3\frac{3}{5} = 3\frac{6}{10}$$

$$5\frac{11}{10} = 5 + \frac{11}{10} = 5 + 1\frac{1}{10} = 6\frac{1}{10}$$ ■

Subtracting Mixed Numbers

To subtract mixed numbers, find the LCD of the fractions. Subtract the fractions, then subtract the whole numbers and simplify.

◆ Example 14   Subtract: $8\dfrac{2}{3}$ from $13\dfrac{3}{4}$.

$$13\frac{3}{4} = 13\frac{9}{12}$$ First change the proper fractions to the LCD, 12.

$$8\frac{2}{3} = 8\frac{8}{12}$$

$$5\frac{1}{12}$$ ■

If the larger of the two mixed numbers does not also have the larger proper fraction, borrow 1 from the whole number. Then add it to the proper fraction before subtracting.

◆ **Example 15**   Subtract: $2\frac{3}{5}$ from $4\frac{1}{2}$.

$$4\frac{1}{2} = 4\frac{5}{10} = 3\frac{15}{10}$$
$$2\frac{3}{5} = 2\frac{6}{10} = 2\frac{6}{10}$$
$$\overline{\phantom{2\frac{3}{5} = 2\frac{6}{10} = }}1\frac{9}{10}$$

First change the proper fractions to the LCD, 10. Then borrow 1 from 4 and add $\frac{10}{10}$ to $\frac{5}{10}$.   ■

◆ **Example 16**   Subtract: $2\frac{3}{7}$ from $8\frac{1}{4}$.

$$8\frac{1}{4} = 8\frac{7}{28} = 7\frac{35}{28}$$
$$2\frac{3}{7} = 2\frac{12}{28} = 2\frac{12}{28}$$
$$\overline{\phantom{2\frac{3}{7} = 2\frac{12}{28} = }}5\frac{23}{28}$$

First change the proper fractions to the LCD, 28. Then borrow 1 from 8 and add $\frac{28}{28}$ to $\frac{7}{28}$.   ■

◆ **Example 17**   Subtract: $12 - 4\frac{3}{8}$.

$$12 = 11\frac{8}{8}$$
$$4\frac{3}{8} = 4\frac{3}{8}$$
$$\overline{\phantom{4\frac{3}{8} = 1}}7\frac{5}{8}$$

First change the whole number 12 to the LCD, 8, as shown. Then borrow 1 from 12.   ■

Applications Involving Addition and Subtraction of Fractions

An electrical circuit with more than one path for the current to flow is called a *parallel circuit*. See Figure 1.25. The current in a parallel circuit is divided among the branches in the circuit. How it is divided depends on the resistance in each branch. Since the current is divided among the branches, the total current (I_T) of the circuit is the same as the current from the source. This equals the sum of the currents through the individual branches of the circuit. That is, $I_T = I_1 + I_2 + I_3 + \cdots$.

Parallel circuit

FIGURE 1.25

◆ Example 18   Find the total current in the parallel circuit in Figure 1.26.

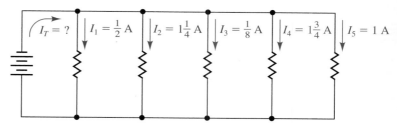

FIGURE 1.26

$$I_T = I_1 + I_2 + I_3 + I_4 + I_5$$

$\dfrac{1}{2}$ A $= \dfrac{4}{8}$ A   First change the proper fractions to the LCD, 8.

$1\dfrac{1}{4}$ A $= 1\dfrac{2}{8}$ A

$\dfrac{1}{8}$ A $= \dfrac{1}{8}$ A

$1\dfrac{3}{4}$ A $= 1\dfrac{6}{8}$ A

$\underline{1\ \ \text{A} = 1\ \ \text{A}}$

$3\dfrac{13}{8}$ A $= 4\dfrac{5}{8}$ A   ■

◆ Example 19   Find the missing dimension in Figure 1.27.

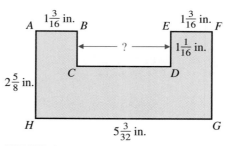

FIGURE 1.27

First, note that the length of side *HG* equals the sum of the lengths of sides *AB*, *CD*, and *EF*. To find the length of the missing dimension, subtract the sum of side *AB* and side *EF* from side *HG*.

That is, first add *AB* + *EF*:   Then subtract *HG* − (*AB* + *EF*):

AB: $1\dfrac{3}{16}$ in.   *HG*: $5\dfrac{3}{32}$ in. $= 5\dfrac{3}{32}$ in. $= 4\dfrac{35}{32}$ in.

EF: $\underline{1\dfrac{3}{16}}$ in.   *AB* + *EF*: $\underline{2\dfrac{3}{8}}$ in. $= \underline{2\dfrac{12}{32}}$ in. $= \underline{2\dfrac{12}{32}}$ in.

AB + *EF*: $2\dfrac{6}{16}$ in., or $2\dfrac{3}{8}$ in. $= 2\dfrac{23}{32}$ in.

Therefore, the missing dimension is $2\dfrac{23}{32}$ in.   ■

The **perimeter** of a geometric figure is the sum of the lengths of its sides.

 Example 20

Find the perimeter of Figure 1.27.

$$AB: \quad 1\frac{3}{16} \text{ in.} = 1\frac{6}{32} \text{ in.}$$

$$BC: \quad 1\frac{1}{16} \text{ in.} = 1\frac{2}{32} \text{ in.}$$

$$CD: \quad 2\frac{23}{32} \text{ in.} = 2\frac{23}{32} \text{ in.}$$

$$DE: \quad 1\frac{1}{16} \text{ in.} = 1\frac{2}{32} \text{ in.}$$

$$EF: \quad 1\frac{3}{16} \text{ in.} = 1\frac{6}{32} \text{ in.}$$

$$FG: \quad 2\frac{5}{8} \text{ in.} = 2\frac{20}{32} \text{ in.}$$

$$GH: \quad 5\frac{3}{32} \text{ in.} = 5\frac{3}{32} \text{ in.}$$

$$HA: \quad 2\frac{5}{8} \text{ in.} = 2\frac{20}{32} \text{ in.}$$

Perimeter: $\quad 15\frac{82}{32} \text{ in.} = 15 + 2\frac{18}{32} \text{ in.} = 17\frac{9}{16} \text{ in.}$ ■

Using a Calculator to Add and Subtract Fractions

Example 21

Add: $\dfrac{2}{3} + \dfrac{7}{12}$.

2 [A⅟c] 3 [+] 7 [A⅟c] 12 [=]

| 1⌐1⌐4 |

Thus, $\dfrac{2}{3} + \dfrac{7}{12} = 1\dfrac{1}{4}$.   ■

Example 22

Add: $7\dfrac{3}{4} + 5\dfrac{11}{12}$.

7 [A⅟c] 3 [A⅟c] 4 [+] 5 [A⅟c] 11 [A⅟c] 12 [=]

| 13⌐2⌐3 |

Thus, $7\dfrac{3}{4} + 5\dfrac{11}{12} = 13\dfrac{2}{3}$.   ■

Example 23   Subtract: $8\frac{1}{4} - 3\frac{5}{12}$.

8 $A\frac{b}{c}$ 1 $A\frac{b}{c}$ 4 − 3 $A\frac{b}{c}$ 5 $A\frac{b}{c}$ 12 =

$$4\Gamma5\Gamma6$$

Thus, $8\frac{1}{4} - 3\frac{5}{12} = 4\frac{5}{6}$. ■

Exercises 1.7

Find the LCD of each set of fractions:

1. $\frac{1}{2}, \frac{1}{8}, \frac{1}{16}$ **2.** $\frac{1}{3}, \frac{2}{5}, \frac{3}{7}$

3. $\frac{1}{6}, \frac{3}{10}, \frac{1}{14}$ **4.** $\frac{1}{9}, \frac{1}{15}, \frac{5}{21}$

5. $\frac{1}{3}, \frac{1}{16}, \frac{7}{8}$ **6.** $\frac{1}{5}, \frac{3}{14}, \frac{4}{35}$

Perform the indicated operations and simplify:

7. $\frac{2}{3} + \frac{1}{6}$ **8.** $\frac{1}{2} + \frac{3}{8}$

9. $\frac{1}{16} + \frac{3}{32}$ **10.** $\frac{5}{6} + \frac{1}{18}$

11. $\frac{2}{7} + \frac{3}{28}$ **12.** $\frac{1}{9} + \frac{2}{45}$

13. $\frac{3}{8} + \frac{5}{64}$ **14.** $\frac{3}{10} + \frac{7}{100}$

15. $\frac{1}{5} + \frac{3}{20}$ **16.** $\frac{3}{4} + \frac{3}{16}$

17. $\frac{4}{5} + \frac{1}{2}$ **18.** $\frac{2}{3} + \frac{4}{9}$

19. $\frac{1}{3} + \frac{1}{6} + \frac{3}{16} + \frac{1}{12}$ **20.** $\frac{3}{16} + \frac{1}{8} + \frac{1}{3} + \frac{1}{4}$

21. $\frac{1}{20} + \frac{1}{30} + \frac{1}{40}$ **22.** $\frac{1}{14} + \frac{1}{15} + \frac{1}{6}$

23. $\frac{3}{10} + \frac{1}{14} + \frac{4}{15}$ **24.** $\frac{5}{36} + \frac{11}{72} + \frac{5}{6}$

25. $\frac{7}{8} - \frac{3}{4}$ **26.** $\frac{9}{64} - \frac{2}{128}$

27. $\frac{4}{5} - \frac{3}{10}$ **28.** $\frac{7}{16} - \frac{1}{3}$

29. $\frac{9}{14} - \frac{3}{42}$ **30.** $\frac{8}{9} - \frac{5}{24}$

31. $\frac{9}{16} - \frac{13}{32} - \frac{1}{8}$ **32.** $\frac{7}{8} - \frac{2}{9} - \frac{1}{12}$

33. $2\frac{1}{2} + 4\frac{3}{4}$ **34.** $3\frac{5}{8} + 5\frac{3}{4}$

35. $3 - \frac{3}{8}$ **36.** $8 - 5\frac{3}{4}$

37. $8\frac{3}{16} - 3\frac{7}{16}$ **38.** $5\frac{3}{8} + 2\frac{3}{4}$

39. $7\frac{3}{16} - 4\frac{7}{8}$ **40.** $8\frac{1}{4} - 4\frac{7}{16}$

41. $3\frac{4}{5} + 9\frac{8}{9}$ **42.** $4\frac{5}{12} + 6\frac{17}{20}$

43. $3\frac{9}{16} + 4\frac{7}{12} + 3\frac{1}{6}$ **44.** $5\frac{2}{5} + 3\frac{7}{10} + 4\frac{7}{15}$

45. $16\frac{5}{8} - 4\frac{7}{12} - 2\frac{1}{2}$ **46.** $12\frac{9}{16} - 3\frac{1}{6} + 2\frac{1}{4}$

47. Find the perimeter of the triangular plot in Illustration 1.

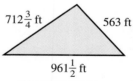
$712\frac{3}{4}$ ft ... 563 ft ... $961\frac{1}{2}$ ft
ILLUSTRATION 1

48. A welder has four pieces of scrap steel angle of lengths $3\frac{1}{4}$ ft, $2\frac{3}{8}$ ft, $3\frac{1}{8}$ ft, and $4\frac{3}{16}$ ft. If they are welded together, how long is the welded piece?

 49. A welder has two pieces of half-inch pipe, one of length $2\frac{3}{8}$ ft and another of length $3\frac{7}{8}$ ft. **a.** What is the total length of the two welded together? **b.** If she needs a total length of $4\frac{3}{4}$ ft, how much must be cut off?

50. What is the difference in the size (diameter) of 6011 welding rods of diameter $\frac{1}{8}$ in. and Super Strength 100 rods of diameter $\frac{3}{32}$ in.?

51. A pilot flies a small plane on a cross-country trip to two cities and begins with a full tank of fuel from the home airport. Upon arrival at the first location, the plane required $13\frac{3}{4}$ gal of 100*LL* aviation fuel. At the next stop, the plane required $11\frac{2}{5}$ gal. Upon return to the home airport, the plane took $10\frac{2}{5}$ gal to fill the tank. How much total fuel was used on the trip?

52. A Piper Warrior holds 50 gal of 100*LL* aviation fuel. A pilot takes off and lands at another airport and fills up the tank, which takes $17\frac{1}{2}$ gal. The pilot then flies to a second airport, which requires $20\frac{3}{8}$ gal. Had the pilot made the trip to the two airports without refueling each time, how much fuel would have been left in the tank?

53. A pilot flies to an island off the coast of North Carolina and uses $25\frac{1}{4}$ gal of fuel. The return trip only uses $23\frac{3}{4}$ gal. The difference is due to the wind. What is the difference in the fuel used?

54. Oil is changed in three cars. They hold $4\frac{1}{2}$ qt, $4\frac{6}{10}$ qt, and $4\frac{1}{16}$ qt. How much total oil is used?

55. A mechanic spent $\frac{1}{3}$ h changing spark plugs, $\frac{1}{4}$ h changing an air filter, and $\frac{1}{4}$ h changing the oil and oil filter. How much total time was spent servicing this car?

56. A heating and cooling specialist needs two pieces of duct $3\frac{3}{4}$ ft and $2\frac{1}{4}$ ft in length. There are two pieces in stock that are each 4 ft long. After these two lengths are cut off, what excess will be left?

57. The cooling requirements for the three separate incubation rooms are $\frac{1}{3}$ ton, $\frac{3}{4}$ ton, and $\frac{9}{16}$ ton. If a central HVAC unit will be installed, how many tons are required?

58. A finished product consists of four components that will be assembled and packaged for shipment. The box manufacturer has requested the total product weight be on the drawing so that the appropriate strength cardboard is used. What is the product weight? (1 lb = 16 oz)

Part	Weight each
1	$3\frac{1}{2}$ oz
2	$33\frac{1}{8}$ oz
3	6 lb
4	$10\frac{1}{3}$ oz

59. What is the distance between points *A* and *B* in Illustration 2?

ILLUSTRATION 2

*Find **a.** the length of the missing dimension and **b.** the perimeter of each figure.*

60.

61.

62.

63.

64. The perimeter of a triangle is $59\frac{9}{32}$ in. One side is $19\frac{5}{8}$ in., and a second side is $17\frac{13}{16}$ in. How long is the remaining side?

Find the total current in each parallel circuit:

65.

66.

67.

68.

69. Find the length of the shaft in Illustration 3.

ILLUSTRATION 3

70. Find the distance between the centers of the two end-holes of the plate in Illustration 4.

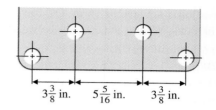

$3\frac{3}{8}$ in. $5\frac{5}{16}$ in. $3\frac{3}{8}$ in.

ILLUSTRATION 4

71. In Illustration 5, find **a.** the length of the tool and **b.** the length of diameter A.

ILLUSTRATION 5

72. A rod $13\frac{13}{16}$ in. long has been cut as shown in Illustration 6. Assume that the waste in each cut is $\frac{1}{16}$ in. What is the length of the remaining piece?

ILLUSTRATION 6

73. Find **a.** the length and **b.** the diameter of the shaft in Illustration 7.

ILLUSTRATION 7

74. Find the missing dimension of the shaft in Illustration 8.

ILLUSTRATION 8

75. Floor joists are spaced 16 in. OC (on center) and are $1\frac{5}{8}$ in. thick. What is the distance between them?

76. If no tap drill chart is available, the correct drill size (*TDS*) can be found by using the formula $TDS = OD - P$, where *OD* is the outside diameter and *P* is the pitch of the thread (the distance between successive threads). Find the tap drill size for a $\frac{3}{8}$-in. outside diameter if the pitch is $\frac{1}{16}$ in.

77. How much must the diameter of a $\frac{7}{8}$-in. shaft be reduced so that its diameter will be $\frac{51}{64}$ in.?

78. What is the difference in thickness between a $\frac{7}{16}$-in. steel plate and a $\frac{5}{8}$-in. steel plate?

79. A planer takes a $\frac{3}{32}$-in. cut on a plate that is $1\frac{7}{8}$ in. thick. What is the thickness of the plate after one cut? What is the thickness of the plate after three cuts?

80. A home is built on a $65\frac{3}{4}$-ft-wide lot. The house is $5\frac{5}{12}$ ft from one side of the lot and is $43\frac{5}{6}$ ft wide. (See Illustration 9.) What is the distance from the house to the other side of the lot?

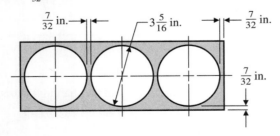

ILLUSTRATION 9

81. See Illustration 10. What width and length steel strip is needed in order to drill three holes of diameter $3\frac{5}{16}$ in.? Allow $\frac{7}{32}$ in. between and on each side of the holes.

ILLUSTRATION 10

82. Find length *x* in Illustration 11.

ILLUSTRATION 11

83. Find the total length of the valve in Illustration 12.

ILLUSTRATION 12

84. Find the total length of the shaft in Illustration 13.

ILLUSTRATION 13

85. A mechanic needs the following lengths of $\frac{3}{8}$-in. copper tubing: $15\frac{3}{8}$ in., $7\frac{3}{4}$ in., $11\frac{1}{2}$ in., $7\frac{7}{32}$ in., and $10\frac{5}{16}$ in. What is the total length of tubing needed?

86. An end view and side view of a shaft are shown in Illustration 14. **a.** Find the diameter of the largest part of the shaft. **b.** Find dimension *A* of the shaft.

Side view
End view

ILLUSTRATION 14

87. A homeowner burns three truckloads of firewood that contain $\frac{2}{3}$, $\frac{3}{4}$, and $\frac{2}{3}$ of a cord, respectively, during the winter. How many cords of firewood did she burn?

88. A homeowner has a lot that measures $1\frac{1}{2}$ acres. Of that area, $\frac{1}{2}$ acre is wooded, $\frac{1}{6}$ acre is covered by the house and shrubbery, and $\frac{1}{3}$ acre is covered by a driveway and groundcover. The rest is planted in grass for lawns all around the house. How much lawn does the homeowner have?

89. A recreational hiker walks a trail with signs that indicate the following distances between points: $1\frac{1}{2}$ mi, $2\frac{3}{4}$ mi, $\frac{3}{4}$ mi, and $\frac{1}{2}$ mi. How far did she walk on the trail?

1.8 Multiplication and Division of Fractions

Multiplying Fractions

$$\frac{a}{b} \times \frac{c}{d} = \frac{a \cdot c}{b \cdot d} \qquad (b \neq 0, d \neq 0)$$

To multiply fractions, multiply the numerators and multiply the denominators. Then reduce the resulting fraction to lowest terms.

Example 1 Multiply: $\frac{5}{9} \times \frac{3}{10}$.

$$\frac{5}{9} \times \frac{3}{10} = \frac{5 \cdot 3}{9 \cdot 10} = \frac{15}{90} = \frac{\cancel{15} \cdot 1}{\cancel{15} \cdot 6} = \frac{1}{6}$$

To simplify the work, consider the following alternative method:

$$\frac{\overset{1}{\cancel{5}}}{\underset{3}{\cancel{9}}} \times \frac{\overset{1}{\cancel{3}}}{\underset{2}{\cancel{10}}} = \frac{1 \cdot 1}{3 \cdot 2} = \frac{1}{6}$$

This method divides the numerator by 15, or (5×3), and the denominator by 15, or (5×3). It does not change the value of the fraction. ■

Example 2 Multiply: $\frac{18}{25} \times \frac{7}{27}$.

As a shortcut, divide a numerator by 9 and a denominator by 9. Then multiply:

$$\frac{\overset{2}{\cancel{18}}}{25} \times \frac{7}{\underset{3}{\cancel{27}}} = \frac{2 \cdot 7}{25 \cdot 3} = \frac{14}{75}$$ ■

Any mixed number must be replaced by an equivalent improper fraction before multiplying or dividing two or more fractions.

Example 3   Multiply: $8 \times 3\frac{3}{4}$.

$$8 \times 3\frac{3}{4} = \frac{\overset{2}{\cancel{8}}}{1} \times \frac{15}{\underset{1}{\cancel{4}}} = \frac{30}{1} = 30 \qquad \text{First change } 3\frac{3}{4} \text{ to an improper fraction.} \qquad ■$$

 Example 4   Multiply: $\frac{9}{16} \times \frac{5}{22} \times \frac{4}{7} \times 3\frac{2}{3}$.

$$\frac{9}{16} \times \frac{5}{22} \times \frac{4}{7} \times 3\frac{2}{3} = \frac{\overset{3}{\cancel{9}}}{\underset{4}{\cancel{16}}} \times \frac{5}{\underset{2}{\cancel{22}}} \times \frac{\overset{1}{\cancel{4}}}{7} \times \frac{\overset{1}{\cancel{11}}}{\underset{1}{\cancel{3}}} = \frac{15}{56} \qquad ■$$

Note: Whenever you multiply several fractions, you may simplify the computation by dividing *any* numerator and *any* denominator by the same number.

 Dividing Fractions

$$\frac{a}{b} \div \frac{c}{d} = \frac{a}{b} \times \frac{d}{c} = \frac{a \cdot d}{b \cdot c} \qquad (b \neq 0, c \neq 0, d \neq 0)$$

To divide a fraction by a fraction, invert the fraction (interchange numerator and denominator) that follows the division sign ($\div$). Then multiply the resulting fractions.

 Example 5   Divide: $\frac{5}{6} \div \frac{2}{3}$.

$$\frac{5}{6} \div \frac{2}{3} = \frac{5}{\underset{2}{\cancel{6}}} \times \frac{\overset{1}{\cancel{3}}}{2} = \frac{5 \cdot 1}{2 \cdot 2} = \frac{5}{4} \text{ or } 1\frac{1}{4} \qquad \text{Invert } \tfrac{2}{3} \text{ and multiply.} \qquad ■$$

 Example 6   Divide: $7 \div \frac{2}{5}$.

$$7 \div \frac{2}{5} = \frac{7}{1} \times \frac{5}{2} = \frac{35}{2} \text{ or } 17\frac{1}{2} \qquad \text{Invert } \tfrac{2}{5} \text{ and multiply.} \qquad ■$$

 Example 7   Divide: $\frac{3}{5} \div 4$.

$$\frac{3}{5} \div 4 = \frac{3}{5} \div \frac{4}{1} = \frac{3}{5} \times \frac{1}{4} = \frac{3}{20} \qquad \text{Invert 4 and multiply.} \qquad ■$$

Example 8

Divide: $\dfrac{9}{10} \div 5\dfrac{2}{5}$.

$$\frac{9}{10} \div 5\frac{2}{5} = \frac{9}{10} \div \frac{27}{5} = \frac{\overset{1}{\cancel{9}}}{\underset{2}{\cancel{10}}} \times \frac{\overset{1}{\cancel{5}}}{\underset{3}{\cancel{27}}} = \frac{1}{6}$$ Invert $\frac{27}{5}$ and multiply. ■

When both multiplication and division of fractions occur, invert only the first fraction that follows a division sign ($\div$). Then proceed according to the rules for multiplying fractions.

Example 9

Perform the indicated operations and simplify: $\dfrac{2}{5} \times \dfrac{1}{3} \div \dfrac{3}{4}$.

$$\frac{2}{5} \times \frac{1}{3} \div \frac{3}{4} = \frac{2}{5} \times \frac{1}{3} \times \frac{4}{3} = \frac{2 \cdot 1 \cdot 4}{5 \cdot 3 \cdot 3} = \frac{8}{45}$$ ■

Example 10

Perform the indicated operations and simplify: $7\frac{1}{3} \div 4 \times 2$.

$$7\frac{1}{3} \div 4 \times 2 = \frac{\overset{11}{\cancel{22}}}{3} \times \frac{1}{\underset{2}{\cancel{4}}} \times \frac{\overset{1}{\cancel{2}}}{1} = \frac{11}{3} \quad \text{or} \quad 3\frac{2}{3}$$ ■

Applications Involving Multiplication and Division of Fractions

Lumber is usually measured in board feet. One *board foot* is the amount of wood contained in a piece of wood that measures one inch thick and one square foot in area, or its equivalent. (See Figure 1.28.) The number of board feet in lumber may be found by the formula

12 in.   1 in.
     1 ft

One board foot
FIGURE 1.28

$$\text{bd ft} = \frac{\overset{\text{number of}}{\underset{\text{boards}}{}} \times \overset{\text{thickness}}{\underset{\text{(in in.)}}{}} \times \overset{\text{width}}{\underset{\text{(in in.)}}{}} \times \overset{\text{length}}{\underset{\text{(in ft)}}{}}}{12}$$

The 12 in the denominator comes from the fact that the simplest form of one board foot can be thought of as a board that is 1 in. thick × 12 in. wide × 1 ft long.

Lumber is either rough or finished. *Rough stock* is lumber that is not planed or dressed; *finished stock* is planed on one or more sides. When measuring lumber, we compute the full size. That is, we compute the measure of the rough stock that is required to make the desired finished piece. When lumber is finished or planed, $\frac{1}{16}$ in. is taken off each side when the lumber is less than $1\frac{1}{2}$ in. thick. If the lumber is $1\frac{1}{2}$ in. or more in thickness, $\frac{1}{8}$ in. is taken off each side. (**Note:** Lumber for framing houses usually measures $\frac{1}{2}$ in. less than the name that we call the piece. For example, a "two-by-four," a piece 2 in. by 4 in., actually measures $1\frac{1}{2}$ in. by $3\frac{1}{2}$ in.)

Example 11

Find the number of board feet contained in 6 pieces of lumber 2 in. × 8 in. × 16 ft (Figure 1.29).

16 ft

2 in.

8 in.

FIGURE 1.29

$$\text{bd ft} = \frac{\overset{\text{number of}}{\underset{\text{boards}}{}} \times \overset{\text{thickness}}{\underset{\text{(in in.)}}{}} \times \overset{\text{width}}{\underset{\text{(in in.)}}{}} \times \overset{\text{length}}{\underset{\text{(in ft)}}{}}}{12}$$

$$= \frac{6 \times 2 \times 8 \times 16}{12} = 128 \text{ bd ft} \qquad ■$$

Example 12

Energy in the form of electrical power is used by industry and consumers alike. Power (in watts, W) equals the voltage (in volts, V) times the current (in amperes, or amps, A). A soldering iron draws a current of $7\frac{1}{2}$ A on a 110-V circuit. Find the wattage, or power, rating of this soldering iron.

$$\text{Power} = (\text{voltage}) \times (\text{current})$$

$$= 110 \quad \times \quad 7\frac{1}{2}$$

$$= 110 \quad \times \quad \frac{15}{2}$$

$$= 825 \text{ W}$$

Power may also be found by computing the product of the square of the current (in amps, A) and the resistance (in ohms, Ω). ■

Example 13

To give $\frac{1}{5}$ grain of Myleran from $\frac{1}{30}$-grain tablets, how many tablets would be given?

To find how many tablets would be given, we divide the amount to be given by the amount in each tablet.

$$\frac{1}{5} \div \frac{1}{30} = \frac{1}{\cancel{5}} \times \frac{\overset{6}{\cancel{30}}}{1}$$

$$= 6 \text{ tablets} \qquad ■$$

Example 14

One form of Ohm's law states that the current I (in amps, A) in a simple circuit equals the voltage E (in volts, V) divided by the resistance R (in ohms, Ω). What current is required for a heating element with a resistance of $7\frac{1}{2}$ Ω operating in a 12-V circuit?

$$\text{Current} = (\text{voltage}) \div (\text{resistance})$$

$$= 12 \quad \div \quad 7\frac{1}{2}$$

$$= 12 \quad \div \quad \frac{15}{2}$$

$$= \overset{4}{\cancel{12}} \quad \times \quad \frac{2}{\underset{5}{\cancel{15}}}$$

$$= \frac{8}{5} \text{ or } 1\frac{3}{5} \text{ A}$$

Using a Calculator to Multiply and Divide Fractions

◆ Example 15

Multiply: $2\frac{5}{6} \times 4\frac{1}{2}$.

2 A$\frac{b}{c}$ 5 A$\frac{b}{c}$ 6 $\times$ 4 A$\frac{b}{c}$ 1 A$\frac{b}{c}$ 2 $=$

$$12 \Gamma 3 \Gamma 4$$

Thus, $2\frac{5}{6} \times 4\frac{1}{2} = 12\frac{3}{4}$.

◆ Example 16

Divide: $5\frac{5}{7} \div 8\frac{1}{3}$.

5 A$\frac{b}{c}$ 5 A$\frac{b}{c}$ 7 $\div$ 8 A$\frac{b}{c}$ 1 A$\frac{b}{c}$ 3 $=$

$$24 \Gamma 35$$

Thus, $5\frac{5}{7} \div 8\frac{1}{3} = \frac{24}{35}$.

Exercises 1.8

Perform the indicated operations and simplify:

1. $\frac{2}{3} \times 18$

2. $8 \times \frac{1}{2}$

3. $\frac{3}{4} \times 12$

4. $3\frac{1}{2} \times \frac{2}{5}$

5. $1\frac{3}{4} \times \frac{5}{16}$

6. $\frac{1}{3} \times \frac{1}{3} \times \frac{1}{3}$

7. $\frac{16}{21} \times \frac{7}{8}$

8. $\frac{7}{12} \times \frac{45}{56}$

9. $\frac{2}{7} \times 35$

10. $\frac{9}{16} \times \frac{2}{3} \times 1\frac{6}{15}$

11. $\frac{5}{8} \times \frac{7}{10} \times \frac{2}{7}$

12. $\frac{9}{16} \times \frac{5}{9} \times \frac{4}{25}$

13. $2\frac{1}{3} \times \frac{5}{8} \times \frac{6}{7}$

14. $\frac{5}{28} \times \frac{3}{5} \times \frac{2}{3} \times \frac{2}{9}$

15. $\dfrac{6}{11} \times \dfrac{26}{35} \times 1\dfrac{9}{13} \times \dfrac{7}{12}$

16. $\dfrac{3}{8} \div \dfrac{1}{4}$

17. $\dfrac{3}{5} \div \dfrac{10}{12}$

18. $\dfrac{10}{12} \div \dfrac{3}{5}$

19. $4\dfrac{1}{2} \div \dfrac{1}{4}$

20. $18\dfrac{2}{3} \div 6$

21. $15 \div \dfrac{3}{8}$

22. $\dfrac{77}{6} \div 6$

23. $\dfrac{7}{11} \div \dfrac{3}{5}$

24. $7 \div 3\dfrac{1}{8}$

25. $\dfrac{2}{5} \times 3\dfrac{2}{3} \div \dfrac{3}{4}$

26. $\dfrac{7}{8} \times \dfrac{1}{2} \div \dfrac{2}{7}$

27. $\dfrac{16}{5} \times \dfrac{3}{2} \times \dfrac{10}{4} \div 5\dfrac{1}{3}$

28. $6 \times 6 \times \dfrac{21}{7} \div 48$

29. $\dfrac{7}{9} \times \dfrac{3}{8} \div \dfrac{28}{81}$

30. $2\dfrac{1}{3} \times \dfrac{5}{8} \div \dfrac{10}{4}$

31. $\dfrac{2}{7} \times \dfrac{5}{9} \times \dfrac{3}{10} \div 6$

32. $\dfrac{9}{4} \times \dfrac{9}{4} \times \dfrac{21}{7} \div 81$

33. $\dfrac{7}{16} \div \dfrac{3}{8} \times \dfrac{1}{2}$

34. $\dfrac{5}{8} \div \dfrac{25}{64} \times \dfrac{5}{6}$

35. A barrel has a capacity of 42 gal. How many gallons does it contain when it is $\dfrac{3}{4}$ full?

36. a. Find the area of a rectangle with length $6\dfrac{1}{3}$ ft and width $3\dfrac{3}{4}$ ft. (Area = length × width.)
   b. Find its perimeter.

37. A welder uses seven 6011 welding rods to weld two metal slabs. If each rod makes a $6\dfrac{1}{2}$-in. weld, find the total length of the weld.

38. A welder has $6\dfrac{2}{3}$ ft of $\dfrac{1}{2}$-in. pipe. How many pieces of pipe, each of length $1\dfrac{3}{4}$ ft, can be obtained from the original pipe?

39. A small aircraft flew a total of $684\dfrac{1}{4}$ mi. If it took the plane $5\dfrac{2}{3}$ h to make the trip, how fast was the aircraft flying?

40. On a Civil Air Patrol mission, five search planes searched for $3\dfrac{1}{4}$ h for a missing aircraft. How many total hours did they search?

41. Nine pieces of 8-in. × 12-in. duct that is $3\dfrac{2}{3}$ ft in length is needed for a building. What is the total length needed?

42. The HVAC supply duct is 17 ft long. Our truck bed can only carry ducts $4\dfrac{1}{2}$ ft long. How many pieces must the 17-ft duct be cut into and how long is each piece, assuming most of the pieces will be $4\dfrac{1}{2}$ ft?

Find the number of board feet in each quantity of lumber:

43. 10 pieces 2 in. × 4 in. × 12 ft

44. 24 pieces 4 in. × 4 in. × 16 ft

45. 175 pieces 1 in. × 8 in. × 14 ft

46. Find the total length of eight pieces of steel each $5\dfrac{3}{4}$ in. long.

47. The outside diameter (OD) of a pipe is $4\dfrac{9}{32}$ in. The walls are $\dfrac{7}{32}$ in. thick. Find the inside diameter. (See Illustration 1.)

ILLUSTRATION 1

48. Two strips of metal are riveted together in a straight line, with nine rivets equally spaced $2\dfrac{5}{16}$ in. apart. What is the distance between the first and last rivet?

49. Two metal strips are riveted together in a straight line, with 16 equally spaced rivets. The distance between the first and last rivet is $28\dfrac{1}{8}$ in. Find the distance between any two consecutive rivets.

50. Find length x, the distance between centers, in Illustration 2.

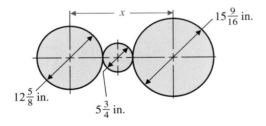

ILLUSTRATION 2

51. From a steel rod 36 in. long, the following pieces are cut:

   3 pieces $2\dfrac{1}{8}$ in. long

   2 pieces $5\dfrac{3}{4}$ in. long

   6 pieces $\dfrac{7}{8}$ in. long

   1 piece $3\dfrac{1}{2}$ in. long

Assume $\dfrac{1}{16}$ in. of waste in each cut. Find the length of the remaining piece.

52. A piece of drill rod 2 ft 6 in. long is to be cut into pins, each $2\frac{1}{2}$ in. long.

    a. Assume no loss of material in cutting. How many pins will you get?

    b. Assume $\frac{1}{16}$ in. waste in each cut. How many pins will you get?

53. The cutting tool on a lathe moves $\frac{3}{128}$ in. along the piece being turned for each revolution of work. The piece revolves at 45 revolutions per minute. How long will it take to turn a length of $9\frac{9}{64}$-in. stock in one operation?

54. Three vents are equally spaced along a wall 26 ft 6 in. (318 in.) long, as shown in Illustration 3. Find dimension *d*.

ILLUSTRATION 3

55. A concrete pad for mounting a condensing unit is 4 ft long, $2\frac{2}{3}$ ft wide, and 3 in. ($\frac{1}{4}$ ft) thick. Find its volume in cubic feet.

56. How many lengths of radiator hose, each $5\frac{1}{4}$ in. long, can be cut from a 6-ft roll?

57. A car dealership received six cars that needed to be detailed. If the service staff took $7\frac{1}{2}$ h to detail the cars, how long did each car take?

58. Four tires can be replaced on a car in $\frac{3}{4}$ h. If 11 cars needed their tires changed, how long would it take?

59. Find the load of a circuit that takes $12\frac{1}{2}$ A at 220 V. (See Example 12.)

60. An electric iron requires $4\frac{1}{4}$ A and has a resistance of $24\frac{1}{2}$ Ω. What voltage does it require to operate? ($V = IR$.)

61. An electric hand drill draws $3\frac{3}{4}$ A and has a resistance of $5\frac{1}{3}$ Ω. What power does it use? ($P = I^2R$.)

62. A wiring job requires the following lengths of BX cable:

    12 pieces $8\frac{1}{2}$ ft long

    7 pieces $18\frac{1}{2}$ ft long

    24 pieces $1\frac{3}{4}$ ft long

    12 pieces $6\frac{1}{2}$ ft long

    2 pieces $34\frac{1}{4}$ ft long

How much cable is needed to complete the job?

63. What current is required for a heating element with a resistance of $10\frac{1}{2}$ Ω operating in a 24-V circuit? (See Example 14.)

64. How many lengths of wire, each $3\frac{3}{4}$ in. long, can be cut from a 25-ft roll?

65. A total of 19 ceiling outlets are to be equally spaced in a straight line between two points that are $130\frac{1}{2}$ ft apart in a hallway. How far apart will the ceiling outlets be, center to center?

66. Tom needs to apply $1\frac{3}{4}$ gal of herbicide per acre of soybeans. How many gallons of herbicide are needed for 120 acres?

67. An airplane sprayer tank holds 60 gal. If $\frac{3}{4}$ gal of water and $\frac{1}{2}$ lb of pesticide are applied per acre, how much pesticide powder is needed per tankful?

68. If 1 ft³ of cotton weighs $22\frac{1}{2}$ lb, how many cubic feet are contained in a bale of cotton weighing 500 lb? In 15 tons of cotton?

69. A test plot of $\frac{1}{20}$ acre produces 448 lb of shelled corn. Find the yield in bushels per acre. (1 bu of shelled corn weighs 56 lb.)

70. A farmer wishes to concrete his rectangular feed lot, which measures 120 ft by 180 ft. He wants to have a base of 4 in. of gravel covered with $3\frac{1}{2}$ in. of concrete.

    a. How many cubic yards of each material must he purchase?

    b. What is his total materials cost rounded to the nearest dollar? Concrete costs $90/yd³ delivered, and gravel costs $11/ton delivered. (1 yd³ of gravel weighs approximately 2500 lb.)

71. A medicine contains $\frac{1}{5}$ alcohol. A bottle holds $2\frac{1}{2}$ oz of this medicine. How many ounces of alcohol does the bottle contain?

72. The doctor orders 45 mg of prednisone. Each tablet contains 10 mg. How many tablets are given to the patient?

73. To give 50 mg of ascorbic acid from 100-mg tablets, how many tablets should be given?

74. To give 500 mg of ascorbic acid from 200-mg tablets, how many tablets should be given?

75. A patient is given $\frac{1}{4}$ of a 5-grain aspirin tablet. How much aspirin does the patient receive?

76. To give 1 grain of digitalis from $1\frac{1}{2}$-grain tablets, how many tablets should be given?

77. A patient is given $\frac{3}{4}$ grain of codeine from $\frac{3}{8}$-grain tablets. How many tablets are given?

78. If you give $\frac{2}{3}$ of a $7\frac{1}{2}$-grain tablet, how many grains does the patient receive?

79. A patient is given $\frac{1}{2}$ grain of Valium from $\frac{1}{6}$-grain tablets. How many tablets are given?

80. Six pieces of pipe are to be welded together with a flat $\frac{1}{4}$-in. plate between them to form guides. The pipes are each $6\frac{1}{8}$ in. long. What would be the overall length l of the assembly shown in Illustration 4?

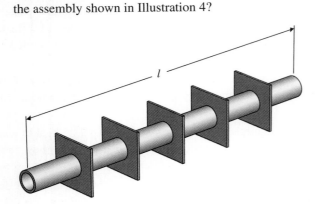

ILLUSTRATION 4

81. A drawing is lacking dimension A. It is critical to make the channel 1 ft long with a cross section as shown in Illustration 5. Find **a.** dimension A and **b.** the volume of metal in the channel.

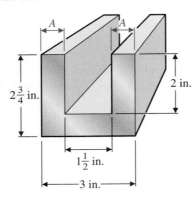

ILLUSTRATION 5

82. A CAD drawing in Illustration 6 shows a bar 2 in. by 4 in. and 36 in. long. How many $3\frac{1}{8}$-in. pieces can be cut if each saw cut is $\frac{1}{8}$-in. scrap?

ILLUSTRATION 6

In a parallel circuit, the total resistance (R_T) is given by the formula*

$$R_T = \cfrac{1}{\cfrac{1}{R_1} + \cfrac{1}{R_2} + \cfrac{1}{R_3} + \cdots}$$

Note: The three dots mean that you should use as many fractions in the denominator as there are resistances in the circuit.

Find the total resistance in each parallel circuit:

83.

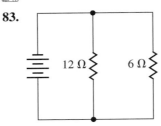

12 Ω   6 Ω

84.

40 Ω   60 Ω   80 Ω

85.

6 Ω   12 Ω   24 Ω   48 Ω

86. Each time a board is cut, the saw blade removes an additional $\frac{1}{8}$ in. of wood that is referred to as the *kerf*. What is the shortest beginning length of board that will allow you to cut five pieces, each 18 in. long, allowing for the kerf? Express your answer in feet and inches.

87. A horticulturist crosses two plants of the same species: one has a red flower and the other has a white flower. He knows from genetics that $\frac{1}{4}$ of the resulting seeds will produce plants with white flowers and $\frac{3}{4}$ will produce plants with red flowers. He harvests seeds and grows out 300 plants. How many of those would he expect to produce white flowers? Red flowers?

88. A tree harvested for firewood has a trunk that is 27 ft long. The woodcutter plans to cut it into $1\frac{1}{2}$-ft lengths before splitting. How many lengths will result?

---

*A calculator approach to working with such equations is shown in Section 6.9.

1.9 The U.S. System of Weights and Measures

Centuries ago, the thumb, hand, foot, and length from nose to outstretched fingers were used as units of measurement. These methods, of course, were not very satisfactory because people's sizes varied. In the 14th century, King Edward II proclaimed the length of the English inch to be the same as three barleycorn grains laid end to end. (See Figure 1.30.) This proclamation helped some, but it did not eliminate disputes over the length of the English inch.

(a) One old way to define one yard.

(b) Three barley corns laid end to end used to define one inch.

FIGURE 1.30

Each of these methods provides rough estimations of measurements. Actually, **measurement** is the comparison of an *observed* quantity with a *standard unit* quantity. In the estimation above, there is no one standard unit. A standard unit that is constant, accurate, and accepted by all is needed for technical measurements. Today, nations have bureaus to set national standards for all measures.

The U.S. system of weights and measures, which is derived from and sometimes called the English system, is a combination of makeshift units of Anglo-Saxon, Roman, and French-Norman weights and measures. The metric system, which is now used by international industry and business, all major U.S. industries, and most federal agencies, is presented in Chapter 3.

The U.S. system requires us to understand and be able to use fractions in everyday life. After we have converted to the metric system, the importance of fractional computations will be greatly reduced.

Take a moment to review the table of U.S. weights and measures on your reference card. Become familiar with this table, because you will use it when changing units.

Example 1

Change 5 ft 9 in. to inches.

$$1 \text{ ft} = 12 \text{ in., so } 5 \text{ ft} = 5 \times 12 \text{ in.} = 60 \text{ in.}$$
$$5 \text{ ft } 9 \text{ in.} = 60 \text{ in.} + 9 \text{ in.} = 69 \text{ in.}$$ ■

To change from one unit or set of units to another, we use what is commonly called a conversion factor. We know that we can multiply any number or quantity by 1 (one) without changing its value. We also know that any fraction whose numerator and denominator are the same is equal to 1. For example, $\frac{5}{5} = 1$, $\frac{16}{16} = 1$, and $\frac{7 \text{ ft}}{7 \text{ ft}} = 1$. Also, since 12 in. = 1 ft, $\frac{12 \text{ in.}}{1 \text{ ft}} = 1$, and likewise, $\frac{1 \text{ ft}}{12 \text{ in.}} = 1$, because the numerator equals the denominator. We call such names for 1 **conversion factors** (or *unit conversion factors*). The information necessary for forming conversion factors is found in tables, many of which are provided on the reference card included with this book.

◆ **Choosing Conversion Factors**

The correct choice for a given conversion factor is the one in which the old units are in the numerator of the original expression and in the denominator of the conversion factor, or the old units are in the denominator of the original expression and in the numerator of the conversion factor. That is, set up the conversion factor so that the old units cancel each other.

Example 2

Change 19 ft to inches.

Since 1 ft = 12 in., the two possible conversion factors are $\frac{1\,ft}{12\,in.} = 1$ and $\frac{12\,in.}{1\,ft} = 1$. We want to choose the one whose numerator is expressed in the new units (in.) and whose denominator is expressed in the old units (ft); that is, $\frac{12\,in.}{1\,ft}$. Therefore,

$$19\,\cancel{ft} \times \frac{12\,in.}{1\,\cancel{ft}} = 19 \times 12\,in. = 228\,in.$$
$$\underset{\text{conversion factor}}{\uparrow}$$ ■

Example 3

Change 8 yd to feet.

$$3\,ft = 1\,yd,$$

$$\text{so}\quad 8\,\cancel{yd} \times \frac{3\,\cancel{ft}}{1\,\cancel{yd}} = 8 \times 3\,ft = 24\,ft$$
$$\underset{\text{conversion factor}}{\uparrow}$$ ■

Example 4

Change 76 oz to pounds.

$$76\,\cancel{oz} \times \frac{1\,lb}{16\,\cancel{oz}} = \frac{76}{16}\,lb = \frac{19}{4}\,lb = 4\frac{3}{4}\,lb$$
$$\underset{\text{conversion factor}}{\uparrow}$$ ■

Sometimes it is necessary to use more than one conversion factor.

Example 5

Change 6 mi to yards.

In the table on your reference card, there is no expression equating miles with yards. Thus, it is necessary to use two conversion factors.

$$6\,\cancel{mi} \times \frac{5280\,\cancel{ft}}{1\,\cancel{mi}} \times \frac{1\,yd}{3\,\cancel{ft}} = \frac{6 \times 5280}{3}\,yd = 10{,}560\,yd$$
$$\underset{\text{conversion factors}}{\uparrow\qquad\uparrow}$$ ■

Example 6

How could a technician mixing chemicals express 4800 fluid ounces (fl oz) in gallons?

No conversions between fluid ounces and gallons are given in the tables. Use the conversion factors for (a) fluid ounces to pints (pt); (b) pints to quarts (qt); and (c) quarts to gallons (gal).

$$4800\,\cancel{fl\,oz} \times \frac{1\,\cancel{pt}}{16\,\cancel{fl\,oz}} \times \frac{1\,\cancel{qt}}{2\,\cancel{pt}} \times \frac{1\,gal}{4\,\cancel{qt}} = \frac{4800}{16 \times 2 \times 4}\,gal = 37.5\,gal$$

Conversion
factors for   **(a)**   **(b)**   **(c)**
$$\qquad\qquad\quad\uparrow\quad\ \ \uparrow\quad\ \ \uparrow$$ ■

The use of a conversion factor is especially helpful for units with which you are unfamiliar, such as rods, chains, or fathoms.

Example 7

Change 561 ft to rods.

Given 1 rod = 16.5 ft, proceed as follows:

$$561\,\cancel{ft} \times \frac{1\text{ rod}}{16.5\,\cancel{ft}} = 34\text{ rods}$$ ■

Example 8

Change 320 ft/min to ft/h.

Here, choose the conversion factor whose denominator is expressed in the new units (hours) and whose numerator is expressed in the old units (minutes).

$$320\,\frac{ft}{\cancel{min}} \times \frac{60\,\cancel{min}}{1\text{ h}} = 19,200\,ft/h$$ ■

The following example shows how to use multiple conversion factors in more complex units.

Example 9

Change 60 mi/h to ft/s.

This requires a series of conversions as follows: (a) from hours to minutes; (b) from minutes to seconds; and (c) from miles to feet.

$$60\frac{\cancel{mi}}{\cancel{h}} \times \frac{1\,\cancel{h}}{60\,\cancel{min}} \times \frac{1\,\cancel{min}}{60\,s} \times \frac{5280\,ft}{1\,\cancel{mi}} = 88\,ft/s$$

Conversion
factors for   (a)   (b)   (c) ■

Exercises 1.9

Fill in each blank:

1. 3 ft 7 in. = _____ in. **2.** 6 yd 4 ft = _____ ft

3. 5 lb 3 oz = _____ oz

4. 7 yd 3 ft 6 in. = _____ in.

5. 4 qt 1 pt = _____ pt **6.** 6 gal 3 qt = _____ pt

7. 2 bu 2 pecks (pk) = _____ pk

8. 5 bu 3 pk = _____ pk

9. 8 ft = _____ in. **10.** 5 yd = _____ ft

11. 3 qt = _____ pt **12.** 4 mi = _____ ft

13. 96 in. = _____ ft **14.** 72 ft = _____ yd

15. 10 pt = _____ qt **16.** 54 in. = _____ ft

17. 88 oz = _____ lb **18.** 32 fl oz = _____ pt

19. 14 qt = _____ gal **20.** 3 bu = _____ pk

21. 56 fl oz = _____ pt **22.** 7040 ft = _____ mi

23. 92 ft = _____ yd **24.** 9000 lb = _____ tons

25. 2 mi = _____ yd **26.** 6000 fl oz = _____ gal

27. 500 fl oz = _____ qt **28.** 3 mi = _____ rods

29. A door is 80 in. in height. Find its height in feet and inches.

30. A plane is flying at 22,000 ft. How many miles high is it?

31. Change the length of a shaft $12\frac{3}{4}$ ft long to inches.

32. A machinist has 15 wrought-iron rods to mill. Each rod weighs 24 oz. What is the total weight of the rods in pounds?

33. The instructions on a carton of chemicals call for mixing 144 fl oz of water, 24 fl oz of chemical No. 1, and 56 fl oz of chemical No. 2. How many quarts are contained in the final mixture?

34. The resistance of 1 ft of No. 32-gauge copper wire is $\frac{4}{25}$ Ω. What is the resistance of 15 yd of this wire?

35. A farmer wishes to wire a shed that is 1 mi from the electricity source in his barn. He uses No. 0-gauge copper wire, which has a resistance of $\frac{1}{10}$ ohm (Ω) per 1000 ft. What is the resistance for the mile of wire?

36. To mix an order of feed, the following quantities of feed are combined: 4200 lb, 600 lb, 5800 lb, 1300 lb, and 2100 lb. How many tons are in the final mixture?

37. A piece of sheet metal has dimensions $3\frac{3}{4}$ ft $\times$ $4\frac{2}{3}$ ft. What is the area in square inches?

38. Three pieces of steel angle of lengths 72 in., 68 in., and 82 in. are welded together. **a.** What is the total length in feet? **b.** Find the total length in yards.

39. An airport runway is 2 mi long. How long is it in **a.** feet and **b.** yards?

40. A given car holds $17\frac{1}{2}$ gal of gas. How many **a.** quarts and **b.** pints is this?

41. A small window air conditioner is charged with 3 lb of refrigerant. How many ounces is this?

42. Air flows through a metal duct at 2200 cubic feet per minute (CFM). Find this airflow in cubic feet per second.

43. A CAD drawing survey sheet shows a property road frontage as 153 ft. How many yards is this?

44. A tank that exists on a property is 3 ft by 6 ft by 4 ft deep. How many gallons of water will this tank hold? (Water weighs 62.4 lb/ft^3; 1 gal of water weighs 8.34 lb.)

45. Given 1 chain = 66 ft, change 561 ft to chains.

46. Given 1 fathom = 6 ft, change 12 fathoms to feet.

47. Given 1 dram = $27\frac{17}{50}$ grains, change 15 drams to grains.

48. Given 1 ounce = 8 drams, change 96 drams to ounces.

49. Change 4500 ft/h to ft/min.

50. Change 28 ft/s to ft/min.

51. Change $1\frac{1}{5}$ mi/s to mi/min.

52. Change 7200 ft/min to ft/s.

53. Change 40 mi/h to ft/s.

54. Change 64 ft/s to mi/h.

55. Change 24 in./s to ft/min.

56. Change 36 in./s to mi/h.

57. Add: 6 yd 2 ft 11 in.
        2 yd 1 ft 8 in.
        5 yd 2 ft 9 in.
        1 yd   6 in.

58. Subtract: 8 yd 1 ft 3 in.
            2 yd 2 ft 6 in.

59. A fish farmer sells three truckloads of fish that weigh an average of 1.5 tons each. How many pounds of fish did she sell?

60. A homeowner estimates that she has 34,850 ft^2 of lawn. Convert that area of lawn into acres.

61. To estimate the height of a tree using a device called a Biltmore stick, a forester must stand 4 rods away from the base of the tree. If the forester has an average pace length of 3 ft, how many paces must he walk from the tree to be at that approximate distance from the tree?

Unit 1B   Review

Simplify:

1. $\dfrac{9}{15}$ **2.** $\dfrac{48}{54}$

3. Change $\dfrac{27}{6}$ to a mixed number in simplest form.

4. Change $3\dfrac{2}{5}$ to an improper fraction.

Perform the indicated operations and simplify:

5. $\dfrac{5}{6} + \dfrac{2}{3}$ **6.** $5\dfrac{3}{8} - 2\dfrac{5}{12}$

7. $\dfrac{5}{12} \times \dfrac{16}{25}$ **8.** $\dfrac{3}{4} \div 1\dfrac{5}{8}$

9. $1\dfrac{2}{3} + 3\dfrac{5}{6} - 2\dfrac{1}{4}$ **10.** $4\dfrac{2}{3} \div 3\dfrac{1}{2} \times 1\dfrac{1}{2}$

11. Find the missing dimension in Illustration 1.

ILLUSTRATION 1

12. A pipe is 72 in. long. Cut three pieces of the following lengths from the pipe: $16\frac{3}{4}$ in., $24\frac{7}{8}$ in., and $12\frac{5}{16}$ in. Assume $\frac{1}{16}$ in. waste in each cut. What length of pipe is left?

13. Find the perimeter of a rectangle with length $6\frac{1}{4}$ in. and width $2\frac{2}{3}$ in.

14. Find the area of a rectangle with length $6\frac{1}{4}$ in. and width $2\frac{2}{3}$ in.

15. Change 4 ft to inches.

16. Change 24 ft to yards.

17. Change 3 lb to ounces.

18. Change 20 qt to gallons.

19. Change 60 mi/h to ft/s.

20. Subtract: 14 ft 4 in.

 8 ft 8 in.

Unit 1C

REVIEW OF OPERATIONS WITH DECIMAL FRACTIONS AND PERCENT

1.10 Addition and Subtraction of Decimal Fractions

Introduction to Decimals

A fraction whose denominator is 10, 100, 1000, or any power of 10 is called a **decimal fraction**. Decimal calculations and measuring instruments calibrated in decimals are the basic tools for measurement in the metric system. The common use of the calculator makes a basic understanding of decimal principles necessary.

Recall the place values of the digits of a whole number from Section 1.1. Each digit to the left of the decimal point represents a multiple of a power of 10. Each digit to the right of the decimal point represents a multiple of a power of $\frac{1}{10}$. Study Table 1.2, which shows place values for decimals.

Note that $10^0 = 1$. (See Section 2.5.)

Table 1.2 Place Values for Decimals

Number	Words	Product form	Exponential form
1,000,000	One million	$10 \times 10 \times 10 \times 10 \times 10 \times 10$	10^6
100,000	One hundred thousand	$10 \times 10 \times 10 \times 10 \times 10$	10^5
10,000	Ten thousand	$10 \times 10 \times 10 \times 10$	10^4
1,000	One thousand	$10 \times 10 \times 10$	10^3
100	One hundred	10×10	10^2
10	Ten	10	10^1
1	One	1	10^0
0.1	One tenth	$\frac{1}{10}$	$\left(\frac{1}{10}\right)^1$ or 10^{-1}
0.01	One hundredth	$\frac{1}{10} \times \frac{1}{10}$	$\left(\frac{1}{10}\right)^2$ or 10^{-2}
0.001	One thousandth	$\frac{1}{10} \times \frac{1}{10} \times \frac{1}{10}$	$\left(\frac{1}{10}\right)^3$ or 10^{-3}
0.0001	One ten-thousandth	$\frac{1}{10} \times \frac{1}{10} \times \frac{1}{10} \times \frac{1}{10}$	$\left(\frac{1}{10}\right)^4$ or 10^{-4}
0.00001	One hundred-thousandth	$\frac{1}{10} \times \frac{1}{10} \times \frac{1}{10} \times \frac{1}{10} \times \frac{1}{10}$	$\left(\frac{1}{10}\right)^5$ or 10^{-5}
0.000001	One millionth	$\frac{1}{10} \times \frac{1}{10} \times \frac{1}{10} \times \frac{1}{10} \times \frac{1}{10} \times \frac{1}{10}$	$\left(\frac{1}{10}\right)^6$ or 10^{-6}

Example 1   In the number 123.456, find the place value of each digit and the number it represents.

Digit	Place value	Number represented
1	Hundreds	1×10^2
2	Tens	2×10^1
3	Ones or units	3×10^0 or 3×1
4	Tenths	$4 \times \left(\dfrac{1}{10}\right)^1$ or 4×10^{-1}
5	Hundredths	$5 \times \left(\dfrac{1}{10}\right)^2$ or 5×10^{-2}
6	Thousandths	$6 \times \left(\dfrac{1}{10}\right)^3$ or 6×10^{-3}

Recall that place values to the left of the decimal point are powers of 10 and place values to the right of the decimal point are powers of $\frac{1}{10}$. ■

Example 2   Write each decimal in words: 0.05; 0.0006; 24.41; 234.001207.

Decimal	Word form
0.05	Five hundredths
0.0006	Six ten-thousandths
24.41	Twenty-four *and* forty-one hundredths
234.001207	Two hundred thirty-four *and* one thousand two hundred seven millionths

Note that the decimal point is read *and.* ■

Example 3   Write each number as a decimal and as a common fraction.

Number	Decimal	Common fraction
One hundred four and seventeen hundredths	104.17	$104\dfrac{17}{100}$
Fifty and three thousandths	50.003	$50\dfrac{3}{1000}$
Five hundred eleven hundred-thousandths	0.00511*	$\dfrac{511}{100,000}$

*This book follows the common practice of writing a zero before the decimal point in a decimal smaller than 1.

■

Often, common fractions are easier to use if they are expressed as decimal equivalents. Every common fraction can be expressed as a decimal. A *repeating decimal* is one in which a digit or a group of digits repeats again and again; it may be written as a common fraction.

A bar over a digit or group of digits means that this digit or group of digits is repeated without ending. Each of the following numbers is a repeating decimal:

0.33333 . . . is written $0.\overline{3}$

72.64444 . . . is written $72.6\overline{4}$

0.21212121 . . . is written $0.\overline{21}$

6.33120120120 . . . is written $6.33\overline{120}$

A *terminating decimal* is a decimal number with a given number of digits. Examples are 0.75, 12.505, and 0.000612.

Changing a Common Fraction to a Decimal

To change a common fraction to a decimal, divide the numerator of the fraction by the denominator.

Example 4 Change $\dfrac{3}{4}$ to a decimal.

$$
\begin{array}{r}
0.75 \\
4\overline{)3.00} \\
\underline{2\,8} \\
20 \\
\underline{20} \\
\end{array}
$$

Divide the numerator by the denominator.

$\dfrac{3}{4} = 0.75$ (a terminating decimal) ■

Example 5 Change $\dfrac{8}{15}$ to a decimal.

$$
\begin{array}{r}
0.5333 \\
15\overline{)8.0000} \\
\underline{7\,5} \\
50 \\
\underline{45} \\
50 \\
\underline{45} \\
50 \\
\underline{45} \\
5 \\
\end{array}
$$

Divide the numerator by the denominator.

$\dfrac{8}{15} = 0.5333\ldots = 0.5\overline{3}$ (a repeating decimal)

The result could be written $0.53\overline{3}$ or $0.5\overline{3}$. It is not necessary to continue the division once it has been established that the quotient has begun to repeat. ■

Since a decimal fraction can be written as a common fraction with a denominator that is a power of ten, it is easy to change a decimal fraction to a common fraction. Simply use the digits that appear to the right of the decimal point (disregarding beginning zeros) as the numerator. Use the place value of the last digit as the denominator. Any digits to the left of the decimal point will be the whole-number part of the resulting mixed number.

 Example 6   Change each decimal to a common fraction or a mixed number.

Decimal	Common fraction or mixed number
a. 0.3	$\dfrac{3}{10}$
b. 0.17	$\dfrac{17}{100}$
c. 0.25	$\dfrac{25}{100} = \dfrac{1}{4}$
d. 0.125	$\dfrac{125}{1000} = \dfrac{1}{8}$
e. 0.86	$\dfrac{86}{100} = \dfrac{43}{50}$
f. 8.1	$8\dfrac{1}{10}$
g. 13.64	$13\dfrac{64}{100} = 13\dfrac{16}{25}$
h. 5.034	$5\dfrac{34}{1000} = 5\dfrac{17}{500}$

■

In on-the-job situations, it is often more convenient to add, subtract, multiply, and divide measurements that are in decimal form rather than in fractional form. Except for the placement of the decimal point, the four arithmetic operations are the same for decimal fractions as they are for whole numbers.

 Example 7   Add 13.2, 8.42, and 120.1.

a. Using decimal fractions:

$$
\begin{array}{r}
13.2 \\
8.42 \\
\underline{120.1} \\
141.72
\end{array}
$$

b. Using common fractions:

$$13\frac{2}{10} = 13\frac{20}{100}$$

$$8\frac{42}{100} = 8\frac{42}{100}$$

$$120\frac{1}{10} = \underline{120\frac{10}{100}}$$

$$141\frac{72}{100} = 141.72 \quad ■$$

Adding or Subtracting Decimal Fractions

Step 1 Write the decimals so that the digits having the same place value are in vertical columns. (Make certain that the decimal points are also lined up vertically.)

Step 2 Add or subtract as with whole numbers.

Step 3 Place the decimal point between the ones digit and the tenths digit of the sum or the difference. (Be certain the decimal point is in the same vertical line as the other decimal points.)

Example 8   Subtract 1.28 from 17.9.

$$\begin{array}{r} 17.90 \\ \underline{1.28} \\ 16.62 \end{array}$$ Zeros can be supplied after the last digit at the right of the decimal point without changing the value of a number. Therefore, 17.9 = 17.90.

Example 9   Add 24.1, 26, and 37.02.

$$\begin{array}{r} 24.10 \\ 26.00 \\ \underline{37.02} \\ 87.12 \end{array}$$ A decimal point can be placed at the right of any whole number, and zeros can be supplied without changing the value of the number.

Example 10   Perform the indicated operations: 51.6 − 2.45 + 7.3 − 14.92.

$$\begin{array}{r} 51.60 \\ \underline{-2.45} \\ 49.15 \\ \underline{+7.30} \\ 56.45 \\ \underline{-14.92} \\ 41.53 \end{array}$$

49.15   difference

56.45   sum

41.53   final difference

Example 11   Find the missing dimension in Figure 1.31.

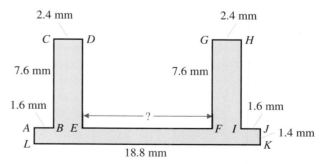

FIGURE 1.31

The missing dimension *EF* equals the sum of the lengths *AB*, *CD*, *GH*, and *IJ* subtracted from the length *LK*. That is, add

AB:	1.6 mm
CD:	2.4 mm
GH:	2.4 mm
IJ:	1.6 mm
	8.0 mm

Then subtract

LK:	18.8 mm
	−8.0 mm
	10.8 mm

That is, length *EF* = 10.8 mm.   ■

Example 12

As we saw in Unit 1B, the total current in a parallel circuit equals the sum of the currents in the branches of the circuit. Find the total current in the parallel circuit shown in Figure 1.32.

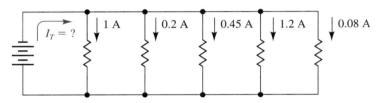

FIGURE 1.32

```
1    A
0.2  A
0.45 A
1.2  A
0.08 A
2.93 A
```
■

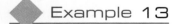

Using a Calculator to Add and Subtract Decimal Fractions

Example 13

Add: $14.62 + $0.78 + $1.40 + $0.05.

14.62 (+) .78 (+) 1.4 (+) .05 (=)

16.85

The sum is $16.85.   ■

Combinations of addition and subtraction may be performed on a calculator as follows.

Example 14   Do as indicated: $74.6 - 8.57 + 5 - 0.0031$.

74.6 $\boxed{-}$ 8.57 $\boxed{+}$ 5 $\boxed{-}$ .0031 $\boxed{=}$

$$\boxed{71.0269}$$

The result is 71.0269.   ■

Exercises 1.10 _____

Write each decimal in words:

1. 0.004   **2.** 0.021   **3.** 0.0005

4. 7.1   **5.** 1.00421   **6.** 1042.007

7. 6.092   **8.** 8.1461

Write each number both as a decimal and as a common fraction or mixed number:

9. Five and two hundredths

10. One hundred twenty-three and six thousandths

11. Seventy-one and twenty-one ten-thousandths

12. Sixty-five thousandths

13. Forty-three and one hundred one ten-thousandths

14. Five hundred sixty-three millionths

Change each common fraction to a decimal:

15. $\dfrac{3}{8}$ **16.** $\dfrac{16}{25}$ **17.** $\dfrac{11}{15}$ **18.** $\dfrac{2}{5}$

19. $\dfrac{17}{50}$ **20.** $\dfrac{11}{9}$ **21.** $\dfrac{14}{11}$ **22.** $\dfrac{128}{25}$

23. $\dfrac{128}{7}$ **24.** $\dfrac{603}{24}$ **25.** $\dfrac{308}{9}$ **26.** $\dfrac{230}{6}$

Change each decimal to a common fraction or a mixed number:

27. 0.7   **28.** 0.6   **29.** 0.11

30. 0.75   **31.** 0.8425   **32.** 3.14

33. 10.76   **34.** 148.255

Find each sum:

35. 137.64
      7.14
      0.008
      6.1

36.   63
      4.7
     19.45
   120.015

37. $147.49 + 7.31 + 0.004 + 8.4$

38. $47 + 6.3 + 20.71 + 170.027$

Subtract:

39. 72.4 from 159   **40.** 3.12 from 4.7

41. $64.718 - 49.41$ **42.** $140 - 16.412$

Perform the indicated operations:

43. $18.4 - 13.72 + 4$ **44.** $34.14 - 8.7 - 16.5$

45. $0.37 + 4.5 - 0.008$

46. $51.7 - 1.11 - 4.6 + 84.1$

47. $1.511 + 14.714 - 6.1743$

48. $0.0056 + 0.023 - 0.00456 + 0.9005$

49. A piece of $\frac{1}{4}$-in. flat steel is 6.25 ft long by 4.2 ft wide. If you cut off two equal pieces of length 2.4 ft and width 4.2 ft, what size piece will be left?

50. A welder needs to weld together pipes of lengths 10.25 ft, 15.4 ft, and 14.1 ft. What is the total length of the new pipe?

51. A crop duster flew 2.3 h on Monday, 3.1 h on Wednesday, and 5.4 h on Friday and Saturday combined. What was the total flying time for the week?

52. An ultralight aircraft flew 125.5 mi to a small airport, then another 110.3 mi to another airport. After spending the night, it flew 97.8 mi to yet another airport. What was the total mileage for the trip?

53. A car needs new tires. The tread on the old tires measures $\frac{1}{16}$ in. If the new tires have a tread of $\frac{3}{8}$ in., what is the difference in the tread written as a decimal?

54. What is the total cost for one piece of 8-in. by 16-in. metal duct at \$17.33 and two pieces of 8-in. by 12-in. metal duct at \$11.58?

55. Find the missing dimensions in Illustration 1.

56. Find the perimeter of the figure in Illustration 1.

3.45 cm

1.87 cm

a

b

2.69 cm

8.32 cm

ILLUSTRATION 1

57. Find the length of the shaft shown in Illustration 2.

4.17 in.   1.30 in.   1.00 in.   1.47 in.

ILLUSTRATION 2

58. The perimeter of the hexagon in Illustration 3 is 6.573 in. Find the length of side *x*.

x

0.938 in.

1.501 in.

0.688 in.

1.313 in.   0.625 in.

ILLUSTRATION 3

59. A steel axle is being designed and drawn as in Illustration 4. It has a $\frac{1}{8}$-in.-diameter hole drilled in the center of its length. If the axle is 9.625 in. long, how far from the end should the center of the hole be dimensioned?

$\frac{1}{8}$ in. dia.

9.625 in.

L

ILLUSTRATION 4

60. What is the internal diameter of a circular tube having an outside diameter (OD) of 1.125 in. and a wall thickness of 0.046 in.?

61. Find the total current in the parallel circuit in Illustration 5.

$I_T = ?$ 0.3 A   0.105 A   0.45 A   0.93 A   0.27 A   0.55 A

ILLUSTRATION 5

62. As we saw in Unit 1A, the total resistance in a series circuit is equal to the sum of the resistances in the circuit. Find the total resistance in the series circuit in Illustration 6.

21.5 Ω   42.6 Ω

62.3 Ω

19.8 Ω

32.2 Ω

ILLUSTRATION 6

63. Find the total resistance in the series circuit in Illustration 7.

15.7 Ω   40 Ω   25.5 Ω

0.6 Ω   1200 Ω   115 Ω

ILLUSTRATION 7

64. In a series circuit, the voltage of the source equals the sum of the separate voltage drops in the circuit. Find the voltage of the source in the circuit in Illustration 8.

3.2 V   5.1 V   0.45 V

0.03 V

0.8 V

2 V   0.007 V

ILLUSTRATION 8

65. Find the difference between the diameters of the circular ends of the taper shown in Illustration 9.

1.625 in.   1.093 in.

ILLUSTRATION 9

66. Find the missing dimension in each figure in Illustration 10.

13.47 cm

6.74 cm   4.89 cm a

0.98 cm

c

c

b

1.23 cm   1.79 cm   2.62 cm

ILLUSTRATION 10

67. Find the wall thickness of the pipe in Illustration 11.

1.50 in.

1.94 in.

ILLUSTRATION 11

68. Find the length, l, of the socket in Illustration 12. Also, find the length of diameter A.

2.375 in.

3.375 in.

l

3.750 in.

1.250 in. A 1.250 in.

ILLUSTRATION 12

69. For a valve to be seated properly in an automobile engine, the factory part measuring 1.732 in. must be ground 0.005 in. Find the size of the valve after it is ground.

70. The standard width of a new piston ring is 0.1675 in. The used ring measures 0.1643 in. How much has it worn?

71. According to the United Nations, the human population in year 2000 was 6.11 billion and it was expected to be about 11.20 billion by the year 2100. If the estimate is accurate, by how much will the human population increase during this century?

72. A homeowner fertilizes his lawn in March at a cost of $114.57, in June at a cost of $145.36, and in September at a cost of $99.21. How much did he spend on fertilizer for his lawn over the course of the year?

73. According to the *Oil and Gas Journal,* proven reserves of petroleum in billions of barrels (bbl) as of January 1, 2007, were estimated by region as follows:

North America	213.3 bbl
Central and South America	102.8 bbl
Europe	15.8 bbl
Middle East	739.2 bbl
Africa	114.1 bbl
Asia and Oceania	33.4 bbl
Unspecified sources	98.9 bbl

If these estimates are accurate, what was the total worldwide petroleum reserve as of that date?

1.11   Rounding Numbers

We often need to make an estimate of a number or a measurement. When a truck driver makes a delivery from one side of a city to another, he or she can only estimate the time it will take to make the trip. An automobile technician must estimate the cost of a repair job and the number of mechanics to assign to that job. On such occasions, estimates are *rounded*.

Earlier, you found that $\frac{1}{3} = 0.33\overline{3}$. There is no exact decimal value to use in a calculation. You must round $0.33\overline{3}$ to a certain number of decimal places, depending on the accuracy needed in a given situation.

There are many rounding procedures in general use today. Some are complicated, and others are simple. We use one of the simplest methods, which will be outlined in the next examples and then stated in the form of a rule.

Example 1

Round 25,348 to the nearest thousand.

Note that 25,348 is more than 25,000 and less than 26,000.

26,000
25,900
25,800
25,700
25,600
25,500
25,400
25,348   As you can see, 25,348 is closer to 25,000 than to 26,000. Therefore, 25,348
25,300   rounded to the nearest thousand is 25,000.
25,200
25,100
25,000   ■

Example 2

Round 2.5271 to the nearest hundredth.

Note that 2.5271 is more than 2.5200 but less than 2.5300.

2.5300 = 2.53
2.5290
2.5280
2.5271   2.5271 is nearer to 2.53 than to 2.52. Therefore, 2.5271 rounded to the
2.5270   nearest hundredth is 2.53.
2.5260
2.5250
2.5240
2.5230
2.5220
2.5210
2.5200 = 2.52

Note: If a number is exactly halfway between two numbers, round up to the larger number.   ■

Rounding Numbers to a Particular Place Value

To round a number to a particular place value:

1. If the digit in the next place to the right is less than 5, drop that digit and all other following digits. Use zeros to replace any whole-number places dropped.
2. If the digit in the next place to the right is 5 or greater, add 1 to the digit in the place to which you are rounding. Drop all other following digits. Use zeros to replace any whole-number places dropped.

  Example 3   Round each number in the left column to the place indicated in each of the other columns.

Number	Hundred	Ten	Unit	Tenth	Hundredth	Thousandth
a. 158.6147	200	160	159	158.6	158.61	158.615
b. 4,562.7155	4,600	4,560	4,563	4,562.7	4,562.72	4,562.716
c. 7.12579	0	10	7	7.1	7.13	7.126
d. 63,576.15	63,600	63,580	63,576	63,576.2	63,576.15	—
e. 845.9981	800	850	846	846.0	846.00	845.998

■

Instead of rounding a number to a particular place value, we often need to round a number to a given number of significant digits. **Significant digits** are those digits in a number we are reasonably sure of being able to rely on in a measurement. Here we present a brief introduction to significant digits. (An in-depth discussion of accuracy and significant digits is given in Section 4.1.)

Significant Digits

The following digits in a number are significant:

• All nonzero digits (258 has three significant digits)
• All zeros between significant digits (2007 has four significant digits)
• All zeros at the end of a decimal number (2.000 and 0.09500 have four significant digits)

The following digits in a number are not significant:

• All zeros at the beginning of a decimal number less than 1 (0.00775 has three significant digits)
• All zeros at the end of a whole number (36,000 has two significant digits)

Rounding Numbers to a Given Number of Significant Digits

To round a number to a given number of significant digits:

1. Count the given number of significant digits from left to right, starting with the first nonzero digit.
2. If the next digit to the right is less than 5, drop that digit and all other following digits. Use zeros to replace any whole-number places dropped.
3. If the next digit to the right is 5 or greater, add 1 to the digit in the place to which you are rounding. Drop all other following digits. Use zeros to replace any whole-number places dropped.

Example 4   Round each number to three significant digits.

a. 74,123   Count three digits from left to right, which is the digit 1. Since the next digit to its right is less than 5, replace the next two digits with zeros. Thus, 74,123 rounded to three significant digits is 74,100.

b. 0.002976401   Count three nonzero digits from left to right, which is the digit 7. Since the next digit to its right is greater than 5, increase the digit 7 by 1 and drop the next four digits. Thus, 0.002976401 rounded to three significant digits is 0.00298.   ■

Example 5   Round each number to the given number of significant digits.

Number	Given number of significant digits	Rounded number
a. 2571.88	3	2570
b. 2571.88	4	2572
c. 345,175	2	350,000
d. 345,175	4	345,200
e. 0.0030162	2	0.0030
f. 0.0030162	3	0.00302
g. 24.00055	3	24.0
h. 24.00055	5	24.001

■

Exercises 1.11

*Round each number to **a.** the nearest hundred and **b.** the nearest ten:*

1. 1652   **2.** 1760   **3.** 3125.4

4. 73.82   **5.** 18,675   **6.** 5968

*Round each number to **a.** the nearest tenth and **b.** the nearest thousandth:*

7. 3.1416   **8.** 0.161616   **9.** 0.05731

10. 0.9836   **11.** 0.07046   **12.** 3.7654

Round each number in the left column to the place indicated in each of the other columns:

	Number	Hundred	Ten	Unit	Tenth	Hundredth	Thousandth
13.	636.1825						
14.	1,451.5254						
15.	17,159.1666						
16.	8.171717						
17.	1,543,679						
18.	41,892.1565						
19.	10,649.83						
20.	84.00659						
21.	649.8995						
22.	147.99545						

Round each number to three significant digits:

23. 236,534   **24.** 202.505   **25.** 0.03275

Round each number to two significant digits:

26. 63,914   **27.** 71.613   **28.** 0.03275

Round each number to four significant digits:

29. 1,462,304   **30.** 23.2347   **31.** 0.000337567

Round each number to three significant digits:

32. 20,714   **33.** 1.00782   **34.** 0.00118952

1.12 Multiplication and Division of Decimal Fractions

Multiplying Two Decimal Fractions

1. Multiply the numbers as you would whole numbers.
2. Count the total number of digits to the right of the decimal points in the two numbers being multiplied. Then place the decimal in the product so that it has that same total number of digits to the right of the decimal point.

 Example 1   Multiply: 42.6 × 1.73.

```
   42.6
   1.73
  12 78
 298 2
 426
 73.698
```

Note that 42.6 has one digit to the *right* of the decimal point and 1.73 has two digits to the *right* of the decimal point. The product has three digits to the *right* of the decimal point. ■

Example 2

Multiply: 30.6×4200.

$$
\begin{array}{r}
30.6 \\
4200 \\
\hline
61\,200 \\
1224 \\
\hline
128520.0
\end{array}
$$

This product has one digit to the *right* of the decimal point. ■

Dividing Two Decimal Fractions

Step 1 Use the same form as in dividing two whole numbers.

Step 2 Multiply both the dividend and the divisor (numerator and denominator) by a power of ten that makes the divisor a whole number.

Step 3 Divide as you would with whole numbers, and place the decimal point in the quotient directly above the decimal point in the dividend.

 Example 3

Divide 24.32 by 6.4.

Method 1:

$$
\frac{24.32}{6.4} \times \frac{10}{10} = \frac{243.2}{64}
$$

$$
\begin{array}{r}
3.8 \\
64\overline{)243.2} \\
192 \\
\hline
512 \\
512 \\
\hline
\end{array}
$$

Method 2:

$$
\begin{array}{r}
3.8 \\
6.4\overline{)24.3.2} \\
19\,2 \\
\hline
5\,1\,2 \\
5\,1\,2 \\
\hline
\end{array}
$$

Moving the decimal point one place to the right in both the dividend and divisor here is the same as multiplying numerator and denominator by 10 in Method 1.

■

 Example 4

Divide 75.1 by 1.62 and round to the nearest hundredth.

To round to the nearest hundredth, you must carry the division out to the thousandths place and then round to hundredths. We show two methods.

Method 1:

$$
\frac{75.1}{1.62} \times \frac{100}{100} = \frac{7510}{162}
$$

$$
\begin{array}{r}
46.358 \\
162\overline{)7510.000} \\
648 \\
\hline
1030 \\
972 \\
\hline
580 \\
486 \\
\hline
940 \\
810 \\
\hline
1300 \\
1296 \\
\hline
4
\end{array}
$$

Method 2:

$$
\begin{array}{r}
46.358 \\
1.62.\overline{)75.10.000} \\
64\,8 \\
\hline
10\,30 \\
9\,72 \\
\hline
58\,0 \\
48\,6 \\
\hline
9\,40 \\
8\,10 \\
\hline
1\,300 \\
1\,296 \\
\hline
4
\end{array}
$$

Moving the decimal point two places to the right in both the dividend and divisor here is the same as multiplying both numerator and denominator by 100 in Method 1.

In both methods, you need to add zeros after the decimal point and carry the division out to the thousandths place. Then round to the nearest hundredth. This gives 46.36 as the result.

■

Example 5

A gasoline station is leased for $1155 per month. How much gasoline must be sold each month to make the cost of the lease equal to 3.5¢ ($0.035) per gallon?

Divide the cost of the lease per gallon into the cost of the lease per month.

$$
\begin{array}{r}
33\,000. \\
0.035.\overline{)1155.000.} \\
\underline{105} \\
105 \\
\underline{105}
\end{array}
$$

That is, 33,000 gal of gasoline must be sold each month. ■

 Example 6

A sprayer tank holds 350 gal. Suppose 20 gal of water and 1.25 gal of pesticide are applied to each acre.

    a. How many acres can be treated on one tankful?

    b. How much pesticide is needed per tankful?

      a. To find the number of acres treated on one tankful, divide the number of gallons of water *and* pesticide into the number of gallons of a full tank.

$$
\begin{array}{r}
16.4 \quad\text{or approximately 16 acres/tankful} \\
21.25.\overline{)350.00.0} \\
\underline{212\,5} \\
137\,50 \\
\underline{127\,50} \\
10\,00\,0 \\
\underline{8\,50\,0} \\
1\,50\,0
\end{array}
$$

    b. To find the amount of pesticide needed per tankful, multiply the number of gallons of pesticide applied per acre times the number of acres treated on one tankful.

$$
\begin{array}{r}
1.25 \\
\underline{16} \\
7\,50 \\
12\,5 \\
\hline
20.00 \quad\text{or approximately 20 gal/tankful.}
\end{array}
$$
■

Using a Calculator to Multiply and Divide Decimal Fractions

 Example 7

Multiply: 8.23 × 65 × 0.4.

8.23 ⊠ 65 ⊠ .4 ⊜

213.98

The product is 213.98. ■

To divide numbers using a calculator, follow the steps in the following example.

Example 8   Divide: $3.69 \div 8.2$.

 3.69 8.2 =

0.45

The quotient is 0.45.   ■

Now we expand the order of operations from Section 1.2 to include how to treat the fraction bar.

 Example 9   Evaluate: $\dfrac{4 + (9 - 3)^2}{4^3 - 2 \cdot 12}$

$$= \frac{4 + (6)^2}{4^3 - 2 \cdot 12}$$ Subtract within parentheses.

$$= \frac{4 + 36}{64 - 2 \cdot 12}$$ Evaluate the powers.

$$= \frac{4 + 36}{64 - 24}$$ Multiply.

$$= \frac{40}{40}$$ Add the numbers in the numerator and subtract the numbers in the denominator. When a problem contains a fraction bar, treat the numerator and the denominator separately before dividing.

$$= 1$$ Divide.   ■

 Example 10   The inductive reactance (in ohms, Ω) in an ac circuit equals the product of 2π times the frequency (in hertz, Hz, that is, cycles/second) times the inductance (in henries, H). Find the inductive reactance in the ac circuit in Figure 1.33. (Use the π key on your calculator, or use $\pi = 3.14$.)

The inductive reactance is

$$2\pi \times \text{frequency} \times \text{inductance}$$

FIGURE 1.33 $2 \times \pi \times \quad 60 \quad \times \quad 0.15 \quad = 56.5 \; \Omega \text{ (rounded to 3 significant digits)}$ ■

 Example 11   The effect of both resistance and inductance in a circuit is called *impedance*. Ohm's law for an ac circuit states that the current (in amps, A) equals the voltage (in volts, V) divided by the impedance (in ohms, Ω). Find the current in a 110-V ac circuit that has an impedance of 65 Ω.

$$\text{current} = \text{voltage} \div \text{impedance}$$
$$= 110 \quad \div \quad 65$$
$$= 1.69 \text{ A (rounded to 3 significant digits)}$$ ■

Exercises 1.12

Multiply:

1. 3.7
    0.15

2. 14.1
    1.7

3. 25.03
    0.42

4. 4000
    6.75

5. 5800
    1600

6. 90,000
    0.00705

Divide:

7. $36 \div 1.2$ **8.** $5.1 \div 1.7$

9. $0.6 \div 0.04$ **10.** $14.356 \div 0.74$

Divide and round to the nearest hundredth:

11. $17,500 \div 70.5$ **12.** $7900 \div 1.52$

13. $75,000 \div 20.4$ **14.** $1850 \div 0.75$

Evaluate each expression following the order of operations:

15. $\dfrac{8^2 - 6^2}{4 \cdot 8 + (7 + 9)}$ **16.** $\dfrac{148 - 3 \cdot 4^2}{5^3 - 2 \cdot 5^2}$

17. $\dfrac{4 \cdot 5 \cdot 6 - 5 \cdot 2^3}{4^2 \cdot 5 + 5 \cdot 2^2}$ **18.** $\dfrac{2^3 + (2 + 3 \cdot 6)^2}{(2 \cdot 5 - 4)^2 + 3 \cdot 5}$

19. A 3.6-ft piece of steel angle is to be divided into 3 equal parts. What is the length of each piece?

20. A 7-ft I-beam is to be divided into 4 equal parts. What is the length of each piece?

21. A small plane flew 321.3 mi in 2.7 h. How fast did the plane fly?

22. To fill the tanks of a small plane after a trip cost $104.06. If it took 24.2 gal of gas, find the price per gallon.

23. A car can travel 475 mi on a full tank of gas. If the car holds 17.12 gal of gas, how many miles per gallon can it get?

24. A set of four new tires cost $465.60. What was the price per tire?

25. A stair detail has 12 risers of $8\frac{7}{8}$ in. each. The owner wants only 11 risers, but the total height must be as before. What is the new riser height dimension (in a decimal fraction) for the drawing in Illustration 1?

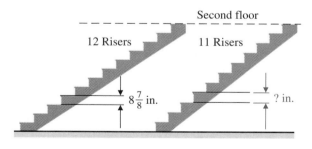

Second floor

12 Risers   11 Risers

$8\frac{7}{8}$ in.   ? in.

ILLUSTRATION 1

26. If the cost of 8-in. by 20-in. metal duct is $24.96 for 4 ft, how much is it per foot? Per inch?

27. In Illustration 2, find **a.** the perimeter of the outside square and **b.** the length of the center line, l.

4.72 m

ILLUSTRATION 2

28. In Illustration 3, find the perimeter of the octagon, which has eight equal sides.

4.75 mm

ILLUSTRATION 3

29. The pitch p of a screw is the reciprocal of the number of threads per inch n; that is, $p = \frac{1}{n}$. If the pitch is 0.0125, find the number of threads per inch.

30. A 78-ft cable is to be cut into 3.25-ft lengths. Into how many such lengths can the cable be cut?

31. A steel rod 32.63 in. long is to be cut into 8 pieces. Each piece is 3.56 in. long. Each cut wastes 0.15 in. of rod in shavings. How many inches of the rod are left?

32. How high is a pile of 32 metal sheets if each sheet is 0.045 in. thick?

33. How many metal sheets are in a stack that measures 18 in. high if each sheet is 0.0060 in. thick?

34. A building measures 45 ft 3 in. by 64 ft 6 in. inside. How many square feet of possible floor space does it contain?

35. The cost of excavation is $4.50/yd³. Find the cost of excavating a basement 87 ft long, 42 ft wide, and 8 ft deep.

36. Each cut on a lathe is 0.018 in. deep. How many cuts would be needed to turn down 2.640-in. stock to 2.388 in.?

37. Find the total length of the crankshaft shown in Illustration 4.

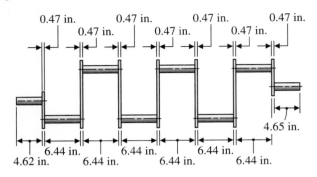

0.47 in.   0.47 in.   0.47 in.   0.47 in.
0.47 in.   0.47 in.   0.47 in.

4.65 in.

6.44 in.   6.44 in.   6.44 in.
4.62 in.   6.44 in.   6.44 in.   6.44 in.

ILLUSTRATION 4

38. A shop foreman may spend $535 on overtime to complete a job. Overtime pay is $26.75 per hour. How many hours of overtime may he use?

39. Find the total piston displacement of a six-cylinder engine if each piston displaces a volume of 56.25 in³.

40. Find the total piston displacement of a six-cylinder engine if each piston displaces 0.9 litres (L).

41. A four-cylinder engine has a total displacement of 2.0 L. Find the displacement of each piston.

42. An eight-cylinder engine has a total displacement of 318 in³. Find the displacement of each piston.

43. The diameter of a new piston is 4.675 in. The average wear per 10,000 mi is 0.007 in. *uniformly over the piston.* **a.** Find the average wear after 80,000 mi. **b.** Find the average diameter of the piston after 100,000 mi.

44. A certain job requires 500 person-hours to complete. How many days will it take for five people working 8 hours per day to complete the job?

45. How many gallons of herbicide are needed for 150 acres of soybeans if 1.6 gal/acre are applied?

46. Suppose 10 gal of water and 1.7 lb of pesticide are to be applied per acre. **a.** How much pesticide would you put in a 300-gal spray tank? **b.** How many acres can be covered with one tankful? (Assume the pesticide dissolves in the water and has no volume.)

47. A cattle feeder buys some feeder cattle, which average 550 lb at $115/hundredweight (that is, $115 per hundred pounds, or $1.15/lb). The price he receives when he sells them as slaughter cattle is $86/hundredweight. If he plans to make a profit of $120 per head, what will be his cost per pound for a 500-lb weight gain?

48. An insecticide is to be applied at a rate of 2 pt/100 gal of water. How many pints are needed for a tank that holds 20 gal? 60 gal? 150 gal? 350 gal? (Assume that the insecticide dissolves in the water and has no volume.)

Find the inductive reactance in each ac circuit (see Example 10):

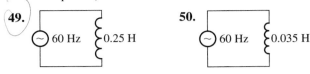

49. 60 Hz   0.25 H

50. 60 Hz   0.035 H

Power (in watts, W) equals voltage times current. Find the power in each circuit:

51. 0.045 A   6.4 V

52. 0.0065 A   0.95 V

53. Find the current in a 220-V ac circuit with impedance 35.5 Ω. (See Example 11.)

54. A flashlight bulb is connected to a 1.5-V dry cell. If it draws 0.25 A, what is its resistance? (Resistance equals voltage divided by current.)

55. A lamp that requires 0.84 A of current is connected to a 115-V source. What is the lamp's resistance?

56. A heating element operates on a 115-V line. If it has a resistance of 18 Ω, what current does it draw? (Current equals voltage divided by resistance.)

57. A nurse gives three tablets of glyceryl trinitrate, containing 0.150 grain each. How many grains are given?

58. A nurse gives two tablets of ephedrine, containing 0.75 grain each. How many grains are given?

59. An order reads 0.5 mg of digitalis, and each tablet contains 0.1 mg. How many tablets should be given?

60. An order reads 1.25 mg of digoxin, and the tablets on hand are 0.25 mg. How many tablets should be given?

61. A statute mile is 5280 ft. A nautical mile used in aviation is 6080.6 ft. This gives the conversion 1 statute mile = 0.868 nautical miles. If a plane flew 350 statute miles, how many nautical miles were flown?

62. Five lathes and four milling machines are to be on one circuit. If each lathe uses 16.0 A and each milling machine uses 13.8 A, what is the amperage requirement for this circuit?

63. A steel plate 1.00 in. thick weighs 40.32 lb/ft². Find the weight of a 4.00 ft × 8.00 ft sheet.

64. Municipal solid waste (MSW) consists basically of trash and recycle that is produced by nonindustrial and nonagricultural sources. According to Environmental Protection Agency estimates, as of 2006, each person in the United States generated an average of 4.6 lb of MSW each day. If you are an average American, how much MSW did you generate in that year?

65. According to U.S. Census Bureau projections, the U.S. population was 305,852,427 as of 3 p.m. EST on December 10, 2008. If that is accurate and if the average person in the United States actually generated 4.6 lb of MSW on that day, how many tons of MSW (rounded to the nearest thousand) were generated in the United States on December 10, 2008?

66. One U.S. bushel contains 1.2445 ft³. A silo that has a capacity of 10,240 ft³ can store how many bushels of corn? Round to the nearest bushel.

67. A rick of firewood is 4 ft by 8 ft by the average length of the stick of firewood. If the firewood in a rick is cut to 16-in. lengths, what is the volume of the rick? Round to the nearest tenth of a cubic foot.

1.13 Percent

Percent is the comparison of any number of parts to 100 parts. The word *percent* means "per hundred." The symbol for percent is %.

You wish to put milk in a pitcher so that it is 25% "full" (Figure 1.34a). First, imagine a line drawn down the side of the pitcher. Then imagine the line divided into 100 equal parts. Each mark shows 1%: that is, each mark shows one out of 100 parts.

(a) This pitcher is 25% full. (b) This pitcher is 83% full. (c) This pitcher is 100% full.

FIGURE 1.34
How full is each pitcher?

FIGURE 1.35

Finally, count 25 marks from the bottom. The amount of milk below the line is 25% of what the pitcher will hold. Note that 100% is a full, or one whole, pitcher of milk.

One dollar equals 100 cents or 100 pennies. Then, 36% of one dollar equals 36 of 100 parts, or 36 cents or 36 pennies. (See Figure 1.35.)

To save 10% of your salary, you would have to save $10 out of each $100 earned.

When the U.S. government spends 15% of its budget on its debt, interest payments are taking $15 out of every $100 the government collects.

A salesperson who earns a commission of 8% receives $8 out of each $100 of goods he or she sells.

A car's radiator holds a mixture that is 25% antifreeze. That is, in each hundred parts of mixture, there are 25 parts of pure antifreeze.

A state charges a 5% sales tax. That is, for each $100 of goods that you buy, a tax of $5 is added to your bill. The $5, a 5% tax, is then paid to the state.

Just remember: *percent* means "per hundred."

Changing a Percent to a Decimal

Percent means the number of parts per 100 parts. Any percent can be written as a fraction with 100 as the denominator.

Example 1   Change each percent to a fraction and then to a decimal.

 a. $75\% = \dfrac{75}{100} = 0.75$ 75 hundredths

 b. $45\% = \dfrac{45}{100} = 0.45$ 45 hundredths

 c. $16\% = \dfrac{16}{100} = 0.16$ 16 hundredths

 d. $7\% = \dfrac{7}{100} = 0.07$ 7 hundredths   ■

Changing a Percent to a Decimal

To change a percent to a decimal, move the decimal point two places to the *left* (divide by 100). Then remove the percent sign (%).

Example 2   Change each percent to a decimal.

 a. $44\% = 0.44$ Move the decimal point two places to the *left* and remove
 b. $24\% = 0.24$ the percent sign (%).
 c. $115\% = 1.15$
 d. $5.7\% = 0.057$
 e. $0.25\% = 0.0025$
 f. $100\% = 1$ ■

If the percent contains a fraction, write the fraction as a decimal. Then proceed as described above.

Example 3   Change each percent to a decimal.

 a. $12\tfrac{1}{2}\% = 12.5\% = 0.125$ Write the fraction part as a decimal and then
                                   change the percent to a decimal.

 b. $6\tfrac{3}{4}\% = 6.75\% = 0.0675$

 c. $165\tfrac{1}{4}\% = 165.25\% = 1.6525$

 d. $\tfrac{3}{5}\% = 0.6\% = 0.006$ ■

For problems involving percents, we must use the decimal form of the percent, or its equivalent fractional form.

Changing a Decimal to a Percent

Changing a decimal to a percent is the reverse of what we did in Example 1.

Example 4   Write 0.75 as a percent.

$$0.75 = \frac{75}{100} \qquad \text{75 hundredths}$$

$$= 75\% \qquad \textit{hundredeths} \text{ means percent} \qquad \blacksquare$$

Changing a Decimal to a Percent

To change a decimal to a percent, move the decimal point two places to the *right* (multiply by 100). Write the percent sign (%) after the number.

Example 5   Change each decimal to a percent.

a. $0.38 = 38\%$ Move the decimal point two places to the *right*. Write
b. $0.42 = 42\%$ the percent sign (%) after the number.
c. $0.08 = 8\%$
d. $0.195 = 19.5\%$
e. $1.25 = 125\%$
f. $2 = 200\%$ ■

Changing a Fraction to a Percent

In some problems, we need to change a fraction to a percent.

Changing a Fraction to a Percent

1. First, change the fraction to a decimal.
2. Then change this decimal to a percent.

Example 6   Change $\frac{3}{5}$ to a percent.

$$\begin{array}{r} 0.6 \\ 5\overline{)3.0} \\ \underline{3\,0} \end{array}$$ First, change $\frac{3}{5}$ to a decimal by dividing the numerator by the denominator.

$$0.6 = 60\%$$ Then change 0.6 to a percent by moving the decimal point two places to the right. Write the percent sign (%) after the number.

So $\frac{3}{5} = 0.6 = 60\%$.   ■

Example 7   Change $\dfrac{3}{8}$ to a percent.

$$\begin{array}{r} 0.375 \\ 8\overline{)3.000} \\ \underline{2\,4} \\ 60 \\ \underline{56} \\ 40 \\ \underline{40} \end{array}$$ First, change $\frac{3}{8}$ to a decimal.

$0.375 = 37.5\%$ Then change 0.375 to a percent.

So $\dfrac{3}{8} = 0.375 = 37.5\%$.   ■

  Example 8   Change $\dfrac{5}{6}$ to a percent.

$$\begin{array}{r} 0.83 \text{ r } 2 \text{ or } 0.83\frac{2}{6} = 0.83\frac{1}{3} \\ 6\overline{)5.00} \\ \underline{4\,8} \\ 20 \\ \underline{18} \\ 2 \end{array}$$ First, change $\frac{5}{6}$ to a decimal.

Note: When the division is carried out to the hundredths place and the remainder is not zero, write the remainder in fraction form, with the remainder over the divisor.

$0.83\dfrac{1}{3} = 83\dfrac{1}{3}\%$ Then change $0.83\frac{1}{3}$ to a percent.

So $\dfrac{5}{6} = 0.83\dfrac{1}{3} = 83\dfrac{1}{3}\%$.   ■

  Example 9   Change $1\dfrac{2}{3}$ to a percent.

$$\begin{array}{r} 0.66 \text{ r } 2 \text{ or } 0.66\frac{2}{3} \\ 3\overline{)2.00} \\ \underline{1\,8} \\ 20 \\ \underline{18} \\ 2 \end{array}$$ First, change $1\frac{2}{3}$ to a decimal.

That is, $1\dfrac{2}{3} = 1.66\dfrac{2}{3}$.

$1.66\dfrac{2}{3} = 166\dfrac{2}{3}\%$ Then change $1.66\frac{2}{3}$ to a percent.

So $1\dfrac{2}{3} = 1.66\dfrac{2}{3} = 166\dfrac{2}{3}\%$.   ■

Changing a Percent to a Fraction

Changing a Percent to a Fraction

1. Change the percent to a decimal.
2. Then change the decimal to a fraction in lowest terms.

◆ Example 10

Change 25% to a fraction in lowest terms.

$$25\% = 0.25$$ First, change 25% to a decimal by moving the decimal point two places to the *left*. Remove the percent sign (%).

$$0.25 = \frac{25}{100} = \frac{1}{4}$$ Then change 0.25 to a fraction. Reduce it to lowest terms.

So $25\% = 0.25 = \frac{1}{4}$.   ■

◆ Example 11

Change 215% to a mixed number.

$$215\% = 2.15$$ First, change 215% to a decimal.

$$2.15 = 2\frac{15}{100} = 2\frac{3}{20}$$ Then change 2.15 to a mixed number in lowest terms.

So $215\% = 2.15 = 2\frac{3}{20}$.   ■

◆ **Changing a Percent That Contains a Mixed Number to a Fraction**

1. Change the mixed number to an improper fraction.
2. Then multiply this result by $\frac{1}{100}$* and remove the percent sign (%).

_____

*Multiplying by $\frac{1}{100}$ is the same as dividing by 100. This is what we do to change a percent to a decimal.

◆ Example 12

Change $33\frac{1}{3}\%$ to a fraction.

$$33\frac{1}{3}\% = \frac{100}{3}\%$$ First, change the mixed number to an improper fraction.

$$\frac{100}{3}\% \times \frac{1}{100} = \frac{\overset{1}{\cancel{100}}}{3} \times \frac{1}{\underset{1}{\cancel{100}}} = \frac{1}{3}$$ Then multiply this result by $\frac{1}{100}$ and remove the percent sign (%).

So $33\frac{1}{3}\% = \frac{1}{3}$.   ■

◆ Example 13

Change $83\frac{1}{3}\%$ to a fraction.

First, $83\frac{1}{3}\% = \frac{250}{3}\%$.

Then $\frac{250}{3}\% \times \frac{1}{100} = \frac{\overset{5}{\cancel{250}}}{3} \times \frac{1}{\underset{2}{\cancel{100}}} = \frac{5}{6}$.

So $83\frac{1}{3}\% = \frac{5}{6}$.   ■

Exercises 1.13 _____

Change each percent to a decimal:

1. 27%
2. 15%
3. 6%
4. 5%
5. 156%
6. 232%
7. 29.2%
8. 36.2%
9. 8.7%
10. 128.7%
11. 947.8%
12. 68.29%
13. 0.28%
14. 0.78%
15. 0.068%
16. 0.0093%
17. $4\frac{1}{4}\%$
18. $9\frac{1}{2}\%$
19. $\frac{3}{8}\%$
20. $50\frac{1}{3}\%$

Change each decimal to a percent:

21. 0.54   **22.** 0.25   **23.** 0.08

24. 0.02   **25.** 0.62   **26.** 0.79

27. 2.17   **28.** 0.345   **29.** 4.35

30. 0.225   **31.** 0.185   **32.** 6.25

33. 0.297   **34.** 7.11   **35.** 5.19

36. 0.815   **37.** 0.0187   **38.** 0.0342

39. 0.0029   **40.** 0.00062

Change each fraction to a percent:

41. $\dfrac{4}{5}$ **42.** $\dfrac{3}{4}$ **43.** $\dfrac{1}{8}$ **44.** $\dfrac{2}{5}$

45. $\dfrac{1}{6}$ **46.** $\dfrac{1}{3}$ **47.** $\dfrac{4}{9}$ **48.** $\dfrac{3}{7}$

49. $\dfrac{3}{5}$ **50.** $\dfrac{5}{6}$ **51.** $\dfrac{13}{40}$ **52.** $\dfrac{17}{50}$

53. $\dfrac{7}{16}$ **54.** $\dfrac{15}{16}$ **55.** $\dfrac{96}{40}$ **56.** $\dfrac{100}{16}$

57. $1\dfrac{3}{4}$ **58.** $2\dfrac{1}{3}$ **59.** $2\dfrac{5}{12}$ **60.** $5\dfrac{3}{8}$

Change each percent to a fraction or a mixed number in lowest terms:

61. 75%   **62.** 45%   **63.** 16%

64. 80%   **65.** 60%   **66.** 15%

67. 93%   **68.** 32%   **69.** 275%

70. 325%   **71.** 125%   **72.** 150%

73. $10\dfrac{3}{4}\%$ **74.** $13\dfrac{2}{5}\%$ **75.** $10\dfrac{7}{10}\%$

76. $40\dfrac{7}{20}\%$ **77.** $17\dfrac{1}{4}\%$ **78.** $6\dfrac{1}{3}\%$

79. $16\dfrac{1}{6}\%$ **80.** $72\dfrac{1}{8}\%$

81. Complete the following table with the equivalents:

Fraction	Decimal	Percent
$\dfrac{3}{8}$	_____	_____
_____	0.45	_____
_____	_____	18%
$1\dfrac{2}{5}$	_____	_____
_____	1.08	_____
_____	_____	$16\dfrac{3}{4}\%$

1.14   Rate, Base, and Part

Any percent problem calls for finding one of three things:

1. the rate (percent),

2. the base, or

3. the part.

Such problems are solved using one of three percent formulas. In these formulas, we let

R = the rate (percent)

B = the base

P = the part or amount (sometimes called the percentage)

The following may help you identify which letter stands for each given number and the unknown in a problem:

1. The rate, R, usually has either a percent sign (%) or the word *percent* with it.

2. The base, B, is usually the whole (or entire) amount. The base is often the number that follows the word *of*.

3. The part, P, is usually some fractional part of the base, B. If you identify R and B first, then P will be the number that is not R or B.

Note: The base and the part should have the same unit(s) of measure.

Example 1   Given: 25% of $80 is $20. Identify R, B, and P.

> R is 25%.   25 is the number with a percent sign. Remember to change 25% to the decimal 0.25 for use in a formula.
>
> B is $80. $80 is the whole amount. It also follows the word *of*.
>
> P is $20. $20 is the part. It is also the number that is not R or B. ■

 Example 2   Given: 72% of the 75 students who took this course last year are now working; find how many are now working. Identify R, B, and P.

> R is 72%.   72 is the number with a percent sign.
>
> B is 75 students.   75 is the whole amount. It also follows the word *of*.
>
> P is the unknown.   The unknown is the number that is some fractional part of the base. It is also the number that is not R or B. ■

Percent Problems: Finding the Part

After you have determined which two numbers are known, you find the third or unknown number by using one of three formulas.

> ### Formulas for Finding Part, Base, and Rate
>
> **1.** $P = BR$ Use to find the part.
>
> **2.** $B = \dfrac{P}{R}$ Use to find the base.
>
> **3.** $R = \dfrac{P}{B}$ Use to find the rate or percent.

Note: After you have studied algebra later in the text, you will need to remember only the first formula.

 Example 3   Find 75% of 180.

> $R = 75\% \doteq 0.75$ 75 is the number with a percent sign.
> $B = 180$ 180 is the whole amount and follows the word *of*.
> $P = $ the unknown   Use Formula 1.
> $P = BR$
> $P = (180)(0.75)$
> $= 135$ ■

 Example 4 $45 is $9\frac{3}{4}\%$ of what amount?

> $R = 9\frac{3}{4}\% = 9.75\% = 0.0975$ $9\frac{3}{4}\%$ is the number with a percent sign.
>
> $B = $ the unknown   Use Formula 2.
> $P = \$45$ $45 is the part.
>
> $B = \dfrac{P}{R}$
>
> $B = \dfrac{\$45}{0.0975}$
>
> $= \$461.54$ ■

Example 5

What percent of 20 metres is 5 metres?

R = the unknown Use Formula 3.
B = 20 m 20 m is the whole amount and follows the word *of.*
P = 5 m 5 m is the part.

$$R = \frac{P}{B}$$

$$R = \frac{5 \text{ m}}{20 \text{ m}}$$

$$= 0.25 = 25\%$$ ■

Example 6

Aluminum is 12% of the mass of a given car. This car has 186 kg of aluminum in it. What is the total mass of the car?

R = 12% = 0.12 12 is the number with a percent sign.
B = the unknown Use Formula 2.
P = 186 kg 186 kg is the part.

$$B = \frac{P}{R}$$

$$B = \frac{186 \text{ kg}}{0.12}$$

$$= 1550 \text{ kg}$$ ■

Example 7

A fuse is a safety device with a core. When too much current flows, the core melts and breaks the circuit. The size of a fuse is the number of amperes of current the fuse can safely carry. A given 50-amp (50-A) fuse blows at 20% overload. What is the maximum current the fuse will carry?

First, find the amount of current overload:

R = 20% = 0.20 20 is the number with a percent sign.
B = 50 A 50 A is the base.
P = the unknown Use Formula 1.

$$P = BR$$

$$P = (50 \text{ A})(0.20)$$

$$= 10 \text{ A}$$

The maximum current the fuse will carry is the normal current plus the overload:

$$50 \text{ A} + 10 \text{ A} = 60 \text{ A}$$ ■

Example 8

Georgia's salary was $600 per week. Then she was given a raise of $50 per week. What percent raise did she get?

R = the unknown Use Formula 3.
B = $600 $600 is the base.
P = $50 $50 is the part.

$$R = \frac{P}{B}$$

$$R = \frac{\$50}{\$600}$$

$$= 0.08\frac{1}{3} = 8\frac{1}{3}\%$$ ■

Example 9

Castings are listed at $9.50 each. A 12% discount is given if 50 or more are bought at one time. We buy 60 castings.

    a. What is the discount on one casting?
    b. What is the cost of one casting?
    c. What is the total cost?

    a. Discount equals 12% of $9.50.

$$R = 12\% = 0.12$$
$$B = \$9.50$$
$$P = \text{the unknown (the discount)}$$
$$P = BR$$
$$P = (\$9.50)(0.12)$$
$$= \$1.14 \text{ (the discount on one casting)}$$

    b. Cost (of one casting) = list − discount
$$= \$9.50 - \$1.14$$
$$= \$8.36$$

    c. Total cost = cost of one casting times the number of castings
$$= (\$8.36)(60)$$
$$= \$501.60$$ ■

You may also need to find the percent increase or decrease in a given quantity.

 Example 10

Mary's hourly wages changed from $18.40 to $19.55. Find the percent increase in her wages.

First, let's find the change in her wages.

$$\$19.55 - \$18.40 = \$1.15$$

Then, this change is what percent of her original wage?

$$R = \frac{P}{B} = \frac{\$1.15}{\$18.40} = 0.0625 = 6.25\%$$ ■

The process of finding the percent increase or percent decrease may be summarized by the following formula:

$$\text{percent increase (or percent decrease)} = \frac{\text{the change}}{\text{the original value}} \times 100\%$$

 Example 11

Normal ac line voltage is 115 volts (V). Find the percent decrease if the line voltage drops to 109 V.

$$\text{percent decrease} = \frac{\text{the change}}{\text{the original value}} \times 100\%$$
$$= \frac{115 \text{ V} - 109 \text{ V}}{115 \text{ V}} \times 100\%$$
$$= 5.22\%$$ ■

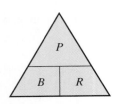

FIGURE 1.36

The triangle in Figure 1.36 can be used to help you remember the three percent formulas, as follows:

1. $P = BR$ To find the part, cover P; B and R are next to each other on the same line, as in multiplication.

2. $B = \dfrac{P}{R}$ To find the base, cover B; P is over R, as in division.

3. $R = \dfrac{P}{B}$ To find the rate, cover R; P is over B, as in division.

Exercises 1.14

Identify the rate (R), the base (B), and the part (P) in each statement 1–10 (do not solve the problem):

1. 60 is 25% of 240.
2. $33\frac{1}{3}\%$ of $300 is $100.
3. 40% of 270 is 108.
4. 72 is 15% of 480.
5. At plant A, 4% of the tires made were defective. Plant A made 28,000 tires. How many tires were defective?
6. On the last test, 25 of the 28 students earned passing grades. What percent of students passed?
7. A girls' volleyball team won 60% of its games. The team won 21 games. How many games did it play?
8. A rancher usually loses 10% of his herd every winter due to weather. He has a herd of 15,000. How many does he expect to lose this winter?
9. An electronics firm finds that 6% of the resistors it makes are defective. There were 2050 defective resistors. How many resistors were made?
10. The interest on a $500 loan is $90. What is the rate of interest?

When finding the percent, round to the nearest tenth of a percent when necessary:

11. What percent of $2080 is $208?
12. The number 2040 is 7.5% of what number?
13. What percent of 5280 ft is 880 yd?
14. 0.35 mi is 4% of what amount?
15. $72 is 4.5% of what amount?
16. What percent of 7.15 is 3.5?
17. Find 235% of 48.
18. What percent of $\frac{1}{8}$ is $\frac{1}{15}$?
19. Find 28% of 32 volts (V).
20. Find 110% of 50.

21. A welder needs to complete 130 welds. If 97 have been completed so far, what is the percent completed?
22. A welder makes high-quality welds 92% of the time. Out of 115 welds, how many are expected to be of high quality?
23. A small airport has a Cessna 172 rental plane. In one month, 24 h of the 65 total rental hours were for lessons. What percent of the total rental time was the plane rented for lessons?
24. On a cross-country trip, 1.5 h were flown under VFR (Visual Flight Rules), and 0.4 h was flown under IFR (Instrument Flight Rules). What percent of the trip was flown under IFR?
25. A car oil filter holds 0.3 qt of oil. The car holds 4.5 qt of oil including the filter. What percent of the oil is in the filter?
26. Air enters an air conditioner at the rate of 75 lb/h, and the unit can remove 1.5 lb/h of moisture. If the air entering contains 2 lb/h moisture, what percent of the moisture is removed?
27. Air flows through a duct at 2400 cubic feet per minute (CFM). After several feet and a few vents, the airflow decreases to 1920 CFM. What is the percent drop that has occurred?
28. A building being designed will have fixed windows. Including the frame, the windows are 2 ft wide and 6 ft high. The south wall is 78 ft 6 in. wide by 12 ft 2 in. high. Local codes allow only 20% window area on south walls. How many windows can you draw on this wall?
29. The embankment leading to a bridge must have a maximum 3% slope. The change in elevation shown in Illustration 1 must be dimensioned to meet these criteria. Find dimension A to complete the drawing.

Embankment   3%

A

100 ft

Concrete abutment

ILLUSTRATION 1

30. A rectangular tank is being designed with an internal catwalk around the inside as shown in Illustration 2. For functional reasons, the walkway cannot be more than 3 in. above the liquid. The liquid level in the tank will be maintained $\frac{3}{4}$ full at all times. What height dimension for the walkway should be put on the drawing?

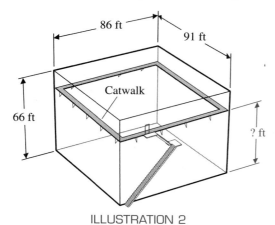

86 ft

91 ft

Catwalk

66 ft

? ft

ILLUSTRATION 2

31. The application rate of a chemical is $2\frac{3}{4}$ lb/acre. How many pounds are needed for 160 acres of corn? If the chemical contains 80% active ingredients by weight, how many pounds of active ingredients will be applied? How many pounds of inert ingredients will be applied?

32. U.S. soybeans average 39% protein. A bushel of soybeans weighs 60 lb. How many pounds of protein are in a bushel? A 120-acre field yields 45 bu/acre. How many pounds of protein does that field yield?

33. A dairy cow produced 7310 lb of milk in a year. A gallon of milk weighs 8.6 lb. How many gallons of milk did the cow produce? The milk tested at 4.2% butterfat. How many gallons of butterfat did the cow produce?

34. You need 15% of a 60-mg tablet. How many mg would you take?

35. Mary needs to give 40% of a 0.75-grain tablet. How many grains does she give?

36. You need 0.15% of 2000 mL. How many millilitres do you need?

37. What percent of 0.600 grain is 0.150 grain?

38. During a line voltage surge, the normal ac voltage increased from 115 V to 128 V. Find the percent increase.

39. During manufacturing, the pressure in a hydraulic line increases from 75 lb/in^2 to 115 lb/in^2. What is the percent increase in pressure?

40. The value of Caroline's house decreased from $93,500 to $75,400 when the area's major employer closed the local plant and moved to another state. Find the percent decrease in the value of her house.

41. Due to wage concessions, Bill's hourly wages dropped from $25.50 to $21.88. Find the percent decrease in his wages.

42. A building has 28,000 ft^2 of floor space. When an addition of 6500 ft^2 is built, what is the percent increase in floor space?

43. Two different items both originally selling for $100.00 are on sale. One item is marked down 55%. The second item is first marked down 40%, then an additional 15%. Find the final sale price for each item.

44. A machinist is hired at $22.15 per hour. After a 6-month probationary period, the wage will increase by 32%. If the machinist successfully completes the apprenticeship, what will the pay be per hour?

45. A homeowner harvests a tree to use for firewood. The entire tree weighs 1640 lb. Of that, 95% is cut and split into sticks of firewood. The rest is leaves and branches too small to use for firewood. How much did the firewood weigh?

46. A fisherman catches a total of 125 lb of fish. When the fish are cleaned, 59 lb of fillet remain and the rest is discarded. What percent of the fish was usable as fillets?

47. A flock of mallards (ducks) is called a sord. One particular mallard sord has 250 live mallards when the last hatchling emerges in the spring. At the end of the following winter, the sord has 187 birds remaining before the first egg hatches. What was the survival rate for the sord?

48. In a local community, wildlife biologists estimate a deer population of 135 on January 1. Over the following 12 months, there are 42 live births of deer fawn, 7 are killed by vehicles on the highway, 3 fawns are killed by dogs, 5 are killed by hunters, and 10 die of disease or other injury. As of December 31, what was the deer population and what was the percentage change?

49. Populations of any organism increase when births exceed deaths. In a suburban area in the upper Midwest, the large number of deer was becoming a problem. A Deer Task Force survey suggested in 2006 that 20 deer per square mile might be an acceptable population level for the citizens in that area. Assume a current population density of 25 deer per square mile and a population growth rate (births minus deaths) of 40% per year. **a.** If there are no significant predators and hunting is not allowed, how many deer can the town planners expect per square mile in the following year? **b.** How many deer can the town planners expect per square mile in the following second year with the same population growth rate?

50. A community has a goal to decrease its municipal solid waste (MSW) by 25% over a 5-year period. Assuming the community has 75,000 residents who each average 4.6 lb of MSW each day, **a.** how much MSW would each resident average each day if the goal were met?

b. How many tons of MSW would the community generate annually if the goal were met? **c.** Another nearby larger community had decreased its annual MSW by 30% to 73,500 tons. How much was its previous annual amount of MSW?

51. An *invoice* is an itemized list of goods and services specifying the price and terms of sale. Illustration 3 shows an invoice for parts and labor for an addition to a home for the week indicated. Complete the invoice.

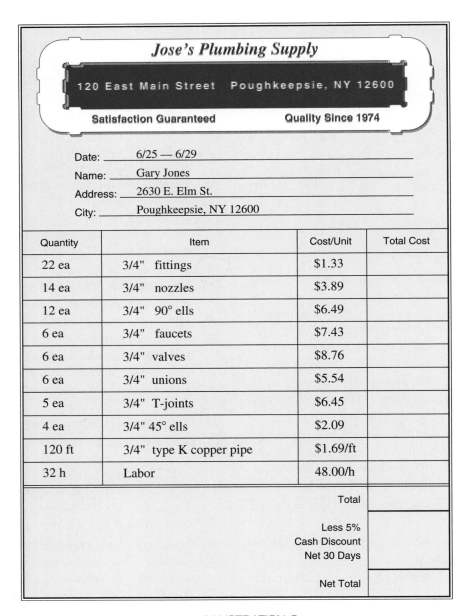

Jose's Plumbing Supply

120 East Main Street   Poughkeepsie, NY 12600

Satisfaction Guaranteed   Quality Since 1974

Date: _____6/25 — 6/29_____

Name: _____Gary Jones_____

Address: _____2630 E. Elm St._____

City: _____Poughkeepsie, NY 12600_____

Quantity	Item	Cost/Unit	Total Cost
22 ea	3/4" fittings	$1.33	
14 ea	3/4" nozzles	$3.89	
12 ea	3/4" 90° ells	$6.49	
6 ea	3/4" faucets	$7.43	
6 ea	3/4" valves	$8.76	
6 ea	3/4" unions	$5.54	
5 ea	3/4" T-joints	$6.45	
4 ea	3/4" 45° ells	$2.09	
120 ft	3/4" type K copper pipe	$1.69/ft	
32 h	Labor	48.00/h	
		Total	
		Less 5% Cash Discount Net 30 Days	
		Net Total	

ILLUSTRATION 3

52. Illustration 4 shows an invoice for grain sold at a local elevator. Complete the invoice.

FARMERS TERMINAL GRAIN COMPANY

Beardstown, Illinois 62618   Since 1893

Customer name: Shaw Farms, Inc.   Account No. 3786

Date	Gross wt–pounds	Weight of empty truck	Net wt–pounds	Type of grain	No. of bushels*	Price/bu	Amount
7/2	21560	9160	12400	Wheat	207	$ 5.67	$1173.69
7/3	26720	9240		Wheat		5.71	
7/5	20240	7480		Wheat		5.74	
7/6	28340	9200		Wheat		5.81	
7/8	26760	9160		Wheat		5.76	
7/8	17880	7485		Wheat		5.76	
10/1	25620	9080		Soybeans		11.72	
10/1	21560	7640		Soybeans		11.69	
10/2	26510	9060		Soybeans		11.68	
10/2	22630	7635		Soybeans		11.65	
10/4	22920	9220		Soybeans		11.72	
10/5	20200	7660		Soybeans		11.81	
10/6	25880	9160		Soybeans		11.90	
10/7	21300	7675		Soybeans		11.84	
10/8	18200	7665		Soybeans		11.79	
10/12	26200	9150		Corn		4.68	
10/12	22600	7650		Corn		4.65	
10/13	27100	9080		Corn		4.66	
10/15	22550	7635		Corn		4.61	
10/15	23600	7680		Corn		4.59	
10/17	26780	9160		Corn		4.63	
10/18	28310	9200		Corn		4.69	
10/21	21560	7665		Corn		4.67	
10/22	25750	9160		Corn		4.65	
						TOTAL	$ 46,363.83

*Round to the nearest bushel. *Note:* Corn weighs 56 lb/bu; soybeans weigh 60 lb/bu; wheat weighs 60 lb/bu.

ILLUSTRATION 4

53. Many lumberyards write invoices for their lumber by the piece. (See Illustration 5.) Complete the invoice, which is for the rough framing of the shell of a home.

KURT'S LUMBER 400 WEST OAK AKRON, OHIO 44300			
SOLD TO Robert Bennett 32 Park Pl E. Akron 44305		DATE 5/16	
QUANTITY	DESCRIPTION	UNIT PRICE	TOTAL
66	2 x 4 " x 16', fir, plate material	$ 7.97	
30	2" x 4" x 10', fir, plate studs	3.95	
14	2" x 4" x 8', fir, knee wall studs	3.39	
17	2" x 6" x 12', fir, kit. ceiling joists	6.59	
4	2" x 12" x 12', fir, kitchen girders	12.10	
9	2" x 6" x 10', fir, kitchen rafters	5.39	
7	2" x 4" x 12', fir, collar beams	4.97	
10	2" x 8" x 12', fir, 2nd floor joists	11.97	
6	2" x 8" x 16', fir, 2nd floor joists	16.89	
11	2" x 8" x 20', fir, 2nd floor joists	18.55	
15	4' x 8' x 3/4", T & G plywood	24.25	
27	2" x 8" x 18', fir, kitchen and living room rafters	16.95	
7	2" x 8" x 16', fir, kitchen and living room rafters	14.39	
1	2" x 10" x 22', fir, kitchen and living room ridge	24.96	
10	1" x 8" x 14', #2 white pine, sub facia	10.37	
27	2" x 8" x 22', fir, bedroom rafters	19.85	
7	2" x 8" x 16', fir, dormer rafters	12.25	
1	2" x 10" x 20', fir, bedroom ridge	17.85	
7	2" x 6" x 20', fir, bedroom ceiling joists	12.19	
8	2" x 6" x 8', fir, bedroom ceiling joists	3.49	
3	2" x 12" x 14', fir, stair stringers	17.65	
80	4' x 8' x 1/2", roof decking	17.29	
7	rolls #15 felt building paper	20.65	
1	50 lb #16 cement nails	33.59	
3	30 lb galvanized roofing nails	34.97	
250	precut fir studs	2.18	

KURT'S LUMBER

Subtotal _____

Less 2% cash discount _____

Subtotal _____

5 3/4% sales tax _____

NET TOTAL _____

ILLUSTRATION 5

54. Complete the electronics parts invoice shown in Illustration 6.

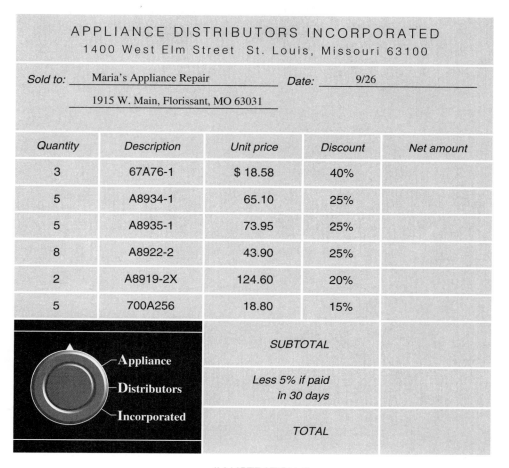

APPLIANCE DISTRIBUTORS INCORPORATED
1400 West Elm Street St. Louis, Missouri 63100

Sold to: Maria's Appliance Repair *Date:* 9/26

1915 W. Main, Florissant, MO 63031

Quantity	Description	Unit price	Discount	Net amount
3	67A76-1	$ 18.58	40%	
5	A8934-1	65.10	25%	
5	A8935-1	73.95	25%	
8	A8922-2	43.90	25%	
2	A8919-2X	124.60	20%	
5	700A256	18.80	15%	
			SUBTOTAL	
			Less 5% if paid in 30 days	
			TOTAL	

Appliance Distributors Incorporated

ILLUSTRATION 6

1.15 Powers and Roots

The **square of a number** is the product of that number times itself. The square of 3 is $3 \cdot 3$ or 3^2 or 9. The square of a number may be found with a calculator as follows.

Example 1

Find 73.6^2 rounded to three significant digits.

73.6 x^2 $=$

```
5416.96
```

Thus, $73.6^2 = 5420$ rounded to three significant digits. ■

Example 2   Find 0.135^2 rounded to three significant digits.

.135 x^2 =

> 0.018225

Thus, $0.135^2 = 0.0182$ rounded to three significant digits.   ■

The **square root of a number** is that positive number which, when multiplied by itself, gives the original number. The square root of 25 is 5 and is written as $\sqrt{25}$. The symbol $\sqrt{}$ is called a *radical*.

 Example 3   Find the square roots of **a.** 16, **b.** 64, **c.** 100, and **d.** 144.

    a. $\sqrt{16} = 4$ because $4 \cdot 4 = 16$
    b. $\sqrt{64} = 8$ because $8 \cdot 8 = 64$
    c. $\sqrt{100} = 10$ because $10 \cdot 10 = 100$
    d. $\sqrt{144} = 12$ because $12 \cdot 12 = 144$

Numbers whose square roots are whole numbers are called **perfect squares**. For example, 1, 4, 9, 16, 25, 36, 49, and 64 are perfect squares.   ■

The square root of a number may be found with a calculator as follows.

 Example 4   Find $\sqrt{21.4}$ rounded to three significant digits.

 $\sqrt{}$ 21.4 =

> 4.626013402

Thus, $\sqrt{21.4} = 4.63$ rounded to three significant digits.   ■

 Example 5   Find $\sqrt{0.000594}$ rounded to three significant digits.

 $\sqrt{}$ .000594   =

> 0.024372115

Thus, $\sqrt{0.000594} = 0.0244$ rounded to three significant digits.   ■

The **cube of a number** is the product of that number times itself three times. The cube of 5 is $5 \cdot 5 \cdot 5$ or 5^3 or 125.

◆ **Example 6**　Find the cubes of **a.** 2, **b.** 3, **c.** 4, and **d.** 10.

　　a. $2^3 = 2 \cdot 2 \cdot 2 = 8$
　　b. $3^3 = 3 \cdot 3 \cdot 3 = 27$
　　c. $4^3 = 4 \cdot 4 \cdot 4 = 64$
　　d. $10^3 = 10 \cdot 10 \cdot 10 = 1000$　　　　　　　　　■

The cube of a number may be found with a calculator as follows:

◆ **Example 7**　Find 12^3.

12 [∧]* 3 [=]

| 1728 |

Thus, $12^3 = 1728$.

*Some calculators use the [y^x] button to find a power.　　　　　　■

◆ **Example 8**　Find 4.25^3 rounded to three significant digits.

4.25 [∧] 3 [=]

| 76.765625 |

Thus, $4.25^3 = 76.8$ rounded to three significant digits.　　　　　■

The **cube root of a number** is that number which, when multiplied by itself three times, gives the original number. The cube root of 8 is 2 and is written as $\sqrt[3]{8}$. (**Note:** $2 \cdot 2 \cdot 2 = 8$. The small ^{3} in the radical is called the *index*.)

◆ **Example 9**　Find the cube roots of **a.** 8, **b.** 27, and **c.** 125.

　　a. $\sqrt[3]{8} = 2$ because $2 \cdot 2 \cdot 2 = 8$
　　b. $\sqrt[3]{27} = 3$ because $3 \cdot 3 \cdot 3 = 27$
　　c. $\sqrt[3]{125} = 5$ because $5 \cdot 5 \cdot 5 = 125$　　　　　　　■

Numbers whose cube roots are whole numbers are called **perfect cubes**. For example, 1, 8, 27, 64, 125, and 216 are perfect cubes.

The cube root of a number may be found with a calculator as follows.

◆ **Example 10**　Find $\sqrt[3]{512}$.

　　a. If your calculator has a $\sqrt[3]{}$ button,

[$\sqrt[3]{}$] 512 [=]

| 8 |

Thus, $\sqrt[3]{512} = 8$.

b. If your calculator has a $\sqrt[x]{}$ button,

3* $\boxed{\sqrt[x]{}}$ 512 $\boxed{=}$

$$\boxed{\quad\quad 8 \quad\quad}$$

Thus, $\sqrt[3]{512} = 8$.

_____

*Here you need to enter the index of the root first. ■

◆ Example 11   Find $\sqrt[3]{4532}$ rounded to three significant digits.

a. If your calculator has a $\sqrt[3]{}$ button,

$\boxed{\sqrt[3]{}}$ 4532 $\boxed{=}$

$$\boxed{16.5486778}$$

Thus, $\sqrt[3]{4532} = 16.5$ rounded to three significant digits.

b. If your calculator has a $\sqrt[x]{}$ button,

3 $\boxed{\sqrt[x]{}}$ 4532 $\boxed{=}$

$$\boxed{16.5486778}$$

Thus, $\sqrt[3]{4532} = 16.5$ rounded to three significant digits. ■

In general, in a **power** of a number, the *exponent* indicates the number of times the *base* is used as a factor. For example, the 4th power of 3 is written 3^4, which means that 3 is used as a factor 4 times ($3^4 = 3 \cdot 3 \cdot 3 \cdot 3 = 81$).

◆ Example 12   Find 2.24^5 rounded to three significant digits.

2.24 $\boxed{\wedge}$ *5 $\boxed{=}$

$$\boxed{56.39493386}$$

Thus, $2.24^5 = 56.4$ rounded to three significant digits.

_____

*Some calculators use the $\boxed{y^x}$ button to find a power. ■

Exercises 1.15

Find each power rounded to three significant digits:

1. 15^2

2. 25^2

3. 14.9^2

4. 0.0279^2

5. 0.00257^2

6. $54,200^2$

7. 9^3

8. 14^3

9. 8.25^3

10. 0.0225^3

11. 0.169^3

12. 24.8^3

13. 2.75^5

14. 3.5^{10}

Find each root rounded to three significant digits:

15. $\sqrt{8.75}$

16. $\sqrt{12,500}$

17. $\sqrt{4750}$

18. $\sqrt{0.0065}$

19. $\sqrt[3]{75,975}$

20. $\sqrt[3]{9.59}$

21. $\sqrt[3]{0.00777}$

22. $\sqrt[3]{675.88}$

1.16 Applications Involving Percent: Personal Finance (Optional)

When money is loaned, the borrower pays a fee for using the money to the lender; this fee is called *interest*. Similarly, when someone deposits money into a savings account, the bank or other financial institution pays an interest fee for using the money to the depositor. There are several types of interest problems. We will start with simple interest, the interest paid only on the original principal. Simple interest problems are based on the following formula:

Simple Interest

$$i = prt$$

where i = the amount of interest paid

p = the principal, which is the original amount of money borrowed or deposited

r = the interest rate written as a decimal

t = the time in years the money is being used

Example 1

A student obtains a simple interest loan to purchase a computer for $1200 at 12.5% over 2 years. How much interest is paid?

Here, p = $1200

r = 12.5% = 0.125, written as a decimal

t = 2 years

Then, $i = prt$

$i = (\$1200)(0.125)(2)$

$= \$300$

Thus, the student pays an extra $300 fee to be able to use the computer while the money is being paid. ■

Savings accounts paid with simple interest are calculated similarly.

When money is borrowed, the total amount to be paid back is the amount borrowed (the principal) plus the interest. The money is usually paid back in regular monthly or weekly installments. The following formula is used to determine the regular payment amount:

Payment Amount

$$\text{payment amount} = \frac{\text{principal} + \text{interest}}{\text{loan period in months or weeks}}$$

Example 2

Find the monthly payment amount for the computer purchase in Example 1.

$$\text{payment amount} = \frac{\text{principal} + \text{interest}}{\text{loan period in months or weeks}}$$

$$\text{payment amount} = \frac{\$1200 + \$300}{24} \qquad \text{2 years} \times \text{12 months}$$

$$= \$62.50 \text{ per month}$$ ■

You probably have heard the term *compound interest* used. What is the difference between simple interest and compound interest? *Simple interest* is the amount of money paid or earned on the principal (initial amount) without the interest added to the initial amount. For example, if you accepted a loan of $1000 with 1% simple interest per month, you would owe

$1000 + ($1000)(0.01) = $1000 + $10 = $1010 at the end of one month,
$1000 + $10 + ($1000)(0.01) = $1000 + $10 + $10 = $1020 at the end of two months,
$1000 + $20 + ($1000)(0.01) = $1000 + $20 + $10 = $1030 at the end of three months, etc.

Compound interest is the amount of money paid or earned on the accumulated interest plus the principal with the accumulated interest for the given period then added to the principal. For example, if you accepted a loan of $1000 with 1% interest compounded monthly, you would owe

$1000 + ($1000)(0.01) = $1000 + $10 = $1010 at the end of one month,
$1010 + ($1010)(0.01) = $1010 + $10.10 = $1020.10 at the end of two months,
$1020.10 + ($1020.10)(0.01) = $1020.10 + $10.20 = $1030.30 at the end of three months, etc.

The following formula is used to compute compound interest for any given number of times the interest is compounded annually. Basic examples would include finding the amount of money owed on a loan after a given period of time with no payments made or the amount of money in a savings account after a given period of time with no additional deposits made.

Compound Interest

$$A = P\left(1 + \frac{r}{n}\right)^{nt}$$

where A = the amount after time t
P = the principal or initial amount
r = the annual interest rate written as a decimal
n = the number of times the interest is compounded each year
t = the number of years

Example 3

Find the amount of money owed at the end of 3 years if $10,000 is borrowed at 8% per year compounded quarterly and no payments are made on the loan.

Here, $P = \$10,000$

$r = 8\% = 0.08$

$n = 4$ (compounded quarterly means at the end of each of four quarters)

$t = 3$ years

$$A = P\left(1 + \frac{r}{n}\right)^{nt}$$

$$= \$10,000\left(1 + \frac{0.08}{4}\right)^{(4)(3)}$$

$$= \$10,000\,(1.02)^{12}$$

$$= \$12,682.42$$

Using a scientific calculator, we have

10000 × (1 + .08 ÷ 4) ∧ (4 × 3) =

```
12682.41795
```

Thus, the answer is $12,682.42 rounded to the nearest cent. ■

A variety of formulas can be used to calculate interest. You are advised to always ask how your interest is calculated.

A *mortgage* is a legal document that pledges a house or other real estate as security for the repayment of a loan. This enables a person to buy and use property without having the funds on hand to pay for the house and property outright; a significant down payment is almost always required before a loan is approved. If the borrower fails to repay the loan, the lender may foreclose, which means that the house and property would be sold so that the lender can recover the amount of the loan remaining. A car loan works similarly. If the buyer fails to repay the car loan, the car is reclaimed so that the lender can recover the amount of the car loan remaining. Most mortgage and car loans are calculated with compound interest.

Amortization is the distribution of a single given amount of money into smaller regular installment loan repayments. Each installment consists of both principal and interest. The following amortization formula is used to calculate the periodic (often monthly) installment payment amount for a given loan amount. Although the amount of each installment payment remains the same, some goes toward the interest and some goes toward the principal, with early payments loaded toward interest and later payments loaded toward principal.

Amortization Formula

$$A = P\left(\frac{i(1 + i)^n}{(1 + i)^n - 1}\right)$$

where A = the periodic payment amount

P = the amount of the principal

i = the interest rate as an annual percentage rate (APR) as a decimal divided by the number of annual interest payments (divide by 12 if monthly or by 52 if weekly)

n = the total number of payments

Note: Interest rate i must be given in terms of APR, annual percentage rate, and not APY, annual percentage yield.

Example 4

You need a home loan of $120,000 after your down payment. How much would your monthly house payment be if the bank charges 6% APR for a loan of 30 years?

Here, $P = \$120{,}000$

$i = 0.06/12 = 0.005$

$n = 12 \text{ payments/year} \times 30 \text{ years} = 360 \text{ monthly payments}$

$$A = P\left(\frac{i(1 + i)^n}{(1 + i)^n - 1}\right)$$

$$= \$120{,}000\left(\frac{0.005(1 + 0.005)^{360}}{(1 + 0.005)^{360} - 1}\right)$$

$$= \$120{,}000\left(\frac{0.005(1.005)^{360}}{(1.005)^{360} - 1}\right)$$

$$= \$719.46$$

Using a scientific calculator, we have

That is, you would pay $719.46 rounded to the nearest cent each month for 30 years to repay this loan.   ■

Example 5

You need a car loan of $18,000 after the trade-in and down payment. You are trying to decide between two choices. (a) Receive a car loan and pay 1.5% APR for 36 months through the car dealership or (b) pay cash to receive a $2500 discount and obtain a loan for the difference at a bank at 6% APR for 36 months.

(a) Here, $P = \$18{,}000$

$i = 0.015/12 = 0.00125$

$n = 36$

$$A = P\left(\frac{i(1 + i)^n}{(1 + i)^n - 1}\right)$$

$$= \$18{,}000\left(\frac{0.00125(1 + 0.00125)^{36}}{(1 + 0.00125)^{36} - 1}\right)$$

$$= \$511.65/\text{month} \text{ or } \$511.65 \times 36 = \$18{,}419.40 \text{ total}$$

(b) Here, $P = \$18{,}000 - \$2500 = \$15{,}500$

$i = 0.06/12 = 0.005$

$n = 36$

$$A = P\left(\frac{i(1 + i)^n}{(1 + i)^n - 1}\right)$$

$$= \$15{,}500\left(\frac{0.005(1 + 0.005)^{36}}{(1 + 0.005)^{36} - 1}\right)$$

$$= \$471.54/\text{month} \text{ or } \$471.54 \times 36 = \$16{,}975.44 \text{ total}$$

It pays to do some good financial research!   ■

Exercises 1.16

1. One family member loans $2000 to another family member for 3 years at 5% simple interest. **a.** How much interest is paid? **b.** Find the monthly payment.

2. Kim deposits $2500 in a savings account that pays 4.5% simple interest. How much interest does she earn in 2 years?

3. Find the amount of money owed at the end of 4 years if $7500 is borrowed at 6.5% per year compounded quarterly and no payments are made on the loan.

4. Find the amount of money owed at the end of 6 years if $10,500 is borrowed at 5.75% per year compounded semiannually and no payments are made on the loan.

5. Mary Lou received $15,000 from her grandparents for her college education 8 years prior to her enrolling in college. Mary Lou invested the money at 5.5% compounded semiannually. How much money would she have in her savings account when she is ready to enroll in college?

6. Ted invests $6000 at 7.5% compounded quarterly for 5 years. How much money does he have at the end of 5 years?

7. You need a home loan of $150,000 after your down payment. How much would your monthly house payment be if the bank charges 6.5% APR for a loan of 30 years?

8. You need a home loan of $75,000 after your down payment. How much would your monthly house payment be if the bank charges 6.25% APR for a loan of 15 years?

9. A farmer purchased 275 acres of land for $4100/acre. He paid 25% down and obtained a loan for the balance at 6.75% APR over a 20-year period. How much is the monthly payment?

10. Denny purchased a new truck for a price of $45,500. He received a rebate of $4500 and paid a sales tax of 6.5% after the rebate. He paid a down payment of 20% and obtained a loan for the balance at 7.25% APR over a 5-year period. How much is the monthly payment?

11. You need a car loan of $24,000 after the trade-in and down payment. You are trying to decide between two choices. **a.** Receive a car loan and pay 0.75% APR for 36 months through the car dealership. Find the monthly payment and the total amount paid. **b.** Pay cash to receive a $1500 discount and obtain a loan for the difference at a bank at 8.5% APR for 36 months. Find the monthly payment and the total amount paid.

12. You need a car loan of $19,500 after the trade-in and down payment. You are trying to decide between two choices. **a.** Receive a car loan and pay 1.75% APR for 36 months through the car dealership. Find the monthly payment and the total amount paid. **b.** Pay cash to receive a $2500 discount and obtain a loan for the difference at a bank at 6.5% APR for 36 months. Find the monthly payment and the total amount paid.

13. Larry purchased a new combine that cost $175,500 less a rebate of $4500, a trade-in of $9500, and a down payment of $5000. He takes out a loan for the balance at 8% APR over 4 years. Find the monthly payment and the total amount Larry paid on the balance.

14. An agricultural equipment dealer bought a tractor for $115,500 and then includes markups of 3.5% to cover expenses and 0.95% for profit. Fred buys this tractor less a trade-in of $7500 and a down payment of $10,000. He takes out a loan for the balance at 7.25% APR over 5 years. Find the monthly payment and the total amount Fred paid on the balance.

Unit 1C Review

1. Change $1\frac{5}{8}$ to a decimal.

2. Change 0.45 to a common fraction in lowest terms.

Perform the indicated operations:

3. $4.206 + 0.023 + 5.9$ 4. $120 - 3.065$

5. $12.1 - 6.25 + 0.004$

6. Find the missing dimension in the figure in Illustration 1.

7. Find the perimeter of the figure in Illustration 1.

ILLUSTRATION 1

8. Round 45.0649 to **a.** the nearest tenth and **b.** the nearest hundredth.

9. Round 45.0649 to **a.** three significant digits and **b.** four significant digits.

10. Multiply: 4.606×0.025

11. Divide: $45.24 \div 2.4$

12. A cable 18.5 in. long is to be cut into lengths of 2.75 in. each. How many cables of this length can be cut? How much of the cable is left?

13. Change 25% to a decimal.

14. Change 0.724 to a percent.

15. Find 16.5% of 420.

16. 240 is 12% of what number?

17. What percent of 240 yd is 96 yd?

18. Jean makes $16.50/h. If she receives a raise of 6%, find her new wage.

Find each power or root rounded to three significant digits:

19. 45.9^2

20. $\sqrt[3]{831}$

Chapter 1 Group Activities

1. Certificates of Deposit

   In small groups, go to some banks and research certificates of deposits, or CDs. Find the following information about their CDs. (You could surf the Internet, too.)

   a. Ask what types of CDs they offer: 3-month or 91-day, 6-month, 12-month, 15-month, or longer-term CDs.

   b. Find the rates for each CD.

   c. Ask if there are minimum deposit requirements.

   d. Ask how the interest is compounded: monthly, quarterly, semiannually, or other.

   e. Ask how the CD works.

   f. Write a report and make a chart using the information from banks you visited that offer different interest rates and different methods of compounding. Select two methods of compounding the same interest rate. Compare them by showing the difference in interest earned over a period of three years. (Deposit the same amount into your two CDs and then record the monthly account balance for 36 months.)

   For example, bank A requires a minimum deposit of $1000 with an annual percentage yield of 3.75% compounded monthly. If you have a 6-month CD with the interest compounded monthly, you must divide the interest rate by 12: $0.0375 \div 12 = 0.003125$.

   Initial deposit: $1000

   1st month: $1000[1 + (0.0375 \div 12)] = \$1003.13$

   2nd month: $\$1003.13(1 + 0.003125) = \$1006.26$

   3rd month: $\$1006.26(1 + 0.003125) = \$1009.40$

   4th month: $\$1009.40(1 + 0.003125) = \$1012.55$

   5th month: $\$1012.55(1 + 0.003125) = \$1015.71$

   6th month: $\$1015.71(1 + 0.003125) = \$1018.88$

   If compounded quarterly then the same example would be

   Initial deposit: $1000

   1st month: $1000

   2nd month: $1000

   3rd month: $\$1000[1 + (0.0375 \div 4)] = \$1009.38$

   4th month: $1009.38

   5th month: $1009.38

   6th month: $\$1009.38(1 + 0.009375) = \$1018.84$

   Determine which bank offers the best CD rate and the best method of interest compounding.

2. Design and sketch a sandbox where the perimeter of the open top is at least 2 ft more than the perimeter of the base. Maximize the use of a 4-ft by 8-ft cardboard sheet. Find the surface area of the five sides and the volume of your sandbox. (The most efficient design could be built with plywood and donated to Head Start.)

3. Design and sketch a picnic table that could be used on campus. Research designs either by visiting local stores or on the Internet. Price the materials necessary to construct one unit and ten units.

Chapter 1 Summary

Glossary of Basic Terms

Area. The number of square units of measure contained in a plane geometric figure. (p. 14)

Common fraction. The ratio or quotient of two integers in the form $\frac{a}{b}$, where $b \neq 0$. The integer above the line is called the numerator; the integer below the line is called the denominator. (p. 28)

Conversion factor. A fraction whose numerator equals its denominator (equal to 1) but with different units to change from one unit or set of units to another. (p. 53)

Cube of a number. The product of that number times itself three times. (p. 90)

Cube root of a number. That number which, when multiplied by itself three times, gives the original number. (p. 91)

Decimal fraction. A fraction whose denominator is 10, 100, 1000, or any power of 10. (p. 57)

Difference. The result of subtracting numbers. (p. 3)

Divisible. One number is divisible by a second number if, when you divide the first number by the second number, you get a zero remainder. (p. 23)

Even integer. An integer divisible by 2. (p. 23)

Formula. A statement of a rule using letters to represent the relationship of certain quantities. (p. 19)

Fraction reduced to lowest terms. A fraction whose numerator and denominator have no common factors. (p. 29)

Grouping symbols. Often parentheses ( ) or brackets [ ] that help to clarify the meaning of mathematical expressions. (p. 11)

Improper fraction. A fraction whose numerator is greater than or equal to its denominator. (p. 30)

Least common denominator (LCD). The smallest positive integer that has all the denominators as divisors. (p. 33)

Measurement. The comparison of an observed quantity with a standard unit quantity. (p. 53)

Mixed number. An integer plus a proper fraction. (p. 30)

Odd integer. An integer that is not divisible by 2. (p. 23)

Percent. The comparison of any number of parts to 100 parts; *percent* means "per hundred." (p. 75)

Perfect cubes. Numbers whose cube roots are whole numbers. (p. 91)

Perfect squares. Numbers whose square roots are whole numbers. (p. 90)

Perimeter. The sum of the lengths of the sides of a geometric figure. (p. 40)

Positive integers. The numbers 1, 2, 3, ... (p. 2)

Power. A number, called the base, and an exponent, which indicates the number of times the base is used as a factor. (p. 92)

Prime factorization. The process of finding the prime factors of a positive integer. (p. 23)

Prime factors of a positive integer. Those prime numbers whose product equals the given positive integer. (p. 23)

Prime number. An integer greater than 1 that has no divisors except itself and 1; the first ten prime numbers are 2, 3, 5, 7, 11, 13, 17, 19, 23, and 29. (p. 23)

Product. The result of multiplying numbers. (p. 6)

Proper fraction. A fraction whose numerator is less than its denominator. (p. 30)

Quotient. The result of dividing numbers. (p. 6)

Significant digits. Those digits in a number we are reasonably sure of being able to rely on in a measurement. (p. 67)

Square of a number. The product of that number times itself. (p. 89)

Square root of a number. That positive number which, when multiplied by itself, gives the original number. (p. 90)

Sum. The result of adding numbers. (p. 3)

Volume. The number of cubic units of measure contained in a solid geometric figure. (p. 15)

Whole numbers. The numbers 0, 1, 2, 3, ... (p. 2)

1.2 Order of Operations

1. **Order of Operations:**
   a. Always do the operations within parentheses or other grouping symbols first.
   b. Then evaluate each power, if any.
   c. Next, perform multiplications and divisions in the order in which they appear as you read from left to right.
   d. Finally, perform additions and subtractions in the order in which they appear as you read from left to right. (p. 12)

1.3 Area and Volume

1. **Area of a rectangle:** $A = lw$ (p. 15)

2. **Volume of a rectangular solid:** $V = lwh$ (p. 16)

1.4   Formulas

1. **Formulas from geometry:**

   a. Area of a triangle: $A = \dfrac{1}{2}bh$ (p. 21)

   b. Area of a parallelogram: $A = bh$ (p. 21)

   c. Area of a trapezoid: $A = \left(\dfrac{a+b}{2}\right)h$ (p. 22)

1.5   Prime Factorization

1. **Divisibility tests:**

   a. *Divisibility by 2:* If a number ends with an even digit, then the number is *divisible by 2*. (p. 25)

   b. *Divisibility by 3:* If the sum of the digits of a number is divisible by 3, then the number itself is *divisible by 3*. (p. 25)

   c. *Divisibility by 5:* If a number has 0 or 5 as its last digit, then the number is *divisible by 5*. (p. 25)

1.6   Introduction to Fractions

1. **Equal or equivalent fractions:**

   a. The numerator and denominator of any fraction may be multiplied or divided by the same number (except zero) without changing the value of the fraction.

   b. Two fractions $\dfrac{a}{b}$ and $\dfrac{c}{d}$ are equal or equivalent if $ad = bc$, where $b \neq 0$ and $d \neq 0$. (p. 28)

2. **Simplifying special fractions:**

   a. Any number (except zero) divided by itself equals 1.

   b. Any number divided by 1 equals itself.

   c. Zero divided by any number (except zero) equals zero.

   d. Any number divided by zero is not meaningful and is called *undefined*. (pp. 29–30)

3. **Changing an improper fraction to a mixed number:** To change an improper fraction to a mixed number, divide the numerator by the denominator. The quotient is the whole-number part. The remainder over the divisor is the proper fraction part of the mixed number. (p. 30)

1.7   Addition and Subtraction of Fractions

1. **Finding the least common denominator:** To find the least common denominator (LCD) of a set of fractions:

   a. Factor each denominator into its prime factors.

   b. Write each prime factor the number of times it appears *most* in any *one* denominator in step (a).

The LCD is the product of these prime factors. (p. 34)

2. **Adding fractions:** To add two or more fractions with the same denominator, first add the numerators. Then place the sum over the common denominator and simplify. (p. 33)

3. **Subtracting fractions:** To subtract two or more fractions with a common denominator, subtract their numerators and place the difference over the common denominator and simplify. (p. 36)

4. **Adding mixed numbers:** To add mixed numbers, find the LCD of the fractions. Add the fractions, then add the whole numbers. Finally, add these two results and simplify. (p. 37)

5. **Subtracting mixed numbers:** To subtract mixed numbers, find the LCD of the fractions. Subtract the fractions, then subtract the whole numbers and simplify. (p. 37)

1.8   Multiplication and Division of Fractions

1. **Multiplying fractions:** To multiply fractions, multiply the numerators and multiply the denominators. Then reduce the resulting fraction to lowest terms. (p. 45)

2. **Dividing fractions:** To divide a fraction by a fraction, invert the fraction that follows the division sign. Then multiply the resulting fractions as described above. (p. 46)

1.9   The U.S. System of Weights and Measures

1. **Choosing conversion factors:** The correct choice for a given conversion factor is the one in which the old units are in the numerator of the original expression and in the denominator of the conversion factor or the old units are in the denominator of the original expression and in the numerator of the conversion factor. That is, set up the conversion factor so that the old units cancel each other. (p. 54)

1.10   Addition and Subtraction of Decimal Fractions

1. **Place values for decimals:** Review Table 1.2 on page 57.

2. **Changing a common fraction to a decimal:** To change a common fraction to a decimal, divide the numerator of the fraction by the denominator. (p. 59)

3. Adding or subtracting decimal fractions: To add or subtract decimal fractions,

   a. Write the decimals so that the digits having the same place value are in vertical columns. (Make certain that the decimal points are also lined up vertically.)

   b. Add or subtract as with whole numbers.

   c. Place the decimal point between the ones digit and the tenths digit of the sum or the difference. (Be certain that the decimal point is in the same vertical line as the other decimal points.) (p. 61)

1.11 Rounding Numbers

1. Rounding numbers to a particular place value: To round a number to a particular place value:

   a. If the digit in the next place to the right is less than 5, drop that digit and all other following digits. Use zeros to replace any whole-number places dropped.

   b. If the digit in the next place is 5 or greater, add 1 to the digit in the place to which you are rounding. Drop all other following digits. Use zeros to replace any whole-number digits dropped. (p. 67)

2. Significant digits:

   a. The following digits in a number are *significant:*
- All nonzero digits.
- All zeros between significant digits.
- All zeros at the end of a decimal number.

   b. The following digits in a number are *not significant:*
- All zeros at the beginning of a decimal number less than 1.
- All zeros at the end of a whole number. (p. 67)

3. Rounding a number to a given number of significant digits: To round a number to a given number of significant digits:

   a. Count the given number of significant digits from left to right, starting with the first nonzero digit.

   b. If the next digit to the right is less than 5, drop that digit and all other following digits. Use zeros to replace any whole-number places dropped.

   c. If the next digit to the right is 5 or greater, add 1 to the digit in the place to which you are rounding. Drop all other following digits. Use zeros to replace any whole-number places dropped. (p. 68)

1.12 Multiplication and Division of Decimal Fractions

1. Multiplying two decimal fractions: To multiply two decimal fractions:

   a. Multiply the numbers as you would whole numbers.

   b. Count the total number of digits to the right of the decimal points in the two numbers being multiplied. Then place the decimal in the product so that it has that same total number of digits to the right of the decimal point. (p. 69)

2. Dividing two decimal fractions: To divide two decimal fractions:

   a. Use the same form as in dividing two whole numbers.

   b. Multiply both the divisor and the dividend (denominator and numerator) by a power of 10 that makes the divisor a whole number.

   c. Divide as you would whole numbers, and place the decimal point in the quotient directly above the decimal point in the dividend. (p. 70)

1.13 Percent

1. Changing a percent to a decimal: To change a percent to a decimal, move the decimal point two places to the *left* (divide by 100). Then remove the percent sign (%). (p. 76)

2. Changing a decimal to a percent: To change a decimal to a percent, move the decimal point two places to the *right* (multiply by 100). Write the percent sign (%) after the number. (p. 77)

3. Changing a fraction to a percent: To change a fraction to a percent:

   a. First, change the fraction to a decimal.

   b. Then change this decimal to a percent. (p. 77)

4. Changing a percent to a fraction: To change a percent to a fraction:

   a. Change the percent to a decimal.

   b. Then change the decimal to a fraction in lowest terms. (p. 78)

5. Changing a percent that contains a mixed number to a fraction: To change a percent that contains a mixed number to a fraction:

   a. Change the mixed number to an improper fraction.

   b. Then multiply this result by $\frac{1}{100}$ and remove the percent sign (%). (p. 79)

1.14 Rate, Base, and Part

1. Percent problems: Any percent problem calls for finding one of three things:

   a. The rate, R, usually has either a percent sign (%) or the word *percent* with it.

   b. The base, B, is usually the whole (or entire) amount. The base is often the number that follows the word *of.*

c. The part, P, is usually some fractional part of the base, B. If you identify R and B first, then P will be the number that is not R or B. (p. 80)

2. Formulas for finding part, base, and rate:

a. $P = BR$ Use to find the part.

b. $B = \dfrac{P}{R}$ Use to find the base.

c. $R = \dfrac{P}{B}$ Use to find the rate or percent. (p. 81)

3. Percent increase (or percent decrease): The process for finding the percent increase or percent decrease may be summarized by the following formula:

$$\begin{array}{c}\text{percent increase} \\ \text{(or percent decrease)}\end{array} = \dfrac{\text{the change}}{\text{the original value}} \times 100\%$$

(p. 83)

1.16 Applications Involving Percent: Personal Finance (Optional)

1. Simple interest:

$$i = prt$$

where i = the amount of interest paid

p = the principal, which is the original amount of money borrowed or deposited

r = the interest rate written as a decimal

t = the time in years the money is being used (p. 93)

2. Payment amount:

$$\text{payment amount} = \dfrac{\text{principal} + \text{interest}}{\begin{array}{c}\text{loan period in months} \\ \text{or weeks}\end{array}}$$

(p. 94)

3. Compound interest:

$$A = P\left(1 + \dfrac{r}{n}\right)^{nt}$$

where A = the amount after time t

P = the principal or initial amount

r = the annual interest rate written as a decimal

n = the number of times the interest is compounded each year

t = the number of years (p. 94)

4. Amortization formula:

$$A = P\left(\dfrac{i(1 + i)^n}{(1 + i)^n - 1}\right)$$

where A = the periodic payment amount

P = the amount of the principal

i = the interest rate as an annual percentage rate (APR) as a decimal divided by the number of annual interest payments (divide by 12 if monthly or by 52 if weekly)

n = the total number of payments (p. 95)

 ## Chapter 1 Review

1. Add: $435 + 2600 + 18 + 5184 + 6$

2. Subtract: $60{,}000 - 4{,}803$

3. Multiply: 7060×1300

4. Divide: $68{,}040 \div 300$

5. Evaluate: $12 - 3(5 - 2)$

6. Evaluate: $(6 + 4)8 \div 2 + 3$

7. Evaluate: $18 \div 2 \times 5 \div 3 - 6 + 4 \times 7$

8. Evaluate: $18/(5 - 3) + (6 - 2) \times 8 - 10$

9. Find the area of the figure in Illustration 1.

ILLUSTRATION 1

10. Find the volume of the figure in Illustration 2.

ILLUSTRATION 2

11. Given the formula $C = \dfrac{5}{9}(F - 32)$ and $F = 50$, find C.

12. Given the formula $P = \dfrac{Fs}{t}$, $F = 600$, $s = 50$, and $t = 10$, find P.

13. Is 460 divisible by 3?

14. Find the prime factorization of 54.

15. Find the prime factorization of 330.

Simplify:

16. $\dfrac{36}{56}$

17. $\dfrac{180}{216}$

Change each to a mixed number in simplest form:

18. $\dfrac{25}{6}$

19. $3\dfrac{18}{5}$

Change each mixed number to an improper fraction:

20. $2\dfrac{5}{8}$

21. $3\dfrac{7}{16}$

Perform the indicated operations and simplify:

22. $\dfrac{3}{8} + \dfrac{7}{8} + \dfrac{6}{8}$

23. $\dfrac{1}{4} + \dfrac{5}{12} + \dfrac{5}{6}$

24. $\dfrac{29}{36} - \dfrac{7}{30}$

25. $5\dfrac{3}{16} + 9\dfrac{5}{12}$

26. $6\dfrac{3}{8} - 4\dfrac{7}{12}$

27. $18 - 6\dfrac{2}{5}$

28. $16\dfrac{2}{3} + 1\dfrac{1}{4} - 12\dfrac{11}{12}$

29. $\dfrac{5}{6} \times \dfrac{3}{10}$

30. $3\dfrac{6}{7} \times 4\dfrac{2}{3}$

31. $\dfrac{3}{8} \div 6$

32. $\dfrac{2}{3} \div 1\dfrac{7}{9}$

33. $1\dfrac{4}{5} \div 1\dfrac{9}{16} \times 11\dfrac{2}{3}$

34. Find dimensions A and B in the figure in Illustration 3.

ILLUSTRATION 3

Fill in each blank:

35. 6 lb 9 oz = _____ oz

36. 168 ft = _____ in.

37. 72 ft = _____ yd

38. 36 mi = _____ yd

Write each common fraction as a decimal:

39. $\dfrac{9}{16}$

40. $\dfrac{5}{12}$

Change each decimal to a common fraction or a mixed number and simplify:

41. 0.45

42. 19.625

Perform the indicated operations:

43. $8.6 + 140 + 0.048 + 19.63$

44. $25 + 16.3 - 18 + 0.05 - 6.1$

45. $86.7 - 18.035$

46. $34 - 0.28$

47. 0.605×5300

48. 18.05×0.106

49. $74.73 \div 23.5$

50. $9.27 \div 0.45$

51. Round 248.1563 to **a.** the nearest hundred, **b.** the nearest tenth, and **c.** the nearest ten.

52. Round 5.64908 to **a.** the nearest tenth, **b.** the nearest hundredth, and **c.** the nearest ten-thousandth.

Change each percent to a decimal:

53. 15%

54. $8\dfrac{1}{4}\%$

Change each decimal to a percent:

55. 0.065

56. 1.2

57. What is $8\dfrac{3}{4}\%$ of \$12,000?

58. Complete the following table with the equivalents:

Fraction	Decimal	Percent
_____	0.25	_____
_____	_____	$37\frac{1}{2}\%$
$\frac{5}{6}$	_____	_____
$8\frac{3}{4}$	_____	_____
_____	2.4	_____
_____	_____	0.15%

59. In a small electronics business, the overhead is $32,000 and the gross income is $84,000. What percent is the overhead?

60. A new tire has a tread depth of $\frac{13}{32}$ in. At 16,000 miles, the tread depth is $\frac{11}{64}$ in. What percent of the tread is left?

61. A farmer bales 60 tons of hay, which contain 20% moisture. How many tons of dry matter does he harvest?

62. Six gears are to be placed on a shaft with a flat $\frac{1}{4}$-in. washer between them as shown in Illustration 4. The gears are $3\frac{1}{16}$ in. thick. What would be the overall length of the shaft if $1\frac{1}{8}$ in. were open at each end of the shaft?

ILLUSTRATION 4

63. A $\frac{7}{8}$-in.-thick board is being used as a base. If holes are drilled $\frac{9}{16}$ in. to insert posts, what is the thickness of the board left below each hole as shown in Illustration 5?

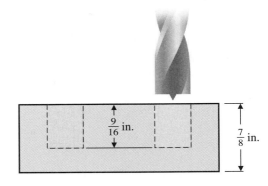

ILLUSTRATION 5

64. A dish manufacturing plant needs a shipping box 20 in. deep with a 10-in. square base. The box company is drawing out the die to cut the cardboard for this box. How large a sheet of cardboard is needed to make one box that allows 1 in. for a glue edge as shown in Illustration 6?

ILLUSTRATION 6

Find each power or root rounded to three significant digits:

65. 15.9^3

66. $\sqrt{19200}$

Chapter 1 Test

1. Add: $47 + 4969 + 7 + 256$

2. Subtract: $4000 - 484$

3. Multiply: 4070×635

4. Divide: $96{,}000 \div 60$

Evaluate each expression:

5. $8 + 2(5 \times 6 + 8)$

6. $15 - 9 \div 3 + 3 \times 4$

7. Find the area of the figure in Illustration 1.

ILLUSTRATION 1

8. Find the volume of the figure in Illustration 2.

10 in.   12 in.
3 in.   4 in.
20 in.

ILLUSTRATION 2

9. Ohm's law states that current (in A) equals voltage (in V) divided by resistance (in Ω). Find the current in the circuit in Illustration 3.

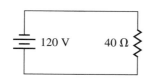

120 V   40 Ω

ILLUSTRATION 3

10. If $P = 2l + 2w$, $l = 20$, and $w = 15$, find P.

11. If $t = \dfrac{d}{r}$, $d = 1050$, and $r = 21$, find t.

12. If $P = 2a + b$, $a = 36$, and $b = 15$, find P.

Find the prime factorization of each number:

13. 90

14. 220

Simplify:

15. $\dfrac{30}{64}$

16. $\dfrac{28}{42}$

17. Change $\dfrac{23}{6}$ to a mixed number.

18. Change $3\dfrac{1}{4}$ to an improper fraction.

19. Add: $\dfrac{3}{8} + \dfrac{1}{4}$

20. Subtract: $\dfrac{5}{16} - \dfrac{5}{32}$

21. Add: $3\dfrac{1}{8}$

$2\dfrac{1}{2}$

$4\dfrac{3}{4}$

22. Subtract: $10\dfrac{1}{8}$

$3\dfrac{5}{16}$

Perform the indicated operations and simplify:

23. $3\dfrac{5}{8} + 2\dfrac{3}{16} - 1\dfrac{1}{4}$

24. $\dfrac{3}{8} \times \dfrac{16}{27}$

25. $\dfrac{3}{8} \div 3\dfrac{5}{16}$

26. $\dfrac{4}{3} \times \dfrac{1}{8} \times \dfrac{9}{20}$

27. $3\dfrac{5}{8} + 1\dfrac{3}{4} \times 6\dfrac{1}{5}$

28. Given the formula $P = 2l + 2w$, $l = 4\dfrac{3}{4}$, and $w = 2\dfrac{1}{2}$, find P.

29. Find the total current in the circuit in Illustration 4.

$I_T = ?$

$3\dfrac{5}{8}$ A $2\dfrac{3}{4}$ A $4\dfrac{5}{16}$ A

ILLUSTRATION 4

Fill in each blank:

30. 120 ft = _____ yd

31. 3 lb 5 oz = _____ oz

32. Express $\dfrac{5}{8}$ as a decimal.

33. Express 2.12 as a mixed number and simplify.

34. Add: 2.147 + 2.04 + 60 + 0.007 + 0.83

35. Subtract: 400 − 2.81

36. Round 27.2847 to the nearest **a.** tenth and **b.** hundredth.

37. Multiply: 6.12 × 1.32

38. Divide: $6.3\overline{)0.315}$

39. 59.45 is 41% of what number?

40. 88 is what percent of 284? (to the nearest tenth)

41. Rachel receives a 6.7% increase in salary. If her salary was $612 per week, what is her new weekly salary?

Find each power or root rounded to three significant digits:

42. 0.235^2

43. $\sqrt[3]{304.8}$

2

Signed Numbers and Powers of 10

Mathematics at Work

Electronics technicians perform a variety of jobs. Electronic engineering technicians apply electrical and electronic theory knowledge to design, build, test, repair, and modify experimental and production electrical equipment in industrial or commercial plants for use by engineering personnel in making engineering design and evaluation decisions.

Other electronics technicians repair electronic equipment such as industrial controls, telemetering and missile control systems, radar systems, and transmitters and antennas using testing instruments. Industrial controls automatically monitor and direct production processes on the factory floor. Transmitters and antennas provide communications links for many organizations. The federal government uses radar and missile control systems for national defense as well as other applications.

Electricians install, maintain, and repair electrical wiring, equipment, and fixtures and ensure that work is in accordance with relevant codes. They also travel to locations to repair equipment and perform preventive maintenance on a regular basis. They use schematics and manufacturers' specifications that show connections and provide instructions on how to locate problems. They also use software programs and testing equipment to diagnose malfunctions. For more information, go to the website listed below.

Lawrence Migdale/StockBoston

Electronics Technician
Electronics technician checking circuitry.

www.cengage.com/mathematics/ewen

Objectives

■ Find the absolute value of a signed number.

■ Add, subtract, multiply, and divide signed numbers.

■ Add, subtract, multiply, and divide signed numbers involving fractions.

■ Use the rules for exponents for powers of 10 to multiply, divide, and raise a power to a power.

■ Work with numbers in scientific notation.

■ Work with numbers in engineering notation.

2.1 Addition of Signed Numbers

Technicians use negative numbers in many ways. In an experiment using low temperatures, for example, you would record 10° below zero as −10°. Or consider sea level as zero altitude. If a submarine dives 75 m, you could consider its depth as −75 m (75 m below sea level). See Figure 2.1.

(a)   (b)

FIGURE 2.1

These measurements indicate a need for numbers other than positive integers, which are the only numbers that we have used up to now. To illustrate the graphical relationship of these numbers, we draw a number line as in Figure 2.2 with a point representing zero and with evenly spaced points that represent the **positive integers** (1, 2, 3, . . .) to the right as shown. Then we mark off similarly evenly spaced points to the left of zero. These points correspond to the **negative integers** (−1, −2, −3, . . .) as shown. The negative integers are preceded by a negative (−) sign; −3 is read "negative 3," and −5 is read "negative 5." Each positive integer corresponds to a negative integer. For example, 3 and −3 are corresponding integers. Note that the distances from 0 to 3 and from 0 to −3 are equal.

FIGURE 2.2
The real number line

The **rational numbers** are defined as those numbers that can be written as the ratio of two integers; that is, *a/b*, where $b \neq 0$. The **irrational numbers** are those numbers that cannot be written as the ratio of two integers, such as $\sqrt{2}$, $-\sqrt{30}$, or the square root of any nonperfect square; π; and several other kinds of numbers that you will study later. The **real numbers** consist of the rational and irrational numbers and are represented on the *real number line* as shown in Figure 2.2. The real number line is dense or full with real numbers; that is, each point on the number line represents a distinct real number, and each real number is represented by a distinct point on the number line. Examples of real numbers are illustrated in Figure 2.3.

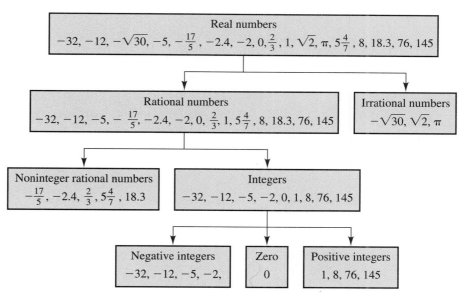

FIGURE 2.3
Examples of real numbers

The **absolute value of a number** is its distance from zero on the number line. Because distance is always considered positive, the absolute value of a number is never negative. We write the absolute value of a number *x* as $|x|$; it is read "the absolute value of *x*." Thus, $|x| \geq 0$. ("$\geq$" means "is greater than or equal to.") For example, $|+6| = 6$, $|4| = 4$, and $|0| = 0$. However, if a number is less than 0 (negative), its absolute value is the corresponding positive number. For example, $|-6| = 6$ and $|-7| = 7$.

Remember:

> The absolute value of a number is never negative.

Example 1

Find the absolute value of each number: **a.** +3, **b.** −5, **c.** 0, **d.** −10, **e.** 15.

a. $|+3| = 3$ The distance between 0 and +3 on the number line is 3 units.

b. $|-5| = 5$ The distance between 0 and −5 on the number line is 5 units.

c. $|0| = 0$ The distance is 0 units.

d. $|-10| = 10$ The distance between 0 and −10 on the number line is 10 units.

e. $|15| = |+15| = 15$ The distance between 0 and +15 on the number line is 15 units.

One number is *larger* than another number if the first number is to the *right* of the second on the number line in Figure 2.2. Thus, 5 is larger than 1, 0 is larger than −3, and 2 is larger than −4. Similarly, one number is *smaller* than another if the first number is to the *left* of the second on the number line in Figure 2.2. Thus, 0 is smaller than 3, −1 is smaller than 4, and −5 is smaller than −2.

The use of **signed numbers** (positive and negative numbers) is one of the most important operations that we will study. Signed numbers are used in work with exponents and certain dial indicators as shown in Section 4.9. Operations with signed numbers are also essential for success in the basic algebra that follows later.

Adding Two Numbers with Like Signs (the Same Signs)

1. To add two positive numbers, add their absolute values. The result is positive. A positive sign may or may not be used before the result. It is usually omitted.
2. To add two negative numbers, add their absolute values and place a negative sign before the result.

Example 2   Add:

    a. $(+2) + (+3) = +5$ **c.** $(+4) + (+5) = +9$

    b. $(-4) + (-6) = -10$ **d.** $(-8) + (-3) = -11$ ■

Adding Two Numbers with Unlike Signs

To add a negative number and a positive number, find the difference of their absolute values. The sign of the number having the larger absolute value is placed before the result.

Example 3   Add:

    a. $(+4) + (-7) = -3$ **d.** $(-8) + (+6) = -2$

    b. $(-3) + (+8) = +5$ **e.** $(-2) + (+5) = +3$

    c. $(+6) + (-1) = +5$ **f.** $(+3) + (-11) = -8$ ■

Adding Three or More Signed Numbers

Step 1 Add the positive numbers.
Step 2 Add the negative numbers.
Step 3 Add the sums from Steps 1 and 2 according to the rules for addition of two signed numbers.

Example 4   Add $(-8) + (+12) + (-7) + (-10) + (+3)$.

Step 1 $(+12) + (+3) = +15$ Add the positive numbers.
Step 2 $(-8) + (-7) + (-10) = -25$ Add the negative numbers.
Step 3 $(+15) + (-25) = -10$ Add the sums from Steps 1 and 2.

Therefore, $(-8) + (+12) + (-7) + (-10) + (+3) = -10$.   ■

 Example 5   Add $(+4) + (-7) + (-2) + (+6) + (-3) + (-5)$.

Step 1 $(+4) + (+6) = +10$
Step 2 $(-7) + (-2) + (-3) + (-5) = -17$
Step 3 $(+10) + (-17) = -7$

Therefore, $(+4) + (-7) + (-2) + (+6) + (-3) + (-5) = -7$.   ■

SIGNED NUMBERS DRILL 1

Being able to work quickly and accurately with signed numbers is as important to success in algebra as being able to work quickly and accurately with the basic number facts in arithmetic. The following oral signed numbers drill is designed to be used in class by the instructor asking students in random order to add a given set of signed numbers and verbally give the answer. A friend may also help you use this drill for extra practice. The answers are on the bottom of page 112 to make it easier for you to drill yourself.

Add the following signed numbers:

1. $(+4) + (+5)$ **2.** $(-6) + (-2)$ **3.** $(+7) + (-3)$
4. $(-8) + (+1)$ **5.** $(-3) + (+9)$ **6.** $(-7) + (-7)$
7. $(-4) + (0)$ **8.** $(0) + (+8)$ **9.** $(+2) + (+1)$
10. $(-5) + (-7)$ **11.** $(-6) + (+2)$ **12.** $(+6) + (-7)$
13. $(-10) + (+8)$ **14.** $(+3) + (-7)$ **15.** $(+6) + (+9)$
16. $(-8) + (-4)$ **17.** $(+4) + (-4)$ **18.** $(-7) + (+8)$
19. $(+1) + (-6)$ **20.** $(-9) + (-7)$ **21.** $(+3) + (-2)$
22. $(-6) + (+9)$ **23.** $(+2) + (-8)$ **24.** $(-7) + (-3)$
25. $(+7) + (-8)$ **26.** $(-3) + (-9)$ **27.** $(+1) + (+5)$
28. $(-2) + (-2)$ **29.** $(-3) + (+4)$ **30.** $(+9) + (-4)$

Exercises 2.1

Find the absolute value of each number:

1. 3   **2.** -4 **3.** -6
4. 0   **5.** $+4$ **6.** $+8$
7. 17   **8.** -37 **9.** -15
10. 49

Add:

11. $(+4) + (+6)$ **12.** $(-5) + (-9)$
13. $(+9) + (-2)$ **14.** $(-10) + (+4)$
15. $(+5) + (-7)$ **16.** $(-4) + (+6)$
17. $(-3) + (-9)$ **18.** $(+4) + (-9)$
19. $(12) + (-6)$ **20.** $(-12) + (6)$
21. $(-4) + (-5)$ **22.** $(+2) + (-11)$

23. $(-3) + (+7)$ **24.** $(+8) + (+2)$
25. $(-5) + (+2)$ **26.** $(-7) + (-6)$
27. $(+7) + (-8)$ **28.** $(8) + (-3)$
29. $(-10) + (6)$ **30.** $(+4) + (-11)$
31. $(-8) + (2)$ **32.** $(+3) + (+7)$
33. $(-2) + (0)$ **34.** $(0) + (+3)$
35. $(9) + (-5)$ **36.** $(+9) + (-9)$
37. $(16) + (-7)$ **38.** $(-19) + (-12)$
39. $(-6) + (+9)$ **40.** $(+20) + (-30)$
41. $(-1) + (-3) + (+8)$ **42.** $(+5) + (-3) + (+4)$
43. $(+1) + (+7) + (-1)$ **44.** $(-5) + (-9) + (-4)$
45. $(-9) + (+6) + (-4)$ **46.** $(+8) + (+7) + (-2)$

47. $(+8) + (-8) + (+7) + (-2)$
48. $(-6) + (+5) + (-8) + (+4)$
49. $(-4) + (-7) + (-7) + (-2)$
50. $(-3) + (-9) + (+5) + (+6)$
51. $(-1) + (-2) + (+9) + (-8)$
52. $(+6) + (+5) + (-7) + (-3)$
53. $(-6) + (+2) + (+7) + (-3)$
54. $(+8) + (-1) + (+9) + (+6)$
55. $(-5) + (+1) + (+3) + (-2) + (-2)$
56. $(+5) + (+2) + (-3) + (-9) + (-9)$
57. $(-5) + (+6) + (-9) + (-4) + (-7)$
58. $(-9) + (+7) + (-6) + (+5) + (-8)$

59. $(+1) + (-4) + (-2) + (+2) + (-9)$
60. $(-1) + (-2) + (-6) + (-3) + (-5)$
61. $(-2) + 8 + (-4) + 6 + (-1)$
62. $14 + (-5) + (-1) + 6 + (-3)$
63. $5 + 6 + (-2) + 9 + (-7)$
64. $(-5) + 4 + (-1) + 6 + (-7)$
65. $(-3) + 8 + (-4) + (-7) + 10$
66. $16 + (-7) + (-5) + 20 + (-5)$
67. $3 + (-6) + 7 + 4 + (-4)$
68. $(-8) + 6 + 9 + (-5) + (-4)$
69. $(-5) + 4 + (-7) + 2 + (-8)$
70. $7 + 9 + (-6) + (-4) + 9 + (-2)$

2.2 Subtraction of Signed Numbers

Subtracting Two Signed Numbers

To subtract two signed numbers, change the sign of the number being subtracted and add according to the rules for addition of signed numbers.

Example 1   Subtract:

a. $(+2) - (+5) = (+2) + (-5)$
  $= -3$

To subtract, change the sign of the number being subtracted, $+5$, and add.

b. $(-7) - (-6) = (-7) + (+6)$
  $= -1$

To subtract, change the sign of the number being subtracted, -6, and add.

c. $(+6) - (-4) = (+6) + (+4)$
  $= +10$

To subtract, change the sign of the number being subtracted, -4, and add.

d. $(+1) - (+6) = (+1) + (-6)$
  $= -5$

To subtract, change the sign of the number being subtracted, $+6$, and add.

e. $(-8) - (-10) = (-8) + (+10) = +2$
f. $(+9) - (-6) = (+9) + (+6) = +15$
g. $(-4) - (+7) = (-4) + (-7) = -11$

■

Subtracting More Than Two Signed Numbers

When more than two signed numbers are involved in subtraction, change the sign of *each* number being subtracted and add the resulting signed numbers.

Answers to Signed Numbers Drill 1

1. $+9$ **2.** -8 **3.** $+4$ **4.** -7 **5.** $+6$ **6.** -14 **7.** -4 **8.** $+8$ **9.** $+3$ **10.** -12 **11.** -4 **12.** -1
13. -2 **14.** -4 **15.** $+15$ **16.** -12 **17.** 0 **18.** $+1$ **19.** -5 **20.** -16 **21.** $+1$ **22.** $+3$ **23.** -6 **24.** -10 **25.** -1
26. -12 **27.** $+6$ **28.** -4 **29.** $+1$ **30.** $+5$

Example 2   Subtract: $(-4) - (-6) - (+2) - (-5) - (+7)$
$$= (-4) + (+6) + (-2) + (+5) + (-7)$$

Change the sign of each number being subtracted and add the resulting signed numbers.

Step 1 $(+6) + (+5) = +11$

Step 2 $(-4) + (-2) + (-7) = -13$

Step 3 $(+11) + (-13) = -2$

Therefore, $(-4) - (-6) - (+2) - (-5) - (+7) = -2$.   ■

Adding and Subtracting Combinations of Signed Numbers

When combinations of additions and subtractions of signed numbers occur in the same problem, change *only* the sign of each number being subtracted. Then add the resulting signed numbers.

 Example 3   Perform the indicated operations:

$$(+4) - (-5) + (-6) - (+8) - (-2) + (+5) - (+1)$$
$$= (+4) + (+5) + (-6) + (-8) + (+2) + (+5) + (-1)$$

Change *only* the sign of each number being subtracted and add the resulting signed numbers.

Step 1 $(+4) + (+5) + (+2) + (+5) = +16$

Step 2 $(-6) + (-8) + (-1) = -15$

Step 3 $(+16) + (-15) = +1$

Therefore, $(+4) - (-5) + (-6) - (+8) - (-2) + (+5) - (+1) = +1$.   ■

 Example 4   Perform the indicated operations:

$$(-12) + (-3) - (-5) - (+6) - (-1) + (+4) - (+3) - (-8)$$
$$= (-12) + (-3) + (+5) + (-6) + (+1) + (+4) + (-3) + (+8)$$

Step 1 $(+5) + (+1) + (+4) + (+8) = +18$

Step 2 $(-12) + (-3) + (-6) + (-3) = -24$

Step 3 $(+18) + (-24) = -6$

Therefore, $(-12) + (-3) - (-5) - (+6) - (-1) + (+4) - (+3) - (-8) = -6$.   ■

SIGNED NUMBERS DRILL 2

The following oral signed numbers drill is designed to be used in class by the instructor asking students in random order to randomly add or subtract a given set of signed numbers by placing a plus or minus sign in the blank and verbally giving the answer. A friend may also help you use this drill for extra practice. The answers are on the bottom of page 114 to make it easier for you to drill yourself.

Add or subtract the following signed numbers:

1. $(+3)$ __ $(+5)$ **2.** (-7) __ (-2) **3.** $(+8)$ __ (-3)

4. (-9) __ $(+1)$ **5.** (-3) __ $(+7)$ **6.** (-7) __ (-7)

7. (-5) __ (0) **8.** (0) __ $(+2)$ **9.** $(+4)$ __ $(+1)$

10. (-5) __ (-8) **11.** (-6) __ $(+3)$ **12.** $(+8)$ __ (-7)

13. (-1) __ $(+7)$ **14.** $(+3)$ __ (-9) **15.** $(+6)$ __ $(+4)$

16. (-8) __ (-3) **17.** $(+5)$ __ (-5) **18.** (-6) __ $(+8)$

19. $(+1)$ __ (-7) **20.** (-9) __ (-5) **21.** $(+3)$ __ (-2)

22. (-2) __ $(+4)$ **23.** $(+2)$ __ (-7) **24.** (-5) __ (-3)

25. $(+4)$ __ (-8) **26.** (-2) __ (-9) **27.** $(+2)$ __ $(+5)$

28. (-3) __ (-3) **29.** $(+2)$ __ (-4) **30.** $(+5)$ __ (-4)

Exercises 2.2

Subtract:

1. $(+4) - (+6)$
2. $(-5) - (-9)$
3. $(+9) - (-2)$
4. $(-10) - (+4)$
5. $(+5) - (-7)$
6. $(-4) - (+6)$
7. $(-3) - (-9)$
8. $(+4) - (-9)$
9. $(12) - (-6)$
10. $(-12) - (6)$
11. $(3) - (9)$
12. $(-2) - (0)$
13. $(0) - (4)$
14. $(-5) - (-10)$
15. $(-7) - (-7)$
16. $(18) - (-18)$
17. $(-4) - (-5)$
18. $(+2) - (-11)$
19. $(-3) - (+7)$
20. $(+6) - (+8)$
21. $(-5) - (+2)$
22. $(-7) - (-6)$
23. $(+7) - (-8)$
24. $(8) - (-3)$
25. $(-10) - (6)$
26. $(+4) - (-11)$
27. $(-8) - (+2)$
28. $(+3) - (+7)$
29. $(-2) - (0)$
30. $(0) - (+3)$
31. $(9) - (-5)$
32. $(+9) - (-9)$
33. $(16) - (-7)$
34. $(-19) - (-12)$
35. $(-6) - (+9)$
36. $(+20) - (-30)$

Perform the indicated operations:

37. $(+6) - (-3) - (+1)$
38. $(+3) - (-7) - (+6)$
39. $(-3) - (-7) - (+8)$
40. $(+3) - (4) - (-9)$
41. $(+3) - (-6) - (+9) - (-8)$
42. $(+10) - (-4) - (6) - (-9)$
43. $(+5) - (-5) + (-8)$
44. $(+1) + (-7) - (-7)$
45. $(-3) + (-5) - (0) - (+7)$
46. $(+4) - (-3) + (+6) - (8)$
47. $(+4) - (-11) - (+12) + (-6)$
48. $(8) - (-6) - (+18) - (4)$
49. $(-7) - (+6) + (-3) - (-2) - (+9)$
50. $(-3) + (-4) + (+7) - (-2) - (+6)$
51. $-9 + 8 - 5 + 6 - 4$
52. $-12 + 2 + 30 - 6$
53. $-8 + 12 - 7 - 4 + 6$
54. $7 + 4 - 8 - 9 + 3$
55. $16 - 18 + 4 - 7 - 2 + 9$
56. $3 - 7 + 5 - 6 - 7 + 2$
57. $8 + 10 - 20 + 4 - 5 - 6 + 1$
58. $5 - 6 - 7 + 2 - 8 + 10$
59. $9 - 7 + 4 + 3 - 8 - 6 - 6 + 1$
60. $-4 + 6 - 7 - 5 + 6 - 7 - 1$

2.3 Multiplication and Division of Signed Numbers

Multiplying Two Signed Numbers

1. If the two numbers have the same sign, multiply their absolute values. This product is always positive.

2. If the two numbers have different signs, multiply their absolute values and place a negative sign before the product.

Answers to Signed Numbers Drill 2

The first answer is for addition and the second answer is for subtraction.

1. $+8, -2$ 2. $-9, -5$ 3. $+5, +11$ 4. $-8, -10$ 5. $+4, -10$ 6. $-14, 0$ 7. $-5, -5$ 8. $+2, -2$ 9. $+5, +3$
10. $-13, +3$ 11. $-3, -9$ 12. $+1, +15$ 13. $+6, -8$ 14. $-6, +12$ 15. $+10, +2$ 16. $-11, -5$ 17. $0, +10$
18. $+2, -14$ 19. $-6, +8$ 20. $-14, -4$ 21. $+1, +5$ 22. $+2, -6$ 23. $-5, +9$ 24. $-8, -2$ 25. $-4, +12$
26. $-11, +7$ 27. $+7, -3$ 28. $-6, 0$ 29. $-2, +6$ 30. $+1, +9$

Example 1   Multiply:

  a. $(+2)(+3) = +6$ Multiply the absolute values of the signed numbers; the product is positive because the two numbers have the same sign.

  b. $(-4)(-7) = +28$ Multiply the absolute values of the signed numbers; the product is positive because the two numbers have the same sign.

  c. $(-2)(+4) = -8$ Multiply the absolute values of the signed numbers; the product is negative because the two numbers have different signs.

  d. $(-6)(+5) = -30$ Multiply the absolute values of the signed numbers; the product is negative because the two numbers have different signs.

  e. $(+3)(+4) = +12$ **g.** $(-5)(+7) = -35$

  f. $(-6)(-9) = +54$ **h.** $(+4)(-9) = -36$ ■

Multiplying More Than Two Signed Numbers

1. If the number of negative factors is even (divisible by 2), multiply the absolute values of the numbers. This product is positive.
2. If the number of negative factors is odd, multiply the absolute values of the numbers and place a negative sign before the product.

 Example 2   Multiply: $(-11)(+3)(-6) = +198$ The number of negative factors is 2, which is even; therefore, the product is positive. ■

 Example 3   Multiply: $(-5)(-4)(+2)(-7) = -280$ The number of negative factors is 3, which is odd; therefore, the product is negative. ■

Dividing Two Signed Numbers

1. If the two numbers have the same sign, divide their absolute values. This quotient is always positive.
2. If the two numbers have different signs, divide their absolute values and place a negative sign before the quotient.

   Since multiplication and division are related operations, the same rules for signed numbers apply to both operations.

 Example 4   Divide:

  a. $\dfrac{+12}{+2} = +6$ Divide the absolute values of the signed numbers; the quotient is positive because the two numbers have the same sign.

  b. $\dfrac{-18}{-6} = +3$ Divide the absolute values of the signed numbers; the quotient is positive because the two numbers have the same sign.

c. $\dfrac{+20}{-4} = -5$ Divide the absolute values of the signed numbers; the quotient is negative because the two numbers have different signs.

d. $\dfrac{-24}{+6} = -4$ Divide the absolute values of the signed numbers; the quotient is negative because the two numbers have different signs.

e. $(+30) \div (+5) = +6$ **g.** $(+16) \div (-4) = -4$

f. $(-42) \div (-2) = +21$ **h.** $(-45) \div (+9) = -5$ ■

SIGNED NUMBERS DRILL 3

The following oral signed numbers drill is designed to be used in class by the instructor asking students in random order to randomly add, subtract, or multiply a given set of signed numbers by placing a plus, minus, or times sign in the blank and verbally giving the answer. A friend may also help you use this drill for extra practice. The answers are on the bottom of page 118 to make it easier for you to drill yourself.

Add, subtract, or multiply the following signed numbers:

1. $(+4) \_\_ (+8)$ **2.** $(-6) \_\_ (-3)$ **3.** $(+8) \_\_ (-3)$
4. $(-9) \_\_ (+1)$ **5.** $(-3) \_\_ (+5)$ **6.** $(-6) \_\_ (-6)$
7. $(-7) \_\_ (0)$ **8.** $(0) \_\_ (-4)$ **9.** $(+3) \_\_ (+1)$
10. $(-5) \_\_ (-8)$ **11.** $(-6) \_\_ (+3)$ **12.** $(+8) \_\_ (-7)$
13. $(-1) \_\_ (+7)$ **14.** $(+3) \_\_ (-3)$ **15.** $(+5) \_\_ (+9)$
16. $(-8) \_\_ (-2)$ **17.** $(+5) \_\_ (-5)$ **18.** $(-7) \_\_ (+6)$
19. $(+1) \_\_ (-4)$ **20.** $(-9) \_\_ (-9)$ **21.** $(+3) \_\_ (-1)$
22. $(-6) \_\_ (+8)$ **23.** $(+2) \_\_ (-7)$ **24.** $(-6) \_\_ (-3)$
25. $(+7) \_\_ (-4)$ **26.** $(-6) \_\_ (-9)$ **27.** $(+7) \_\_ (+3)$
28. $(-10) \_\_ (-10)$ **29.** $(-3) \_\_ (+5)$ **30.** $(+8) \_\_ (-5)$

Exercises 2.3

Multiply:

1. $(+4)(+6)$ **2.** $(-5)(-9)$ **3.** $(+9)(-2)$ **44.** $(-2)(-8)(-3)$ **45.** $(+5)(+2)(+3)$
4. $(-10)(+4)$ **5.** $(+5)(-7)$ **6.** $(-4)(+6)$ **46.** $(-4)(+3)(0)(+3)$ **47.** $(-3)(-2)(4)(-7)$
7. $(-3)(-9)$ **8.** $(+4)(-9)$ **9.** $(12)(-6)$ **48.** $(-3)(-1)(+1)(+2)$ **49.** $(-9)(-2)(+3)(+1)(-3)$
10. $(-12)(6)$ **11.** $(3)(9)$ **12.** $(-2)(0)$ **50.** $(-6)(-2)(-4)(-1)(-2)(+2)$
13. $(0)(4)$ **14.** $(-5)(-10)$ **15.** $(-7)(-7)$
16. $(18)(-18)$ **17.** $(+15)(-20)$ **18.** $(-11)(-3)$ *Divide:*
19. $(1)(-13)$ **20.** $(-22)(+14)$ **21.** $(+3)(-2)$ **51.** $\dfrac{+10}{+2}$ **52.** $\dfrac{-8}{-4}$ **53.** $\dfrac{+27}{-3}$
22. $(+5)(+7)$ **23.** $(-6)(-8)$ **24.** $(-9)(+2)$
25. $(-7)(-3)$ **26.** $(-4)(+4)$ **27.** $(-8)(+2)$ **54.** $\dfrac{-48}{+6}$ **55.** $\dfrac{-32}{-4}$ **56.** $\dfrac{+39}{-13}$
28. $(+5)(-3)$ **29.** $(-6)(-9)$ **30.** $(+4)(+7)$
31. $(-6)(+1)$ **32.** $(+4)(-2)$ **33.** $(-8)(-3)$ **57.** $\dfrac{+14}{+7}$ **58.** $\dfrac{45}{+15}$ **59.** $\dfrac{-54}{-6}$
34. $(-2)(+6)$ **35.** $(-3)(-9)$ **36.** $(8)(-4)$
37. $(-9)(7)$ **38.** $(-8)(0)$ **39.** $(9)(-1)$ **60.** $\dfrac{+72}{-9}$ **61.** $\dfrac{-100}{+25}$ **62.** $\dfrac{+84}{+12}$
40. $(-4)(-10)$ **41.** $(+3)(-2)(+1)$
42. $(-5)(-9)(+2)$ **43.** $(-3)(+3)(-4)$ **63.** $\dfrac{-75}{-25}$ **64.** $\dfrac{+36}{-6}$ **65.** $\dfrac{+85}{+5}$

66. $\dfrac{-270}{+9}$ **67.** $\dfrac{+480}{+12}$ **68.** $\dfrac{-350}{+70}$

69. $\dfrac{-900}{-60}$ **70.** $\dfrac{+4800}{-240}$

71. $(-49) \div (-7)$ **72.** $(+9) \div (-3)$

73. $(+80) \div (+20)$ **74.** $(-60) \div (+12)$

75. $(+45) \div (-15)$ **76.** $(+120) \div (-6)$

77. $(-110) \div (-11)$ **78.** $(+84) \div (+6)$

79. $(-96) \div (-12)$ **80.** $(-800) \div (+25)$

2.4 Signed Fractions

The rules for operations with signed integers also apply to fractions.

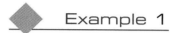 **Example 1** Add: $\left(-\dfrac{1}{4}\right) + \left(-\dfrac{3}{16}\right) = \left(-\dfrac{4}{16}\right) + \left(-\dfrac{3}{16}\right)$ The LCD is 16.

$$= \dfrac{-4 - 3}{16}$$

$$= -\dfrac{7}{16}$$ Combine the numerators. ■

 Example 2 Add: $\dfrac{3}{5} + \left(-\dfrac{2}{3}\right) = \dfrac{9}{15} + \left(-\dfrac{10}{15}\right)$ The LCD is 15.

$$= \dfrac{9 - 10}{15}$$

$$= -\dfrac{1}{15}$$ Combine the numerators. ■

 Example 3 Add: $\left(-\dfrac{4}{9}\right) + \dfrac{2}{3} = \left(-\dfrac{4}{9}\right) + \dfrac{6}{9}$ The LCD is 9.

$$= \dfrac{-4 + 6}{9}$$

$$= \dfrac{2}{9}$$ Combine the numerators. ■

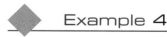 **Example 4** Add: $\left(-2\dfrac{3}{4}\right) + \left(-1\dfrac{5}{6}\right) = \left(-2\dfrac{9}{12}\right) + \left(-1\dfrac{10}{12}\right)$ The LCD is 12.

$$= -3\dfrac{19}{12}$$ Add the signed mixed numbers.

$$= -\left(3 + \dfrac{12 + 7}{12}\right)$$ Change the improper fraction to a mixed number.

$$= -\left(3 + 1\dfrac{7}{12}\right)$$

$$= -4\dfrac{7}{12}$$ ■

Example 5   Subtract: $\left(-\dfrac{5}{9}\right) - \left(-\dfrac{5}{12}\right) = \left(-\dfrac{5}{9}\right) + \left(+\dfrac{5}{12}\right)$ Change the sign of the fraction being subtracted and add.

$$= \left(-\dfrac{20}{36}\right) + \left(+\dfrac{15}{36}\right)$$ The LCD is 36.

$$= \dfrac{-20 + 15}{36}$$

$$= -\dfrac{5}{36}$$ Combine the numerators.   ■

Example 6   Subtract: $\left(-1\dfrac{3}{8}\right) - \left(+2\dfrac{5}{6}\right) = \left(-1\dfrac{3}{8}\right) + \left(-2\dfrac{5}{6}\right)$ Change the sign of the mixed number being subtracted and add.

$$= \left(-1\dfrac{9}{24}\right) + \left(-2\dfrac{20}{24}\right)$$ The LCD is 24.

$$= -3\dfrac{29}{24}$$ Add the mixed numbers.

$$= -\left(3 + \dfrac{24 + 5}{24}\right)$$ Change the improper fraction to a mixed number.

$$= -\left(3 + 1\dfrac{5}{24}\right)$$

$$= -4\dfrac{5}{24}$$ ■

Example 7   Subtract: $\left(-3\dfrac{5}{12}\right) - \left(-1\dfrac{2}{3}\right) = \left(-3\dfrac{5}{12}\right) + \left(+1\dfrac{2}{3}\right)$ Change the sign of the mixed number being subtracted and add.

$$= \left(-3\dfrac{5}{12}\right) + \left(1\dfrac{8}{12}\right)$$ The LCD is 12.

$$= \left(-2\dfrac{17}{12}\right) + \left(1\dfrac{8}{12}\right)$$ Borrow 1 or $\frac{12}{12}$.

$$= -1\dfrac{9}{12}$$ Add the mixed numbers.

$$= -1\dfrac{3}{4}$$ Reduce to lowest terms.   ■

Answers to Signed Numbers Drill 3

The first answer is for addition, the second answer is for subtraction, and the third answer is for multiplication.

1. +12, −4, +32 **2.** −9, −3, +18 **3.** +5, +11, −24 **4.** −8, −10, −9 **5.** +2, −8, −15 **6.** −12, 0, +36
7. −7, −7, 0 **8.** −4, +4, 0 **9.** +4, +2, +3 **10.** −13, −3, +40 **11.** −3, −9, −18 **12.** +1, +15, −56
13. +6, −8, −7 **14.** 0, +6, −9 **15.** +14, −4, +45 **16.** −10, −6, +16 **17.** 0, +10, −25 **18.** −1, −13, −42
19. −3, +5, −4 **20.** −18, 0, +81 **21.** +2, +4, −3 **22.** +2, −14, −48 **23.** −5, +9, −14 **24.** −9, −3, +18
25. +3, +11, −28 **26.** −15, +3, +54 **27.** +10, +4, +21 **28.** −20, 0, +100 **29.** +2, −8, −15 **30.** +3, +13, −40

◆ **Example 8** Multiply: $\left(-\dfrac{2}{5}\right)\left(-\dfrac{5}{8}\right) = \dfrac{10}{40}$ The product is positive.

$\qquad\qquad\qquad\qquad\quad = \dfrac{1}{4}$ Reduce to lowest terms. ■

◆ **Example 9** Multiply: $\left(\dfrac{4}{15}\right)\left(-\dfrac{5}{2}\right) = -\dfrac{20}{30}$ The product is negative.

$\qquad\qquad\qquad\qquad\quad = -\dfrac{2}{3}$ Reduce to lowest terms. ■

◆ **Example 10** Divide: $\left(-\dfrac{3}{7}\right) \div \left(-\dfrac{9}{14}\right) = \left(-\dfrac{\overset{1}{\cancel{3}}}{\cancel{7}_{1}}\right) \times \left(-\dfrac{\overset{2}{\cancel{14}}}{\cancel{9}_{3}}\right)$ Invert and multiply; cancels are shown.

$\qquad\qquad\qquad\qquad\qquad = \dfrac{2}{3}$ The product is positive. ■

◆ **Example 11** Divide: $\left(-\dfrac{11}{15}\right) \div \dfrac{2}{3} = \left(-\dfrac{11}{\cancel{15}_{5}}\right) \times \dfrac{\overset{1}{\cancel{3}}}{2}$ Invert and multiply; cancels are shown.

$\qquad\qquad\qquad\qquad\quad = -\dfrac{11}{10} \text{ or } -1\dfrac{1}{10}$ The product is negative. ■

One more rule about fractions will help you.

Equivalent Signed Fractions

$$\dfrac{a}{-b} = \dfrac{-a}{b} = -\dfrac{a}{b}$$

That is, a negative fraction may be written in three different but equivalent forms. However, the form $-\dfrac{a}{b}$ is the customary form.

For example, $\dfrac{3}{-4} = \dfrac{-3}{4} = -\dfrac{3}{4}$.

Note: $\dfrac{-a}{-b} = \dfrac{a}{b}$, using the rules for dividing signed numbers.

◆ **Example 12**   Add: $\dfrac{3}{-4} + \dfrac{-2}{3} = \left(-\dfrac{3}{4}\right) + \left(-\dfrac{2}{3}\right)$ Change to customary form.

$$= \left(-\dfrac{9}{12}\right) + \left(-\dfrac{8}{12}\right) \qquad \text{The LCD is 12.}$$

$$= \dfrac{-9 + -8}{12}$$

$$= -\dfrac{17}{12} \text{ or } -1\dfrac{5}{12} \qquad \text{Combine the numerators.} \qquad ■$$

◆ **Example 13**   Add: $\dfrac{3}{-4} + \dfrac{2}{3} = \left(-\dfrac{3}{4}\right) + \dfrac{2}{3}$ Change to customary form.

$$= \left(-\dfrac{9}{12}\right) + \left(\dfrac{8}{12}\right) \qquad \text{The LCD is 12.}$$

$$= \dfrac{-9 + 8}{12}$$

$$= -\dfrac{1}{12} \qquad \text{Combine the numerators.} \qquad ■$$

◆ **Example 14**   Subtract: $\dfrac{3}{-4} - \dfrac{2}{3} = \left(-\dfrac{3}{4}\right) + \left(-\dfrac{2}{3}\right)$ Change to customary form, change the sign of the fraction being subtracted, and add.

$$= \left(-\dfrac{9}{12}\right) + \left(-\dfrac{8}{12}\right) \qquad \text{The LCD is 12.}$$

$$= \dfrac{-9 - 8}{12}$$

$$= -\dfrac{17}{12} \text{ or } -1\dfrac{5}{12} \qquad \text{Combine the numerators.} \qquad ■$$

◆ **Example 15**   Multiply: $\left(-\dfrac{1}{4}\right)\left(\dfrac{-3}{5}\right) = \left(-\dfrac{1}{4}\right)\left(-\dfrac{3}{5}\right)$ Change to customary form.

$$= \dfrac{3}{20} \qquad \text{The product is positive.} \qquad ■$$

◆ **Example 16**   Multiply: $\left(\dfrac{-1}{2}\right)\left(\dfrac{3}{4}\right) = \left(-\dfrac{1}{2}\right)\left(\dfrac{3}{4}\right)$ Change to customary form.

$$= -\dfrac{3}{8} \qquad \text{The product is negative.} \qquad ■$$

◆ **Example 17**   Divide: $\left(\dfrac{-2}{3}\right) \div 3 = \left(-\dfrac{2}{3}\right) \div 3$ Change to customary form.

$$= \left(-\dfrac{2}{3}\right)\left(\dfrac{1}{3}\right) \qquad \text{Invert and multiply.}$$

$$= -\dfrac{2}{9} \qquad \text{The product is negative.} \qquad ■$$

◆ Example 18 Divide: $\left(\dfrac{-3}{7}\right) \div \dfrac{-5}{6} = \left(-\dfrac{3}{7}\right) \div \left(-\dfrac{5}{6}\right)$ Change to customary form.

$$= \left(-\dfrac{3}{7}\right)\left(-\dfrac{6}{5}\right)$$ Invert and multiply.

$$= \dfrac{18}{35}$$ The product is positive. ■

Exercises 2.4

Perform the indicated operations and simplify:

1. $\dfrac{1}{8} + \left(-\dfrac{5}{16}\right)$

2. $\left(-\dfrac{2}{3}\right) + \left(-\dfrac{2}{7}\right)$

3. $\dfrac{1}{2} + \left(-\dfrac{7}{16}\right)$

4. $\dfrac{2}{3} + \left(-\dfrac{7}{9}\right)$

5. $\left(-5\dfrac{3}{4}\right) + \left(-6\dfrac{2}{5}\right)$

6. $\left(-1\dfrac{3}{8}\right) + \left(5\dfrac{5}{12}\right)$

7. $\left(-3\dfrac{2}{3}\right) + \left(-\dfrac{4}{9}\right) + \left(4\dfrac{5}{6}\right)$

8. $\left(-\dfrac{3}{4}\right) + \left(-1\dfrac{1}{6}\right) + \left(-1\dfrac{1}{3}\right)$

9. $\left(-\dfrac{1}{4}\right) - \left(-\dfrac{1}{5}\right)$

10. $\left(-\dfrac{2}{9}\right) - \left(+\dfrac{1}{2}\right)$

11. $1\dfrac{3}{8} - \left(+\dfrac{5}{16}\right)$

12. $\left(-\dfrac{1}{3}\right) - \left(+3\dfrac{1}{2}\right)$

13. $\left(+1\dfrac{3}{4}\right) - (-4)$

14. $2\dfrac{3}{4} - \left(-3\dfrac{1}{4}\right)$

15. $\left(-\dfrac{2}{3}\right) + \left(-\dfrac{5}{6}\right) - \dfrac{1}{4}$

16. $\left(-\dfrac{3}{4}\right) - \left(-1\dfrac{2}{3}\right) + \left(-1\dfrac{5}{6}\right)$

17. $-\dfrac{1}{9} \times \dfrac{1}{7}$

18. $\left(-\dfrac{2}{3}\right)\left(-\dfrac{1}{2}\right)$

19. $\left(-3\dfrac{1}{3}\right)\left(-1\dfrac{4}{5}\right)$

20. $\dfrac{21}{8} \times 1\dfrac{7}{9}$

21. $\dfrac{4}{5} \div \left(-\dfrac{8}{9}\right)$

22. $\left(-1\dfrac{1}{4}\right) \div \dfrac{3}{5}$

23. $\left(-\dfrac{7}{9}\right) \div \left(-\dfrac{8}{3}\right)$

24. $2\dfrac{3}{4} \div \left(-3\dfrac{1}{6}\right)$

25. $\left(\dfrac{-1}{4}\right) + \left(\dfrac{1}{-5}\right)$

26. $\left(\dfrac{-4}{5}\right) + \left(-1\dfrac{1}{2}\right)$

27. $\dfrac{3}{4} + \left(\dfrac{-3}{8}\right)$

28. $\left(\dfrac{-3}{2}\right) + \left(\dfrac{-8}{3}\right)$

29. $\dfrac{5}{8} - \left(\dfrac{-5}{8}\right)$

30. $\left(\dfrac{-1}{4}\right) - \left(\dfrac{1}{-5}\right)$

31. $\left(\dfrac{-6}{8}\right) - (-4)$

32. $(-2) - \left(\dfrac{-1}{-4}\right)$

33. $\left(\dfrac{-1}{4}\right)\left(\dfrac{1}{-5}\right)$

34. $(-2)\left(\dfrac{-1}{-4}\right)$

35. $\left(\dfrac{-5}{-8}\right)\left(-5\dfrac{1}{3}\right)$

36. $\left(\dfrac{-3}{-4}\right)(-12)$

37. $32 \div \left(\dfrac{-2}{3}\right)$

38. $\left(\dfrac{-4}{9}\right) \div (-2)$

39. $\left(\dfrac{-2}{-3}\right) \div \left(\dfrac{2}{-3}\right)$

40. $\left(-1\dfrac{3}{5}\right) \div \left(-3\dfrac{1}{5}\right)$

41. $\left(\dfrac{-2}{3}\right) + \left(-\dfrac{5}{6}\right) + \dfrac{1}{4} + \dfrac{1}{8}$

42. $\left(\dfrac{-3}{4}\right) + \left(\dfrac{2}{-3}\right) - \left(\dfrac{-1}{-2}\right) - \left(\dfrac{-5}{6}\right)$

43. $\left(\dfrac{-2}{5}\right)\left(\dfrac{3}{-4}\right)\left(\dfrac{-15}{-18}\right)$

44. $\left(-2\dfrac{3}{4}\right) \div \left(1\dfrac{3}{5}\right)\left(\dfrac{-2}{5}\right)$

45. $\left(\dfrac{-2}{3}\right) + \left(-\dfrac{1}{2}\right)\left(\dfrac{5}{-6}\right)$

46. $\left(\dfrac{-4}{5}\right) \div \left(-1\dfrac{1}{2}\right) - \left(\dfrac{2}{-5}\right)$

2.5 Powers of 10

Multiplying Powers of 10

To multiply two powers of 10, add the exponents as follows:
$$10^a \times 10^b = 10^{a+b}$$

Note: The rules for working with powers of 10 shown in this section also apply to other bases, as shown in Section 5.4.

 Example 1   Multiply: $(10^2)(10^3)$

Method 1: $(10^2)(10^3) = (10 \cdot 10)(10 \cdot 10 \cdot 10) = 10^5$

Method 2: $(10^2)(10^3) = 10^{2+3} = 10^5$ Add the exponents.   ■

 Example 2   Multiply: $(10^3)(10^5)$

Method 1: $(10^3)(10^5) = (10 \cdot 10 \cdot 10)(10 \cdot 10 \cdot 10 \cdot 10 \cdot 10) = 10^8$

Method 2: $(10^3)(10^5) = 10^{3+5} = 10^8$ Add the exponents.   ■

Example 3   Multiply each of the following powers of 10:

    a. $(10^9)(10^{12}) = 10^{9+12} = 10^{21}$ Add the exponents.

    b. $(10^{-12})(10^{-7}) = 10^{(-12)+(-7)} = 10^{-19}$

    c. $(10^{-9})(10^6) = 10^{(-9)+6} = 10^{-3}$

    d. $(10^{10})(10^{-6}) = 10^{10+(-6)} = 10^4$

    e. $10^5 \cdot 10^{-8} \cdot 10^4 \cdot 10^{-3} = 10^{5+(-8)+4+(-3)} = 10^{-2}$ ■

Dividing Powers of 10

To divide two powers of 10, subtract the exponents as follows:
$$10^a \div 10^b = 10^{a-b}$$

 Example 4   Divide: $\dfrac{10^6}{10^2}$

Method 1: $\dfrac{10^6}{10^2} = \dfrac{\cancel{10} \cdot \cancel{10} \cdot 10 \cdot 10 \cdot 10 \cdot 10}{\cancel{10} \cdot \cancel{10}} = 10^4$

Method 2: $\dfrac{10^6}{10^2} = 10^{6-2} = 10^4$ Subtract the exponents.   ■

Example 5   Divide: $\dfrac{10^5}{10^3}$

Method 1: $\dfrac{10^5}{10^3} = \dfrac{\cancel{10} \cdot \cancel{10} \cdot \cancel{10} \cdot 10 \cdot 10}{\cancel{10} \cdot \cancel{10} \cdot \cancel{10}} = 10^2$

Method 2: $\dfrac{10^5}{10^3} = 10^{5-3} = 10^2$ Subtract the exponents.   ■

 Example 6   Divide each of the following powers of 10:

    a. $\dfrac{10^{12}}{10^4} = 10^{12-4} = 10^8$ Subtract the exponents.

    b. $\dfrac{10^{-5}}{10^5} = 10^{(-5)-5} = 10^{-10}$

    c. $\dfrac{10^6}{10^{-9}} = 10^{6-(-9)} = 10^{15}$

    d. $10^{-8} \div 10^{-5} = 10^{-8-(-5)} = 10^{-3}$

    e. $10^5 \div 10^9 = 10^{5-9} = 10^{-4}$ ■

◢ Raising a Power of 10 to a Power

To raise a power of 10 to a power, multiply the exponents as follows:
$$(10^a)^b = 10^{ab}$$

 Example 7   Find the power $(10^2)^3$.

Method 1: $(10^2)^3 = 10^2 \cdot 10^2 \cdot 10^2$
$= 10^{2+2+2}$ Use the product of powers rule.
$= 10^6$

Method 2: $(10^2)^3 = 10^{(2)(3)} = 10^6$ Multiply the exponents.   ■

 Example 8   Find each power of 10:

    a. $(10^4)^3 = 10^{(4)(3)} = 10^{12}$ Multiply the exponents.

    b. $(10^{-5})^2 = 10^{(-5)(2)} = 10^{-10}$

    c. $(10^{-6})^{-3} = 10^{(-6)(-3)} = 10^{18}$

    d. $(10^4)^{-4} = 10^{(4)(-4)} = 10^{-16}$

    e. $(10^{10})^8 = 10^{(10)(8)} = 10^{80}$ ■

In Section 1.10, we stated that $10^0 = 1$. Let's see why. To show this, we use the substitution principle, which states that if $a = b$ and $a = c$, then $b = c$.

$a = b$

$\dfrac{10^n}{10^n} = 10^{n-n}$ To divide powers, subtract the exponents.

$= 10^0$

$a = c$

$\dfrac{10^n}{10^n} = 1$ Any number other than zero divided by itself equals 1.

Therefore, $b = c$; that is, $10^0 = 1$.

Zero Power of 10

$$10^0 = 1$$

We also have used the fact that $10^{-a} = \dfrac{1}{10^a}$. To show this, we start with $\dfrac{1}{10^a}$:

$$\frac{1}{10^a} = \frac{10^0}{10^a} \qquad (1 = 10^0)$$

$$= 10^{0-a} \qquad \text{To divide powers, subtract the exponents.}$$

$$= 10^{-a}$$

Negative Power of 10

$$10^{-a} = \frac{1}{10^a}$$

For example, $10^{-3} = \dfrac{1}{10^3}$ and $10^{-8} = \dfrac{1}{10^8}$.

In a similar manner, we can also show that

$$\frac{1}{10^{-a}} = 10^a$$

For example, $\dfrac{1}{10^{-5}} = 10^5$ and $\dfrac{1}{10^{-2}} = 10^2$.

Combinations of multiplications and divisions of powers of 10 can also be done easily using the rules of exponents.

◆   Example 9   Perform the indicated operations. Express the results using positive exponents.

a.
$$\frac{10^4 \cdot 10^0}{10^{-3}} = \frac{10^{4+0}}{10^{-3}} \qquad \text{Add the exponents in the numerator.}$$

$$= \frac{10^4}{10^{-3}}$$

$$= 10^{4-(-3)} \qquad \text{Subtract the exponents.}$$

$$= 10^7$$

b.
$$\frac{10^{-2} \cdot 10^5}{10^2 \cdot 10^{-5} \cdot 10^8} = \frac{10^{-2+5}}{10^{2+(-5)+8}} \qquad \text{Add the exponents in the numerator and in the denominator.}$$

$$= \frac{10^3}{10^5}$$

$$= 10^{3-5} \qquad \text{Subtract the exponents.}$$

$$= 10^{-2}$$

$$= \frac{1}{10^2} \qquad \text{Express the result using a positive exponent.}$$

c. $\dfrac{10^2 \cdot 10^{-3}}{10^4 \cdot 10^{-7}} = \dfrac{10^{2+(-3)}}{10^{4+(-7)}}$ Add the exponents in the numerator and in the denominator.

$$= \dfrac{10^{-1}}{10^{-3}}$$

$$= 10^{(-1)-(-3)}$$ Subtract the exponents.

$$= 10^2$$

d. $\dfrac{10^{-5} \cdot 10^8 \cdot 10^{-6}}{10^3 \cdot 10^4 \cdot 10^{-1}} = \dfrac{10^{(-5)+8+(-6)}}{10^{3+4+(-1)}}$ Add the exponents in the numerator and in the denominator.

$$= \dfrac{10^{-3}}{10^6}$$

$$= 10^{-3-6}$$ Subtract the exponents.

$$= 10^{-9}$$

$$= \dfrac{1}{10^9}$$ Express the result using a positive exponent.

■

Exercises 2.5

Perform the indicated operations using the laws of exponents. Express the results using positive exponents.

1. $10^4 \cdot 10^9$

2. $10^{-3} \cdot 10^5$

3. $\dfrac{10^4}{10^8}$

4. $10^4 \div 10^{-6}$

5. $\dfrac{1}{10^{-3}}$

6. $\dfrac{1}{10^{-5}}$

7. $(10^4)^3$

8. $(10^3)^{-2}$

9. $10^{-6} \cdot 10^{-4}$

10. $10^{-15} \cdot 10^{10}$

11. $\dfrac{10^{-3}}{10^{-6}}$

12. $10^{-2} \div 10^{-5}$

13. $(10^{-3})^4$

14. $(10^{-3})^{-5}$

15. $\left(\dfrac{10^6}{10^8}\right)^2$

16. $\left(\dfrac{10^{-2}}{10^{-5}}\right)^3$

17. $\dfrac{(10^0)^3}{10^{-2}}$

18. $\left(\dfrac{10^0}{10^{-3}}\right)^2$

19. $10^2 \cdot 10^{-5} \cdot 10^{-3}$

20. $10^{-6} \cdot 10^{-1} \cdot 10^4$

21. $10^3 \cdot 10^4 \cdot 10^{-5} \cdot 10^3$

22. $\dfrac{10^0 \cdot 10^{-3}}{10^{-6} \cdot 10^3}$

23. $\dfrac{10^3 \cdot 10^2 \cdot 10^{-7}}{10^5 \cdot 10^{-3}}$

24. $\dfrac{10^{-2} \cdot 10^{-3} \cdot 10^{-7}}{10^3 \cdot 10^4 \cdot 10^{-5}}$

25. $\dfrac{10^8 \cdot 10^{-6} \cdot 10^{10} \cdot 10^0}{10^4 \cdot 10^{-17} \cdot 10^8}$

26. $\dfrac{(10^{-4})^6}{10^4 \cdot 10^{-3}}$

27. $\dfrac{(10^{-9})^{-2}}{10^{16} \cdot 10^{-4}}$

28. $\left(\dfrac{10^4}{10^{-7}}\right)^3$

29. $\left(\dfrac{10^5 \cdot 10^{-2}}{10^{-4}}\right)^2$

30. $\left(\dfrac{10^{-7} \cdot 10^{-2}}{10^9}\right)^{-3}$

2.6 Scientific Notation

Scientific Notation

Scientific notation is a method that is especially useful for writing very large or very small numbers. To write a number in scientific notation, write it as a product of a number between 1 and 10 and a power of 10.

Example 1   Write 226 in scientific notation.

$$226 = 2.26 \times 10^2$$

Remember that 10^2 is a short way of writing $10 \times 10 = 100$. Note that multiplying 2.26 by 100 gives 226. ■

  Example 2   Write 52,800 in scientific notation.

$$52,800 = 5.28 \times 10,000 = 5.28 \times (10 \times 10 \times 10 \times 10)$$
$$= 5.28 \times 10^4$$ ■

Writing a Decimal Number in Scientific Notation

To write a decimal number in scientific notation,

1. Reading from left to right, place a decimal point after the first nonzero digit.
2. Place a caret (∧) at the position of the original decimal point.
3. If the decimal point is to the *left* of the caret, the exponent of the power of 10 is the same as the number of decimal places from the caret to the decimal point.

$$26,638 = 2.6638. \times 10^{④} = 2.6638 \times 10^4$$
$$\overleftarrow{\;④\;} \; ∧$$

4. If the decimal point is to the *right* of the caret, the exponent of the power of 10 is the same as the negative of the number of places from the caret to the decimal point.

$$0.00986 = 0.009.86 \times 10^{-③} = 9.86 \times 10^{-3}$$
$$∧\overrightarrow{③}$$

5. If the decimal point is already after the first nonzero digit, the exponent of 10 is zero.

$$2.15 = 2.15 \times 10^0$$

  Example 3   Write 2738 in scientific notation.

$$2738 = 2.738 \times 10^{③} = 2.738 \times 10^3$$
$$\overleftarrow{③}∧$$ ■

  Example 4   Write 0.0000003842 in scientific notation.

$$0.0000003842 = 0.0000003.842 \times 10^{-⑦} = 3.842 \times 10^{-7}$$
$$∧\overrightarrow{\;⑦\;}$$ ■

Writing a Number in Scientific Notation in Decimal Form

To change a number in scientific notation to decimal form,

1. Multiply the decimal part by the given *positive* power of 10 by moving the decimal point to the *right* the same number of decimal places as indicated by the exponent of 10. Supply zeros when needed.

2. Multiply the decimal part by the given *negative* power of 10 by moving the decimal point to the *left* the same number of decimal places as indicated by the exponent of 10. Supply zeros when needed.

 Example 5   Write 2.67×10^2 as a decimal.

$$2.67 \times 10^2 = 267$$ Move the decimal point two places to the *right*, since the exponent of 10 is $+2$.   ■

 Example 6   Write 8.76×10^4 as a decimal.

$$8.76 \times 10^4 = 87{,}600$$ Move the decimal point four places to the *right*, since the exponent of 10 is $+4$. It is necessary to write two zeros.   ■

 Example 7   Write 5.13×10^{-4} as a decimal.

$$5.13 \times 10^{-4} = 0.000513$$ Move the decimal point four places to the *left*, since the exponent of 10 is -4. It is necessary to write three zeros.   ■

You may find it useful to note that a number in scientific notation with

a. a positive exponent greater than 1 is *greater than* 10, and
b. a negative exponent is between 0 and 1.

That is, a number in scientific notation with a positive exponent represents a relatively large number. A number in scientific notation with a negative exponent represents a relatively small number.

Scientific notation may be used to compare two positive numbers expressed as decimals. First, write both numbers in scientific notation. The number having the greater power of 10 is the larger. If the powers of 10 are equal, compare the parts of the numbers that are between 1 and 10.

 Example 8   Which is greater, 0.000876 or 0.0004721?

$$0.000876 = 8.76 \times 10^{-4}$$
$$0.0004721 = 4.721 \times 10^{-4}$$

Since the exponents are the same, compare 8.76 and 4.721. Since 8.76 is greater than 4.721, 0.000876 is greater than 0.0004721.   ■

 Example 9   Which is greater, 0.0062 or 0.0382?

$$0.0062 = 6.2 \times 10^{-3}$$
$$0.0382 = 3.82 \times 10^{-2}$$

Since -2 is greater than -3, 0.0382 is greater than 0.0062.   ■

Scientific notation is especially helpful for multiplying and dividing very large and very small numbers. To perform these operations, you must first know some rules for exponents. Many calculators perform multiplication, division, and powers of numbers entered in scientific notation and give the results, when very large or very small, in scientific notation.

Multiplying Numbers in Scientific Notation

To multiply numbers in scientific notation, multiply the decimals between 1 and 10. Then add the exponents of the powers of 10.

◆ **Example 10**

Multiply $(4.5 \times 10^8)(5.2 \times 10^{-14})$. Write the result in scientific notation.

$$(4.5 \times 10^8)(5.2 \times 10^{-14}) = (4.5)(5.2) \times (10^8)(10^{-14})$$
$$= 23.4 \times 10^{-6}$$
$$= (2.34 \times 10^1) \times 10^{-6}$$
$$= 2.34 \times 10^{-5}$$

Note that 23.4×10^{-6} is not in scientific notation, because 23.4 is not between 1 and 10.

To find this product using a calculator that accepts numbers in scientific notation, use the following procedure.

4.5 ⬚EE⬚* 8 ⬚×⬚ 5.2 ⬚EE⬚ ⬚(−)⬚ 14 ⬚=⬚

⬚ 2.34 x 10⁻⁵ ⬚

Notes: 1. You may need to set your calculator in scientific notation mode.

2. The ⬚(−)⬚ or ⬚(+/−)⬚ key is used to enter a negative number.

The product is 2.34×10^{-5}.   ■

Dividing Numbers in Scientific Notation

To divide numbers in scientific notation, divide the decimals between 1 and 10. Then subtract the exponents of the powers of 10.

◆ **Example 11**

Divide $\dfrac{4.8 \times 10^{-7}}{1.6 \times 10^{-11}}$. Write the result in scientific notation.

$$\frac{4.8 \times 10^{-7}}{1.6 \times 10^{-11}} = \frac{4.8}{1.6} \times \frac{10^{-7}}{10^{-11}}$$
$$= 3 \times 10^4$$

Using a calculator, we have

4.8 ⬚EE⬚ ⬚(−)⬚ 7 ⬚÷⬚ 1.6 ⬚EE⬚ ⬚(−)⬚ 11 ⬚=⬚

⬚ 3 x 10⁴ ⬚

The quotient is 3×10^4.   ■

*Key may also be ⬚Exp⬚.

Example 12   Evaluate $\dfrac{(6 \times 10^{-6})(3 \times 10^9)}{(2 \times 10^{-10})(4 \times 10^{-5})}$. Write the result in scientific notation.

$$\frac{(6 \times 10^{-6})(3 \times 10^9)}{(2 \times 10^{-10})(4 \times 10^{-5})} = \frac{(6)(3)}{(2)(4)} \times \frac{(10^{-6})(10^9)}{(10^{-10})(10^{-5})} = \frac{18}{8} \times \frac{10^3}{10^{-15}}$$

$$= 2.25 \times 10^{18}$$

Again, use a calculator.

6 EE (−) 6 × 3 EE 9 ÷ 2 EE (−) 10 ÷ 4 EE (−) 5 =

2.25 × 10¹⁸   Note: Some calculators display this result as 2.25 E18.

The result is 2.25×10^{18}.   ■

Powers of Numbers in Scientific Notation

To find the power of a number in scientific notation, find the power of the decimal between 1 and 10. Then multiply the exponent of the power of 10 by this same power.

Example 13   Find the power $(4.5 \times 10^6)^2$. Write the result in scientific notation.

$$(4.5 \times 10^6)^2 = (4.5)^2 \times (10^6)^2$$
$$= 20.25 \times 10^{12} \qquad \text{Note that 20.25 is not between 1 and 10.}$$
$$= (2.025 \times 10^1) \times 10^{12}$$
$$= 2.025 \times 10^{13}$$

4.5 EE 6 x^2 =

2.025 × 10¹³

The result is 2.025×10^{13}.   ■

Example 14   Find the power $(3 \times 10^{-8})^5$. Write the result in scientific notation.

$$(3 \times 10^{-8})^5 = 3^5 \times (10^{-8})^5$$
$$= 243 \times 10^{-40}$$
$$= (2.43 \times 10^2) \times 10^{-40}$$
$$= 2.43 \times 10^{-38}$$

The ⌃ key is used to raise a number to a power.

3 EE (−) 8 ⌃ or y^x 5 =

2.43 × 10⁻³⁸

The result is 2.43×10^{-38}.   ■

Exercises 2.6

Write each number in scientific notation:

1. 356
2. 15,600
3. 634.8
4. 24.85
5. 0.00825
6. 0.00063
7. 7.4
8. 377,000
9. 0.000072
10. 0.00335
11. 710,000
12. 1,200,000
13. 0.0000045
14. 0.0000007
15. 0.000000034
16. 4,500,000,000
17. 640,000
18. 85,000

Write each number in decimal form:

19. 7.55×10^4
20. 8.76×10^2
21. 5.31×10^3
22. 5.14×10^5
23. 7.8×10^{-2}
24. 9.44×10^{-3}
25. 5.55×10^{-4}
26. 4.91×10^{-6}
27. 6.4×10^1
28. 3.785×10^{-2}
29. 9.6×10^2
30. 7.3×10^3
31. 5.76×10^0
32. 6.8×10^{-5}
33. 6.4×10^{-6}
34. 7×10^8
35. 5×10^{10}
36. 5.05×10^0
37. 6.2×10^{-7}
38. 2.1×10^{-9}
39. 2.5×10^{12}
40. 1.5×10^{11}
41. 3.3×10^{-11}
42. 7.23×10^{-8}

Find the larger number:

43. 0.0037; 0.0048
44. 0.029; 0.0083
45. 0.000042; 0.00091
46. 148,000; 96,988
47. 0.00037; 0.000094
48. 0.8216; 0.792
49. 0.0613; 0.00812
50. 0.0000613; 0.01200

Find the smaller number:

51. 0.008; 0.0009
52. 295,682; 295,681
53. 1.003; 1.0009
54. 21.8; 30.2
55. 0.00000000998; 0.01
56. 0.10108; 0.10102
57. 0.000314; 0.000271
58. 0.00812; 0.0318

Perform the indicated operations (write each result in scientific notation with the decimal part rounded to three significant digits when necessary):

59. $(4 \times 10^{-6})(6 \times 10^{-10})$

60. $(3 \times 10^7)(3 \times 10^{-12})$

61. $\dfrac{4.5 \times 10^{16}}{1.5 \times 10^{-8}}$

62. $\dfrac{1.6 \times 10^6}{6.4 \times 10^{10}}$

63. $\dfrac{(4 \times 10^{-5})(6 \times 10^{-3})}{(3 \times 10^{-10})(8 \times 10^8)}$

64. $\dfrac{(5 \times 10^4)(3 \times 10^{-5})(4 \times 10^6)}{(1.5 \times 10^6)(2 \times 10^{-11})}$

65. $(1.2 \times 10^6)^3$

66. $(2 \times 10^{-9})^4$

67. $(6.2 \times 10^{-5})(5.2 \times 10^{-6})(3.5 \times 10^8)$

68. $\dfrac{(5 \times 10^{-6})^2}{4 \times 10^6}$

69. $\left(\dfrac{2.5 \times 10^{-4}}{7.5 \times 10^8}\right)^2$

70. $\left(\dfrac{2.5 \times 10^{-9}}{5 \times 10^{-7}}\right)^4$

71. $(18,000)(0.00005)$

72. $(4500)(69,000)(150,000)$

73. $\dfrac{2,400,000}{36,000}$

74. $\dfrac{(3500)(0.00164)}{2700}$

75. $\dfrac{84,000 \times 0.0004 \times 142,000}{0.002 \times 3200}$

76. $\dfrac{(0.0025)^2}{3500}$

77. $\left(\dfrac{48,000 \times 0.0144}{0.0064}\right)^2$

78. $\left(\dfrac{0.0027 \times 0.16}{12,000}\right)^3$

79. $\left(\dfrac{1.3 \times 10^4}{(2.6 \times 10^{-3})(5.1 \times 10^8)}\right)^5$

80. $\left(\dfrac{9.6 \times 10^{-3}}{(2.45 \times 10^{-4})(1.1 \times 10^5)}\right)^6$

81. $\left(\dfrac{18.4 \times 2100}{0.036 \times 950}\right)^8$

82. $\left(\dfrac{0.259 \times 6300}{866 \times 0.013}\right)^{10}$

2.7 Engineering Notation

Numbers may also be written in engineering notation, similar to scientific notation, as follows:

> ## Engineering Notation
>
> **Engineering notation** is used to write a number with its decimal part between 1 and 1000 and a power of 10 whose exponent is divisible by 3.

> ## Writing a Decimal Number in Engineering Notation
>
> To write a decimal number in engineering notation,
>
> 1. Move the decimal point *in groups of three digits* until the decimal point indicates a number between 1 and 1000.
> 2. If the decimal point has been moved to the *left*, the exponent of the power of 10 in engineering notation is the same as the number of places the decimal point was moved.
> 3. If the decimal point has been moved to the *right*, the exponent of the power of 10 in engineering notation is the same as the negative of the number of places the decimal point was moved.
>
> In any case, the exponent will be divisible by 3.

 Example 1

Write 48,500 in engineering notation.

$$48{,}500 = 48.5 \times 10^3$$ Move the decimal point in groups of three decimal places until the decimal part is between 1 and 1000.

Check The exponent of the power of 10 must be divisible by 3. ■

 Example 2

Write 375,000,000,000 in engineering notation.

$$375{,}000{,}000{,}000 = 375 \times 10^9$$ Move the decimal point in groups of three decimal places until the decimal part is between 1 and 1000.

Check 9, the exponent of the power of 10, is divisible by 3. ■

 Example 3

Write 0.000000000002045 in engineering notation.

$$0.000000000002045 = 2.045 \times 10^{-12}$$ Move the decimal point in groups of three decimal places until the decimal part is between 1 and 1000.

Check -12, the exponent of the power of 10, is divisible by 3. In this case, this number is also written in scientific notation. ■

Writing a number in engineering notation in decimal form is similar to writing a number in scientific notation in decimal form.

 Example 4   Write 405×10^6 as a decimal.

$$405 \times 10^6 = 405,000,000$$ Move the decimal point six places to the *right,* since the exponent is +6.   ■

 Example 5   Write 87.035×10^{-6} as a decimal.

$$87.035 \times 10^{-6} = 0.000087035$$ Move the decimal point six places to the *left,* since the exponent is −6.   ■

Operations with numbers in engineering notation using a calculator are very similar to operations with numbers in scientific notation. If your calculator has an engineering notation mode, set it in this mode. If not, use scientific notation and convert the result to engineering notation.

 Example 6   Multiply $(26.4 \times 10^6)(722 \times 10^3)$. Write the result in engineering notation.

In this example, we will show an arithmetic step-by-step analysis and then show how to use a calculator to find this product.

$$(26.4 \times 10^6)(722 \times 10^3) = (26.4)(722) \times (10^6)(10^3)$$
$$= 19060.8 \times 10^9$$
$$= (19.0608 \times 10^3) \times 10^9$$
$$= 19.0608 \times 10^{12}$$

To find this product using a calculator that accepts numbers in engineering notation or scientific notation, use the following procedure:

26.4 **EE** 6 **×** 722 **EE** 3 **=**

| 19.0608 X 10¹² |

Note: If you use scientific notation, your result is 1.90608×10^{13}, which, when converted to engineering notation, is 19.0608×10^{12}.   ■

 Example 7   Divide $\dfrac{12.75 \times 10^{-15}}{236 \times 10^{-9}}$. Write the result in engineering notation rounded to three significant digits.

Find this quotient using a calculator as follows:

12.75 **EE** **(−)** 15 **÷** 236 **EE** **(−)** 9 **=**

| 54.02542373 X 10⁻⁹ |

So, the result rounded to three significant digits is 54.0×10^{-9}.

Note: If you use scientific notation, your result is 5.40×10^{-8}, which, when converted to engineering notation, is 54.0×10^{-9}.   ■

Example 8   Find the power $(15.4 \times 10^9)^2$ and write the result in engineering notation rounded to three significant digits.

15.4 **EE** 9 x^2 =

$$237.16 \times 10^{18}$$

The result rounded to three significant digits is 237×10^{18}. ■

 Example 9   Find the square root $\sqrt{740.5 \times 10^{-18}}$ and write the result in engineering notation rounded to three significant digits.

$\sqrt{}$ 740.5 **EE** (−) 18 =

$$27.21212965 \times 10^{-9}$$

The result rounded to three significant digits is 27.2×10^{-9}. ■

For comparison purposes, the following table shows six numbers written in both scientific notation and engineering notation:

Number	Scientific notation	Engineering notation
6,710,000	6.71×10^6	6.71×10^6
805,000	8.05×10^5	805×10^3
34,500,000	3.45×10^7	34.5×10^6
0.000096	9.6×10^{-5}	96×10^{-6}
0.000007711	7.711×10^{-6}	7.711×10^{-6}
0.000000444	4.44×10^{-7}	444×10^{-9}

Exercises 2.7

Write each number in engineering notation:

1. 28,000
2. 135,000
3. 3,450,000
4. 29,000,000
5. 220,000,000,000
6. 7,235,000,000,000,000
7. 0.0066
8. 0.00015
9. 0.0000000765
10. 0.0000000000044
11. 0.975
12. 0.0000000625

Write each number in decimal form:

13. 57.7×10^3
14. 135×10^6
15. 4.94×10^{12}
16. 46×10^9
17. 567×10^6
18. 3.24×10^{18}

19. 26×10^{-6}
20. 751×10^{-3}
21. 5.945×10^{-9}
22. 602.5×10^{-6}
23. 10.64×10^{-12}
24. 6.3×10^{-15}

Perform the indicated operations and write each result in engineering notation rounded to three significant digits:

25. $(35.5 \times 10^6)(420 \times 10^9)$
26. $(9.02 \times 10^{-6})(69.5 \times 10^{-24})$
27. $(2.7 \times 10^9)(27 \times 10^{-6})(270 \times 10^{-12})$
28. $(6 \times 10^{-12})(20 \times 10^{-9})(400 \times 10^{-6})$
29. $\dfrac{70.5 \times 10^6}{120 \times 10^{-9}}$
30. $\dfrac{450 \times 10^{-12}}{51 \times 10^6}$

31. $\dfrac{(5.15 \times 10^9)(65.3 \times 10^{-6})}{(27 \times 10^6)(800 \times 10^{12})}$

32. $\dfrac{(750 \times 10^{-12})(25 \times 10^{-6})(1.5 \times 10^{-3})}{(30 \times 10^{-9})(2 \times 10^{15})}$

33. $\dfrac{1}{2.95 \times 10^{-9}}$ **34.** $\dfrac{1}{55 \times 10^{12}}$

35. $(350 \times 10^9)^2$

36. $(92.5 \times 10^{-12})^2$

37. $\sqrt{80.5 \times 10^{12}}$

38. $\sqrt{750 \times 10^{-18}}$

39. $\dfrac{\sqrt{6.05 \times 10^9}}{(244 \times 10^{-6})^2}$

40. $\dfrac{1}{(24 \times 10^{-9})^2}$

Chapter 2 Group Activities

1. In small groups, go through Section 2.5 and look at all of the highlighted rules for multiplying powers of 10, dividing powers of 10, and raising a power of 10 to a power. Look over the examples and make up five new examples similar to Examples 1 through 8 of the text, methods 1 and 2. After you have done this, go back and change all of the 10s to 2s. Then to 3s, 4s . . . and continue this up to 9. Do the other numbers behave as the tens do? Do the rules still work? In your own words, tell what you have found.

2. In small groups, the members of the group should write what profession they wish to pursue. Ideally, several careers will be listed. Take each career one at a time and think of all the ways in which that profession could use powers of 10 and scientific notation. After this has been completed for each career, compare the careers as to whether there is overlap in the ways powers of ten and scientific notation are used. Do the same with fractions. Do the careers use these numbers similarly? If they do, explain. If they do not, explain how the same type of number is used but in different contexts.

Chapter 2 Summary

Glossary of Basic Terms

Absolute value of a number.   Its distance from zero on the number line. The absolute value of a number is never negative. (p. 109)

Engineering notation.   A number written with its decimal part between 1 and 1000 and a power of 10 whose exponent is divisible by 3. (p. 131)

Irrational numbers.   Those numbers that cannot be written as the ratio of two integers. (p. 109)

Negative integers. $-1, -2, -3, \ldots$, or those integers less than zero. (p. 108)

Positive integers. $1, 2, 3, \ldots$, or those integers greater than zero. (p. 108)

Rational numbers.   Those numbers that can be written as the ratio of two integers; that is, a/b, where $b \neq 0$. (p. 109)

Real numbers.   Those numbers that are either rational or irrational. (p. 109)

Scientific notation.   A number written with its decimal part between 1 and 10 and a power of ten. (p. 125)

Signed numbers.   Positive and negative numbers. (p. 110)

2.1   Addition of Signed Numbers

1. **Adding two numbers with like signs:**
   a. To add two positive numbers, add their absolute values. The result is positive.
   b. To add two negative numbers, add their absolute values and place a negative sign before the result. (p. 110)

2. **Adding two numbers with unlike signs:** To add a negative number and a positive number, find the difference of their absolute values. The sign of the number having the larger absolute value is placed before the result. (p. 110)

3. **Adding three or more signed numbers:** To add three or more signed numbers,
   a. Add the positive numbers.
   b. Add the negative numbers.
   c. Add the sums from steps (a) and (b) according to the rules for addition of two signed numbers. (p. 110)

2.2 Subtraction of Signed Numbers

1. **Subtracting two signed numbers:** To subtract two signed numbers, change the sign of the number being subtracted and add according to the rules for addition of signed numbers. (p. 112)

2. **Subtracting more than two signed numbers:** When more than two signed numbers are involved in subtraction, change the sign of *each* number being subtracted and add the resulting signed numbers. (p. 112)

3. **Adding and subtracting combinations of signed numbers:** When combinations of addition and subtraction of signed numbers occur in the same problem, change *only* the sign of each number being subtracted. Then add the resulting signed numbers. (p. 113)

2.3 Multiplication and Division of Signed Numbers

1. **Multiplying two signed numbers:**
   a. To multiply two numbers with the same sign, multiply their absolute values. This product is positive.
   b. To multiply two numbers with different signs, multiply their absolute values and place a negative sign before the result. (p. 114)

2. **Multiplying more than two signed numbers:** To multiply more than two signed numbers, multiply their absolute values; then
   a. place a positive sign before the result if the number of negative signed numbers is even or
   b. place a negative sign before the result if the number of negative signed numbers is odd. (p. 115)

3. **Dividing two signed numbers:**
   a. To divide two numbers with the same sign, divide their absolute values. This product is positive.
   b. To divide two numbers with different signs, divide their absolute values and place a negative sign before the result. (p. 115)

2.4 Signed Fractions

1. **Equivalent signed fractions:** The following are equivalent signed fractions:

$$\frac{a}{-b} = \frac{-a}{b} = -\frac{a}{b} \quad \text{(p. 119)}$$

2.5 Powers of 10

1. **Multiplying powers of 10:** To multiply two powers of 10, add the exponents: $10^a \times 10^b = 10^{a+b}$ (p. 122)

2. **Dividing powers of 10:** To divide two powers of 10, subtract the exponents: $10^a \div 10^b = 10^{a-b}$ (p. 122)

3. **Raising a power of 10 to a power:** To raise a power of 10 to a power, multiply the exponents: $(10^a)^b = 10^{ab}$ (p. 123)

4. **Zero power of 10:** The zero power of 10 is 1: $10^0 = 1$ (p. 124)

5. **Negative power of 10:** Negative powers of 10 may be written $10^{-a} = \dfrac{1}{10^a}$ or $\dfrac{1}{10^{-a}} = 10^a$ (p. 124)

2.6 Scientific Notation

1. **Writing a decimal number in scientific notation:** To write a decimal number in scientific notation,
   a. Reading from left to right, place a decimal point after the first nonzero digit.
   b. Place a caret (∧) at the position of the original decimal point.
   c. If the decimal point is to the *left* of the caret, the exponent of the power of 10 is the same as the number of decimal places from the caret to the decimal point.
   d. If the decimal point is to the *right* of the caret, the exponent of the power of 10 is the same as the negative of the number of places from the caret to the decimal point.
   e. If the decimal point is already after the first nonzero digit, the exponent of 10 is zero. (p. 126)

2. **Writing a number in scientific notation in decimal form:** To change a number in scientific notation to decimal form,
   a. Multiply the decimal part by the given *positive* power of 10 by moving the decimal point to the *right* the same number of decimal places as indicated by the exponent of 10. Supply zeros when needed.
   b. Multiply the decimal part by the given *negative* power of 10 by moving the decimal point to the *left* the same number of decimal places as indicated by the exponent of 10. Supply zeros when needed. (pp. 126–127)

3. **Useful note:** A number in scientific notation is greater than 10 if it has a positive exponent and between 0 and 1 if it has a negative exponent. (p. 127)

2.7 Engineering Notation

1. **Writing a decimal number in engineering notation:**
   To write a decimal number in engineering notation,
   a. Move the decimal point *in groups of three digits* until the decimal point indicates a number between 1 and 1000.
   b. If the decimal point has been moved to the *left*, the exponent of the power of 10 in engineering

notation is the same as the number of places the decimal point was moved.
   c. If the decimal point has been moved to the *right*, the exponent of the power of 10 in engineering notation is the same as the negative of the number of places the decimal point was moved.
   In any case, the exponent will be divisible by 3. (p. 131)

Chapter 2 Review

Find the absolute value of each number:

1. $+5$ **2.** -16 **3.** 13

Add:

4. $(-4) + (+7)$ **5.** $(-6) + (-2)$

6. $(+5) + (-8)$ **7.** $(-9) + (+2) + (-6)$

Subtract:

8. $3 - 6$ **9.** $(-7) - (+4)$

10. $(+9) - (-10)$ **11.** $(-6) - (+4) - (-8)$

Perform the indicated operations:

12. $(-2) - (+7) + (+4) + (-5) - (-10)$

13. $5 - 6 + 4 - 9 + 4 + 3 - 12 - 8$

Multiply:

14. $(-6)(+4)$ **15.** $(+4)(+9)$ **16.** $(-9)(-8)$

17. $(-2)(-7)(+1)(+3)(-2)$

Divide:

18. $\dfrac{-18}{-3}$ **19.** $(+30) \div (-5)$ **20.** $\dfrac{+45}{+9}$

Perform the indicated operations and simplify:

21. $\left(-\dfrac{6}{7}\right) - \left(\dfrac{5}{-6}\right)$ **22.** $\dfrac{-3}{16} \div \left(-2\dfrac{1}{4}\right)$

23. $\dfrac{-5}{8} + \left(-\dfrac{5}{6}\right) - \left(+1\dfrac{2}{3}\right)$ **24.** $\left(-\dfrac{9}{16}\right)\left(2\dfrac{2}{3}\right)$

Perform the indicated operations using the laws of exponents and express the results using positive exponents:

25. $10^9 \cdot 10^{-14}$ **26.** $10^6 \div 10^{-3}$

27. $(10^{-4})^3$ **28.** $\dfrac{(10^{-3} \cdot 10^5)^3}{10^6}$

Write each number in scientific notation:

29. $476{,}000$ **30.** 0.0014

Write each number in decimal form:

31. 5.35×10^{-5} **32.** 6.1×10^7

Find the larger number:

33. $0.00063; 0.00105$ **34.** $0.056; 0.06$

Find the smaller number:

35. $0.000075; 0.0006$ **36.** $0.04; 0.00183$

Perform the indicated operations and write each result in scientific notation:

37. $(9.5 \times 10^{10})(4.6 \times 10^{-13})$ **38.** $\dfrac{8.4 \times 10^8}{3 \times 10^{-6}}$

39. $\dfrac{(50{,}000)(640{,}000{,}000)}{(0.0004)^2}$ **40.** $(4.5 \times 10^{-8})^2$

41. $(2 \times 10^9)^4$ **42.** $\left(\dfrac{1.2 \times 10^{-2}}{3 \times 10^{-5}}\right)^3$

Write each number in engineering notation:

43. $275{,}000$ **44.** $32{,}000{,}000$ **45.** 0.00045

Write each number in decimal form:

46. 31.6×10^6 **47.** 746×10^{-3}

Perform the indicated operations and write each result in engineering notation rounded to three significant digits when necessary:

48. $(39.4 \times 10^6)(120 \times 10^{-3})$ **49.** $\dfrac{84.5 \times 10^{-9}}{3.48 \times 10^6}$

50. $\dfrac{1}{21.7 \times 10^{-6}}$

Chapter 2 Test

Perform the indicated operations:

1. $-7 + (+9)$ **2.** $-28 + (-11)$ **3.** $18 + (-6)$

4. $-65 + 42$ **5.** $+112 + 241$ **6.** $-6 - (-10)$

7. $+16 - (+18)$ **8.** $(-4)(-7)$ **9.** $(+15)(-22)$

10. $(+17)(-8)$ **11.** $\dfrac{+20}{-4}$

12. $(-160) \div (-8)$

13. $(-2) + (-6) + (+5) + (-9) + (+10)$

14. $(-3)(-2)(+1)(+2)(-1)(+2)(+1)$

15. $(-5) + (-6) - (-7) - (+4) + (+3)$

16. $8 + (-1) + (-5) - (-3) + 10$

17. $-8 + 5 + 2 - 12 + 5 - 3$

Perform the indicated operations and simplify:

18. $\left(-\dfrac{2}{3}\right)(-6) + \left(-1\dfrac{1}{3}\right)$ **19.** $2\dfrac{1}{5} - \left(-1\dfrac{3}{10}\right) + 2\dfrac{3}{5}$

20. $\left(-\dfrac{5}{9}\right)\left(3\dfrac{2}{5}\right)$

21. Write 0.000182 in scientific notation.

22. Write 4.7×10^6 in decimal form.

Perform the indicated operations using the laws of exponents and express each result using positive exponents:

23. $(10^{-3})(10^6)$ **24.** $10^3 \div 10^{-5}$

25. $(10^2)^4$ **26.** $\dfrac{10^8 \cdot 10^{-6}}{(10^{-3})^{-2}}$

27. $\dfrac{(10^{-4})(10^{-8})^2}{(10^4)^{-6}}$

Perform the indicated operations. Write each result in scientific notation rounded to three significant digits when necessary:

28. $\dfrac{(7.6 \times 10^{13})(5.35 \times 10^{-6})}{4.64 \times 10^8}$

29. $\dfrac{(150{,}000)(18{,}000)(0.036)}{(0.0056)(48{,}000)}$

30. $\dfrac{(25{,}000)(0.125)}{(0.05)^3}$

Write each number in engineering notation:

31. 825,000   **32.** 0.0000751

33. Write 880×10^{-6} in decimal form.

Perform the indicated operations and write the result in engineering notation rounded to three significant digits:

34. $(39.4 \times 10^6)(120 \times 10^{-3})(45.0 \times 10^{12})$

35. $\dfrac{(3.03 \times 10^{12})^2}{\sqrt{615 \times 10^{-3}}}$

Cumulative Review

1. Evaluate: $16 \div 8 + 5 \times 2 - 3 + 7 \times 9$
2. Find the area of the figure in Illustration 1.

2 cm   3 cm

3 cm

7 cm

9 cm

ILLUSTRATION 1

3. Find the total resistance in the series circuit in Illustration 2.

R_1 R_2 R_3
520 Ω   55 Ω   60 Ω

R_5 R_4
3040 Ω   75 Ω

ILLUSTRATION 2

4. Is 2306 divisible by 6?
5. Find the prime factorization of 630.
6. Change $\dfrac{32}{9}$ to a mixed number in simplest form.
7. Find the area of a trapezoid with bases of 40 ft and 72 ft and height 80 ft.

Perform the indicated operations and simplify:

8. $\dfrac{5}{16} + \dfrac{1}{16} + \dfrac{7}{16}$ 9. $\dfrac{3}{8} + \dfrac{1}{4} + \dfrac{7}{16}$

10. $6\dfrac{1}{2} - 4\dfrac{5}{8}$ 11. $\dfrac{2}{5} \times \dfrac{1}{8}$

12. 5 lb 3 oz = _____ oz
13. Round 615.2875 to the nearest **a.** hundred **b.** tenth **c.** ten and **d.** thousandth.
14. Change $7\dfrac{2}{5}\%$ to a decimal.
15. Find 28.5% of $14,000.
16. 212 is 32% of what number?
17. To the nearest hundredth, 58 is what percent of 615?
18. A used car is listed to sell at $6800. Joy bought it for $6375. What percent markdown did she receive?
19. Find the value of $-8 + (+9) + (-3) - (-12)$.
20. Multiply: $(-8)(-9)(+3)(-1)(+2)$

Perform the indicated operations and simplify:

21. $-\dfrac{3}{8} + \left(-\dfrac{1}{4}\right) - 2\dfrac{5}{16}$ 22. $-\dfrac{5}{8} \times \dfrac{5}{8}$

23. Write 318,180 in scientific notation.
24. Find the larger number: 0.000618; 0.00213
25. Simplify. Express the result using positive exponents.

$$\dfrac{(10^{-4} \times 10^3)^{-2}}{10^6}$$

Write each number in engineering notation:

26. 4500   27. 0.00027

Write each number in decimal form:

28. 281×10^{-9} 29. 16.3×10^6

Perform the indicated operations and write each result in scientific notation rounded to three significant digits:

30. $(4.62 \times 10^4)(1.52 \times 10^6)$ 31. $\dfrac{5.61 \times 10^7}{1.18 \times 10^{10}}$

32. $\dfrac{(5.62 \times 10^{-3})(6.28 \times 10^6)}{(5.1 \times 10^6)(2 \times 10^{12})}$ 33. $\sqrt{4.28 \times 10^{-6}}$

3 The Metric System

Mathematics at Work

Many health care professionals in the allied health areas provide essential, critical support in a variety of areas. A sampling of health care professional support areas are registered nurses, licensed practical nurses, dental hygienists, dental assistants, occupational therapy assistants, physical therapy assistants, radiologic technologists, respiratory therapy assistants, surgical technologists and technicians, pharmacy technicians, and emergency medical technicians.

The health care professionals support our health care within prescribed duties as outlined by the various specific job descriptions in clinics and hospitals, and in public health, industrial, government, and private settings. Most health care degree and certificate programs are accredited by the corresponding national and/or state health accrediting agency. For more information, go to the website listed below.

Comstock Images/Jupiter Images

Allied Health Care Professionals
Lab technician performing an x-ray on a child.

■ Apply the basic concepts of the metric system, involving the SI prefixes and units of measure.

■ Use conversion factors to change from one unit to another within the metric system of weights and measures using length, mass and weight, volume and area, time, current, and other units.

■ Use the correct formula to change temperature measures from degrees Celsius to degrees Fahrenheit or vice versa.

■ Use conversion factors to change units within the U.S. system, from U.S. system units to metric system units, and from metric system units to U.S. system units.

3.1 Introduction to the Metric System

In early recorded history, parts of the human body were used for standards of measurement. However, these standards were neither uniform nor acceptable to all. So the next step was to define the various standards, such as the inch, the foot, and the rod. But each country introduced or defined its own standards, which were often not related to those in other countries. Then, in 1670, as the need for a single worldwide measurement system became recognized, Gabriel Mouton, a Frenchman, proposed a uniform decimal measurement system. By the 1800s, metric standards were first adopted worldwide. The U.S. Congress legalized the use of the metric system throughout the United States over 140 years ago on July 28, 1866. In 1960, the International System of Units was adopted for the modern metric system. The abbreviation for the International System of Units is **SI** (from the French *Système International d'Unités*) and is commonly called the **metric system**.

Throughout U.S. history, several attempts have been made to convert the nation to the metric system. By the 1970s, the United States was the only nonmetric industrialized nation left in the world, but the U.S. government did little to implement the system. Industry and business, however, found their foreign markets significantly limited because metric products were preferred. Now many segments of U.S. business and industry have independently gone metric because world trade requires it. Metric countries just naturally want metric products. And the inherent simplicity of the metric system of measurement and standardization of weights and measures have led to major cost savings in industries that have converted to it. Most major U.S. industries, such as the automotive, aviation, and farm implement industries, as well as the Department of Defense and other federal agencies, have effectively converted to the metric system. In some industries, you—the student and worker—will need to know and use both systems.

The SI metric system has seven base units, as shown in Table 3.1. Other commonly used metric units are shown in Table 3.2.

Table 3.1 Seven Base Metric Units

Basic unit	SI abbreviation	For measuring
metre*	m	length
kilogram	kg	mass
second	s	time
ampere	A	electric current
kelvin	K	temperature
candela	cd	light intensity
mole	mol	molecular substance

At present, there is some difference of opinion on the spelling of metre and litre. We have chosen the "re" spelling because it is the internationally accepted spelling and because it distinguishes the unit from other meters, such as parking meters and electricity meters.

Table 3.2 Other Commonly Used Metric Units

Unit	SI abbreviations	For measuring
litre*	L	volume
cubic metre	m^3	volume
square metre	m^2	area
newton	N	force
metre per second	m/s	speed
joule	J	energy
watt	W	power
radian	rad	plane angle

See Table 3.1 footnote.

The metric system, a decimal or base 10 system, is very similar to our decimal number system. It is an easy system to use, because calculations are based on the number 10 and its multiples. The SI system has special prefixes that name multiples and submultiples; these can be used with almost all SI units. Table 3.3 shows the prefixes and the corresponding symbols.

Table 3.3 Prefixes for SI Units

Multiple or submultiple* decimal form	Power of 10	Prefix	Prefix symbol	Pronunciation	Meaning
1,000,000,000,000	10^{12}	tera	T	těr′ă	one trillion times
1,000,000,000	10^{9}	giga	G	jĭg′ă	one billion times
1,000,000	10^{6}	mega	M	měg′ă	one million times
1,000	10^{3}	kilo**	k	kĭl′ō or kēl′ō	one thousand times
100	10^{2}	hecto	h	hěk′tō	one hundred times
10	10^{1}	deka	da	děk′ă	ten times
0.1	10^{-1}	deci	d	děs′ĭ	one tenth of
0.01	10^{-2}	centi**	c	sěnt′ĭ	one hundredth of
0.001	10^{-3}	milli**	m	mĭl′ĭ	one thousandth of
0.000001	10^{-6}	micro	μ	mī′krō	one millionth of
0.000000001	10^{-9}	nano	n	năn′ō	one billionth of
0.000000000001	10^{-12}	pico	p	pē′kō	one trillionth of

*Factor by which the unit is multiplied.
**Most commonly used prefixes.

Because the same prefixes are used with most all SI metric units, it is not necessary to memorize long lists or many tables.

Example 1

Write the SI abbreviation for 45 kilometres.

The symbol for the prefix *kilo* is k.
The symbol for the unit *metre* is m.
The SI abbreviation for 45 kilometres is 45 km. ■

Example 2

Write the SI unit for the abbreviation 50 mg.

The prefix for m is *milli*.
The unit for g is *gram*.
The SI unit for 50 mg is 50 milligrams. ■

In summary, the U.S. or English system is an ancient one based on standards initially determined by parts of the human body, which is why there is no consistent relationship between units. In the metric system, however, standard units are subdivided into multiples of 10, similar to our number system, and the names associated with each subdivision have prefixes that indicate a multiple of 10.

Exercises 3.1

Give the metric prefix for each value:

1. 1000
2. 100
3. 0.01
4. 0.1
5. 0.001
6. 10
7. 1,000,000
8. 0.000001

Give the SI symbol for each prefix:

9. hecto
10. kilo
11. deci
12. milli
13. centi
14. deka
15. micro
16. mega

Write the abbreviation for each quantity:

17. 65 milligrams   **18.** 125 kilolitres

19. 82 centimetres   **20.** 205 millilitres

21. 36 microamps   **22.** 75 kilograms

23. 19 hectolitres   **24.** 5 megawatts

Write the SI unit for each abbreviation:

25. 18 m   **26.** 15 L   **27.** 36 kg

28. 85 mm   **29.** 24 ps   **30.** 9 dam

31. 135 mL   **32.** 45 dL   **33.** 45 mA

34. 75 MW

35. The basic SI unit of length is _____.

36. The basic SI unit of mass is _____.

37. The basic SI unit of electric current is _____.

38. The basic SI unit of time is _____.

39. The common SI units of volume are _____ and _____.

40. The common SI unit of power is _____.

3.2   Length

The basic SI unit of length is the **metre (m).** The height of a door knob is about 1 m. (See Figure 3.1.) One metre is a little more than 1 yd. (See Figure 3.2.)

FIGURE 3.1
The height of a doorknob is about 1 m.

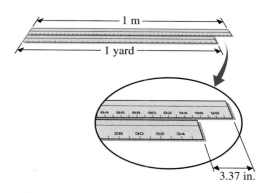

FIGURE 3.2
One metre is a little more than 1 yd.

Long distances are measured in kilometres (km) (1 km = 1000 m). The length of five city blocks is about 1 km. (See Figure 3.3.)

The centimetre (cm) is used to measure short distances, such as the width of this page (about 21 cm), or the width of a board. The width of your small fingernail is about 1 cm. (See Figure 3.4.)

FIGURE 3.3
The length of five city blocks is about 1 km.

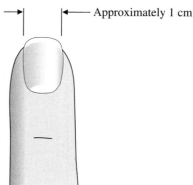

FIGURE 3.4
The width of your small fingernail is about 1 cm.

FIGURE 3.5
The thickness of a dime is about 1 mm.

The millimetre (mm) is used to measure very small lengths, such as the thickness of a sheet of metal or the depth of a tire tread. The thickness of a dime is about 1 mm. (See Figure 3.5.)

Each of the large numbered divisions on the metric ruler in Figure 3.6 marks one centimetre (cm). The smaller lines indicate halves of centimetres, and the smallest divisions show tenths of centimetres, or millimetres.

FIGURE 3.6
Metric ruler

Example 1   Read *A*, *B*, *C*, and *D* on the metric ruler in Figure 3.7. Give each result in millimetres, centimetres, and metres.

FIGURE 3.7

Answers: A = 12 mm, 1.2 cm, 0.012 m
B = 20 mm, 2.0 cm, 0.020 m
C = 25 mm, 2.5 cm, 0.025 m
D = 128 mm, 12.8 cm, 0.128 m   ■

To convert from one metric unit to another, we could use the same conversion factor procedure that we used in the U.S. or English system in Section 1.9.

◆ Choosing Conversion Factors

The correct choice for a given conversion factor is the one in which the old units are in the numerator of the original expression and in the denominator of the conversion factor, or the old units are in the denominator of the original expression and in the numerator of the conversion factor. That is, set up the conversion factor so that the old units cancel each other.

Example 2

Change 3.6 km to metres.

Since *kilo* means 10^3 or 1000, 1 km = 1000 m. The two possible conversion factors are $\frac{1\ km}{1000\ m}$ and $\frac{1000\ m}{1\ km}$. Choose the one whose numerator is expressed in the new units (m) and whose denominator is expressed in the old units (km). This is $\frac{1000\ m}{1\ km}$.

$$3.6\ \cancel{km} \times \frac{1000\ m}{1\ \cancel{km}} = 3600\ m$$

↑—— conversion factor

 Example 3

Change 4 m to centimetres.

First, *centi* means 10^{-2} or 0.01; and 1 cm = 10^{-2} m. Choose the conversion factor with metres in the denominator and centimetres in the numerator.

$$4\ \cancel{m} \times \frac{1\ cm}{10^{-2}\ \cancel{m}} = 400\ cm$$

↑—— conversion factor

Note: Conversions within the metric system only involve moving the decimal point.

Exercises 3.2

Which is longer?

1. 1 metre or 1 millimetre
2. 1 metre or 1 centimetre
3. 1 metre or 1 kilometre
4. 1 millimetre or 1 kilometre
5. 1 centimetre or 1 millimetre
6. 1 kilometre or 1 centimetre

Fill in each blank:

7. 1 m = _____ mm
8. 1 km = _____ m
9. 1 cm = _____ m
10. 1 m = _____ cm
11. 1 m = _____ km
12. 1 hm = _____ m
13. 1 mm = _____ m
14. 1 m = _____ hm
15. 1 cm = _____ mm
16. 1 mm = _____ cm
17. 1 dam = _____ dm
18. 1 dm = _____ m

Which metric unit (km, m, cm, or mm) should you use to measure each item?

19. Diameter of an automobile tire
20. Thickness of sheet metal
21. Metric wrench sizes
22. Length of an auto race
23. Length of a discus throw in track and field
24. Length and width of a table top

25. Distance between Chicago and St. Louis
26. Thickness of plywood
27. Thread size of a pipe
28. Length and width of a house lot

Fill in each blank with the most reasonable unit (km, m, cm, or mm):

29. A common metric wrench set varies in size from 6–19 _____.
30. The diameter of a wheel on a ten-speed bicycle is 56 _____.
31. A jet plane generally flies about 8–9 _____ high.
32. The width of a door in our house is 91 _____.
33. The length of the ridge on our roof is 24 _____.
34. Antonio's waist size is 95 _____.
35. The steering wheel on Brenda's car is 36 _____ in diameter.
36. Jan drives 12 _____ to school.
37. The standard metric size for plywood is 1200 _____ wide and 2400 _____ long.
38. The distance from home plate to the centerfield wall in a baseball park is 125 _____.
39. Read the measurements indicated by the letters on the metric ruler in Illustration 1 and give each result in millimetres and centimetres.

ILLUSTRATION 1

Use a metric ruler to measure each line segment. Give each result in millimetres and centimetres:

40. ├───────────┤

41. ├─────────────┤

42. ├──────────────────┤

43. ├───────────────┤

44. ├───────────┤

45. ├─────────────┤

46. ├──────────┤

47. ├────────────────────┤

48. ├─────┤

49. Change 675 m to km.

50. Change 450 cm to m.

51. Change 1540 mm to m.

52. Change 3.2 km to m.

53. Change 65 cm to m.

54. Change 1.4 m to mm.

55. Change 7.3 m to cm.

56. Change 0.25 km to m.

57. Change 1250 m to km.

58. Change 4.5 m to cm.

59. Change 275 mm to cm.

60. Change 48 cm to mm.

61. Change 125 mm to cm.

62. Change 0.75 m to μm.

63. What is your height in metres and centimetres?

3.3 Mass and Weight

The **mass** of an object is the quantity of material making up the object. One unit of mass in the SI system is the *gram* (g). The gram is defined as the mass contained in 1 cubic centimetre (cm^3) of water, at its maximum density. A common paper clip has a mass of about 1 g. Three aspirin have a mass of about 1 g. (See Figure 3.8.)

(a) A common paperclip has a mass of about 1 g.   (b) Three aspirin have a mass of about 1 g.

FIGURE 3.8

Because the gram is so small, the **kilogram (kg)** is the basic unit of mass in the SI system. One kilogram is defined as the mass contained in 1 cubic decimetre (dm^3) of water at its maximum density.

For very large quantities, such as a trainload of coal or grain or a shipload of ore, the metric ton (1000 kg) is used. The milligram (mg) is used to measure very, very small masses such as medicine dosages. One grain of salt has a mass of about 1 mg.

The **weight** of an object is a measure of the earth's gravitational force—or pull—acting on the object. The SI unit of weight is the *newton* (N). As you are no doubt aware, the terms *mass* and *weight* are commonly used interchangeably by the general public. We have presented them here as technical terms, as they are used in the technical, engineering, and scientific professions. To further illustrate the difference, the mass of an astronaut remains relatively constant while his or her weight varies (the weight decreases as the distance from the earth increases). If the spaceship is in orbit or farther out in space, we say the crew is "weightless," because they seem to float freely in space. Their mass has not changed, although their weight is near zero. (See Figure 3.9.)

FIGURE 3.9
"Weightless" astronaut.

 Example 1

Change 12 kg to grams.

First, *kilo* means 10^3 or 1000; and 1 kg = 1000 g. Choose the conversion factor with kilograms in the denominator and grams in the numerator.

$$12 \text{ kg} \times \frac{1000 \text{ g}}{1 \text{ kg}} = 12{,}000 \text{ g}$$

$\uparrow$ —— conversion factor ■

 Example 2

Change 250 mg to grams.

First, *milli* means 10^{-3} or 0.001; and 1 mg = 10^{-3} g. Choose the conversion factor with milligrams in the denominator and grams in the numerator.

$$250 \text{ mg} \times \frac{10^{-3} \text{ g}}{1 \text{ mg}} = 0.25 \text{ g}$$

$\uparrow$ —— conversion factor ■

Exercises 3.3

Which is larger?

1. 1 gram or 1 milligram
2. 1 gram or 1 kilogram
3. 1 milligram or 1 kilogram
4. 1 metric ton or 1 kilogram
5. 1 milligram or 1 microgram
6. 1 kilogram or 1 microgram

Fill in each blank:

7. 1 g = _____ mg
8. 1 kg = _____ g
9. 1 cg = _____ g
10. 1 mg = _____ g
11. 1 metric ton = _____ kg
12. 1 g = _____ cg
13. 1 mg = _____ μg
14. 1 μg = _____ mg

Which metric unit (kg, g, mg, or metric ton) should you use to measure the mass of each item?

15. A bar of handsoap   **16.** A vitamin capsule

17. A bag of flour   **18.** A four-wheel-drive tractor

19. A pencil   **20.** Your mass

21. A trainload of coal   **22.** A bag of potatoes

23. A contact lens   **24.** An apple

Fill in each blank with the most reasonable unit (kg, g, mg, or metric ton):

25. A slice of bread has a mass of about 25 _____.

26. Elevators in the college have a load limit of 2200 _____.

27. I take 1000 _____ of vitamin C every day.

28. My uncle's new truck can haul a load of 4 _____.

29. Postage rates for letters are based on the number of _____.

30. I take 1 _____ of vitamin C every day.

31. My best friend has a mass of 65 _____.

32. A jar of peanut butter contains 1200 _____.

33. The local grain elevator shipped 20,000 _____ of wheat last year.

34. One common size of aspirin tablets is 325 _____.

35. Change 875 g to kg.   **36.** Change 127 mg to g.

37. Change 85 g to mg.   **38.** Change 1.5 kg to g.

39. Change 3.6 kg to g.   **40.** Change 430 g to mg.

41. Change 270 mg to g.   **42.** Change 1350 g to kg.

43. Change 885 μg to mg.   **44.** Change 18 mg to μg.

45. Change 375 μg to mg.   **46.** Change 6.4 mg to μg.

47. Change 2.5 metric tons to kg.

48. Change 18,000 kg to metric tons.

49. Change 225,000 kg to metric tons.

50. Change 45 metric tons to kg.

51. What is your mass in kilograms?

3.4 Volume and Area

Volume

A common unit of volume in the metric system is the **litre (L).** One litre of milk is a little more than 1 quart. The litre is commonly used for liquid volume. (See Figure 3.10.)

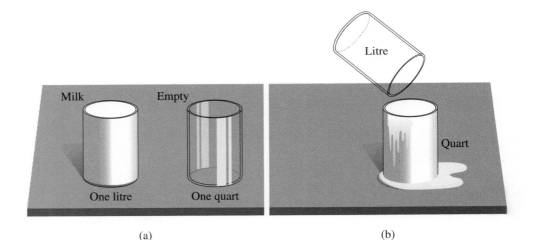

(a)   (b)

FIGURE 3.10
One litre is a little more than 1 qt.

The *cubic metre* (m^3) is used to measure large volumes. One cubic metre is the volume contained in a cube 1 m on an edge. The *cubic centimetre* (cm^3) is used to measure small volumes. It is the volume contained in a cube 1 cm on an edge.

Note: It is important to understand the relationship between the litre and the cubic centimetre. The litre is defined as the volume in 1 cubic decimetre (dm^3). That is, 1 L of liquid fills a cube 1 dm (10 cm) on an edge. (See Figure 3.11.)

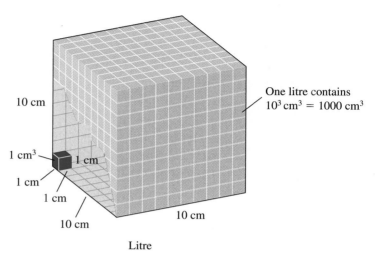

10 cm

One litre contains
$10^3 \, cm^3 = 1000 \, cm^3$

1 cm³

1 cm

1 cm

1 cm

10 cm

10 cm

Litre

FIGURE 3.11
One litre contains 1000 cm³.

The volume of the cube in Figure 3.11 can also be found by the formula

$V = lwh$

$V = (10 \text{ cm})(10 \text{ cm})(10 \text{ cm})$

$\quad = 1000 \text{ cm}^3$ Note: (cm)(cm)(cm) = cm³

Thus, 1 L = 1000 cm³. Dividing each side by 1000, we have

$$\frac{1}{1000} L = 1 \text{ cm}^3$$

or   **1 mL = 1 cm³** $1 \text{ mL} = \frac{1}{1000} L$

Milk, soft drinks, and gasoline are sold by the litre. Liquid medicine and eye drops are sold by the millilitre. Large quantities of liquid are sold by the kilolitre (1000 L).

In Section 3.3, the kilogram was defined as the mass of 1 dm³ of water. Since 1 dm³ = 1 L, 1 litre of water has a mass of 1 kg.

Example 1

Change 0.5 L to millilitres.

$$0.5 \, \cancel{L} \times \frac{1000 \text{ mL}}{1 \, \cancel{L}} = 500 \text{ mL}$$

└── conversion factor ■

Example 2   Change 4.5 cm³ to mm³.

$$4.5 \text{ cm}^3 \times \left(\frac{10 \text{ mm}}{1 \text{ cm}}\right)^3 = 4500 \text{ mm}^3$$

⎣— conversion factor

Use the length conversion factor 1 cm = 10 mm and first form the conversion factor with cm in the denominator and mm in the numerator. Then raise the conversion factor to the third power to obtain cubic units in both numerator and denominator. Since the numerator equals the denominator, both the length conversion factor $\frac{10 \text{ mm}}{1 \text{ cm}}$ and its third power $\left(\frac{10 \text{ mm}}{1 \text{ cm}}\right)^3$ equal 1.

Alternative Method:

$$4.5 \text{ cm}^3 \times \frac{1000 \text{ mm}^3}{1 \text{ cm}^3} = 4500 \text{ mm}^3$$

⎣— conversion factor

The alternative method conversion factor 1 cm³ = 1000 mm³ is taken directly from the metric volume conversion table on the laminated reference card. The first method is preferred, because only the length conversion needs to be remembered or found in a table.   ■

Area

A common unit of area in the metric system is the **square metre (m²),** the area contained in a square whose sides are each 1 m long. The *square centimetre* (cm²) and the *square millimetre* (mm²) are smaller units of area. (See Figure 3.12.) The larger area units are the square kilometre (km²) and the hectare (ha).

FIGURE 3.12
Relative sizes of 1 cm² and 1 mm².

Example 3   Change 2400 cm² to m².

$$2400 \text{ cm}^2 \times \left(\frac{1 \text{ m}}{100 \text{ cm}}\right)^2 = 0.24 \text{ m}^2$$

⎣— conversion factor

Use the length conversion factor 1 m = 100 cm and first form the conversion factor with cm in the denominator and m in the numerator. Then raise the conversion factor to the second power to obtain square units in both numerator and denominator. Since the numerator equals the denominator, both the length conversion factor $\frac{1 \text{ m}}{100 \text{ cm}}$ and its second power $\left(\frac{1 \text{ m}}{100 \text{ cm}}\right)^2$ equal 1.

Alternative Method:

$$2400 \text{ cm}^2 \times \frac{1 \text{ m}^2}{10,000 \text{ cm}^2} = 0.24 \text{ m}^2$$

⎣— conversion factor

The alternative method conversion factor 1 m² = 10,000 cm² is taken directly from the metric area conversion table. The first method is again preferred, because only the length conversion needs to be remembered or found in a table.   ■

Example 4

Change 1.2 km² to m².

$$1.2 \text{ km}^2 \times \left(\frac{1000 \text{ m}}{1 \text{ km}}\right)^2 = 1,200,000 \text{ m}^2$$

Alternative Method:

$$1.2 \text{ km}^2 \times \frac{10^6 \text{ m}^2}{1 \text{ km}^2} = 1,200,000 \text{ m}^2 \qquad ■$$

1 hectare (ha) =
10,000 m² =
1 hm²

100 m

100 m

FIGURE 3.13
Hectare

The **hectare (ha)** is the basic metric unit of land area. The area of 1 hectare equals the area of a square 100 m on a side, whose area is 10,000 m² or 1 square hectometre (hm²). (See Figure 3.13.)

The hectare is used because it is more convenient to say and use than "square hectometre." The metric prefixes are *not* used with the hectare unit. Instead of saying the prefix "kilo" with "hectare," we say "1000 hectares."

Example 5

How many hectares are contained in a rectangular field 360 m by 850 m?

The area in m² is

$$(360 \text{ m})(850 \text{ m}) = 306,000 \text{ m}^2$$

$$306,000 \text{ m}^2 \times \frac{1 \text{ ha}}{10,000 \text{ m}^2} = 30.6 \text{ ha}$$

360 m

850 m

■

Exercises 3.4

Which is larger?

1. 1 litre or 1 millilitre
2. 1 millilitre or 1 kilolitre
3. 1 cubic millimetre or 1 cubic centimetre
4. 1 cubic metre or 1 litre
5. 1 square kilometre or 1 hectare
6. 1 square centimetre or 1 square millimetre

Fill in each blank:

7. 1 L = _____ mL
8. 1 mL = _____ L
9. 1 m³ = _____ cm³
10. 1 mm³ = _____ cm³
11. 1 cm² = _____ mm²
12. 1 km² = _____ ha
13. 1 m³ = _____ L
14. 1 cm³ = _____ mL

Which metric unit (m³, L, mL, m², cm², or ha) should you use to measure the following?

15. Oil in your car's crankcase
16. Cough syrup
17. Floor space in a warehouse
18. Size of a farm
19. Cross-sectional area of a piston
20. Piston displacement in an engine
21. Cargo space in a truck
22. Paint needed to paint a house
23. Eye drops
24. Page size of this book
25. Size of an industrial park
26. Gasoline in your car's gas tank

Fill in each blank with the most reasonable unit (m³, L, mL, m², cm², or ha):

27. Lateesha ordered 12 _____ of concrete for her new driveway.
28. I drink 250 _____ of orange juice each morning.
29. Juan, a farmer, owns a 2500- _____ storage tank for diesel fuel.
30. Dwight planted 75 _____ of wheat this year.
31. Our house has 195 _____ of floor space.
32. We must heat 520 _____ of living space in our house.
33. When I was a kid, I mowed 6 _____ of lawns each week.
34. Our community's water tower holds 650 _____ of water.
35. The cross section of a log is 2500 _____.

36. Darnell bought a 25-_____ tarpaulin for his truck.

37. I need copper tubing with a cross section of 4 _____.

38. We should each drink 2 _____ of water each day.

39. Change 1500 mL to L.

40. Change 0.60 L to mL.

41. Change 1.5 m³ to cm³.

42. Change 450 mm³ to cm³.

43. Change 85 cm³ to mL.

44. Change 650 L to m³.

45. Change 85,000 m² to km².

46. Change 18 m² to cm².

47. Change 85,000 m² to ha.

48. Change 250 ha to km².

49. What is the mass of 500 mL of water?

50. What is the mass of 1 m³ of water?

51. How many hectares are contained in a rectangular field that measures 75 m by 90 m?

52. How many hectares are contained in a rectangular field that measures $\frac{1}{4}$ km by $\frac{1}{2}$ km?

3.5 Time, Current, and Other Units

The basic SI unit of time is the **second (s),** which is the same in all units of measurement. Time is also measured in minutes (min), hours (h), days, and years.

$$1 \text{ min} = 60 \text{ s}$$
$$1 \text{ h} = 60 \text{ min}$$
$$1 \text{ day} = 24 \text{ h}$$
$$1 \text{ year} = 365\tfrac{1}{4} \text{ days (approximately)}$$

 Example 1

Change 4 h 15 min to seconds.

First, $4 \text{ h} \times \dfrac{60 \text{ min}}{1 \text{ h}} = 240 \text{ min}$

And $4 \text{ h } 15 \text{ min} = 240 \text{ min} + 15 \text{ min}$
$$= 255 \text{ min}$$

Then, $255 \text{ min} \times \dfrac{60 \text{ s}}{1 \text{ min}} = 15{,}300 \text{ s}$ ■

Very short periods of time are commonly used in electronics. These are measured in parts of a second, given with the appropriate metric prefix.

 Example 2

What is the meaning of each unit?   **a.** 1 ms   **b.** 1 μs   **c.** 1 ns   **d.** 1 ps

a. 1 ms = 1 millisecond = 10^{-3} s   It means one-thousandth of a second.

b. 1 μs = 1 microsecond = 10^{-6} s   It means one-millionth of a second.

c. 1 ns = 1 nanosecond = 10^{-9} s   It means one-billionth of a second.

d. 1 ps = 1 picosecond = 10^{-12} s   It means one-trillionth of a second.   ■

 Example 3

Change 25 ms to seconds.

First, *milli* means 10^{-3}, and 1 ms = 10^{-3} s. Then

$$25 \text{ ms} \times \dfrac{10^{-3} \text{ s}}{1 \text{ ms}} = 0.025 \text{ s}$$
$$\uparrow$$
conversion factor   ■

Example 4   Change 0.00000025 s to nanoseconds.

First, *nano* means 10^{-9}, and 1 ns = 10^{-9} s. Then

$$0.00000025 \cancel{s} \times \frac{1 \text{ ns}}{10^{-9} \cancel{s}} = 250 \text{ ns}$$

↑—— conversion factor ■

The basic SI unit of electric current is the **ampere (A),** sometimes called the amp. This same unit is used in the U.S. system. The ampere is a fairly large amount of current, so smaller currents are measured in parts of an ampere and are given the appropriate SI prefix.

 Example 5   What is the meaning of each unit?   **a.** 1 mA   **b.** 1 μA

a. 1 mA = 1 milliampere = 10^{-3} A   It means one-thousandth of an ampere.
b. 1 μA = 1 microampere = 10^{-6} A   It means one-millionth of an ampere. ■

 Example 6   Change 275 μA to amperes.

First, *micro* means 10^{-6}, and 1 μA = 10^{-6} A. Then

$$275 \cancel{\mu A} \times \frac{10^{-6} \text{ A}}{1 \cancel{\mu A}} = 0.000275 \text{ A}$$ ■

 Example 7   Change 0.045 A to milliamps.

First, *milli* means 10^{-3}, and 1 mA = 10^{-3} A. Then

$$0.045 \cancel{A} \times \frac{1 \text{ mA}}{10^{-3} \cancel{A}} = 45 \text{ mA}$$ ■

The common metric unit for both electrical and mechanical power is the *watt* (W).

 Example 8   What is the meaning of each unit?   **a.** 1 mW   **b.** 1 kW   **c.** 1 MW

a. 1 mW = 1 milliwatt = 10^{-3} W   It means one-thousandth of a watt.
b. 1 kW = 1 kilowatt = 10^{3} W   It means one thousand watts.
c. 1 MW = 1 megawatt = 10^{6} W   It means one million watts. ■

 Example 9   Change 0.025 W to milliwatts.

First, *milli* means 10^{-3}, and 1 mW = 10^{-3} W. Then

$$0.025 \cancel{W} \times \frac{1 \text{ mW}}{10^{-3} \cancel{W}} = 25 \text{ mW}$$ ■

 Example 10   Change 2.3 MW to watts.

First, *mega* means 10^{6}, and 1 MW = 10^{6} W. Then

$$2.3 \cancel{MW} \times \frac{10^{6} \text{ W}}{1 \cancel{MW}} = 2.3 \times 10^{6} \text{ W or } 2,300,000 \text{ W}$$ ■

A few other units that are commonly used in electronics are listed in Table 3.4. The metric prefixes are used with each of these units in the same way as with the other metric units we have studied.

Table 3.4 Units Commonly Used in Electronics		
Unit	Symbol	Used to measure
volt	V	voltage
ohm	Ω	resistance
hertz	Hz	frequency
farad	F	capacitance
henry	H	inductance
coulomb	C	charge

Exercises 3.5

Which is larger?

1. 1 amp or 1 milliamp
2. 1 microsecond or 1 picosecond
3. 1 second or 1 nanosecond
4. 1 megawatt or 1 milliwatt
5. 1 kilovolt or 1 megavolt
6. 1 volt or 1 millivolt

Write the abbreviation for each unit:

7. 43 kilowatts
8. 7 millivolts
9. 17 picoseconds
10. 1.2 amperes
11. 3.2 megawatts
12. 55 microfarads
13. 450 ohms
14. 70 nanoseconds

Fill in each blank:

15. 1 kW = _____ W
16. 1 mA = _____ A
17. 1 ns = _____ s
18. 1 day = _____ s
19. 1 A = _____ μA
20. 1 F = _____ μF
21. 1 V = _____ MV
22. 1 Hz = _____ kHz
23. Change 0.35 A to mA.
24. Change 18 kW to W.
25. Change 350 ms to s.
26. Change 1 h 25 min 16 s to s.
27. Change 13,950 s to h, min, and s.
28. Change 15 MV to kV.
29. Change 175 μF to mF.
30. Change 145 ps to ns.
31. Change 1500 kHz to MHz.
32. Change 5×10^{12} W to MW.

3.6 Temperature

The basic SI unit for temperature is **kelvin (K),** which is used mostly in scientific and engineering work. Everyday temperatures are measured in *degrees Celsius* (°C). The United States also measures temperatures in degrees Fahrenheit (°F).

On the Celsius scale, water freezes at 0° and boils at 100°. Each degree Celsius is 1/100 of the difference between the boiling temperature and the freezing temperature of water. Figure 3.14 shows some approximate temperature readings in degrees Celsius and Fahrenheit and compares them with a related activity.

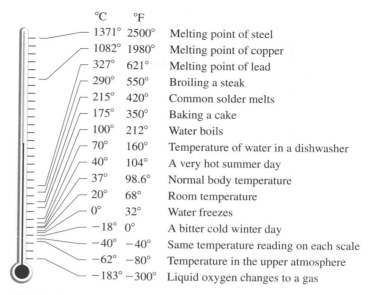

°C	°F	
1371°	2500°	Melting point of steel
1082°	1980°	Melting point of copper
327°	621°	Melting point of lead
290°	550°	Broiling a steak
215°	420°	Common solder melts
175°	350°	Baking a cake
100°	212°	Water boils
70°	160°	Temperature of water in a dishwasher
40°	104°	A very hot summer day
37°	98.6°	Normal body temperature
20°	68°	Room temperature
0°	32°	Water freezes
−18°	0°	A bitter cold winter day
−40°	−40°	Same temperature reading on each scale
−62°	−80°	Temperature in the upper atmosphere
−183°	−300°	Liquid oxygen changes to a gas

FIGURE 3.14
Related temperature readings in degrees Celsius and degrees Fahrenheit

The formulas for changing between degrees Celsius and degrees Fahrenheit are:

$$C = \frac{5}{9}(F - 32°)$$

$$F = \frac{9}{5}C + 32°$$

Important

◆ Example 1   Change 68°F to degrees Celsius.

$$C = \frac{5}{9}(F - 32°)$$

$$C = \frac{5}{9}(68° - 32°)$$

$$= \frac{5}{9}(36°) \qquad \text{First subtract within parentheses.}$$

$$= 20° \qquad \text{Multiply.}$$

Thus, 68°F = 20°C.   ■

◆ Example 2   Change 35°C to degrees Fahrenheit.

$$F = \frac{9}{5}C + 32°$$

$$F = \frac{9}{5}(35°) + 32°$$

$$= 63° + 32° \qquad \text{First multiply.}$$

$$= 95° \qquad \text{Add.}$$

That is, 35°C = 95°F.   ■

Example 3   Change 10°F to degrees Celsius.

$$C = \frac{5}{9}(F - 32°)$$

$$C = \frac{5}{9}(10° - 32°)$$

$$= \frac{5}{9}(-22°)$$

$$= -12.2°$$

So 10°F = −12.2°C.   ■

 Example 4   Change −60°C to degrees Fahrenheit.

$$F = \frac{9}{5}C + 32°$$

$$F = \frac{9}{5}(-60°) + 32°$$

$$= -108° + 32°$$

$$= -76°$$

So −60°C = −76°F.   ■

Exercises 3.6

Use Figure 3.14 to choose the most reasonable answer for each statement:

1. The boiling temperature of water is **a.** 212°C, **b.** 100°C, **c.** 0°C, or **d.** 50°C.

2. The freezing temperature of water is **a.** 32°C, **b.** 100°C, **c.** 0°C, or **d.** −32°C.

3. Normal body temperature is **a.** 100°C, **b.** 50°C, **c.** 37°C, or **d.** 98.6°C.

4. The body temperature of a person who has a fever is **a.** 102°C, **b.** 52°C, **c.** 39°C, or **d.** 37°C.

5. The temperature on a hot summer day in the California desert is **a.** 108°C, **b.** 43°C, **c.** 60°C, or **d.** 120°C.

6. The temperature on a cold winter day in Chicago is **a.** 20°C, **b.** 10°C, **c.** 30°C, or **d.** −10°C.

7. The thermostat in your home should be set at **a.** 70°C, **b.** 50°C, **c.** 19°C, or **d.** 30°C.

8. Solder melts at **a.** 215°C, **b.** 420°C, **c.** 175°C, or **d.** 350°C.

9. Freezing rain is most likely to occur at **a.** 32°C, **b.** 25°C, **c.** −18°C, or **d.** 0°C.

10. The weather forecast calls for a low temperature of 3°C. What should you plan to do? **a.** Sleep with the windows open. **b.** Protect your plants from frost. **c.** Sleep with the air conditioner on. **d.** Sleep with an extra blanket.

Fill in each blank:

11. 77°F = _____ °C

12. 45°C = _____ °F

13. 325°C = _____ °F

14. 140°F = _____ °C

15. −16°C = _____ °F

16. 5°F = _____ °C

17. −16°F = _____ °C

18. −40°C = _____ °F

19. −78°C = _____ °F

20. −10°F = _____ °C

 3.7 Metric and U.S. Conversion

In technical work, you must sometimes change from one system of measurement to another. The approximate conversions between metric units and U.S. units are found in the Metric and U.S. Conversion Table on the laminated reference card. Most numbers are rounded to three or four significant digits. Due to this rounding and your choice of conversion factors, there may be a small difference in the last digit(s) of the answers involving conversion factors. This small difference is acceptable. In this section, round each result to three significant digits, when necessary. You may review significant digits in Section 1.11.

Figure 3.15 shows the relative sizes of each of four sets of common metric and U.S. units of area.

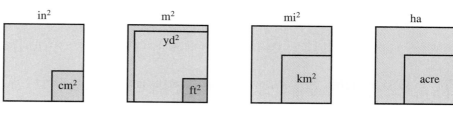

FIGURE 3.15
Relative sizes of some common metric and U.S. units of area

Example 1   Change 17 in. to centimetres.

$$17 \text{ in.} \times \frac{2.54 \text{ cm}}{1 \text{ in.}} = 43.2 \text{ cm}$$
↑ —— conversion factor

Example 2   Change 1950 g to pounds.

$$1950 \text{ g} \times \frac{1 \text{ lb}}{454 \text{ g}} = 4.30 \text{ lb}$$
↑ —— conversion factor

Note: If you choose a different conversion factor, the result may vary slightly due to rounding. For example,

$$1950 \text{ g} \times \frac{0.00220 \text{ lb}}{1 \text{ g}} = 4.29 \text{ lb}$$
↑ —— conversion factor

Example 3   Change 0.85 qt to millilitres.

$$0.85 \text{ qt} \times \frac{0.946 \text{ L}}{1 \text{ qt}} \times \frac{10^3 \text{ mL}}{1 \text{ L}} = 804 \text{ mL}$$
↑   ↑ —— conversion factors

Example 4   Change 5 yd² to ft².

$$5 \text{ yd}^2 \times \left(\frac{3 \text{ ft}}{1 \text{ yd}} \right)^2 = 45 \text{ ft}^2$$

Use the length conversion factor 1 yd = 3 ft and first form the conversion factor with yd in the denominator and ft in the numerator. Then raise the conversion factor to the second power to obtain square units in both numerator and denominator. Since the numerator equals the denominator, both the length conversion factor $\frac{3\,\text{ft}}{1\,\text{yd}}$ and its second power $\left(\frac{3\,\text{ft}}{1\,\text{yd}}\right)^2$ equal 1.

Alternative Method:

$$5\ \cancel{\text{yd}^2} \times \frac{9\ \text{ft}^2}{1\ \cancel{\text{yd}^2}} = 45\ \text{ft}^2$$

The alternative method conversion factor 1 yd² = 9 ft² is taken directly from the U.S. area conversion table on your reference card. The first method is again preferred, because only the length conversion needs to be remembered or found in a table. ■

Example 5   How many square inches are in a metal plate 14 cm² in area?

$$14\ \text{cm}^2 \times \left(\frac{1\ \text{in.}}{2.54\ \text{cm}}\right)^2 = 2.17\ \text{in}^2$$

Alternative Method:

$$14\ \cancel{\text{cm}^2} \times \frac{0.155\ \text{in}^2}{1\ \cancel{\text{cm}^2}} = 2.17\ \text{in}^2 \qquad\qquad ■$$

Example 6   Change 147 ft³ to cubic yards.

$$147\ \text{ft}^3 \times \left(\frac{1\ \text{yd}}{3\ \text{ft}}\right)^3 = 5.44\ \text{yd}^3$$

Alternative Method:

$$147\ \cancel{\text{ft}^3} \times \frac{1\ \text{yd}^3}{27\ \cancel{\text{ft}^3}} = 5.44\ \text{yd}^3 \qquad\qquad ■$$

Example 7   How many cubic yards are in 12 m³?

$$12\ \text{m}^3 \times \left(\frac{1.09\ \text{yd}}{1\ \text{m}}\right)^3 = 15.5\ \text{yd}^3$$

Alternative Method:

$$12\ \cancel{\text{m}^3} \times \frac{1.31\ \text{yd}^3}{1\ \cancel{\text{m}^3}} = 15.7\ \text{yd}^3 \qquad\qquad ■$$

In the U.S. system, the *acre* is the basic unit of land area. Historically, the acre was the amount of ground that a yoke of oxen could plow in one day.

1 acre = 43,560 ft²

1 mi² = 640 acres = 1 section

Example 8   How many acres are in a rectangular field that measures 1350 ft by 2750 ft?

The area in ft² is

$$(1350\ \text{ft})(2750\ \text{ft}) = 3{,}712{,}500\ \text{ft}^2$$

$$3{,}712{,}500\ \cancel{\text{ft}^2} \times \frac{1\ \text{acre}}{43{,}560\ \cancel{\text{ft}^2}} = 85.2\ \text{acres} \qquad\qquad ■$$

Professional journals and publications in nearly all scientific areas, including agronomy and animal science, have been metric for several years, so that scientists around the world can better understand and benefit from U.S. research.

Land areas in the United States are still typically measured in the U.S. system. When converting between metric and U.S. land-area units, use the following relationship:

$$1 \text{ hectare} = 2.47 \text{ acres}$$

A good approximation is

$$1 \text{ hectare} = 2.5 \text{ acres}$$

Example 9   How many acres are in 30.6 ha?

$$30.6 \text{ ha} \times \frac{2.47 \text{ acres}}{1 \text{ ha}} = 75.6 \text{ acres}$$ ■

Example 10   How many hectares are in the rectangular field in Example 8?

$$85.2 \text{ acres} \times \frac{1 \text{ ha}}{2.47 \text{ acres}} = 34.5 \text{ ha}$$ ■

Considerable patience and education will be necessary before the hectare becomes the common unit of land area in the United States. The mammoth task of changing all property documents is only one of many obstacles.

The following example shows how to use multiple conversion factors involving more complex units.

Example 11   Change 165 lb/in² to kg/cm².

This conversion requires a series of conversion factors, as follows:

a. from pounds to kilograms
b. from in² to cm²

$$165 \frac{\text{lb}}{\text{in}^2} \times \frac{1 \text{ kg}}{2.20 \text{ lb}} \times \left(\frac{1 \text{ in.}}{2.54 \text{ cm}} \right)^2 = 11.6 \text{ kg/cm}^2$$

Conversion factors for   ↑   ↑
                          a.   **b.**   ■

Exercises 3.7 _____

Fill in each blank, rounding each result to three significant digits when necessary. (Small differences in the last significant digit of answers are acceptable due to the choice of any conversion factor that has been rounded.):

1. 8 lb = _____ kg
2. 16 ft = _____ m
3. 38 cm = _____ in.
4. 81 m = _____ ft
5. 4 yd = _____ cm
6. 17 qt = _____ L
7. 30 kg = _____ lb
8. 15 mi = _____ km
9. 3.2 in. = _____ mm
10. 2 lb 4 oz = _____ g

11. A road sign reads "75 km to Chicago." What is this distance in miles?

12. A hole is 35 mm wide. How many inches wide is it?

13. The diameter of a bolt is 0.425 in. Express this diameter in mm.

14. Change $3\frac{13}{32}$ in. to cm.

15. A tank contains 8 gal of fuel. How many litres of fuel are in the tank?

16. How many pounds does a 150-kg satellite weigh?

17. An iron bar weighs 2 lb 6 oz. Express its weight in **a.** oz and **b.** kg.

18. A precision part is milled to 1.125 in. in width. What is the width in millimetres?

19. A football field is 100 yd long. What is its length **a.** in feet and **b.** in metres?

20. A micro wheel weighs 0.045 oz. What is its weight in mg?

21. A hole is to be drilled in a metal plate 5 in. in diameter. What is the diameter **a.** in cm and **b.** in mm?

22. A can contains 15 oz of tomato sauce. How many grams does the can contain?

23. Change 3 yd^2 to m^2.

24. Change 12 cm^2 to in^2.

25. How many ft^2 are in 140 yd^2?

26. How many m^2 are in 15 yd^2?

27. Change 18 in^2 to cm^2.

28. How many ft^2 are in a rectangle 12.6 yd long and 8.6 yd wide? $(A = lw)$

29. How many ft^2 are in a rectangle 12.6 m long and 8.6 m wide?

30. Find the area of the figure in Illustration 1 in in^2.

ILLUSTRATION 1

31. Change 15 yd^3 to m^3.

32. Change 5473 in^3 to cm^3.

33. How many mm^3 are in 17 in^3?

34. How many in^3 are in 25 cm^3?

35. Change 84 ft^3 to cm^3.

36. How many cm^3 are in 98 in^3?

37. A commercial lot 80 ft wide and 180 ft deep sold for $32,400. What was the price per square foot? What was the price per frontage foot?

38. A concrete sidewalk is to be built (as shown in Illustration 2) around the outside of a corner lot that measures 140 ft by 180 ft. The sidewalk is to be 5 ft wide. What is the surface area of the sidewalk? The sidewalk is to be 4 in. thick. How many yards (actually, cubic yards)

of concrete are needed? Concrete costs $90/$yd^3$ delivered. How much will the sidewalk cost?

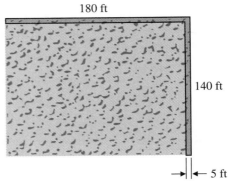

ILLUSTRATION 2

39. How many acres are in a rectangular field that measures 2400 ft by 625 ft?

40. How many acres are in a rectangular field that measures 820 yd by 440 yd?

41. How many hectares are in the field in Exercise 39?

42. How many hectares are in the field in Exercise 40?

43. A house lot measures 145 ft by 186 ft. What part of an acre is the lot?

44. How many acres are in $\frac{1}{4}$ mi^2?

45. How many acres are in $\frac{1}{8}$ section?

46. How many acres are in 520 square rods?

47. A corn yield of 10,550 kg/ha is equivalent to how many lb/acre? To how many bu/acre? (1 bu of corn weighs 56 lb.)

48. A soybean yield of 45 bu/acre is equivalent to how many kg/ha? To how many metric tons/hectare? (1 bu of soybeans weighs 60 lb.)

49. How many acres are in eight 30-in. rows 440 yards long?

50. a. How many rows 30 in. apart can be planted in a rectangular field 3300 ft long and 2600 ft wide? The rows run lengthwise.

b. Suppose seed corn is planted at 20 lb/acre. How many bushels of seed corn will be needed to plant the field?

c. Suppose 1 lb of seed corn contains 1200 kernels. How many bags, each containing 80,000 kernels, will be needed?

51. Change 25.6 kg/cm^2 to lb/in^2.

52. Change 1.5 g/cm^2 to mg/mm^2.

53. Change 65 mi/h to m/s.

54. Change 415 lb/ft^3 to g/cm^3.

Chapter 3 Group Activities

1. Mathematics is used in a variety of places. One location where mathematics is used frequently is in the medical profession. In small groups, brainstorm about the places in a hospital where you think math is used. Think of the different departments and the different professions in the hospital such as radiology, general surgery, etc. After you have thought about this, divide and go to a hospital to check your theory of where and how math is used. Get permission from the proper authorities to ask the employees how they use math. One example is pediatricians who use math in prescribing medication to children. They must be careful to get the weight of a child and use this information to prescribe the proper dosage. The prescription notifies the pharmacist of the amount of medication to give the patient. Make a report on your findings of how math is used in the medical field and make special note of the conversions that doctors and nurses must use. Plan a similar activity for another workplace/profession.

2. Do the following:
   a. Write how old you are to the day. Convert this to days. Convert this to hours and then to minutes.
   b. Write how tall you are. Convert this to feet, to yards, to inches, to metres, and to centimetres.
   c. Write how much you weigh. Convert this to kilograms and to grams.

   Do a little research and see what gravity is on earth and how your weight is determined by gravity. Further research what gravity is on the moon and how your weight would differ on the moon compared to on earth. $(W = mg)$

Chapter 3 Summary

Glossary of Basic Terms

Ampere (A).   The basic SI unit of electric current. (p. 153)

Hectare (ha).   The basic metric unit for land area. (p. 151)

Kelvin (K).   The basic SI unit for temperature; everyday metric temperatures are measured in degrees Celsius (°C). (p. 154)

Kilogram (kg).   The basic SI unit of mass. (p. 146)

Litre (L).   A common SI unit of volume. (p. 148)

Mass.   The quantity of material making up an object. (p. 146)

Metre (m).   The basic SI unit of length. (p. 143)

Second (s).   The basic SI unit of time. (p. 152)

SI.   Abbreviation for the International System of Units (from the French *Système International d'Unités*) and commonly called the **metric system.** (p. 140)

Square metre (m²).   A common SI unit of area. (p. 150)

Weight.   A measure of the earth's gravitational force (pull) acting on an object. (p. 147)

3.1   Introduction to the Metric System

1. **SI base units:** Review the seven SI base units in Table 3.1 on p. 141.

2. **Prefixes for SI units:** Review the prefixes for SI units in Table 3.3 on p. 142.

3.2   Length

1. **Choosing conversion factors:** The correct choice for a given conversion factor is the one in which the old units are in the numerator of the original expression and in the denominator of the conversion factor or the old units are in the denominator of the original expression and in the numerator of the conversion factor. That is, set up the conversion factor so that the old units cancel each other. (p. 144)

3.6   Temperature

1. **Formulas for changing between degrees Celsius and degrees Fahrenheit:**

$$C = \frac{5}{9}(F - 32°)$$

$$F = \frac{9}{5}C + 32°\ \text{(p. 155)}$$

Chapter 3   Review

Give the metric prefix for each value:

1. 0.001

2. 1000

Give the SI abbreviation for each prefix:

3. mega

4. micro

Write the SI abbreviation for each quantity:

5. 42 millilitres

6. 8.3 nanoseconds

Write the SI unit for each abbreviation:

7. 18 km

8. 350 mA

9. 50 μs

Which is larger?

10. 1 L or 1 mL

11. 1 kW or 1 MW

12. $1 km^2$ or 1 ha

13. $1 m^3$ or 1 L

Fill in each blank:

14. 650 m = _____ km

15. 750 mL = _____ L

16. 6.1 kg = _____ g

17. 4.2 A = _____ μA

18. 18 MW = _____ W

19. 25 μs = _____ ns

20. $250 cm^2$ = _____ mm^2

21. $25,000 m^2$ = _____ ha

22. $0.6 m^3$ = _____ cm^3

23. $250 cm^3$ = _____ mL

24. 72°F = _____ °C

25. −25°C = _____ °F

26. Water freezes at _____ °C.

27. Water boils at _____ °C.

28. 180 lb = _____ kg

29. 126 ft = _____ m

30. 360 cm = _____ in.

31. $275 in^2$ = _____ cm^2

32. $18 yd^2$ = _____ ft^2

33. $5 m^3$ = _____ ft^3

34. 15.0 acres = _____ ha

Choose the most reasonable quantity:

35. Jorge and Maria drive **a.** 1600 cm, **b.** 470 m, **c.** 12 km, or **d.** 2400 mm to college each day.

36. Chuck's mass is **a.** 80 kg, **b.** 175 kg, **c.** 14 μg, or **d.** 160 Mg.

37. A car's gas tank holds **a.** 18 L, **b.** 15 kL, **c.** 240 mL, or **d.** 60 L of gasoline.

38. Jamilla, being of average height, is **a.** 5.5 m, **b.** 325 mm, **c.** 55 cm, or **d.** 165 cm tall.

39. A car's average gas consumption is **a.** 320 km/L, **b.** 15 km/L, **c.** 35 km/L, or **d.** 0.75 km/L.

40. On Illinois winter mornings, the temperature sometimes dips to **a.** −50°C, **b.** −30°C, **c.** 30°C, or **d.** −80°C.

41. Abdul drives **a.** 85 km/h, **b.** 50 km/h, **c.** 150 km/h, or **d.** 25 km/h on the interstate highway.

42. Complete the following table of metric system prefixes using the given sample metric unit:

Prefix	Symbol	Power of 10	Sample unit	How many?	How many?
tera	T	10^{12}	m	10^{12} m = 1 Tm	1 m = 10^{-12} Tm
giga	G	10^{9}	W	10^{9} W = 1 GW	1 W = 10^{-9} GW
mega	M	10^{6}	Hz		
kilo	k	10^{3}	g		
hecto	h	10^{2}	Ω		
deka	da	10^{1}	L		
deci	d	10^{-1}	g	10^{-1} g = 1 dg	1 g = 10 dg
centi	c	10^{-2}	m		
milli	m	10^{-3}	A	10^{-3} A = 1 mA	1 A = 10^{3} A
micro	μ	10^{-6}	W		
nano	n	10^{-9}	s		
pico	p	10^{-12}	s		

Chapter 3 Test

1. Give the metric prefix for 1000.
2. Give the metric prefix for 0.01.
3. Which is larger, 200 mg or 1 g?
4. Write the SI unit for the abbreviation 240 μL.
5. Write the abbreviation for 30 hectograms.
6. Which is longer, 1 km or 25 cm?

Fill in each blank:

7. 4.25 km = _____ m
8. 7.28 mm = _____ μm
9. 72 m = _____ mm
10. 256 hm = _____ cm
11. 12 dg = _____ mg
12. 16.2 g = _____ mg
13. 7.236 metric tons = _____ kg
14. 310 g = _____ cg
15. 72 hg = _____ mg
16. 1.52 dL = _____ L
17. 175 L = _____ m^3
18. 2.7 m^3 = _____ cm^3
19. 400 ha = _____ km^2
20. 0.2 L = _____ mL
21. What is the basic SI unit of time?
22. Write the abbreviation for 25 kilowatts.

Fill in each blank:

23. 280 W = _____ kW
24. 13.9 mA = _____ A
25. 720 ps = _____ ns
26. What is the basic SI unit for temperature?
27. What is the freezing temperature of water on the Celsius scale?

Fill in each blank, rounding each result to three significant digits when necessary:

28. 25°C = _____ °F
29. 28°F = _____ °C
30. 98.6°F = _____ °C
31. 100 km = _____ mi
32. 200 cm = _____ in.
33. 1.8 ft^3 = _____ in^3
34. 37.8 ha = _____ acres
35. 80.2 kg = _____ lb

4 Measurement

Mathematics at Work

Science technicians use the principles of science and mathematics to solve problems in research and development and to help invent and improve products and processes. Their jobs are to set up, operate, and maintain laboratory instruments; monitor experiments; make observations; calculate and record results; and keep detailed logs of the work. Biological technicians work with biologists studying living organisms and assist in a variety of medical and biological research projects. Chemical technicians work with chemists and chemical engineers to conduct chemical and physical laboratory tests to assist scientists in making qualitative and quantitative analyses of solids, liquids, and gases in the research and development of new products, new processes, quality control, maintenance of environmental standards, and other work related to the experimental or practical application of chemistry and related sciences. Environmental technicians perform laboratory and field tests to monitor environmental resources and determine contaminants and sources of pollution. Nuclear technicians operate nuclear test and research equipment, monitor radiation, and assist nuclear engineers and physicists in research. Petroleum technicians measure and record physical and geologic conditions in oil or gas wells and collect and examine geologic data or test samples to determine petroleum or mineral content. For more information, go to the website listed below.

Corbis Images/Jupiter Images

Science Technician
Science technician working in a laboratory.

www.cengage.com/mathematics/ewen

Objectives

■ Distinguish the difference between an exact number and an approximate number (measurement).

■ Find the number of significant digits (accuracy) of a measurement.

■ Find the precision and greatest possible error of a measurement.

■ Know the difference between the accuracy and the precision of a measurement.

■ Read metric and U.S. measurements on a vernier caliper and a micrometer caliper.

■ Use the rules for measurement to add, subtract, multiply, and divide measurements.

■ Find the relative error and percent of error of a measurement.

■ Given a measurement and its tolerance, find the upper limit, the lower limit, and the tolerance interval.

■ Using the color code of electrical resistors, find the value of a given resistor as well as its tolerance.

■ Read circular scales on dial indicators and uniform and nonuniform scales on a volt-ohm meter.

4.1 Approximate Numbers and Accuracy

Approximate Numbers (Measurements) versus Exact Numbers

A *tachometer* is used to measure the number of revolutions an object makes with respect to some unit of time. Since the unit of time is usually minutes, a tachometer usually measures revolutions per minute (rpm). This measurement is usually given in integral units, such as 10 rpm or 255 rpm. Tachometers are used in industry to test motors to see whether or not they turn at a specified rate. In the shop, both wood and metal lathes have specified rpm rates. Tachometers are also commonly used in sports cars to help drivers shift gears at the appropriate engine rpm. A tachometer normally measures the spindle speed or the rpm of a shaft, not the surface speed.

Consider the diagram of the tachometer in Figure 4.1. Each of the printed integral values on the dial indicates hundreds of rpm. That is, if the dial indicator is at 10, then the reading is 10 hundred rpm, or 1000 rpm. If the dial indicator is at 70, then the reading is 70 hundred rpm, or 7000 rpm. Each of the subdivisions between 0 and 10, 10 and 20, 20 and 30, and so forth, represents an additional 100 rpm.

FIGURE 4.1
Tachometer

Example 1

Read the tachometer in Figure 4.2.

The indicator is at the sixth division past 30; so the reading is 36 thousand, or 36,000 rpm.

■

FIGURE 4.2

Tachometer readings are only approximate. Tachometers are calibrated in tens, hundreds, or thousands of revolutions per minute, and it is impossible to read the exact number of rpm.

As was noted in Section 1.9, **measurement** is the comparison of an *observed* quantity with a *standard unit* quantity. Consider the measurement of the length of a metal block like the one in Figure 4.3.

1. First, measure the block with ruler A, graduated only in inches. This means measurements will be to the nearest inch. The measured length is 2 in.
2. Measure the same block with ruler B, graduated in half-inches. Measurements now are to the nearest half-inch. The measured length is $2\frac{1}{2}$ in. to the nearest half-inch.
3. Measure the block again with ruler C, graduated in fourths of an inch. To the nearest $\frac{1}{4}$ in., the measurement is $2\frac{1}{4}$ in.
4. Measure the block again with ruler D, graduated in eighths of an inch. To the nearest $\frac{1}{8}$ in., the measurement is $2\frac{3}{8}$ in.

FIGURE 4.3
Measuring the length of a metal block with rulers of
different precision

If you continue this process, by using finer and finer graduations on the ruler, will you ever find the "exact" length? No—since all measurements are only approximations, the "exact" length cannot be found. A measurement is only as good as the measuring instrument you use. It would be rather difficult for you to measure the diameter of a pinhead with a ruler.

Up to this time in your study of mathematics, all measurements have probably been treated as exact numbers. An **exact number** is a number that has been determined as a result of counting—such as 24 students enrolled in a class—or by some definition—such as 1 hour (h) = 60 minutes (min), or 1 in. = 2.54 cm. (These are conversion definitions accepted by all international government bureaus of standards.) Addition, subtraction, multiplication, and division of exact numbers usually make up the main content of elementary-school mathematics.

However, nearly all data of a technical nature involve **approximate numbers**; that is, numbers that have been determined by some measurement process. This process may be direct, as with a ruler, or indirect, as with a surveying transit. Before studying how to perform calculations with approximate numbers (measurements), we must determine the "correctness" of an approximate number. First, we must realize that no measurement can be found exactly. The length of the cover of this book can be found using many instruments. The better the measuring device used, the better the measurement.

In summary,

Exact Versus Approximate Numbers:
1. Only counting numbers are exact.
2. All measurements are approximations.

Accuracy and Significant Digits

The **accuracy of a measurement** means the number of digits, called **significant digits**, that it contains. These indicate the number of units we are reasonably sure of having counted and of being able to rely on in a measurement. The greater the number of significant digits given in a measurement, the better the accuracy, and vice versa.

Example 2

The average distance between the moon and the earth is 239,000 mi. This measurement indicates measuring 239 thousands of miles. Its accuracy is indicated by 3 significant digits. ■

Example 3

A measurement of 10,900 m indicates measuring 109 hundreds of metres. Its accuracy is 3 significant digits. ■

Example 4

A measurement of 0.042 cm indicates measuring 42 thousandths of a centimetre. Its accuracy is 2 significant digits. ■

Example 5

A measurement of 12.000 m indicates measuring 12,000 thousandths of metres. Its accuracy is 5 significant digits. ■

Notice that sometimes a zero is significant and sometimes it is not. Apply the following rules to determine whether a digit is significant or not.

Significant Digits

1. All nonzero digits are significant.

   For example, the measurement 1765 kg has 4 significant digits. (This measurement indicates measuring 1765 units of kilograms.)

2. All zeros between significant digits are significant.

   For example, the measurement 30,060 m has 4 significant digits. (This measurement indicates measuring 3006 tens of metres.)

3. A zero in a whole-number measurement that is specially tagged, such as by a bar above it, is significant.

   For example, the measurement $3\bar{0},000$ ft has 2 significant digits. (This measurement indicates measuring $3\bar{0}$ thousands of feet.)

4. All zeros to the right of a significant digit *and* a decimal point are significant.

   For example, the measurement 6.100 L has 4 significant digits. (This measurement indicates measuring $610\bar{0}$ thousandths of litres.)

5. Zeros to the right in a whole-number measurement that are not tagged are *not* significant.

   For example, the measurement 4600 V has 2 significant digits. (This measurement indicates measuring 46 hundreds of volts.)

6. Zeros to the left in a decimal measurement that is less than 1 are *not* significant. For example, the measurement 0.00960 s has 3 significant digits. (This measurement indicates $96\bar{0}$ hundred-thousandths of a second.)

 Example 6   Determine the accuracy of each measurement.

Measurement	Accuracy (significant digits)
a. 109.006 m	6
b. 0.000589 kg	3
c. 75 V	2
d. 239,000 mi	3
e. 239,00$\bar{0}$ mi	6
f. 239,0$\bar{0}$0 mi	5
g. 0.03200 mg	4
h. 1.20 cm	3
i. 9.020 μA	4
j. 100.050 km	6

■

Exercises 4.1

Determine the accuracy of each measurement; that is, give the number of significant digits for each measurement:

1. 115 V
2. 47,000 lb
3. 7009 ft
4. 420 m
5. 6972 m
6. 320,070 ft
7. 440$\bar{0}$ ft
8. 4400 Ω
9. 44$\bar{0}$0 m
10. 0.0040 g
11. 173.4 m
12. 2070 ft
13. 41,$\bar{0}$00 mi
14. 0.025 A
15. 0.0350 in.
16. 6700 g
17. 173 m
18. 8060 ft
19. 240,$\bar{0}$00 V
20. 2500 g
21. 72,0$\bar{0}$0 mi
22. 137 V
23. 0.047000 A
24. 7.009 g
25. 0.20 mi
26. 69.72 m
27. 32.0070 g
28. 61$\bar{0}$ L
29. 15,0$\bar{0}$0 mi
30. 0.07050 mL
31. 100.020 in.
32. 250.0100 m
33. 900,200 ft
34. 15$\bar{0}$ cm
35. 16,000 W
36. 0.001005 m

 ## 4.2 Precision and Greatest Possible Error

Precision

The **precision of a measurement** is the smallest unit with which the measurement is made; that is, the position of the last significant digit or the smallest unit or calibration on the measuring instrument.

Example 1

The precision of the measurement 239,000 mi is 1000 mi. (The position of the last significant digit is in the thousands place.) ■

Example 2

The precision of the measurement 10,900 m is 100 m. (The position of the last significant digit is in the hundreds place.) ■

Example 3

The precision of the measurement 6.90 L is 0.01 L. (The position of the last significant digit is in the hundredths place.) ■

Example 4

The precision of the measurement 0.0016 A is 0.0001 A. (The position of the last significant digit is in the ten-thousandths place.) ■

Example 5

Determine the precision of each measurement (see Example 6 in Section 4.1).

Measurement	Accuracy (significant digits)	Precision
a. 109.006 m	6	0.001 m
b. 0.000589 kg	3	0.000001 kg
c. 75 V	2	1 V
d. 239,000 mi	3	1000 mi
e. 239,00$\overline{0}$ mi	6	1 mi
f. 239,0$\overline{0}$0 mi	5	10 mi
g. 0.03200 mg	4	0.00001 mg
h. 1.20 cm	3	0.01 cm
i. 9.020 μA	4	0.001 μA
j. 100.050 km	6	0.001 km

■

The precision of a measuring instrument is determined by the smallest unit or calibration on the instrument. The precision of the tachometer in Figure 4.4(a) is 100 rpm. The precision of the tachometer in Figure 4.4(b) is 1000 rpm.

(a) Precision: 100 rpm   (b) Precision: 1000 rpm

FIGURE 4.4
The precision of a measuring instrument is determined by its smallest calibration.

The precision of a ruler graduated in eighths of an inch is $\frac{1}{8}$ in. The precision of a ruler graduated in fourths of an inch is $\frac{1}{4}$ in. However, if a measurement is given as $4\frac{5}{8}$ in., you have no way of knowing what ruler was used. Therefore, you cannot tell whether the precision is $\frac{1}{8}$ in., $\frac{1}{16}$ in., $\frac{1}{32}$ in., or what. The measurement could have been $4\frac{5}{8}$ in., $4\frac{10}{16}$ in., $4\frac{20}{32}$ in., or a similar measurement of some other precision. *Unless stated otherwise, you should assume that the smallest unit used is the one that is recorded.* In this case, the precision is assumed to be $\frac{1}{8}$ in.

Now study closely the enlarged portions of the tachometer (tach) readings in Figure 4.5, given in hundreds of rpm. Note that in each case the measurement is 4100 rpm, although the locations of the pointer are slightly different. Any actual speed between 4050 and 4150 is read 4100 rpm on the scale of this tachometer.

FIGURE 4.5
Any actual speed between 4050 rpm and 4150 rpm is read 4100 rpm on the scale of this tachometer.

The *greatest possible error* is one-half of the smallest unit on the scale on which the measurement is read. We see this in the tach readings in Figure 4.5, where any reading within 50 rpm of 4100 rpm is read as 4100 rpm. Therefore, the greatest possible error is 50 rpm.

If you have a tach reading of 5300 rpm, the greatest possible error is $\frac{1}{2}$ of 100 rpm, or 50 rpm. This means that the actual rpm is between $5300 - 50$ and $5300 + 50$; that is, between 5250 rpm and 5350 rpm.

Next consider the measurements of the three metal rods shown in Figure 4.6. Note that in each case, the measurement of the length is $4\frac{5}{8}$ in., although it is obvious that the rods are of different lengths. Any rod with actual length between $4\frac{9}{16}$ in. and $4\frac{11}{16}$ in. will measure $4\frac{5}{8}$ in. on this scale. The greatest possible error is one-half of the smallest unit on the scale with which the measurement is made.

FIGURE 4.6
Any rod with actual length between $4\frac{9}{16}$ in. and $4\frac{11}{16}$ in. is read $4\frac{5}{8}$ in. on this scale.

If the length of a metal rod is given as $3\frac{3}{16}$ in., the greatest possible error is $\frac{1}{2}$ of $\frac{1}{16}$ in., or $\frac{1}{32}$ in. This means that the actual length of the rod is between $3\frac{3}{16}$ in. $- \frac{1}{32}$ in. and $3\frac{3}{16}$ in. $+ \frac{1}{32}$ in., that is, between $3\frac{5}{32}$ in. and $3\frac{7}{32}$ in.

Greatest Possible Error

The **greatest possible error of a measurement** is equal to one-half its precision.

 Example 6   Find the precision and greatest possible error of the measurement 8.00 kg.

The position of the last significant digit is in the hundredths place; therefore, the precision is 0.01 kg.

The greatest possible error is one-half the precision.

$$(0.5)(0.01 \text{ kg}) = 0.005 \text{ kg}$$ ■

 Example 7   Find the precision and greatest possible error of the measurement 26,000 gal.

The position of the last significant digit is in the thousands place; therefore, the precision is 1000 gal.

The greatest possible error is one-half the precision.

$$\frac{1}{2} \times 1000 \text{ gal} = 500 \text{ gal}$$ ■

 Example 8   Find the precision and greatest possible error of the measurement 0.0460 mg.

Precision: 0.0001 mg
Greatest possible error: $(0.5)(0.0001 \text{ mg}) = 0.00005 \text{ mg}$ ■

Example 9

Find the precision and the greatest possible error of the measurement $7\frac{5}{8}$ in.

The smallest unit is $\frac{1}{8}$ in., which is the precision.
The greatest possible error is $\frac{1}{2} \times \frac{1}{8}$ in. $= \frac{1}{16}$ in., which means that the actual length is within $\frac{1}{16}$ in. of $7\frac{5}{8}$ in. ■

Many calculations with measurements are performed by people who do not make the actual measurements. Therefore, it is often necessary to agree on a method of recording measurements to indicate the precision of the instrument used.

Exercises 4.2

Find **a.** *the precision and* **b.** *the greatest possible error of each measurement:*

1. 2.70 A

2. 13.0 ft

3. 14.00 cm

4. 1.000 in.

5. 15 km

6. 1.010 cm

7. 17.50 mi

8. 6.100 m

9. 0.040 A

10. 0.0001 in.

11. 0.0805 W

12. $10,\overline{0}00$ W

13. $14\overline{0}0$ Ω

14. 301,000 Hz

15. $3\overline{0},000$ L

16. $7,0\overline{0}0,000$ g

17. 428.0 cm

18. 60.0 cm

19. 120 V

20. $30\overline{0}$ km

21. 67.500 m

22. $1\frac{7}{8}$ in.

23. $3\frac{2}{3}$ yd

24. $3\frac{3}{4}$ yd

25. $9\frac{7}{32}$ in.

26. $4\frac{5}{8}$ mi

27. $9\frac{5}{16}$ mi

28. $5\frac{13}{64}$ in.

29. $9\frac{4}{9}$ in^2

30. $18\frac{4}{5}$ in^3

4.3 The Vernier Caliper

In your use of U.S. and metric rulers for making measurements, you have seen that very precise results are difficult to obtain. When more precise measurements are required, you must use a more precise instrument. One such instrument is the **vernier caliper**, which is used by technicians in machine shops, plant assembly lines, and many other workplaces. This instrument is a slide-type measuring instrument used to take precise inside, outside, and depth measurements. It has two metric scales and two U.S. scales. A vernier caliper is shown in Figure 4.7.

FIGURE 4.7

Vernier caliper. The metric scales on this vernier caliper are located above the U.S. or English scales.

To make an outside measurement, the jaws are closed snugly around the outside of an object, as in Figure 4.8(a). For an inside measurement, the smaller jaws are placed inside the object to be measured, as in Figure 4.8(b). For a depth measurement, the depth gauge is inserted into the opening to be measured, as in Figure 4.8(c).

(a) Outside measurement   (b) Inside measurement   (c) Depth measurement

FIGURE 4.8
Measurements with a vernier caliper

Here are some tips for using a vernier caliper (and the micrometer in Section 4.4):

1. Check that the instrument is held perpendicular (that is, at 90°) to the surface of the part being measured.
2. When measuring the diameter of a round piece, check that the *full* diameter is being measured.
3. On rounds, take two readings at approximately 90° to each other. Then average the two readings.

Let's first consider the metric scales (Figure 4.9). One of them is fixed and located on the upper part of the beam. This fixed scale is divided into centimetres and subdivided into millimetres, so record all readings in millimetres (mm). The other metric scale, called the vernier scale, is the upper scale on the slide. The vernier scale is divided into tenths of millimetres (0.10 mm) and subdivided into halves of tenths of millimetres ($\frac{1}{20}$ mm or 0.05 mm). The precision of this vernier scale is therefore 0.05 mm.

Metric vernier scale

Metric fixed scale

FIGURE 4.9
The fixed metric scale on the beam of this vernier caliper is divided into centimetres and further subdivided into millimetres (mm). Its movable metric vernier scale is divided into tenths of millimetres and subdivided into increments of 0.05 mm, which is the precision of this vernier caliper.

The figures in the examples and exercises have been computer generated for easier reading.

◆ Reading a Vernier Caliper in Millimetres

Step 1 Determine the number of whole millimetres in a measurement by counting—on the fixed scale—the number of millimetre graduations that are to the left of the zero graduation on the vernier scale. (Remember that each numbered graduation on the fixed scale represents 10 mm.) The zero graduation on the vernier scale may be directly in line with a graduation on the fixed scale. If so, read the total measurement directly from the fixed scale. Write it in millimetres, followed by a decimal point and two zeros.

Step 2 The zero graduation on the vernier scale may not be directly in line with a graduation on the fixed scale. In that case, find the graduation on the vernier scale that is most nearly in line with any graduation on the fixed scale.

a. If the vernier graduation is a long graduation, it represents the number of tenths of millimetres between the last graduation on the fixed scale and the zero graduation on the vernier scale. Then insert a zero in the hundredths place.

b. If the vernier graduation is a short graduation, add 0.05 mm to the vernier graduation that is on the immediate left of the short graduation.

Step 3 Add the numbers from Steps 1 and 2 to determine the total measurement.

◆ **Example 1**

Read the measurement in millimetres on the vernier caliper in Figure 4.10.

Step 1 The first mark to the left of the zero mark is   45.00 mm

Step 2 The mark on the vernier scale that most nearly lines up with a mark on the fixed scale is   0.20 mm

Step 3 The total measurement is   45.20 mm

■

FIGURE 4.10

FIGURE 4.11

◆ **Example 2**

Read the measurement in millimetres on the vernier caliper in Figure 4.11.

The total measurement is 21.00 mm, because the zero graduation on the vernier scale most nearly lines up with a mark on the fixed scale.

■

Example 3   Read the measurement in millimetres on the vernier caliper in Figure 4.12.

FIGURE 4.12

Step 1   The first mark to the left of the zero mark is   104.00 mm

Step 2   The mark on the vernier scale that most nearly lines up
with a mark on the fixed scale is   0.85 mm

Step 3   Total measurement   104.85 mm

■

Exercises 4.3A

Read the measurement in millimetres shown on each vernier caliper:

1.

2.

3.

4.

5.

6.

7.

8.

9.

10.

11.

12.

13–24. *Read the measurement in millimetres shown on each vernier caliper in Exercises 4.3B* (page 180).

Now consider the two U.S. scales on the vernier caliper in Figure 4.13. One is fixed, the other movable. The fixed scale is located on the lower part of the beam, where each inch is divided into tenths (0.100 in., 0.200 in., 0.300 in., and so on). Each tenth is subdivided into four parts, each of which represents 0.025 in. The vernier scale is divided into 25 parts, each of which represents thousandths (0.001 in.). This means that the precision of this scale is 0.001 in.

FIGURE 4.13
The fixed U.S. or English scale on the beam of this vernier caliper is divided into tenths of inches and further subdivided into increments of 0.025 in. Its movable U.S. vernier scale is divided into increments of 0.001 in., which is the precision of this vernier caliper.

Reading a Vernier Caliper in Thousandths of an Inch

Step 1 Determine the number of inches and tenths of inches by reading the first numbered division that is to the left of the zero graduation on the vernier scale.

Step 2 Add 0.025 in. to the number from Step 1 for each graduation between the last numbered division on the fixed scale and the zero graduation on the vernier scale. (If this zero graduation is directly in line with a graduation on the fixed scale, read the total measurement directly from the fixed scale.)

continued

Step 3   If the zero graduation on the vernier scale is not directly in line with a graduation on the fixed scale, find the graduation on the vernier scale that is most nearly in line with any graduation on the fixed scale. This graduation determines the number of thousandths of inches in the measurement.

Step 4   Add the numbers from Steps 1, 2, and 3 to determine the total measurement.

Example 4   Read the measurement in inches shown on the vernier caliper in Figure 4.14.

Step 1	The first numbered mark to the left of the zero mark is	1.600 in.
Step 2	The number of 0.025-in. graduations is 3; 3 × 0.025 in. =	0.075 in.
Step 3	The mark on the vernier scale that most nearly lines up with a mark on the fixed scale is	0.014 in.
Step 4	Total measurement	1.689 in.

■

FIGURE 4.14   FIGURE 4.15

Example 5   Read the measurement in inches shown on the vernier caliper in Figure 4.15.

Step 1	The first numbered mark to the left of the zero mark is	0.800 in.
Step 2	The number of 0.025-in. graduations is 1; 1 × 0.025 in. =	0.025 in.
Step 3	Total measurement	0.825 in.

Note: The zero graduation is directly in line with a graduation on the fixed scale.

■

Example 6   Read the measurement in inches shown on the vernier caliper in Figure 4.16.

FIGURE 4.16

Step 1	The first numbered mark to the left of the zero mark is	1.200 in.
Step 2	The number of 0.025-in. graduations is 2; 2 × 0.025 in. =	0.050 in.
Step 3	The mark on the vernier scale that most nearly lines up with a mark on the fixed scale is	0.008 in.
Step 4	Total measurement	1.258 in.

■

Digital vernier calipers that provide metric and U.S. readings in millimetres and decimal inches by means of a one-button process are also in common use as shown in Figure 4.17(a) and (b). This display reads down to 0.01 mm or 0.0005 in., which indicates its precision. Before each measurement, carefully close the caliper and check that it reads 0.000 or press the zero button. Then make your outside, inside, or depth measurement as before. Just as we must learn to read clocks with hands as well as digital clocks because both types are in common use, we must learn to use both regular and digital vernier calipers. Some digital vernier calipers have a three-mode display with millimetre, decimal inch, and fractional inch readings. This display in Figure 4.17(c) reads down to 1/128 in., which indicates its precision.

(a)　(b)　(c)

FIGURE 4.17
Digital vernier caliper with displays in (a) millimetre mode; (b) decimal inch mode; and (c) fractional inch mode

Exercises 4.3B

Read the measurement in inches shown on each vernier caliper:

1.

2.

3.

4.

5.

6.

7.

8.

9.

10.

11.

12.

13–24. *Read the measurement in inches shown on each vernier caliper in Exercises 4.3A (pages 176 and 177).*

4.4 **The Micrometer Caliper**

The **micrometer caliper** (micrometer or "mike") is an instrument for measuring very small lengths using the movement of a finely threaded rotating screw, which gives it better precision than a vernier caliper. It is used in technical fields in which fine precision is required. Micrometers are available in metric units and U.S. units. The metric "mike" is graduated and read in hundredths of a millimetre (0.01 mm); the U.S. "mike" is graduated and read in thousandths of an inch (0.001 in.). The parts of a micrometer are labeled in Figure 4.18.

FIGURE 4.18
Basic parts of a micrometer

To use a micrometer properly, place the object to be measured between the anvil and spindle and turn the thimble until the object fits snugly. *Do not force the turning of the thimble, because this may damage the very delicate threads on the spindle that are located inside the thimble.* Some calipers have a ratchet to protect the instrument; the ratchet prevents the thimble from being turned with too much force. A metric micrometer is shown in Figure 4.19, and the basic parts are labeled. The barrels of most metric micrometers are graduated in millimetres. The micrometer in Figure 4.19 also has graduations of halves of millimetres, which are indicated by the lower set of graduations on the barrel. The threads on the spindle are made so that it takes two complete turns of the thimble for the spindle to move precisely one millimetre. The head is divided into 50 equal divisions—each division indicating 0.01 mm, which is the precision.

FIGURE 4.19
Metric micrometer

Reading a Metric Micrometer in Millimetres

Step 1   Find the *whole number* of mm in the measurement by counting the number of mm graduations on the barrel to the left of the head.

Step 2   Find the *decimal part* of the measurement by reading the graduation on the head (see Figure 4.19) that is most nearly in line with the center line on the barrel. Then multiply this reading by 0.01. If the head is at, or immediately to the right of, the half-mm graduation, then add 0.50 mm to the reading on the head.

Step 3   Add the numbers found in Step 1 and Step 2.

 Example 1   Read the measurement shown on the metric micrometer in Figure 4.20.

FIGURE 4.20

Step 1   The barrel reading is   6.00 mm
Step 2   The head reading is   0.24 mm
Step 3   The total measurement is   6.24 mm

◆ **Example 2**   Read the measurement shown on the metric micrometer in Figure 4.21.

Step 1	The barrel reading is	14.00 mm
Step 2	The head reading is	0.12 mm
Step 3	The total measurement is	14.12 mm

FIGURE 4.21   FIGURE 4.22

◆ **Example 3**   Read the measurement shown on the metric micrometer in Figure 4.22.

Step 1	The barrel reading is (Note that the head is past the half-mm mark.)	8.50 mm
Step 2	The head reading is	0.15 mm
Step 3	The total measurement is	8.65 mm

Exercises 4.4A _____

Read the measurement shown on each metric micrometer:

1.

2.

3.

4.

5.

6.

7.

8.

9.

10.

11.

12.

13.

14.

15.

16.

17.

18.

19.

20.

The barrel of the U.S. micrometer shown in Figure 4.23 is divided into tenths of an inch. Each tenth is subdivided into four 0.025-in. parts. The threads on the spindle allow the spindle to move 0.025 in. in one complete turn of the thimble and 4 × 0.025 in., or 0.100 in., in four complete turns. The head is divided into 25 equal divisions—each division indicating 0.001 in., which is the precision.

FIGURE 4.23
U.S. micrometer

Reading a U.S. Micrometer in Thousandths of an Inch

Step 1 Read the last numbered graduation showing on the barrel. Multiply this number by 0.100 in.

Step 2 Find the number of smaller graduations between the last numbered graduation and the head. Multiply this number by 0.025 in.

Step 3 Find the graduation on the head that is most nearly in line with the center line on the barrel. Multiply the number represented by this graduation by 0.001 in.

Step 4 Add the numbers found in Steps 1, 2, and 3.

◆ **Example 4**

FIGURE 4.24

Read the measurement shown on the U.S. micrometer in Figure 4.24.

Step 1	3 numbered divisions on the barrel; 3 × 0.100 in. =	0.300 in.
Step 2	1 small division on the barrel; 1 × 0.025 in. =	0.025 in.
Step 3	The head reading is 17; 17 × 0.001 in. =	0.017 in.
Step 4	The total measurement is	0.342 in.

■

◆   **Example 5**   Read the measurement shown on the U.S. micrometer in Figure 4.25.

Step 1	4 numbered divisions on the barrel; 4 × 0.100 in. =	0.400 in.
Step 2	2 small divisions on the barrel; 2 × 0.025 in. =	0.050 in.
Step 3	The head reading is 21; 21 × 0.001 in. =	0.021 in.
Step 4	The total measurement is	0.471 in.

■

FIGURE 4.25

Exercises 4.4B

Read the measurement shown on each U.S. micrometer:

1.

2.

3.

4.

5.

6.

7.

8.

9.

10.

11.

12.

13.

14.

15.

16.

17.

18.

19.

20.

Other Micrometers

By adding a vernier scale on the barrel of a micrometer (as shown in Figure 4.26), we can increase the precision by one more decimal place. That is, the metric micrometer with vernier scale has a precision of 0.001 mm. The U.S. micrometer with vernier scale has a precision of 0.0001 in. Of course, these micrometers cost more because they require more precise threading than the ones previously discussed. Nevertheless, many jobs require this precision.

FIGURE 4.26
Micrometer with vernier scale

Micrometers are basic, useful, and important tools of the technician. Figure 4.27 shows just a few of their uses.

(a) Measuring a piece of die steel

(b) Measuring the diameter of a crankshaft bearing

(c) Measuring tubing wall thickness with a round anvil micrometer

(d) Checking out-of-roundness on centerless grinding work

(e) Measuring the pitch diameter of a screw thread

(f) Measuring the depth of a shoulder with a micrometer depth gauge

FIGURE 4.27
Examples of how micrometers are used

Digital micrometers that provide metric and U.S. (mm/inch) readings by means of a one-button process are in common use as shown in Figure 4.28. The display in this figure reads down to 0.001 mm or 0.0001 in., which indicates its precision. Before each measurement, carefully close the anvil and spindle and check that it reads 0.000 or press the zero button. Then make your measurement as before.

(a)

(b)

FIGURE 4.28
Digital micrometer with displays in (a) millimetre mode and (b) decimal inch mode

---

4.5 Addition and Subtraction of Measurements

Precision versus Accuracy

Recall that the *precision* of a measurement is the smallest unit with which a measurement is made; that is, the *position* of the last significant digit or the smallest unit or calibration on the measuring instrument. Recall also that the *accuracy* of a measurement is the *number* of digits, called significant digits, which indicate the number of units we are reasonably sure of having counted when making a measurement. Unfortunately, some people tend to use the terms *precision* and *accuracy* interchangeably even though each term expresses a different aspect of a given measurement.

 Example 1

Compare the precision and the accuracy of the measurement 0.0007 mm.

Since the precision is 0.0001 mm, its precision is relatively good. However, since the accuracy is only one significant digit, its accuracy is relatively poor.   ■

Example 2

Given the measurements 13.00 m, 0.140 m, 3400 m, and 0.006 m, find the measurement that is **a.** the least precise, **b.** the most precise, **c.** the least accurate, and **d.** the most accurate.

First, let's find the precision and the accuracy of each measurement.

Measurement	Precision	Accuracy (significant digits)
13.00 m	0.01 m	4
0.140 m	0.001 m	3
3400 m	100 m	2
0.006 m	0.001 m	1

From the table, we find

a. The least precise measurement is 3400 m.

b. The most precise measurements are 0.140 m and 0.006 m.

c. The least accurate measurement is 0.006 m.

d. The most accurate measurement is 13.00 m.   ■

In a series circuit, the electromagnetic force (emf) of the source equals the sum of the separate voltage drops across each resistor in the circuit. Suppose that someone measures the voltage across the first resistor R_1 in Figure 4.29. He uses a voltmeter calibrated in hundreds of volts and measures 15,800 V. Across the second resistor R_2, he uses a voltmeter in thousands of volts and measures 11,000 V. Does the total emf equal 26,800 V? Note that the first voltmeter and its reading indicate a precision of 100 V and a greatest possible error of 50 V. This means that the actual reading lies between 15,750 V and 15,850 V. The second voltmeter and its reading indicate a precision of 1000 V and a greatest possible error of 500 V. The actual reading, therefore, lies between 10,500 V and 11,500 V. This means that we are not very certain of the digit in the hundreds place in the sum 26,800 V.

To be consistent when adding or subtracting measurements of different precision, the sum or difference can be no more *precise* than the least precise measurement.

R_2

11,000 V

R_1

15,800 V

FIGURE 4.29

Adding or Subtracting Measurements of Different Precision

1. Make certain that all measurements are expressed in the same unit. If they are not, change them all to any common unit.

2. Add or subtract.

3. Then round the result to the same precision as the *least precise* measurement.

The total emf in the circuit shown in Figure 4.29 is therefore calculated as follows:

Measurement	*Rounded*
R_1:	15,800 V
R_2:	11,000 V
	26,800 V → 27,000 V

Example 3   Use the rules for addition of measurements to add 13,800 ft, 14,020 ft, 19,864 ft, 2490 ft, and 14,700 ft.

Since all of the measurements are in the same unit (that is, ft), add them together:

13,800 ft
14,020 ft
19,864 ft
 2,490 ft
14,700 ft
64,874 ft → 64,900 ft

Round this sum to the same precision as the least precise measurement. Since the precision of both 13,800 ft and 14,700 ft is 100 ft, round the sum to the nearest 100 ft. Thus, the sum is 64,900 ft. ■

 Example 4   Use the rules for addition of measurements to add 735,000 V, 490,000 V, 86,000 V, 1,300,000 V, and $20\overline{0},000$ V.

Since all of the measurements are in the same unit, add:

  735,000 V
  490,000 V
   86,000 V
1,300,000 V
  $20\overline{0},000$ V
2,811,000 V → 2,800,000 V

The least precise measurement is 1,300,000 V, which has a precision of 100,000 V. Round the sum to the nearest hundred thousand volts: 2,800,000 V. ■

 Example 5   Use the rules for addition of measurements to add 13.8 m, 140.2 cm, 1.853 m, and 29.95 cm.

First, change each measurement to a common unit (say, m) and add:

13.8 m → 13.8 m
140.2 cm → 1.402 m
1.853 m → 1.853 m
29.95 cm → 0.2995 m
             17.3545 m → 17.4 m

The least precise measurement is 13.8 m, which is precise to the nearest tenth of a metre. So round the sum to the nearest tenth of a metre: 17.4 m. ■

 Example 6   Use the rules for subtraction of measurements to subtract 19.352 cm from 41.7 cm.

Since both measurements have the same unit, subtract:

41.7 cm
19.352 cm
22.348 cm → 22.3 cm

The least precise measurement is 41.7 cm, which is precise to the nearest tenth of a cm. Round the difference to the nearest tenth of a cm: 22.3 cm. ■

Exercises 4.5

In each set of measurements, find the measurement that is **a.** *the most accurate and* **b.** *the most precise:*

1. 14.7 in.; 0.017 in.; 0.09 in.
2. 459 ft; 600 ft; 190 ft
3. 0.737 mm; 0.94 mm; 16.01 mm
4. 4.5 cm; 9.3 cm; 7.1 cm
5. 0.0350 A; 0.025 A; 0.00050 A; 0.041 A
6. 134.00 g; 5.07 g; 9.000 g; 0.04 g
7. 145 cm; 73.2 cm; 2560 cm; 0.391 cm
8. 15.2 km; 631.3 km; 20.0 km; 37.7 km
9. 205,000 Ω; 45,000 Ω; 5̄00,000 Ω; 90,000 Ω
10. 1,500,000 V; 65,000 V; 30,0̄00 V; 20,000 V

In each set of measurements, find the measurement that is **a.** *the least accurate and* **b.** *the least precise:*

11. 15.5 in.; 0.053 in.; 0.04 in.
12. 635 ft; 400 ft; 240 ft
13. 43.4 cm; 0.48 cm; 14.05 cm
14. 4.9 kg; 670 kg; 0.043 kg
15. 0.0730 A; 0.043 A; 0.00008 A; 0.91 A
16. 197.0 m; 5.43 m; 4.000 m; 0.07 m
17. 2.1 m; 31.3 m; 461.5 m; 0.6 m
18. 295 m; 91.3 m; 1920 m; 0.360 m
19. 405,000 Ω; 35,000 Ω; 8̄00,000 Ω; 500,000 Ω
20. 1,600,000 V; 36,000 V; 40,0̄00 V; 60,000 V

Use the rules for addition of measurements to find the sum of each set of measurements:

21. 14.7 m; 3.4 m
22. 168 in.; 34.7 in.; 61 in.
23. 42.6 cm; 16.41 cm; 1.417 cm; 34.4 cm
24. 407 g; 1648.5 g; 32.74 g; 98.1 g
25. 26,000 W; 19,600 W; 8450 W; 42,500 W
26. 5420 km; 1926 km; 850 km; 2̄000 km
27. 140,000 V; 76,200 V; 4700 V; 254,000 V; 370,000 V
28. 19,200 m; 8930 m; 50,040 m; 137 m
29. 14 V; 1.005 V; 0.017 V; 3.6 V
30. 120.5 cm; 16.4 cm; 1.417 m
31. 10.555 cm; 9.55 mm; 13.75 cm; 206 mm
32. 1350 cm; 1476 mm; 2.876 m; 4.82 m

Use the rules for subtraction of measurements to subtract the second measurement from the first:

33. 140.2 cm
      13.8 cm

34. 14.02 mm
      13.8 mm

35. 9200 mi
       627 mi

36. 1,900,000 V
       645,000 V

37. 167 mm
       13.2 cm

38. 16.41 oz
      11.372 oz

39. 98.1 g
      32.743 g

40. 4.000 in.
      2.006 in.

41. 0.54361 in.
      0.214   in.

42. If you bolt four pieces of metal with thicknesses 0.136 in., 0.408 in., 1.023 in., and 0.88 in. together, what is the total thickness?

43. If you clamp five pieces of metal with thicknesses 2.38 mm, 10.5 mm, 3.50 mm, 1.455 mm, and 8.200 mm together, what is the total thickness?

44. What is the current going through R_5 in the circuit in Illustration 1? (*Hint:* In a parallel circuit, the current is divided among its branches. That is, $I_T = I_1 + I_2 + I_3 + \cdots.$)

ILLUSTRATION 1

45. A welder cuts several pieces of steel angle of lengths 3.32 ft, 2.15 ft, 3.2 ft, and 4.0 ft. What is the total length of the pieces?

46. A welder weighed some bins of scrap metal. The bins weighed 266 lb, 620 lb, and 1200 lb, respectively. What was the total weight of scrap metal in all three bins?

47. A pilot loads baggage in the baggage compartment of a small plane. The baggage weighs 23.25 lb, 18.6 lb, and 25 lb. What is the total weight of the baggage?

48. To compensate for too much cargo on a plane with 38.35 gal of fuel, fuel is drained. First 8.2 gal are drained, then another 6.33 gal. After this, how much fuel is left?

49. The antifreeze of a car leaks. One day it lost 0.52 gal, a second day it lost 0.4 gal, and a third day it lost 0.34 gal. What was the total antifreeze lost over the three-day period?

50. On a long trip a car is driven 340 mi the first day and 400 mi the second day, and the trip is finished with 253 mi the last day. What are the total miles driven?

51. A furnace burned 23.52 gal of gas in September, 25.8 gal in October, 33.24 gal in November, and 41 gal in December. What was the total gas burned over the 4-month period?

52. A 6-room building has the following supply air requirements. $A = 120$ ft^3/min, $B = 265$ ft^3/min, $C = 61$ ft^3/min, and $D = 670$ ft^3/min. What is the required HVAC unit supply air flow?

53. In making a specific CAD drawing, the pictorial representation must be precise to the nearest thousandth, and the dimensions must be identified with a precision of five decimal places. The shaft overall length must be shown on the drawing. Calculate the total length and show the dimension that you would put on the drawing in Illustration 2.

54. As part of an environmental science class, four families are selected to weigh their trash and recycling for a week using their bathroom scales, which have different precisions. The results are shown below in pounds:

	Trash	Recycle	Can/bottles for refund	Precision of scale
Family 1	35.3	21.5	4.9	0.1 lb
Family 2	14.4	28.6	3.8	0.2 lb
Family 3	18.5	36.0	2.5	0.5 lb
Family 4	46	12	4	1 lb

Find the total amount of each and the percentage of material that was recycled and returned for refund.

55. A fisherman brought home a cooler of fish and ice that weighed a total of 17.4 lb on his bathroom scale. He knows that the empty cooler weighs 3.6 lb. He added a 5-lb bag of ice and nothing else except the fish. How much did the fish weigh?

2.00000
TYP

1.33333
TYP

ILLUSTRATION 2

4.6 Multiplication and Division of Measurements

Suppose that you want to find the area of a rectangular plot of ground that measures 206 m by 84 m. The product, 17,304 m^2, shows five significant digits. The original measurements have three and two significant digits, respectively.

To be consistent when multiplying or dividing measurements, the product or quotient can be no more *accurate* than the least accurate measurement.

Multiplying or Dividing Measurements

1. First, multiply and/or divide the measurements.

2. Then round the result to the same number of significant digits as the measurement that has the least number of significant digits. That is, round the result to the same accuracy as the *least accurate* measurement.

Using this procedure, the area of the plot of ground [(206 m)(84 m) = 17,304 m^2] is rounded to 17,000 m^2.

Example 1

Use the rules for multiplication of measurements: (20.41 g)(3.5 cm).

Step 1   (20.41 g)(3.5 cm) = 71.435 g cm

Step 2   Round this product to two significant digits, which is the accuracy of the least accurate measurement, 3.5 cm. That is,

(20.41 g)(3.5 cm) = 71 g cm   ■

Example 2

Use the rules for multiplication of measurements: (125 m)(345 m)(204 m).

Step 1   (125 m)(345 m)(204 m) = 8,797,500 m^3

Step 2   Round this product to three significant digits, which is the accuracy of the least accurate measurement (which is the accuracy of each measurement in this example). That is,

(125 m)(345 m)(204 m) = 8,8$\overline{0}$0,000 m^3 ■

Example 3

Use the rules for division of measurements to divide 288,000 ft^3 by 216 ft.

Step 1 $\dfrac{288,000 \text{ ft}^3}{216 \text{ ft}} = 1333.333 \ldots \text{ ft}^2$

Step 2   Round this quotient to three significant digits, which is the accuracy of the least accurate measurement (which is the accuracy of each measurement in this example). That is,

$$\frac{288,000 \text{ ft}^3}{216 \text{ ft}} = 1330 \text{ ft}^2$$ ■

Example 4

Use the rules for multiplication and division of measurements to evaluate

$$\frac{(4750 \text{ N})(4.82\text{m})}{1.6 \text{ s}}$$

Step 1 $\dfrac{(4750 \text{ N})(4.82 \text{ m})}{1.6 \text{ s}} = 14,309.375 \dfrac{\text{N m}}{\text{s}}$

Step 2   Round this result to two significant digits, which is the accuracy of the least accurate measurement, 1.6 s. That is,

$$\frac{(4750 \text{ N})(4.82 \text{ m})}{1.6 \text{ s}} = 14,000 \frac{\text{N m}}{\text{s}} \text{ or } 14,000 \text{ N m/s}$$ ■

There are even more sophisticated methods for dealing with the calculations of measurements. The method that one uses (and indeed, whether one should even follow any given procedure) depends on the number of measurements and the sophistication needed for a particular situation.

The procedures for addition, subtraction, multiplication, and division of measurements are based on methods followed and presented by the American Society for Testing and Materials.

Note: To multiply or divide measurements, the units do not need to be the same. (They must be the same in addition and subtraction of measurements.) Also note that the units are multiplied and/or divided in the same manner as the corresponding numbers.

Exercises 4.6

Use the rules for multiplication and/or division of measurements to evaluate:

1. (126 m)(35 m)

2. (470 mi)(1200 mi)

3. (1463 cm)(838 cm)

4. (2.4 A)(3600 Ω)

5. (18.7 m)(48.2 m)

6. (560 cm)(28.0 cm)

7. (4.7 Ω)(0.0281 A)

8. (5.2 km)(6.71 km)

9. (24.2 cm)(16.1 cm)(18.9 cm)

10. (0.045 m)(0.0292 m)(0.0365 m)

11. (2460 m)(960 m)(1970 m)

12. $(46\overline{0}$ in.)(235 in.)(368 in.)

13. $(0.480 \text{ A})^2(150 \ \Omega)$

14. $360 \text{ ft}^2 \div 12 \text{ ft}$

15. $62{,}500 \text{ in}^3 \div 25 \text{ in.}$

16. $9180 \text{ yd}^3 \div 36 \text{ yd}^2$

17. $1520 \text{ m}^2 \div 40 \text{ m}$

18. $18.4 \text{ m}^3 \div 9.2 \text{ m}^2$

19. $4800 \text{ V} \div 14.2 \text{ A}$

20. $\dfrac{4800 \text{ V}}{6.72 \ \Omega}$

21. $\dfrac{5.63 \text{ km}}{2.7 \text{ s}}$

22. $\dfrac{0.497 \text{ N}}{(1.4 \text{ m})(8.0 \text{ m})}$

23. $\dfrac{(120 \text{ V})^2}{47.6 \ \Omega}$

24. $\dfrac{(19 \text{ kg})(3.0 \text{ m/s})^2}{2.46 \text{ m}}$

25. $\dfrac{140 \text{ g}}{(3.2 \text{ cm})(1.7 \text{ cm})(6.4 \text{ cm})}$

26. Find the area of a rectangle measured as 6.5 cm by 28.3 cm. ($A = lw$)

27. $V = lwh$ is the formula for the volume of a rectangular solid, where l = length, w = width, and h = height. Find the volume of a rectangular solid when $l = 16.4$ ft, $w = 8.6$ ft, and $h = 6.4$ ft.

28. Find the volume of a cube measuring 8.10 cm on each edge. ($V = e^3$, where e is the length of each edge.)

29. The formula $s = 4.90t^2$ gives the distance, s, in metres, that an object falls in a given time, t. Find the distance a ball falls in 2.4 seconds.

30. Given K.E. $= \frac{1}{2}mv^2$, $m = 2.87 \times 10^6$ kg, and $v = 13.4$ m/s. Find K.E.

31. A formula for finding the horsepower of an engine is $p = \frac{d^2 n}{2.50}$, where d is the diameter of each cylinder in inches and n is the number of cylinders. What is the horsepower of an 8-cylinder engine if each cylinder has a diameter of 3.00 in.? (*Note:* Eight is an exact number. Ignore the number of significant digits in an exact number when determining the number of significant digits in a product or quotient.)

32. Six pieces of metal, each 2.48 mm in thickness, are fitted together. What is the total thickness of the 6 pieces?

33. Find the volume of a cylinder with radius 6.2 m and height 8.5 m. The formula for the volume of a cylinder is $V = \pi r^2 h$.

34. In 2000, the United States harvested 10,200,000,000 bu of corn from 73,100,000 acres. In 2004, 11,800,000,000 bu were harvested from 73,600,000 acres. What was the yield in bu/acre for each year? What was the increase in yield between 2000 and 2004?

35. A room 24 ft long and 14 ft wide, with a ceiling height of 8.0 ft, has its air changed six times per hour. What are its ventilation requirements in CFM (ft³/min)?

36. A welder welds two pieces of pipe together and uses 2.25 rods. If this same weld is done 6 times, how many rods are used?

37. A rectangular metal storage bin has been welded together. Its dimensions are 13.5 in., 17.25 in., and $2\overline{0}$ in. What is the volume of such a storage bin?

38. A plane flew 1.8 h for each of 4 lessons. How many hours has it flown?

39. A plane flew 3.4 h and used 32.65 gal of gas. How many gallons per hour did it use?

40. A plane flies 60.45 mi due north, then flies 102.3 mi due east. What would be the area of a rectangle formed by these dimensions?

41. What is the area of a windshield of a car if it measures 55.3 in. by 28.25 in.?

42. A vehicle traveled 620 mi and used 24.2 gal of gas. How many miles per gallon did the vehicle get?

43. The trunk space of a small car measures 3.0 ft in width, 4.2 ft in length, and 1.5 ft in depth. Find the volume of the trunk.

44. An old furnace measures 26.5 in. wide, 35 in. long, and 70 in. high. How much space does the furnace occupy?

45. 52.6 ft of duct is needed to put a furnace in a house. If the duct comes only in 6-ft sections, how many sections should be ordered?

46. The weather forecast for tonight calls for about 1 inch of rain per hour. The local shopping center has a paved parking lot that measures 2.50 acres and has two storm sewers to handle the runoff. **a.** Assuming 100% runoff, how many gallons of water per hour must the storm sewers handle tonight to avoid flooding if 1.00 in. of rain per hour falls? (*Note:* The volume of 1 acre of water 1 inch deep is 27,150 gallons.) **b.** If each storm sewer is rated at 25,000 gal/h, can we expect flooding from the parking lot?

47. According to climatologists, the carbon dioxide (CO_2) levels in the atmosphere in 2008 were the highest levels in 650,000 years, standing at 387 parts per million (ppm). The current rate of increase is 2.1 ppm per year. If that rate of increase remains constant from 2008 until 2100, what would be the expected CO_2 level in our atmosphere by the end of the year 2100?

48. According to the Environmental Protection Agency, a total of 245 million tons of solid waste was generated in the United States in 2005. Of that, 11.7% was food scraps. If all of those food scraps had been composted instead of put into landfills, how many pounds of food scraps would have been kept out of our landfills?

49. There is so much water in the world that we normally measure large bodies of water in cubic miles or cubic kilometres instead of gallons. Cayuga Lake in upstate New York contains an estimated 9.4 km³. How many cubic miles of water does Cayuga Lake contain?

4.7 Relative Error and Percent of Error

Technicians must determine the importance of measurement error, which may be expressed in terms of relative error. The **relative error of a measurement** is found by comparing the greatest possible error with the measurement itself.

$$\text{relative error} = \frac{\text{greatest possible error}}{\text{measurement}}$$

 Example 1   Find the relative error of the measurement 0.08 cm.

The precision is 0.01 cm. The greatest possible error is one-half the precision, which is 0.005 cm.

$$\text{relative error} = \frac{\text{greatest possible error}}{\text{measurement}} = \frac{0.005 \ \text{cm}}{0.08 \ \text{cm}} = 0.0625$$

Note that the units will always cancel, which means that the relative error is expressed as a unitless decimal. When this decimal is expressed as a percent, we have the percent of error. ■

> The **percent of error of a measurement** is the relative error expressed as a percent.

Percent of error can be used to compare different measurements because, being a percent, it compares each error in terms of 100. (The percent of error in Example 1 is 6.25%.)

 Example 2   Find the relative error and percent of error of the measurement 13.8 m.

The precision is 0.1 m and the greatest possible error is then 0.05 m. Therefore,

$$\text{relative error} = \frac{\text{greatest possible error}}{\text{measurement}} = \frac{0.05 \ \text{m}}{13.8 \ \text{m}} = 0.00362$$

$$\text{percent of error} = 0.362\%$$ ■

◆ **Example 3**   Compare the measurements $3\frac{3}{4}$ in. and 16 mm. Which one is better? (Which one has the smaller percent of error?)

Measurement	$3\dfrac{3}{4}$ in.	16 mm
Precision	$\dfrac{1}{4}$ in.	1 mm
Greatest possible error	$\dfrac{1}{2} \times \dfrac{1}{4}$ in. $= \dfrac{1}{8}$ in.	$\dfrac{1}{2} \times 1$ mm $= 0.5$ mm

Relative error

$$\frac{\frac{1}{8}\ \text{in.}}{3\frac{3}{4}\ \text{in.}} = \frac{1}{8} \div 3\frac{3}{4} \qquad \frac{0.5\ \cancel{mm}}{16\ \cancel{mm}} = 0.03125$$

$$= \frac{1}{8} \div \frac{15}{4}$$

$$= \frac{1}{8} \times \frac{4}{15}$$

$$= \frac{1}{30}$$

$$= 0.0333$$

Percent of error	3.33%	3.125%

Therefore, 16 mm is the better measurement, because its percent of error is smaller. ■

Tolerance

Lower limit

Upper limit

FIGURE 4.30
The tolerance interval is the difference between the upper limit and the lower limit.

In industry, the **tolerance** of a part or component is the acceptable amount that the part or component may vary from a given size. For example, a steel rod may be specified as $14\frac{3}{8}$ in. $\pm \frac{1}{32}$ in. The symbol "$\pm$" is read "plus or minus." This means that the rod may be as long as $14\frac{3}{8}$ in. $+ \frac{1}{32}$ in.; that is, $14\frac{13}{32}$ in. This is called the *upper limit*. Or it may be as short as $14\frac{3}{8}$ in. $- \frac{1}{32}$ in.; that is, $14\frac{11}{32}$ in. This is called the *lower limit*. Therefore, the specification means that any rod between $14\frac{11}{32}$ in. and $14\frac{13}{32}$ in. would be acceptable. We say that the tolerance is $\pm \frac{1}{32}$ in. The **tolerance interval**—the difference between the upper limit and the lower limit—is $\frac{2}{32}$ in., or $\frac{1}{16}$ in.

A simple way to check the tolerance of the length of a metal rod would be to carefully mark off lengths that represent the lower limit and upper limit, as shown in Figure 4.30. To check the acceptability of a rod, place one end of the rod flush against the metal barrier on the left. If the other end is between the upper and lower limit marks, the part is acceptable. If the rod is longer than the upper limit, it can then be cut to the acceptable limits. If the rod is shorter than the lower limit, it must be rejected. (It can be melted down for another try.)

Example 4

The specifications for a stainless steel cylindrical piston are given as follows:

Diameter: 10.200 cm ± 0.001 cm
Height: 14.800 cm ± 0.005 cm

Find the upper limit, the lower limit, and the tolerance interval for each dimension.

	Given length	Tolerance	Upper limit	Lower limit	Tolerance interval
Diameter:	10.200 cm	±0.001 cm	10.201 cm	10.199 cm	0.002 cm
Height:	14.800 cm	±0.005 cm	14.805 cm	14.795 cm	0.010 cm

■

Tolerance may also be expressed as a percent. For example, resistors are color coded to indicate the tolerance of a given resistor. If the fourth band is silver, this indicates that the acceptable tolerance is ±10% of the given resistance. If the fourth band is gold, this indicates that the acceptable tolerance is ±5% of the given resistance. This is fully discussed in the next section.

Many times, bids may be accepted under certain conditions. They may be accepted, for example, when they are less than 10% over the architect's estimate.

Example 5

If the architect's estimate for a given project is $356,200 and bids may be accepted if they are less than 10% over the estimate, what is the maximum acceptable bid?

10% of $356,200 = (0.10)($356,200) = $35,620

The upper limit or maximum acceptable bid is $356,200 + $35,620 = $391,820.　■

Exercises 4.7

For each measurement, find the precision, the greatest possible error, the relative error, and the percent of error (to the nearest hundredth percent):

1. 1400 lb
2. 240,000 Ω
3. 875 rpm
4. 12,500 V
5. 0.085 g
6. 0.188 cm
7. 2 g
8. 2.2 g
9. 2.22 g
10. 18,0$\overline{0}$0 W
11. 1.00 kg
12. 1.0 kg
13. 0.041 A
14. 0.08 ha
15. $11\frac{7}{8}$ in.
16. $1\frac{3}{4}$ in.
17. 12 ft 8 in.
18. 4 lb 13 oz

Compare each set of measurements by indicating which measurement is better or best:

19. 13.5 cm; $8\frac{3}{4}$ in.
20. 364 m; 36.4 cm
21. 16 mg; 19.7 g; $12\frac{3}{16}$ oz
22. 68,000 V; 3450 Ω; 3.2 A

Complete the table:

	Given measurement	Tolerance	Upper limit	Lower limit	Tolerance interval
23.	$3\frac{1}{2}$ in.	$\pm\frac{1}{8}$ in.	$3\frac{5}{8}$ in.	$3\frac{3}{8}$ in.	$\frac{1}{4}$ in.
24.	$5\frac{3}{4}$ in.	$\pm\frac{1}{16}$ in.			
25.	$6\frac{5}{8}$ in.	$\pm\frac{1}{32}$ in.			
26.	$7\frac{7}{16}$ in.	$\pm\frac{1}{32}$ in.			
27.	$3\frac{7}{16}$ in.	$\pm\frac{1}{64}$ in.			
28.	$\frac{9}{64}$ in.	$\pm\frac{1}{128}$ in.			
29.	$3\frac{3}{16}$ in.	$\pm\frac{1}{128}$ in.			
30.	$9\frac{3}{16}$ mi	$\pm\frac{1}{32}$ mi			
31.	1.19 cm	±0.05 cm			
32.	1.78 m	±0.05 m			
33.	0.0180 A	±0.0005 A			
34.	9.437 L	±0.001 L			
35.	24,000 V	±2000 V			
36.	375,000 W	±10,000 W			
37.	10.31 km	±0.05 km			
38.	21.30 kg	±0.01 kg			

Complete the table:

	Architect's estimate	Maximum rate above estimate	Maximum acceptable bid
39.	$48,250	10%	
40.	$259,675	7%	
41.	$1,450,945	8%	
42.	$8,275,625	5%	

4.8 Color Code of Electrical Resistors

The resistance of an electrical resistor is often given in a color code. A series of four colored bands is painted on the resistor. Each color on any of the first three bands stands for a digit or number as given in the following table.

Color on any of the first three bands	Digit or number
Black	0
Brown	1
Red	2
Orange	3
Yellow	4
Green	5
Blue	6
Violet	7
Gray	8
White	9
Gold on the third band	Multiply the value by 0.1
Silver on the third band	Multiply the value by 0.01

The fourth band indicates the tolerance of the resistor as given in the following table.

Color of the fourth band	Tolerance
Gold	±5%
Silver	±10%
Black or no fourth band	±20%

The value of each resistor is in ohms, Ω, and is given in two significant digits. The color bands are read from left to right when the resistor is in the position shown in Figure 4.31.

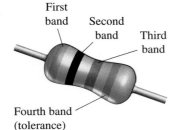

First band
Second band
Third band
Fourth band (tolerance)

Electrical resistor

FIGURE 4.31

Finding the Value of a Resistor

Step 1   The digit corresponding to the color of the first band is the first digit of the resistance.

Step 2   The digit corresponding to the color of the second band is the second digit of the resistance.

Step 3   **a.** The third band indicates the number of zeros to be written after the first two digits from Steps 1 and 2.

b. If the third band is gold, multiply the number corresponding to the digits from Steps 1 and 2 by 0.1. That is, place the decimal point *between* the two digits.

      c. If the third band is silver, multiply the number corresponding to the digits from Steps 1 and 2 by 0.01. That is, place the decimal point *before* the two digits.

Step 4   The fourth band indicates the tolerance written as a percent. The tolerance is

      a. ±5% if the fourth band is gold

      b. ±10% if the fourth band is silver

      c. ±20% if the fourth band is black or if there is no fourth band.

 Example 1

Find the resistance of the resistor shown in Figure 4.32.

Step 1   The first digit is 4—the digit that corresponds to *yellow*.

Step 2   The second digit is 5—the digit that corresponds to *green*.

Step 3A   *Orange* on the third band means that three zeros should be written after the digits from Steps 1 and 2.

So the resistance is 45,000 Ω.

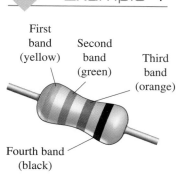

First band (yellow) Second band (green) Third band (orange)

Fourth band (black)

FIGURE 4.32

 Example 2

Find the tolerance, the upper limit, the lower limit, and the tolerance interval for the resistor shown in Figure 4.32.

The black fourth band indicates a tolerance of ±20%.
20% of 45,000 Ω = (0.20)(45,000 Ω) = 9000 Ω.
That is, the tolerance is ±9000 Ω.

The upper limit is 45,000 Ω + 9000 Ω = 54,000 Ω.
The lower limit is 45,000 Ω − 9000 Ω = 36,000 Ω.
The tolerance interval is then 18,000 Ω.

 Example 3

Find the resistance of the resistor shown in Figure 4.33.

Step 1   The first digit is 3—the digit that corresponds to *orange*.

Step 2   The second digit is 0—the digit that corresponds to *black*.

Step 3A   *Red* on the third band means that two zeros should be written after the digits from Steps 1 and 2.

So the resistance is 3000 Ω.

First band (orange) Second band (black) Third band (red)

Fourth band (silver)

FIGURE 4.33

Example 4   Find the tolerance, the upper limit, the lower limit, and the tolerance interval for the resistor shown in Figure 4.33.

The silver fourth band indicates a tolerance of ±10%.
10% of 3000 Ω = (0.10)(3000 Ω) = 300 Ω.
That is, the tolerance is ±300 Ω.

The upper limit is 3000 Ω + 300 Ω = 3300 Ω.
The lower limit is 3000 Ω − 300 Ω = 2700 Ω.
The tolerance interval is then 600 Ω.   ■

 Example 5   Find the resistance of the resistor shown in Figure 4.34.

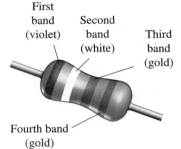

First
band
(violet)   Second
band
(white)   Third
band
(gold)

Fourth band
(gold)

FIGURE 4.34

Step 1   The first digit is 7—the digit that corresponds to *violet*.

Step 2   The second digit is 9—the digit that corresponds to *white*.

Step 3B   *Gold* on the third band means to place the decimal point *between* the digits from Steps 1 and 2.

So the resistance is 7.9 Ω.   ■

 Example 6   Find the resistance of the resistor shown in Figure 4.35.

First
band
(gray)   Second
band
(brown)   Third
band
(black)

Fourth band
(silver)

FIGURE 4.35

Step 1   The first digit is 8—the digit that corresponds to *gray*.

Step 2   The second digit is 1—the digit that corresponds to *brown*.

Step 3A   *Black* on the third band means that no zeros are to be included after the digits from Steps 1 and 2.

So the resistance is 81 Ω.   ■

 Example 7   Find the resistance of the resistor shown in Figure 4.36.

First
band
(blue)   Second
band
(green)   Third
band
(silver)

Fourth band
(gold)

FIGURE 4.36

Step 1   The first digit is 6—the digit that corresponds to *blue*.

Step 2   The second digit is 5—the digit that corresponds to *green*.

Step 3C   *Silver* on the third band means to place the decimal point *before* the digits from Steps 1 and 2.

So the resistance is 0.65 Ω.   ■

Example 8

A serviceperson needs a 680,000-Ω resistor. What color code on the first three bands is needed?

Step 1 The color that corresponds to the first digit, 6, is *blue*.

Step 2 The color that corresponds to the second digit, 8, is *gray*.

Step 3A The color that corresponds to four zeros is *yellow*.

So the colors that the serviceperson is looking for are blue, gray, and yellow.   ■

Exercises 4.8

For each resistor shown, find the resistance and the tolerance, written as a percent:

1.   First band (orange)   Second band (blue)   Third band (brown)   Fourth band (silver)

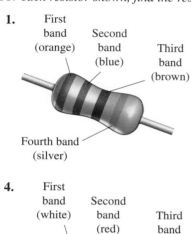

2.   First band (green)   Second band (gray)   Third band (gold)   Fourth band (gold)

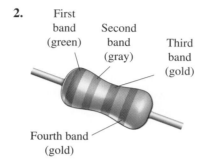

3.   First band (gray)   Second band (orange)   Third band (yellow)   Fourth band (black)

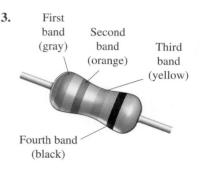

4.   First band (white)   Second band (red)   Third band (red)   Fourth band (silver)

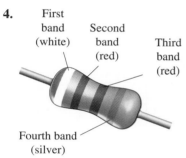

5.   First band (brown)   Second band (yellow)   Third band (green)   Fourth band (black)

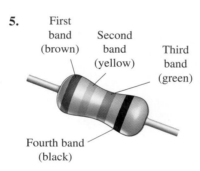

6.   First band (red)   Second band (red)   Third band (silver)   Fourth band (gold)

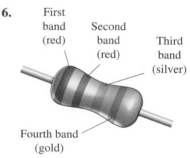

7.   First band (violet)   Second band (black)   Third band (black)   Fourth band (gold)

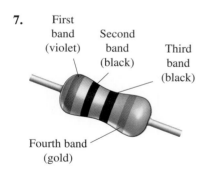

8.   First band (white)   Second band (orange)   Third band (silver)   Fourth band (silver)

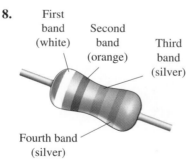

9.   First band (green)   Second band (black)   Third band (yellow)   Fourth band (black)

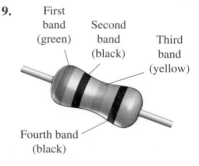

10.   First band (blue)   Second band (black)   Third band (gold)   Fourth band (silver)

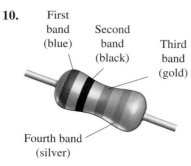

11.   First band (brown)   Second band (black)   Third band (blue)   Fourth band (black)

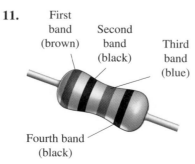

12.   First band (violet)   Second band (green)   Third band (yellow)   Fourth band (black)

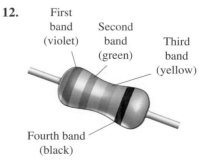

What color code on the first three bands is needed for each resistance?

*Find **a.** the tolerance in ohms, Ω, **b.** the upper limit, **c.** the lower limit, and **d.** the tolerance interval for each resistor:*

13. 4800 Ω   **14.** 95 Ω   **15.** 72,000 Ω

16. 3.1 Ω   **17.** 650,000 Ω   **18.** 100 Ω

19. 0.25 Ω   **20.** 9000 Ω   **21.** 4,500,000 Ω

22. 40 Ω   **23.** 7.6 Ω   **24.** 0.34 Ω

25. Exercise 1   **26.** Exercise 2   **27.** Exercise 3

28. Exercise 5   **29.** Exercise 7   **30.** Exercise 12

4.9 Reading Scales

Circular Scales

In reading circular scales, you must first determine the basic unit of each scale being used. Circular *dial indicators* or dial gauges are useful for making very precise comparisons between a known measurement and some measurement that must be checked for precision. They are used for inspection operations, in toolrooms, and in machine shops in a wide variety of applications. Some of these uses are shown in Figure 4.37.

(a) To ensure positive vise alignment on milling machine, operator uses indicator against parallel clamped in vise jaws.

(b) Precise alignment of cutting bar is assured by using a dial indicator.

(c) Planer operator uses dial test indicator to check depth of cut on flat casting.

(d) Lathe operator uses dial indicator to check total indicator runout (TIR).

FIGURE 4.37
Uses of a dial indicator

Let's first study the metric dial indicator shown in Figure 4.38(a). Each graduation represents 0.01 mm. If the needle deflects six graduations to the right (+) of zero, the object being measured is 6 × 0.01 mm. This is 0.06 mm larger than the desired measurement. If the needle deflects 32 graduations to the left (−) of zero, the object being measured is 32 × 0.01 mm. This is 0.32 mm smaller than the desired measurement.

(a) Metric dial indicator   (b) U.S. dial indicator

FIGURE 4.38

Note the smaller dial on the lower left portion of the dial in Figure 4.38(a). This small needle records the number of complete revolutions that the large needle makes. Each complete revolution of the large needle corresponds to 1.00 mm.

◆ Example 1

FIGURE 4.39

Read the metric dial indicator in Figure 4.39.

The small needle reads +3   × 1.00 mm = +3.00 mm
The large needle reads +36 × 0.01 mm = +0.36 mm
The total reading is   +3.36 mm

This measurement is 3.36 mm more than the desired measurement.   ■

Now look closely at the U.S. dial indicator in Figure 4.38(b). Each graduation represents 0.001 in. If the needle deflects 7 graduations to the right (+) of zero, the object being measured is 7 × 0.001 in. This is 0.007 in. larger than the desired measurement. If the needle deflects to the left (−) 14 graduations, the object being measured is 14 × 0.001 in. This is 0.014 in. smaller than the desired measurement.

Note the smaller dial on the lower left portion of the dial in Figure 4.38(b). This small needle records the number of complete revolutions that the large needle makes. Each complete revolution of the large needle corresponds to 0.100 in. Other dials are read in a similar manner.

FIGURE 4.40

Example 2

Read the U.S. dial indicator in Figure 4.40.

The small needle reads -2×0.100 in. $= -0.200$ in.
The large needle reads -23×0.001 in. $= \underline{-0.023 \text{ in.}}$
The total reading is -0.223 in.

This measurement is 0.223 in. less than the desired measurement.   ■

Digital dial indicators that provide metric and U.S. (mm/inch) readings by means of a one-button process are also in common use as shown in Figure 4.41. These displays read down to 0.01 mm or 0.001 in., which indicates its precision. Before each measurement, check that it reads zero or press the zero button and then make your measurement as before.

(a)   (b)

FIGURE 4.41
Digital dial indicators with display in (a) mm mode and (b) decimal inch mode

Uniform Scales

Figure 4.42 shows some of the various scales that may be found on a *volt-ohm meter* (VOM). This instrument is used to measure voltage (measured in volts, V) and resistance (measured in ohms, Ω) in electrical circuits. Note that the voltage scales are uniform, while the resistance scale is nonuniform. On a given voltage scale, the graduations are equally spaced and each subdivision represents the same number of volts. On the resistance scale, the graduations are not equally spaced and subdivisions on various intervals represent different numbers of ohms. To make things clear in the examples and exercises that follow, we show only one of the VOM scales at a time in a given figure.

FIGURE 4.42
Volt-ohm meter (VOM) scales

The first uniform voltage scale that we study is shown in Figure 4.43. This scale has a range of 0–10 V. There are 10 large divisions, each representing 1 V. Each large division is divided into 5 equal subdivisions. Each subdivision is $\frac{1}{5}$ V, or 0.2 V.

Example 3 Read the scale shown in Figure 4.43.

FIGURE 4.43

The needle is on the third graduation to the right of 8. Each subdivision is 0.2 V. Therefore, the reading is 8.6 V. ■

Figure 4.44 shows a voltage scale that has a range of 0–2.5 V. There are 5 large divisions, each representing 0.5 V. Each division is divided into 5 subdivisions. Each subdivision is $\frac{1}{5} \times 0.5$ V $= 0.1$ V.

Example 4 Read the scale shown in Figure 4.44.

FIGURE 4.44

The needle is on the second graduation to the right of 1.5. Each subdivision is 0.1 V. Therefore, the reading is 1.7 V. ■

Figure 4.45 shows a voltage scale that has a range of 0–250 V. There are 10 large divisions, each representing 25 V. Each division is divided into 5 subdivisions. Each subdivision is $\frac{1}{5} \times 25$ V = 5 V.

◆ **Example 5**   Read the scale shown in Figure 4.45.

FIGURE 4.45

The needle is on the first graduation to the right of 150. Each subdivision is 5 V. Therefore, the reading is 155 V.   ■

Nonuniform Scales

Figure 4.46 shows a nonuniform ohm scale usually found on a VOM. First, consider that part of the scale between 0 and 5. Each large division represents 1 ohm (Ω). Each large division is divided into 5 subdivisions. Therefore, each subdivision represents $\frac{1}{5} \times 1$ Ω, or 0.2 Ω.

FIGURE 4.46

- Between 5 and 10, each subdivision is divided into 2 sub-subdivisions. Each sub-subdivision represents $\frac{1}{2} \times 1$ Ω, or 0.5 Ω.
- Between 10 and 20, each division represents 1 Ω.
- Between 20 and 100, each large division represents 10 Ω. Between 20 and 30, there are 5 subdivisions. Therefore, each subdivision represents $\frac{1}{5} \times 10$ Ω, or 2 Ω.
- Between 30 and 100, each large division has 2 subdivisions. Each subdivision represents $\frac{1}{2} \times 10$ Ω, or 5 Ω.
- Between 100 and 200, each large division represents 50 Ω.
- Between 100 and 150, there are 5 subdivisions. Each subdivision represents $\frac{1}{5} \times 50$ Ω, or 10 Ω.
- Between 200 and 500, there are 3 subdivisions. Each subdivision represents $\frac{1}{3} \times 300$ Ω, or 100 Ω.

The subdivisions on each part of the ohm scale may be summarized as follows:

Range	Each subdivision represents:
0–5 Ω	0.2 Ω
5–10 Ω	0.5 Ω
10–20 Ω	1 Ω
20–30 Ω	2 Ω
30–100 Ω	5 Ω
100–150 Ω	10 Ω
200–500 Ω	100 Ω

 Example 6

Read the scale shown in Figure 4.46.

The needle is on the second subdivision to the left of 2, where each subdivision represents 0.2 Ω. Therefore, the reading is 2.4 Ω. ■

Example 7

Read the scale shown in Figure 4.47.

FIGURE 4.47

The needle is on the subdivision between 70 and 80. A subdivision represents 5 Ω on this part of the scale. Therefore, the reading is 75 Ω. ■

Exercises 4.9

Read each metric dial indicator (the arrow near the zero indicates the direction of deflection of the needle):

1.

2.

3.

Read each U.S. dial indicator (the arrow near the zero indicates the direction of deflection of the needle):

Read each scale:

19.

20.

21.

22.

23.

24.

25.

26.

27.

28.

29.

30.

31.

32.

33.

34.

35.

36.

37.

38.

39.

40.

41.

42.

Chapter 4   Group Activities

1. In industry, mathematics is used in many places. One area is measurement, specifically with regard to precision and accuracy. Usually, an industrial plant has a quality assurance department whose responsibility is to ensure that all the measurement instruments, such as a vernier caliper, are within a certain specification. In your groups, do the following. Each member measure once with a yardstick five or six objects around the room, e.g., book, top of desk, chalkboard, room door, length of room, and pencil. Do not report your measurements until each member of the group has measured all the objects. Come back together and compare your measurements. Are all the measurements of each object the same? How different are they? Now measure the objects again, using a ruler. Again, did each member of the group get the same measurements? If you converted the yardstick measurements to feet or inches, would the measurements be the same? Explain any differences in the measurements. Think about all the factors that could have made the measurements different. Note the smallest unit of measure on each measuring device. Does this pose a problem? Discuss this in your group and report your findings. To find out more, consult a local manufacturer or industry quality assurance department. Ask about SPS (statistical process control) and gauge R&R (gauge repeatability and reproducibility). The library or Internet resources can help explain the importance of gauge R&R and SPS in industry.

2. Take inside and outside measurements for a portion of one of your school's buildings as indicated by your instructor. Choose an appropriate scale and show the numerical values of all the dimensions. Compare the dimensions from your drawings with those in a full set of architectural plans.

Chapter 4   Summary

Glossary of Basic Terms

Accuracy of a measurement.   The number of significant digits that a measurement contains. These indicate the number of units we are reasonably sure of having counted when making the measurement. The greater the number of significant digits given in a measurement, the better the accuracy, and vice versa. (p. 168)

Approximate number.   A number that has been determined by some measurement process. (p. 167)

Exact number.   A number that has been determined as a result of counting or by some definition. (p. 167)

Greatest possible error of a measurement.   One-half of the smallest unit on the scale on which the measurement is read and equal to one-half of the measurement's precision. (p. 172)

Measurement.   The comparison of an *observed* quantity with a *standard unit* quantity. (p. 167)

Micrometer caliper.   An instrument for measuring very small lengths using the movement of a finely threaded rotating screw, which gives it better precision than a vernier caliper. It is used in technical fields in which fine precision is required. (See Figure 4.18 on p. 181.)

Percent of error of a measurement.   The relative error expressed as a percent. (p. 196)

Precision of a measurement.   The smallest unit with which the measurement is made; that is, the position of the last significant digit or the smallest unit or calibration on the measuring instrument. (p. 169)

Relative error of a measurement.   The greatest possible error divided by the measurement itself. (p. 196)

Significant digits.   Those digits in a number that we are reasonably sure of having counted and of being able to rely on in a measurement. (p. 168)

Tolerance.   The acceptable amount that a given part or component may vary from a given size. (p. 197)

Tolerance interval.   The difference between the upper limit and the lower limit. (p. 197)

Vernier caliper.   A slide-type measuring instrument used to take precise inside, outside, and depth measurements. (See Figure 4.7 on p. 173.)

4.1 Approximate Numbers and Accuracy

1. **Exact versus approximate numbers:**
   a. Only counting numbers are exact.
   b. All measurements are approximations. (p. 168)

2. **Significant digits:**
   a. The following digits are *significant:*
      • All nonzero digits
      • All zeros between significant digits
      • A zero in a whole-number measurement that is specially tagged, such as by a bar above it
      • All zeros to the right of a significant digit *and* a decimal point
   b. The following digits are *not significant:*
      • Zeros to the right in a whole-number measurement that are not tagged
      • Zeros to the left in a decimal measurement that is less than 1 (pp. 168–169)

4.3 The Vernier Caliper

1. **Review this section to read the various vernier calipers presented. (p. 173)**

4.4 The Micrometer Caliper

1. **Review this section to read the various micrometer calipers presented. (p. 181)**

4.5 Addition and Subtraction of Measurements

1. **Adding or subtracting measurements of different precision:** To add or subtract measurements of different precision,
   a. Make certain that all measurements are expressed in the same unit. If they are not, change them all to any common unit.
   b. Add or subtract.
   c. Then round the result to the same precision as the *least precise* measurement. (p. 190)

4.6 Multiplication and Division of Measurements

1. **Multiplying or dividing measurements:** To multiply or divide measurements,
   a. First, multiply and/or divide the measurements.
   b. Then round the result to the same number of significant digits as the measurement that has the least number of significant digits. That is, round the result to the same accuracy as the *least accurate* measurement. (p. 193)

4.7 Relative Error and Percent of Error

1. **Relative error:**

   $$\text{relative error} = \frac{\text{greatest possible error}}{\text{measurement}} \quad (\text{p. 196})$$

2. **Percent of error:**

   percent of error = the relative error expressed as a percent (p. 196)

4.8 Color Code of Electrical Resistors

1. **Review this section to read the color code of electrical resistors, which gives the size and tolerance of a given resistor. (p. 200)**

4.9 Reading Scales

1. **Review this section to read the various circular scales, uniform scales, and nonuniform scales presented. (p. 204)**

 # Chapter 4 Review

Give the number of significant digits (the accuracy) of each measurement:

1. 4.06 kg
2. 24,000 mi
3. $36\overline{0}0$ V
4. 5.60 cm
5. 0.0070 W
6. 0.0651 s
7. 20.00 m
8. 20.050 km

*Find **a.** the precision and **b.** the greatest possible error of each measurement:*

9. 6.05 m
10. 15.0 mi
11. 160,500 L
12. 2300 V
13. 17.00 cm
14. $13,0\overline{0}0,000$ V
15. $1\frac{5}{8}$ in.
16. $10\frac{3}{16}$ mi

Read the measurement shown on the vernier caliper in Illustration 1 in

17. Metric units   **18.** U.S. units

ILLUSTRATION 1

19. Read the measurement shown on the metric micrometer in Illustration 2.

ILLUSTRATION 2

20. Read the measurement shown on the U.S. micrometer in Illustration 3.

ILLUSTRATION 3

21. Find the measurement that is **a.** the most accurate and **b.** the most precise:

2500 V; 36,500 V; 60,000 V; 9.6 V; 120 V

22. Find the measurement that is **a.** the least accurate and **b.** the least precise:

0.0005 A; 0.0060 A; 0.425 A; 0.0105 A; 0.0055 A

Use the rules for addition of measurements to find the sum of each set of measurements:

23. 18,000 W; 260,000 W; 2300 W; 45,500 W; 398,000 W

24. 16.8 cm; 19.7 m; 0.14 km; 240 m

25. Use the rules for subtraction of measurements to subtract:

1,500,000 V − 1,125,000 V

Use the rules for multiplication and/or division of measurements to evaluate:

26. 15.6 cm × 18.5 cm × 6.5 cm

27. $\dfrac{98.2 \text{ m}^3}{16.7 \text{ m}}$ **28.** $\dfrac{239 \text{ N}}{(24.8 \text{ m})(6.7 \text{ m})}$ **29.** $\dfrac{(220 \text{ V})^2}{365 \text{ }\Omega}$

*Find **a.** the relative error and **b.** the percent of error (to the nearest hundredth percent) for each measurement:*

30. $5\dfrac{7}{16}$ in.   **31.** 15.60 cm

32. Given a resistor of 2000 Ω with a tolerance of ±10%, find the upper and lower limits.

For each resistor find its resistance and its tolerance written as a percent:

33. First band (brown)   Second band (red)   Third band (yellow)

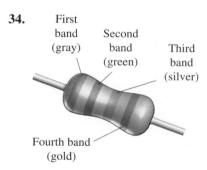

Fourth band (black)

34. First band (gray)   Second band (green)   Third band (silver)

Fourth band (gold)

Read each scale:

35.

36.

37.

 Chapter 4   Test

Give the number of significant digits in each measurement:

1. 1.806 g   **2.** 7.00 L   **3.** 0.00015 A

*Find **a.** the precision and **b.** the greatest possible error of each measurement:*

4. 6.13 mm   **5.** 2400 Ω   **6.** $5\frac{3}{4}$ in.

Read the measurement shown on the vernier caliper in Illustration 1 in

7. Metric units   **8.** U.S. units

ILLUSTRATION 1

9. Read the measurement shown on the metric micrometer in Illustration 2.

ILLUSTRATION 2

10. Read the measurement shown on the U.S. micrometer in Illustration 3.

ILLUSTRATION 3

11. Find the measurement that is **a.** the most accurate, **b.** the most precise, **c.** the least accurate, and **d.** the least precise:

208 m; 17,060 m; 25.9 m; 0.067 m

12. Find the measurement that is **a.** the most accurate, **b.** the most precise, **c.** the least accurate, and **d.** the least precise:

360 V; 0.5 V; 125,000 V; 600,000 V

13. Use the rules of measurement to multiply:

(4.0 m)(12 m)(0.60 m)

14. Use the rules of measurement to add:

12.9 L + 341 L + 2104 L

15. Use the rules of measurement to subtract:

108.07 g − 56.1 g

16. Use the rules of measurement to divide:

6.28 m^2 ÷ 25 m

17. Use the rules of measurement to evaluate:

$$\frac{(56.3 \text{ m})(25 \text{ m})(112.5 \text{ m})}{(21.275 \text{ m})^2}$$

18. Find **a.** the relative error and **b.** the percent of error (to the nearest hundredth percent) for the measurement 5.20 m

Read each scale:

19.

20.

Cumulative Review

1. Evaluate: $1\frac{3}{8} - \frac{1}{2} \times \frac{3}{4} + 1\frac{5}{8} \div \frac{1}{16}$

2. Round 32,518.612 to nearest **a.** ten **b.** hundredth.

3. 18.84 is 31.4% of what number?

4. Evaluate: $(-4)(5) + (-6)(-4) - 7(-4) \div 2(7)$

5. Perform the indicated operations and simplify. Express the result using positive exponents. $\dfrac{(10^3 \cdot 10^{-2})^3}{10^3 \cdot 10^5}$

6. Perform the indicated operations and write the result in scientific notation. $\dfrac{(62.3 \times 10^3)(4.18 \times 10^{-5})}{(17.3 \times 10^{-4})^2}$

7. Give the SI abbreviation for milli.

8. Write the abbreviation for 25 kilograms.

9. Write the SI unit for 250 μs.

10. Which is larger: 1 amp or 1 mega-amp?

Round each result to three significant digits when necessary.

11. Change 120 km to m. 12. Change 250 cm to m.

13. Change 50 g to kg.

14. Change 4060 kg to metric tons.

15. Change 86°C to °F. 16. Change 50°F to °C.

17. Change 163 in² to cm². 18. Change 120 m to km.

19. Change 10 L to mL.

Give the number of significant digits in each measurement:

20. 0.25 Ω 21. 7.002 m

*Find **a.** the precision and **b.** the greatest possible error of each measurement:*

22. 14.28 mm 23. 62.3 lb

Read the measurement on the vernier caliper in Illustration 1 in

24. Metric units. 25. U.S. units.

ILLUSTRATION 1

26. Read the measurement shown on the metric micrometer in Illustration 2.

ILLUSTRATION 2

27. Read the measurement shown on the U.S. micrometer in Illustration 3.

ILLUSTRATION 3

Use the rules for addition and subtraction of measurements to evaluate:

28. Add: 6120 km, 1743 km, 1400 km, 25,608 km

29. Subtract: 98.2 L − 52.16 L

Use the rules for multiplication and/or division of measurements to evaluate:

30. (283 cm)(150 cm)   **31.** 583 ft^2 ÷ 17.28 ft

32. Find **a.** the precision, **b.** the greatest possible error, **c.** the relative error, and **d.** the percent of error to nearest hundredth percent of the measurement 2.135 cm.

Read each scale:

33.

ILLUSTRATION 4

34.

ILLUSTRATION 5

5

Polynomials: An Introduction to Algebra

Mathematics at Work

The nation's construction industry depends on a technical and competent workforce. This workforce includes, but is not limited to, carpenters who cut, fit, and assemble wood and other materials in construction projects; plumbers, pipefitters, and steamfitters who install, maintain, and repair many different types of pipe systems that carry water, steam, air, and other liquids; painters who apply paint, stain, varnish, and other finishes to buildings and other structures; electricians who install, maintain, and repair electrical wiring, equipment, and fixtures; bricklayers and stonemasons who build walls and other structures with bricks, blocks, stones, and other masonry materials; and structural and reinforcing metal workers who use materials made from iron, steel, and other materials to construct highways, bridges, buildings, and towers.

Construction trade workers often learn their own trade through apprenticeship programs administered by local joint union–management committees or through community college or trade school programs, some of which are offered in partnership with the local joint union–management committees. For more information, go to the website listed below.

BrandXPictures/Jupiter Images

Construction Trades
Carpenter framing a building.

www.cengage.com/mathematics/ewen

Objectives

■ Apply the rules for order of operations to evaluate expressions with numbers and to evaluate algebraic expressions when the values of the letters are given.

■ Simplify algebraic expressions by removing parentheses and combining like terms.

■ Add and subtract polynomials.

■ Multiply monomials.

■ Multiply polynomials.

■ Divide a monomial and a polynomial by a monomial.

■ Divide a polynomial by a polynomial.

5.1   Fundamental Operations

In arithmetic, we perform mathematical operations with specific numbers. In algebra, we perform these same basic mathematical operations with numbers and **variables**—letters that represent unknown quantities. Algebra allows us to express and solve general as well as specific problems that cannot be solved using only arithmetic. As a result, employers in technical and scientific areas require a certain level of skill and knowledge of algebra. Your problem-solving skills will increase significantly as your algebra skills increase.

To begin our study of algebra, some basic mathematical principles that you will apply are listed below. Most of them you probably already know; the rest will be discussed. Note that "$\neq$" means "is not equal to."

Basic Mathematical Principles

1. $a + b = b + a$ (Commutative Property for Addition)
2. $ab = ba$ (Commutative Property for Multiplication)
3. $(a + b) + c = a + (b + c)$ (Associative Property for Addition)
4. $(ab)c = a(bc)$ (Associative Property for Multiplication)
5. $a(b + c) = ab + ac$, or $(b + c)a = ba + ca$ (Distributive Property)
6. $a + 0 = a$
7. $a \cdot 0 = 0$
8. $a + (-a) = 0$ (Additive Inverse)
9. $a \cdot 1 = a$
10. $a \cdot \dfrac{1}{a} = 1$ $(a \neq 0)$ (Multiplicative Inverse)

In mathematics, letters are often used to represent numbers. Thus, it is necessary to know how to indicate arithmetic operations and carry them out using letters.

Addition: $x + y$ means add x and y.

Subtraction: $x - y$ means subtract y from x or add the negative of y to x; that is, $x + (-y)$.

Multiplication: xy or $x \cdot y$ or $(x)(y)$ or $(x)y$ or $x(y)$ means multiply x by y.

Division: $x \div y$ or $\dfrac{x}{y}$ means divide x by y, or find a number z such that $zy = x$.

Exponents: $xxxx$ means use x as a factor 4 times, which is abbreviated by writing x^4. In the expression x^4, x is called the *base*, and 4 is called the **exponent**. For example, 2^4 means $2 \cdot 2 \cdot 2 \cdot 2 = 16$.

Order of Operations

1. Perform all operations inside parentheses first. If the problem contains a fraction bar, treat the numerator and the denominator separately.
2. Evaluate all powers, if any. For example, $6 \cdot 2^3 = 6 \cdot 8 = 48$.
3. Perform any multiplications or divisions in order, from left to right.
4. Do any additions or subtractions in order, from left to right.

 Example 1

Evaluate: $4 - 9(6 + 3) \div (-3)$.

$$= 4 - 9(9) \div (-3) \qquad \text{Add within parentheses.}$$
$$= 4 - 81 \div (-3) \qquad \text{Multiply.}$$
$$= 4 - (-27) \qquad \text{Divide.}$$
$$= 31 \qquad \text{Subtract.}$$

■

 Example 2

Evaluate: $(-6) + 5(-2)^2(-9) - 7(3 - 5)^3$.

$$= (-6) + 5(-2)^2(-9) - 7(-2)^3 \qquad \text{Subtract within parentheses.}$$
$$= (-6) + 5(4)(-9) \quad - 7(-8) \qquad \text{Evaluate the powers.}$$
$$= (-6) - 180 \qquad + 56 \qquad \text{Multiply.}$$
$$= -130 \qquad \text{Add and subtract.}$$

■

To **evaluate an expression**, replace the letters with given numbers; then do the arithmetic using the order of operations. The result is the value of the expression.

Example 3

Evaluate: $\dfrac{x^2 - y + 5}{2x - 2}$, if $x = 4$ and $y = 3$.

$$\frac{x^2 - y + 5}{2x - 2} = \frac{4^2 - 3 + 5}{2(4) - 2} \qquad \text{Replace } x \text{ with 4 and } y \text{ with 3.}$$

$$= \frac{16 - 3 + 5}{8 - 2} \qquad \text{Evaluate the power and multiply.}$$

$$= \frac{18}{6} \qquad \text{Add and subtract.}$$

$$= 3 \qquad \text{Divide.}$$

Note: In a fraction, the line between the numerator and denominator serves as parentheses for both. That is, do the operations in both numerator and denominator before doing the division.

■

Example 4   Evaluate: $\dfrac{ab}{3c} + c$, if $a = 6$, $b = 10$, and $c = -5$.

$$\frac{ab}{3c} + c = \frac{6 \cdot 10}{3(-5)} + (-5) \qquad \text{Replace } a \text{ with 6, } b \text{ with 10, and } c \text{ with } -5.$$

$$= \frac{60}{-15} + (-5) \qquad \text{Multiply in the numerator and in the denominator.}$$

$$= -4 + (-5) \qquad \text{Divide.}$$

$$= -9 \qquad \text{Add.} \qquad ■$$

Exercises 5.1

Evaluate each expression:

1. $3(-5)^2 - 4(-2)$

2. $(-2)(-3)^2 + 3(-2) \div 6$

3. $4(-3) \div (-6) - (-18) \div 3$

4. $48 \div (-2)(-3) + (-2)^2$

5. $(-72) \div (-3) \div (-6) \div (-2) - (-4)(-2)(-5)$

6. $28 \div (-7)(2)^2 + 3(-4 - 2)^2 - (-3)^2$

7. $[(-2)(-3) + (-24) \div (-2)] \div [-10 + 7(-1)^2]$

8. $(-9)^2 \div 3^3(6) + [3(-2) - 5(-3)]$

9. $[(-2)(-8)^2 \div (-2)^3] - [-4 + (-2)^4]^2$

10. $[(-2)(3) + 5(-2)][5(-4) - 8(-3)]^2$

In Exercises 11–16, let $x = 2$ and $y = 3$, and evaluate each expression:

11. $2x - y$

12. $x - 2y$

13. $x^2 - y^2$

14. $5y^2 - x^2$

15. $\dfrac{3x + y}{3 + y}$

16. $\dfrac{2(x + y) - 2x}{2(y - x)}$

In Exercises 17–26, let $x = -1$ and $y = 5$, and evaluate each expression:

17. $xy^2 - x$

18. $4x^3 - y^2$

19. $\dfrac{2y}{x} - \dfrac{2x}{y}$

20. $3 + 4(x + y)$

21. $3 - 4(x + y)$

22. $1.7 - 5(2x - y)$

23. $\dfrac{1}{x} - \dfrac{1}{y} + \dfrac{2}{xy}$

24. $(2.4 - x)(x - xy)$

25. $\dfrac{y - 4x}{3x - 6xy}$

26. $\dfrac{(y - x)^2 - 4y}{4x^2 + 2}$

In Exercises 27–32, let $x = -3$, $y = 4$, and $z = 6$. Evaluate each expression:

27. $(2xy^2z)^2$

28. $(x^2 - y^2)z$

29. $(y^2 - 2x^2)z^2$

30. $\left(\dfrac{x + 3y}{z}\right)^2$

31. $\dfrac{(7 - x)^2}{z - y}$

32. $(2x + 3y)(y + z)$

In Exercises 33–40, let $x = -1$, $y = 2$, and $z = -3$. Evaluate each expression:

33. $(2x + 6)(3y - 4)$

34. $z^2 - 5yx^2$

35. $(3x + 5)(2y - 1)(5z + 2)$

36. $(3x - 4z)(2x + 3z)$

37. $(x - xy)^2(z - 2x)$

38. $3x^2(y - 3z)^2 - 6x$

39. $(x^2 + y^2)^2$

40. $(3x^2 - z^2)^2$

41. $\dfrac{x^2 + (z - y)^2}{4x^2 + z^2}$

42. $\dfrac{(3x^2 + 2)^2 - y^2}{6 - 3x^2y^2}$

5.2 Simplifying Algebraic Expressions

Parentheses are often used to clarify the order of operations when the order of operations is complicated or may be ambiguous. Sometimes it is easier to simplify such an expression by first removing the parentheses—before doing the indicated operations. Two rules for removing parentheses are as follows

> ### Removing Parentheses
>
> 1. Parentheses preceded by a plus sign may be removed without changing the signs of the terms within. Think of using the Distributive Property, $a(b + c) = ab + ac$, from Section 5.1 and multiplying each term inside the parentheses by 1. That is,
>
> $$3w + (4x + y) = 3w + 4x + y$$
>
> 2. Parentheses preceded by a minus sign may be removed if the signs of *all* the terms within the parentheses are changed; then the minus sign that preceded the parentheses is dropped. Think of using the Distributive Property, $a(b + c) = ab + ac$, from Section 5.1 and multiplying each term inside the parentheses by -1. That is,
>
> $$3w - (4x - y) = 3w - 4x + y$$

(Notice that the sign of the term $4x$ inside the parentheses is not written. It is therefore understood to be plus.)

Example 1

Remove the parentheses from the expression $5x - (-3y + 2z)$.

$$5x - (-3y + 2z) = 5x + 3y - 2z$$ Change the signs of *all* of the terms within
parentheses; then drop the minus sign that
precedes the parentheses.   ■

Example 2

Remove the parentheses from the expression $7x + (-y + 2z) - (w - 4)$.

$$7x + (-y + 2z) - (w - 4) = 7x - y + 2z - w + 4$$

Drop the plus sign before the first set of parentheses and do not change any of the signs within its parentheses. Change the signs of *all* of the terms within the second set of parentheses; then drop the minus sign that precedes its parentheses.   ■

A **term** is a single number or a product of a number and one or more letters raised to powers. The following are examples of terms:

$$5x, \quad 8x^2, \quad -4y, \quad 15, \quad 3a^2b^3, \quad t$$

The **numerical coefficient** is the numerical factor of a term. The numerical factor of the term $16x^2$ is 16. The numerical coefficient of the term $-6a^2b$ is -6. The numerical coefficient of y is 1.

Terms are parts of an algebraic expression separated by plus and minus signs. For example, $3xy + 2y + 8x^2$ is an expression consisting of three terms.

$$3xy \;+\; 2y \;+\; 8x^2$$

1st	2nd	3rd
term	term	term

Like Terms

Terms with the same variables with exactly the same exponents are called **like terms**. For example, $4x$ and $11x$ have the same variables and are like terms. The terms $-5x^2y^3$ and $8x^2y^3$ have the same variables with the same exponents and are like terms. The terms $8m$ and $5n$ have different variables, and the terms $7x^2$ and $4x^3$ have different exponents, so these are **unlike terms**.

Example 3 The following table gives examples of like terms and unlike terms.

Like Terms	Unlike Terms	
a. $2x$ and $3x$	**e.** $2x^2$ and $3x$	Different exponents
b. $2ax$ and $5ax$	**f.** $2ax$ and $5bx$	Different variables
c. $2x^3$ and $18x^3$	**g.** $2x^3$ and $18x^2$	Different exponents
d. $2a^2x^4$, a^2x^4, and $11a^2x^4$	**h.** $2a^2x^4$, $3ax^4$, and $11a^2x^3$	Different exponents

■

Like terms that occur in a single expression can be combined into one term by combining coefficients (using the Distributive Property from Section 5.1). Thus, $ba + ca = (b + c)a$.

 Example 4 Combine the like terms $2x + 3x$.

$$2x + 3x = (2 + 3)x$$
$$= 5x$$

■

 Example 5 Combine the like terms $4ax + 6ax$.

$$4ax + 6ax = (4 + 6)ax$$
$$= 10ax$$

■

 Example 6 Combine the like terms $2a^2x^4 + a^2x^4 + 11a^2x^4$.

$$2a^2x^4 + a^2x^4 + 11a^2x^4 = 2a^2x^4 + 1a^2x^4 + 11a^2x^4$$
$$= (2 + 1 + 11)a^2x^4$$
$$= 14a^2x^4$$

■

 Example 7 Combine the like terms $9a^3b^4 + 2a^2b^3 + 7a^3b^4$.

$$9a^3b^4 + 2a^2b^3 + 7a^3b^4 = (9 + 7)a^3b^4 + 2a^2b^3$$
$$= 16a^3b^4 + 2a^2b^3$$

■

Some expressions contain parentheses that must be removed before combining like terms. Follow the order of operations.

 Example 8 Simplify: $4x - (x - 2)$.

$$4x - (x - 2) = 4x - x + 2 \qquad$$ Remove the parentheses by changing the signs of both terms within parentheses; then drop the minus sign that precedes the parentheses.

$$= 3x + 2 \qquad$$ Combine like terms.

■

 Example 9 Simplify: $4x - (-2x - 3y) + 5y$.

$$4x - (-2x - 3y) + 5y = 4x + 2x + 3y + 5y$$

Remove the parentheses by changing the signs of both terms within parentheses; then drop the minus sign that precedes the parentheses.

$$= 6x + 8y \qquad$$ Combine like terms.

■

ample 10   Simplify: $(7 - 2x) + (5x + 1)$.

$$(7 - 2x) + (5x + 1) = 7 - 2x + 5x + 1$$

Drop the implied plus sign before the first set of parentheses and do not change any of the signs within its parentheses. Drop the plus sign before the second set of parentheses and do not change any of the signs within its parentheses. Remove the parentheses.

$$= 3x + 8$$ Combine like terms.   ■

From Section 5.1, $a(b + c) = ab + ac$. The Distributive Property is applied to remove parentheses when a number, a letter, or some product precedes the parentheses.

ample 11   Remove the parentheses from each expression.

a. $3(6x + 5) = (3)(6x) + (3)(5)$ Apply the Distributive Property by multiplying each term within the parentheses by 3.

$$= 18x + 15$$ Multiply.

b. $-5(2a - 7) = (-5)(2a) - (-5)(7)$ Apply the Distributive Property by multiplying each term within the parentheses by -5.

$$= -10a + 35$$ Multiply.

c. $\frac{1}{2}(10x^2 + 28x) = \left(\frac{1}{2}\right)(10x^2) + \left(\frac{1}{2}\right)(28x)$ Apply the Distributive Property by multiplying each term within the parentheses by $\frac{1}{2}$.

$$= 5x^2 + 14x$$ Multiply.   ■

ample 12   Simplify: $3x + 5(x - 3)$.

$$3x + 5(x - 3) = 3x + 5x - 15$$ Apply the Distributive Property by multiplying each term within the parentheses by 5.

$$= 8x - 15$$ Combine like terms.   ■

ample 13   Simplify: $4y - 6(-y + 2)$.

$$4y - 6(-y + 2) = 4y + 6y - 12$$ Apply the Distributive Property by multiplying each term within the parentheses by -6.

$$= 10y - 12$$ Combine like terms.   ■

ses 5.2

parentheses from each expression:

$\vdash c)$ **2.** $a - (b + c)$

$) - c)$ **4.** $a - (-b + c)$

$) - c)$ **6.** $x + (y + z + 3)$

$+ z - 3)$ **8.** $x - (-y - z + 3)$

$- z + 3)$ **10.** $x + (-y - z - 3)$

$+ (3y + 4r)$ **12.** $(2x + 4) - (3y + 4r)$

13. $(3x - 5y + 8) + (6z - 2w + 3)$

14. $(4x + 6y - 9) + (-2z + 5w + 3)$

15. $(-5x - 3y - 2) - (6z - 3w - 5)$

16. $(-9x + 6) - (3z + 3w - 1)$

17. $(2x + 3y - 5) + (-z - w + 2) - (-3r + 2s + 7)$

18. $(5x - 11y - 2) - (7z + 3) + (3r + 7) - (4s - 2)$

19. $-(2x - 3y) - (z + 4w) - (4r - s)$

20. $-(3x + y) - (2z + 7w) - (3r - 5s + 2)$

Combine the like terms:

21. $b + b$

22. $4h + 6h$

23. $x^2 + 2x^2 + 3x + 7x$

24. $9k + 3k$

25. $5m - 2m$

26. $4x + 6x - 5x$

27. $3a + 5b - 2a + 7b$

28. $11 + 2m - 6 + m$

29. $6a^2 + a + 1 - 2a$

30. $5x^2 + 3x^2 - 8x^2$

31. $2x^2 + 16x + x^2 - 13x$

32. $13x^2 + 14xy + 6y^2 - 3y^2 + x^2$

33. $1.3x + 5.6x - 13.2x + 4.5x$

34. $2.3x^2 - 4.7x + 0.92x^2 - 2.13x$

35. $\dfrac{5}{9}x + \dfrac{1}{4}y + \dfrac{1}{3}x - \dfrac{3}{8}y$

36. $\dfrac{1}{2}x - \dfrac{2}{3}y - \dfrac{3}{4}x + \dfrac{5}{6}y$

37. $4x^2y - 2xy - y^2 - 3x^2 - 2x^2y + 3y^2$

38. $3x^2 - 5x - 2 + 4x^2 + x - 4 + 5x^2 - x + 2$

39. $2x^3 + 4x^2y - 4y^3 + 3x^3 - x^2y + y - y^3$

40. $4x^2 - 5x - 7x^2 - 3x - y^2 + 2x^2 + 3xy - 2y^2$

Simplify by first removing the parentheses and then combining the like terms:

41. $y - (y - 1)$

42. $x + (2x + 1)$

43. $4x + (4 - x)$

44. $5x - (2 - 3x)$

45. $10 - (5 + x)$

46. $x - (-x - y) + 2y$

47. $2y - (7 - y)$

48. $-y - (y + 3)$

49. $(5y + 7) - (y + 2)$

50. $(2x + 4) - (x - 7)$

51. $(4 - 3x) + (3x + 1)$

52. $10 - (y + 6) + (3y - 2)$

53. $-5y + 9 - (-5y + 3)$

54. $0.5x + (x - 1) - (0.2x + 8)$

55. $0.2x - (0.2x - 28)$

56. $(0.3x - 0.5) - (-2.3x + 1.4)$

57. $\left(\dfrac{1}{2}x - \dfrac{2}{3}\right) - \left(2 - \dfrac{3}{4}x\right)$

58. $\left(\dfrac{3}{4}x - 1\right) + \left(-\dfrac{1}{2}x - \dfrac{2}{3}\right)$

59. $4(3x + 9y)$

60. $6(-2a + 8b)$

61. $-12(3x^2 - 4y^2)$

62. $-3(-a^2 - 4a)$

63. $(5x + 13) - 3(x - 2)$

64. $(7x + 8) - 5(x - 6)$

65. $-9y - 0.5(8 - y)$

66. $12(x + 1) - 3(4x - 2)$

67. $2y - 2(y + 21)$

68. $3x - 3(6 - x)$

69. $6n - (2n - 8)$

70. $14x - 8(2x - 8)$

71. $0.8x - (-x + 7)$

72. $-(x - 3) - 3(4 + x)$

73. $4(2 - 3n) - 2(5 - 3n)$

74. $(x + 4) - 2(2x - 7)$

75. $\dfrac{2}{3}(6x - 9) - \dfrac{3}{4}(12x - 16)$

76. $13\left(7x - 2\dfrac{1}{2}\right) - 9\left(8x + 9\dfrac{2}{3}\right)$

77. $0.45(x + 3) - 0.75(2x + 13)$

78. $0.6(0.5x^2 - 0.9x) + 0.4(x^2 + 0.4x)$

5.3 Addition and Subtraction of Polynomials

A **monomial**, or *term*, is any algebraic expression that contains only products of numbers and variables, which have nonnegative integer exponents. The following expressions are examples of monomials:

$$2x, \quad 5, \quad -3b, \quad \frac{3}{4}a^2bw, \quad \sqrt{315}\,mn$$

A **polynomial** is either a monomial or the sum or difference of unlike monomials. We consider two special types of polynomials. A **binomial** is a polynomial that is the sum or difference of two unlike monomials. A **trinomial** is the sum or difference of three unlike monomials.

The following table shows examples of monomials, binomials, and trinomials.

Monomials	$3x$	$4ab^2$	$-15x^2y^3$	one term
Binomials	$a + b$	$5a^2 + 3$	$7xy^2 - 4x^2y$	two terms
Trinomials	$a + 3b - 2c$	$8x^2 - 3x + 12$	$2a^3b + 3a^2b^2 + ab^3$	three terms

Expressions that contain variables in the denominator are *not* polynomials. For example,

$$\frac{3}{4x}, \qquad \frac{8x}{3x-5}, \qquad \text{and} \qquad \frac{33}{4x^2} + \frac{8}{x-1}$$

are *not* polynomials.

> The **degree of a monomial in one variable** is the same as the exponent of the variable.

Example 1

Find the degree of each monomial: **a.** $-7m$, **b.** $6x^2$, **c.** $5y^3$, **d.** 5.

a. $-7m$ has degree 1.   The exponent of m is 1.

b. $6x^2$ has degree 2.   The exponent of x is 2.

c. $5y^3$ has degree 3.   The exponent of y is 3.

d. 5 has degree 0   5 may be written as $5x^0$.   ■

> The **degree of a polynomial in one variable** is the same as the highest-degree monomial contained in the polynomial.

Example 2

Find the degree of each polynomial: **a.** $5x^4 + x^2$ and **b.** $6y^3 + 4y^2 - y + 1$.

a. $5x^4 + x^2$ has degree 4, the highest-degree monomial.

degree 4 ⌐   ⌐ degree 2

b. $6y^3 + 4y^2 - y + 1$ has degree 3, the highest-degree monomial.

degree 3 ⌐

degree 2   degree 1

degree 0   ■

A polynomial is in **decreasing order** if each term is of some degree less than the preceding term. The following polynomial is written in decreasing order:

$$4x^5 - 3x^4 - 4x^2 - x + 5$$

exponents decrease

A polynomial is in **increasing order** if each term is of some degree larger than the preceding term. The following polynomial is written in increasing order:

$$5 - x - 4x^2 - 3x^4 + 4x^5$$

exponents increase

Adding Polynomials

To add polynomials, add their like terms.

Example 3

Add: $(3x + 4) + (5x - 7)$.

$$(3x + 4) + (5x - 7) = (3x + 5x) + [4 + (-7)] \qquad \text{Add the like terms.}$$
$$= 8x - 3$$ ■

Example 4   Add: $(5x^2 + 6x - 8) + (4x^2 - 3)$.

$$(5x^2 + 6x - 8) + (4x^2 - 3) = (5x^2 + 4x^2) + 6x + [(-8) + (-3)] \qquad \text{Ad}$$
$$\text{ter}$$
$$= 9x^2 + 6x - 11$$

 Example 5   Add: $(5a + 2b - 3c + 4) + (-2a - 4b + 7c - 3) + (6a + b - 4c)$.

$$(5a + 2b - 3c + 4) + (-2a - 4b + 7c - 3) + (6a + b - 4c)$$
$$= [5a + (-2a) + 6a] + [2b + (-4b) + b] + [(-3c) + 7c + (-4c)] +$$
$$\text{Add th}$$

$$= 9a - 1b + 0c + 1$$
$$= 9a - b + 1$$

We sometimes find it easier to find the sum of polynomials by writing th
in columns and then adding the columns, as shown in the next example. Here,
mials are also written in decreasing order, which is also common as an organiz

 Example 6   Add: $(2x^2 - 5x) + (3x^2 + 2x - 4) + (-4x^2 + 5)$.

$$\begin{array}{ll}
2x^2 - 5x & \text{Write the like terms in columns.} \\
3x^2 + 2x - 4 & \\
\underline{-4x^2 \qquad + 5} & \\
x^2 - 3x + 1 & \text{Add the like terms.}
\end{array}$$

◢ Subtracting Polynomials

To subtract two polynomials, change all the signs of the terms of the secon
polynomial and then add the two resulting polynomials.

 Example 7   Subtract: $(5a - 9b) - (2a - 4b)$.

$$(5a - 9b) - (2a - 4b) = (5a - 9b) + (-2a + 4b) \qquad \text{Change all th}$$
$$\text{the terms of}$$
$$\text{polynomial a}$$

$$= [5a + (-2a)] + [(-9b) + 4b] \qquad \text{Add the like}$$
$$= 3a - 5b$$

Example 8   Subtract: $(5x^2 - 3x - 4) - (2x^2 - 5x + 6)$.

$$(5x^2 - 3x - 4) - (2x^2 - 5x + 6) = (5x^2 - 3x - 4) + (-2x^2 + 5x - 6)$$
$$\text{Change all the signs of the te}$$
$$\text{second polynomial and add.}$$

$$= [5x^2 + (-2x^2)] + [(-3x) + 5x] + [(-$$
$$\text{Add the like terms.}$$

$$= 3x^2 + 2x - 10$$

Subtraction can also be done in columns; the subtraction in long division
nomial by another polynomial is usually done using columns.

Example 9   Find the difference: $(5x^2 - 3x - 4) - (2x^2 - 7x + 5)$.

$$\text{Subtract:} \quad \begin{array}{r} 5x^2 - 3x - 4 \\ 2x^2 - 7x + 5 \end{array} \quad \rightarrow \quad \text{Add:} \quad \begin{array}{r} 5x^2 - 3x - 4 \\ -2x^2 + 7x - 5 \\ \hline 3x^2 + 4x - 9 \end{array}$$

The arrow indicates the change of the subtraction problem to an addition problem. Do this by changing the signs of each of the terms in the second polynomial. ■

Exercises 5.3

Classify each expression as a monomial, a binomial, or a trinomial:

1. $3m + 27$

2. $4a^2bc^3$

3. $-5x - 7y$

4. $2x^2 + 7y + 3z^2$

5. $-5xy$

6. $a + b + c$

7. $2x + 3y - 5z$

8. $2a - 3b^3$

9. $-42x^3 - y^4$

10. $15x^{14} - 3x^2 + 5x$

Rearrange each polynomial in decreasing order and state its degree:

11. $1 - x + x^2$

12. $2x^3 - 3x^4 + 2x$

13. $4x + 7x^2 - 1$

14. $y^3 - 1 + y^2$

15. $-4x^2 + 5x^3 - 2$

16. $3x^3 + 6 - 2x + 4x^5$

17. $7 - 3y + 4y^3 - 6y^2$

18. $1 - x^5$

19. $x^3 - 4x^4 + 2x^2 - 7x^5 + 5x - 3$

20. $360x^2 - 720x - 120x^3 + 30x^4 + 1 - 6x^5 + x^6$

Add the following polynomials:

21. $(5a^2 - 7a + 5) + (2a^2 - 3a - 4)$

22. $(2b - 5) + 3b + (-4b - 7)$

23. $(6x^2 - 7x + 5) + (3x^2 + 2x - 5)$

24. $(4x - 7y - z) + (2x - 5y - 3z) + (-3x - 6y - 4z)$

25. $(2a^3 - a) + (4a^2 + 7a) + (7a^3 - a - 5)$

26. $(5y - 7x + 4z) + (3z - 6y + 2x) + (13y + 7z - 6x)$

27. $(3x^2 + 4x - 5) + (-x^2 - 2x + 2) + (-2x^2 + 2x + 7)$

28. $(-x^2 + 6x - 8) + (10x^2 - 13x + 3) +$
$(-12x^2 - 14x + 3)$

29. $(3x^2 + 7) + (6x - 7) + (2x^2 + 5x - 13) + (7x - 9)$

30. $(5x + 3y) + (-3x - 3y) + (-x - 6y) + (3x - 4y)$

31. $(5x^3 - 11x - 1) + (11x^2 + 3) + (3x + 7) + (2x^2 - 2)$

32. $(3x^4 - 5x^2 + 4) + (6x^4 - 6x^2 + 1) + (2x^4 - 7x^2)$

33. $\quad \begin{array}{r} 4y^2 - 3y - 15 \\ 7y^2 - 6y + 8 \\ -3y^2 + 4y + 13 \end{array}$

34. $\quad \begin{array}{r} 129a - 13b - 56c \\ -13a - 52b + 21c \\ 44a \qquad\quad + 11c \end{array}$

35. $\quad \begin{array}{r} 3a^3 + 2a^2 \qquad + 5 \\ a^3 \qquad - 7a - 2 \\ -5a^2 + 4a \\ - 2a - 3 \\ \hline \end{array}$

36. $\quad \begin{array}{r} 4x^2 - 3xy \qquad + 5x - 6y \\ 9xy - 4y^2 \qquad + 6y - 4 \\ x^2 \qquad + y^2 - x \qquad + 3 \\ -8x^2 + xy - 2y^2 + 3x \qquad - 10 \\ \hline \end{array}$

Find each difference:

37. $(3x^2 + 4x + 7) - (x^2 - 2x + 5)$

38. $(2x^2 + 5x - 9) - (3x^2 - 4x + 7)$

39. $(3x^2 - 5x + 4) - (6x^2 - 7x + 2)$

40. $(1 - 3x - 2x^2) - (-1 - 5x + x^2)$

41. $(3a - 4b) - (2a - 7b)$

42. $(-13x^2 - 3y^2 - 4y) - (-5x - 4y + 5y^2)$

43. $(7a - 4b) - (3x - 4y)$

44. $(-16y^3 - 42y^2 - 3y - 5) - (12y^2 - 4y + 7)$

45. $(12x^2 - 3x - 2) - (11x^2 - 7)$

46. $(14z^3 - 6y^3) - (2y^2 + 4z^3)$

47. $(20w^2 - 17w - 6) - (13w^2 + 7w)$

48. $(y^2 - 2y + 1) - (2y^2 + 3y + 5)$

49. $(2x^2 - 5x - 2) - (x^2 - x + 8)$

50. $(3 - 5z + 3z^2) - (14 + z - 2z^2)$

51. Subtract $4x^2 + 2x - 7$ from $8x^2 - 2x + 5$.

52. Subtract $-6x^2 - 3x + 4$ from $2x^2 - 6x - 2$.

53. Subtract $9x^2 + 6$ from $3x^2 + 2x - 4$.

54. Subtract $-4x^2 - 6x + 2$ from $4x^2 + 6$.

Subtract the following polynomials:

55. $\quad \begin{array}{r} 2x^3 + 4x - 1 \\ x^3 + x + 2 \end{array}$

56. $\quad \begin{array}{r} 7x^4 + 3x^3 + 5x \\ -2x^4 + x^3 - 6x + 6 \end{array}$

57. $\quad \begin{array}{r} 12x^5 - 13x^4 + 7x^2 \\ 4x^5 + 5x^4 + 2x^2 - 1 \end{array}$

58. $\quad \begin{array}{r} 8x^3 + 6x^2 - 15x + 7 \\ 14x^3 + 2x^2 + 9x - 1 \end{array}$

5.4 Multiplication of Monomials

Earlier, we used exponents to write products of repeated number factors as fol

$$6^2 = 6 \cdot 6 = 36$$
$$5^3 = 5 \cdot 5 \cdot 5 = 125$$
$$2^4 = 2 \cdot 2 \cdot 2 \cdot 2 = 16$$

A power with a variable base may also be used as follows:

$$x^3 = x \cdot x \cdot x$$
$$c^4 = c \cdot c \cdot c \cdot c$$

In the expression 2^4, the number 2 is called the *base,* and 4 is called the *ex* expression may also be called *the fourth power of 2.* In x^3, the letter x is called 3 is called the exponent.

Example 1   Multiply:

    a. $x^3 \cdot x^2 = (x \cdot x \cdot x)(x \cdot x) = x^5$

    b. $a^4 \cdot a^2 = (a \cdot a \cdot a \cdot a)(a \cdot a) = a^6$

    c. $c^5 \cdot c^3 = (c \cdot c \cdot c \cdot c \cdot c)(c \cdot c \cdot c) = c^8$

Rule 1 for Exponents: Multiplying Powers

$$a^m \cdot a^n = a^{m+n}$$

That is, to multiply powers with the same base, add the exponents.

To multiply two monomials, multiply their numerical coefficients and c variable factors according to Rule 1 for exponents.

Example 2   Multiply: $(2x^3)(5x^4)$.

$$(2x^3)(5x^4) = 2 \cdot 5 \cdot x^3 \cdot x^4 = 10x^{3+4} \qquad \text{Add the exponents.}$$
$$= 10x^7$$

Example 3   Multiply: $(3a)(-15a^2)(4a^4b^2)$.

$$(3a)(-15a^2)(4a^4b^2) = (3)(-15)(4)(a)(a^2)(a^4)(b^2)$$
$$= -180a^7b^2 \qquad \text{Add the exponents}$$

A *special note* about the meaning of $-x^2$ is needed here. Note that x is $-x$. That is,

$$-x^2 = -(x \cdot x)$$

If $-x$ is squared, we have $(-x)^2 = (-x)(-x) = x^2$.

Rule 2 for Exponents: Raising a Power to a Power

$$(a^m)^n = a^{mn}$$

That is, to raise a power to a power, multiply the exponents.

Example 4

Find $(x^3)^5$.

By Rule 1:

$$(x^3)^5 = x^3 \cdot x^3 \cdot x^3 \cdot x^3 \cdot x^3 = x^{15}$$

By Rule 2:

$$(x^3)^5 = x^{3 \cdot 5} = x^{15} \qquad \text{Multiply the exponents.}$$ ■

Example 5

Find $(x^5)^9$.

$$(x^5)^9 = x^{45} \qquad \text{Multiply the exponents.}$$ ■

Rule 3 for Exponents: Raising a Product to a Power

$$(ab)^m = a^m b^m$$

That is, to raise a product to a power, raise each factor to that same power.

Example 6

Find $(xy)^3$.

$$(xy)^3 = x^3 y^3$$ ■

Example 7

Find $(2x^3)^2$.

$$(2x^3)^2 = 2^2(x^3)^2$$
$$= 4x^6 \qquad \text{Square each factor.}$$ ■

Example 8

Find $(-3x^4)^5$.

$$(-3x^4)^5 = (-3)^5(x^4)^5$$
$$= -243x^{20} \qquad \text{Raise each factor to the fifth power.}$$ ■

Example 9

Find $(ab^2c^3)^4$.

$$(ab^2c^3)^4 = a^4(b^2)^4(c^3)^4$$
$$= a^4 b^8 c^{12} \qquad \text{Raise each factor to the fourth power.}$$ ■

Example 10

Find $(2a^2bc^3)^2$.

$$(2a^2bc^3)^2 = 2^2(a^2)^2(b)^2(c^3)^2$$
$$= 4a^4 b^2 c^6$$ ■

Example 11   Evaluate $(4a^2)(-5ab^2)$ when $a = 2$ and $b = 3$.

$$(4a^2)(-5ab^2) = 4(-5)(a^2)(a)(b^2) \qquad \text{First multiply.}$$
$$= -20a^3b^2$$
$$= -20(2)^3(3)^2 \qquad \text{Substitute.}$$
$$= -20(8)(9)$$
$$= -1440$$

Exercises 5.4

Find each product:

1. $(3a)(-5)$ **2.** $(7x)(2x)$

3. $(4a^2)(7a)$ **4.** $(4x)(6x^2)$

5. $(-9m^2)(-6m^2)$ **6.** $(5x^2)(-8x^3)$

7. $(8a^6)(4a^2)$ **8.** $(-4y^4)(-9y^3)$

9. $(13p)(-2pq)$ **10.** $(4ab)(10a)$

11. $(6n)(5n^2m)$ **12.** $(-9ab^2)(6a^2b^3)$

13. $(-42a)\left(-\dfrac{1}{2}a^3b\right)$ **14.** $(28m^3)\left(\dfrac{1}{4}m^2\right)$

15. $\left(\dfrac{2}{3}x^2y^2\right)\left(\dfrac{9}{16}xy^2\right)$ **16.** $\left(-\dfrac{5}{6}a^2b^6\right)\left(\dfrac{9}{20}a^5b^4\right)$

17. $(8a^2bc)(3ab^3c^2)$ **18.** $(-4xy^2z^3)(4x^5z^3)$

19. $\left(\dfrac{2}{3}x^2y\right)\left(\dfrac{9}{32}xy^4z^3\right)$ **20.** $\left(\dfrac{3}{5}m^4n^7\right)\left(\dfrac{20}{9}m^2nq^3\right)$

21. $(32.6mnp^2)(-11.4m^2n)$ **22.** $(5.6a^2b^3c)(6.5a^4b^5)$

23. $(5a)(-17a^2)(3a^3b)$ **24.** $(-4a^2b)(-5ab^3)(-2a^4)$

Use the rules for exponents to simplify:

25. $(x^3)^2$ **26.** $(xy)^4$ **27.** $(x$

28. $(2x^2)^5$ **29.** $(-3x^4)^2$ **30.** $(5$

31. $(-x^3)^3$ **32.** $(-x^2)^4$ **33.** $(x$

34. $(3x)^4$ **35.** $(x^5)^6$ **36.** $(-$

37. $(-5x^3y^2)^2$ **38.** $(-x^2y^4)^5$ **39.** $(1$

40. $(-7w^2)^3$ **41.** $(25n^4)^3$ **42.** $(3$

43. $(3x^2 \cdot x^4)^2$ **44.** $(16x^4 \cdot x^5)^3$ **45.** $(2$

46. $(-4a^2b^3c^4)^4$ **47.** $(-2h^3k^6m^2)^5$ **48.** $(4$

Evaluate each expression when $a = 2$ and $b =$

49. $(4a)(17b)$ **50.** $(3a)(-5b^2)$ **51.** $(9$

52. $(a^2)(ab)$ **53.** $(41a^3)(-2b^3)$ **54.** $(a$

55. $(a^2)^2$ **56.** $(3a)^2$ **57.** $(4$

58. $(-2ab^2)^2$ **59.** $(5a^2b^3)^2$ **60.** $(-$

61. $(5ab)(a^2b^2)$ **62.** $(-3ab)^3$ **63.** $(9$

64. $(-2a^2)(6ab)$ **65.** $-a^2b^4$ **66.** $(-$

67. $(-a)^4$ **68.** $-b^4$

5.5 Multiplication of Polynomials

To multiply a polynomial by a monomial, multiply each term of the polyn[o]
monomial, and then add the products as shown in the following examples.

Example 1   Multiply: $3a(a^2 - 2a + 1)$.

$$3a(a^2 - 2a + 1) = 3a(a^2) + 3a(-2a) + 3a(1) \qquad \text{Multiply each term of}$$
$$= 3a^3 - 6a^2 + 3a \qquad \text{polynomial by } 3a.$$

Example 2   Multiply: $(-5a^3b)(3a^2 - 4ab + 5b^3)$.

$$(-5a^3b)(3a^2 - 4ab + 5b^3)$$
$$= (-5a^3b)(3a^2) + (-5a^3b)(-4ab) + (-5a^3b)(5b^3) \qquad \text{Multiply each}$$
$$= -15a^5b + 20a^4b^2 - 25a^3b^4 \qquad \text{polynomial b}$$

To multiply a polynomial by another polynomial, multiply each term of the first polynomial by each term of the second polynomial. Then add the products. Arrange the work as shown in Example 3.

 Example 3   Multiply: $(5x - 3)(2x + 4)$.

Step 1	Write each polynomial in decreasing order, one under the other.	$5x - 3$
		$\underset{2\ \ \ 3}{\times}$
		$\underline{2x + 4}$
Step 2	Multiply each term of the upper polynomial by the first term in the lower one.	$10x^2 - 6x \qquad \leftarrow 2x(5x - 3)$
Step 3	Multiply each term of the upper polynomial by the second term in the lower one. Place like terms in the same columns.	$\underline{\qquad 20x - 12} \quad \leftarrow 4(5x - 3)$
Step 4	Add the like terms.	$10x^2 + 14x - 12 \qquad$ Add.

To multiply two binomials, such as $(x + 3)(2x + 5)$, you may think of finding the area of a rectangle with sides $(x + 3)$ and $(2x + 5)$ as shown in Figure 5.1.

FIGURE 5.1

Note that the total area is $2x^2 + 6x + 5x + 15$ or $2x^2 + 11x + 15$. Using the four-step process as above, we have

Step 1 $2x + 5$
        $\underline{x + 3}$
Step 2 $2x^2 + 5x \qquad \leftarrow x(2x + 5)$
Step 3 $\underline{\qquad 6x + 15} \quad \leftarrow 3(2x + 5)$
Step 4 $2x^2 + 11x + 15 \qquad$ Add.

 Example 4   Multiply: $(x + 3)(x^2 + 2x - 4)$.

Step 1 $x^2 + 2x - 4$
        $\underline{\qquad x + 3}$
Step 2 $x^3 + 2x^2 - 4x \qquad \leftarrow x(x^2 + 2x - 4)$
Step 3 $\underline{\qquad 3x^2 + 6x - 12} \quad \leftarrow 3(x^2 + 2x - 4)$
Step 4 $x^3 + 5x^2 + 2x - 12 \qquad$ Add.

Example 5

Multiply: $(3a + b)(c + 2d)$.

Step 1 $3a + b$
             $\underline{c + 2d}$

Step 2 $3ac + bc$ $\leftarrow c(3a + b)$

Step 3 $\underline{ 6ad + 2bd}$ $\leftarrow 2d(3a + b)$

Step 4 $3ac + bc + 6ad + 2bd$ Add.

Note that there are no like terms in Steps 2 and 3.

Exercises 5.5

Find each product.

1. $4(a + 6)$

2. $3(a^2 - 5)$

3. $-6(3x^2 + 2y)$

4. $-5(8x - 4y^2)$

5. $a(4x^2 - 6y + 1)$

6. $c(2a + b + 3c)$

7. $x(3x^2 - 2x + 5)$

8. $y(3x + 2y^2 + 4y)$

9. $2a(3a^2 + 6a - 10)$

10. $5x(8x^2 - x + 5)$

11. $-3x(4x^2 - 7x - 2)$

12. $-6x(8x^2 + 5x - 9)$

13. $4x(-7x^2 - 3y + 2xy)$

14. $7a(2a + 3b - 4ab)$

15. $3xy(x^2y - xy^2 + 4xy)$

16. $-2ab(3a^2 + 4ab - 2b^2)$

17. $-6x^3(1 - 6x^2 + 9x^4)$

18. $5x^4(2x^3 + 8x^2 - 1)$

19. $5ab^2(a^3 - b^3 - ab)$

20. $7w^2y(w^2 - 4y^2 + 6w^2y^3)$

21. $\dfrac{2}{3}m(14n - 12m)$

22. $\dfrac{1}{2}a^2b(8ab^2 - 2a^2b)$

23. $\dfrac{4}{7}yz^3\left(28y - \dfrac{2}{5}z\right)$

24. $-\dfrac{1}{8}rs(3s - 16t)$

25. $-4a(1.3a^5 + 2.5a^2 + 1)$

26. $1.28m(2.3m^2 + 4.7n^2)$

27. $417a(3.2a^2 + 4a)$

28. $1.2m^2n^3(9.7m + 6.5mn - 13n^2)$

29. $4x^2y(6x^2 - 4xy + 5y^2)$

30. $x^2y^3z(x^4 - 3x^2y - 3yz + 4z^2)$

31. $\dfrac{2}{3}ab^3\left(\dfrac{3}{4}a^2 - \dfrac{1}{2}ab^2 + \dfrac{5}{6}b^3\right)$

32. $-\dfrac{5}{9}a^2b^4\left(\dfrac{3}{7}a^3b^2 - \dfrac{3}{5}ab - \dfrac{15}{16}b^4\right)$

33. $3x(x - 4) + 2x(1 - 5x) - 6x(2x - 3)$

34. $x(x - 2) - 3x(x + 8) - 2(x^2 + 3x - 5)$

35. $xy(3x + 2xy - y^2) - 2xy^2(2x - xy + 3y)$

36. $ab^2(2a - 3a^2b + b) - a^2b(1 + 2ab^2 - 4b$

37. $(x + 1)(x + 6)$

38. $(x + 10)($

39. $(x + 7)(x - 2)$

40. $(x - 3)(x$

41. $(x - 5)(x - 8)$

42. $(x + 9)(x$

43. $(3a - 5)(a - 4)$

44. $(5x - 2)($

45. $(6a + 4)(2a - 3)$

46. $(3x + 5)($

47. $(4a + 8)(6a + 9)$

48. $(5x - 4)($

49. $(3x - 2y)(5x + 2y)$

50. $(4x - 6y$

51. $(2x - 3)(2x - 3)$

52. $(5m - 9)$

53. $(2c - 5d)(2c + 5d)$

54. $(3a + 2b$

55. $(-7m - 3)(-13m + 1)$

56. $(w - r)($

57. $(x^5 - x^2)(x^3 - 1)$

58. $(7w^4 - 6r^2)(7w^4 + 5r^2)$

59. $(2y^2 - 4y - 8)(5y - 2)$

60. $(m^2 + 2m + 4)(m - 2)$

61. $(4x - 2y - 13)(6x + 3y)$

62. $(4y - 3z)(2y^2 - 5yz + 6z^2)$

63. $(g + h - 6)(g - h + 3)$

64. $(2x - 3y + 4)(4x - 5y - 2)$

65. $(8x - x^3 + 2x^4 - 1)(x^2 + 2 + 5x^3)$

66. $(y^5 - y^4 + y^3 - y^2 + y - 1)(y + 1)$

5.6 Division by a Monomial

To divide a monomial by a monomial, first write the quotient in fraction form
both numerator and denominator into prime factors. Reduce to lowest term
both the numerator and the denominator by their common factors. The remain
the numerator and the denominator give the quotient.

Example 1 Divide: $6a^2 \div 2a$.

$$6a^2 \div 2a = \frac{6a^2}{2a}$$

$$= \frac{\overset{1}{\cancel{2}} \cdot 3 \cdot \overset{1}{\cancel{a}} \cdot a}{\underset{1}{\cancel{2}} \cdot \underset{1}{\cancel{a}}} = \frac{3a}{1} = 3a$$

■

 Example 2 Divide: $4a \div 28a^3$.

$$4a \div 28a^3 = \frac{4a}{28a^3}$$

$$= \frac{\overset{1}{\cancel{2}} \cdot \overset{1}{\cancel{2}} \cdot \overset{1}{\cancel{a}}}{\underset{1}{\cancel{2}} \cdot \underset{1}{\cancel{2}} \cdot 7 \cdot \underset{1}{\cancel{a}} \cdot a \cdot a} = \frac{1}{7a^2}$$

■

Example 3 Divide: $\dfrac{6a^2bc^3}{10ab^2c}$.

$$\frac{6a^2bc^3}{10ab^2c} = \frac{\cancel{2} \cdot 3 \cdot \cancel{a} \cdot a \cdot \cancel{b} \cdot \cancel{c} \cdot c \cdot c}{\cancel{2} \cdot 5 \cdot \cancel{a} \cdot \cancel{b} \cdot b \cdot \cancel{c}} = \frac{3ac^2}{5b}$$

■

Note: Since division by zero is undefined, we will assume that there are no zero denominators here and for the remainder of this chapter.

To divide a polynomial by a monomial, divide each term of the polynomial by the monomial. Simplify each result when possible.

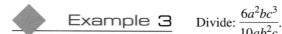 **Example 4** Divide: $\dfrac{15x^3 - 3x^2 + 21x}{3x}$.

$$\frac{15x^3 - 3x^2 + 21x}{3x} = \frac{15x^3}{3x} - \frac{3x^2}{3x} + \frac{21x}{3x} \quad \text{Divide each term of the polynomial by } 3x.$$

$$= \frac{\cancel{3} \cdot 5 \cdot \cancel{x} \cdot x \cdot x}{\cancel{3} \cdot \cancel{x}} - \frac{\cancel{3} \cdot \cancel{x} \cdot x}{\cancel{3} \cdot \cancel{x}} + \frac{\cancel{3} \cdot 7 \cdot \cancel{x}}{\cancel{3} \cdot \cancel{x}}$$

$$= 5x^2 - x + 7$$

■

 Example 5 Divide: $(30d^4y - 28d^2y^2 + 12dy^2) \div (-6dy^2)$.

$$\frac{30d^4y - 28d^2y^2 + 12dy^2}{-6dy^2}$$

$$= \frac{30d^4y}{-6dy^2} + \frac{-28d^2y^2}{-6dy^2} + \frac{12dy^2}{-6dy^2} \quad \text{Divide each term of the polynomial by } -6dy^2.$$

$$= \frac{\cancel{6} \cdot 5 \cdot \cancel{d} \cdot d \cdot d \cdot d \cdot \cancel{y}}{-\cancel{6} \cdot \cancel{d} \cdot \cancel{y} \cdot y} + \frac{-\cancel{2} \cdot 14 \cdot \cancel{d} \cdot d \cdot \cancel{y} \cdot \cancel{y}}{-\cancel{2} \cdot 3 \cdot \cancel{d} \cdot \cancel{y} \cdot \cancel{y}} + \frac{\cancel{6} \cdot 2 \cdot \cancel{d} \cdot \cancel{y} \cdot \cancel{y}}{-\cancel{6} \cdot \cancel{d} \cdot \cancel{y} \cdot \cancel{y}}$$

$$= \frac{5d^3}{-y} + \frac{14d}{3} + \frac{2}{-1}$$

$$= -\frac{5d^3}{y} + \frac{14d}{3} - 2$$

■

Exercises 5.6

Divide:

1. $\dfrac{9x^5}{3x^3}$ **2.** $\dfrac{15x^6}{5x^4}$ **3.** $\dfrac{18x^{12}}{12x^4}$ **4.** $\dfrac{20x^7}{4x^5}$

5. $\dfrac{18x^3}{3x^5}$ **6.** $\dfrac{4x^2}{12x^6}$ **7.** $\dfrac{8x^2}{12x}$ **8.** $\dfrac{-6x^3}{2x}$

9. $\dfrac{x^2y}{xy}$ **10.** $\dfrac{xy^2}{x^2y}$

11. $(15x) \div (6x)$ **12.** $(14x^2) \div (2x^3)$

13. $(15a^3b) \div (3ab^2)$ **14.** $(-13a^5) \div (7a)$

15. $(16m^2n) \div (2m^3n^2)$ **16.** $(-108m^4n) \div (27m^3n)$

17. $0 \div (113w^2r^3)$ **18.** $(-148wr^3) \div (148wr^3)$

19. $(207p^3) \div (9p)$ **20.** $(42x^2y^3) \div (-14x^2y^4)$

21. $\dfrac{92mn}{-46mn}$ **22.** $\dfrac{-132rs^3}{-33r^2s^2}$ **23.** $\dfrac{252}{7r^2}$

24. $\dfrac{118a^3}{-2a^4}$ **25.** $\dfrac{92x^3y}{-28xy^3}$ **26.** $\dfrac{45x^6}{-72x^3y^2}$

27. $\dfrac{-16a^5b^2}{-14a^8b^4}$ **28.** $\dfrac{35a^3b^4c^6}{63a^3b^2c^8}$

29. $(-72x^3yz^4) \div (-162xy^2)$

30. $(-144x^2z^3) \div (216x^5y^2z)$

31. $(4x^2 - 8x + 6) \div 2$

32. $(18y^3 + 12y^2 + 6y) \div 6$

33. $(x^4 + x^3 + x^2) \div x^2$

34. $(20r^2 - 16r - 12) \div (-4)$

35. $(ax - ay - az) \div a$

36. $(14c^3 - 28c^2 - 2c) \div (-2c)$

37. $\dfrac{24a^4 - 16a^2 - 8a}{8}$ **38.** $\dfrac{88x^5 - 110\ldots}{11\imath}$

39. $\dfrac{b^{12} - b^9 - b^6}{b^3}$ **40.** $\dfrac{27a^3 - 18a}{-9c}$

41. $\dfrac{bx^4 - bx^3 + bx^2 - 4bx}{-bx}$

42. $\dfrac{4a^5 - 32a^4 + 8a^3 - 12a^2}{-4a^2}$

43. $\dfrac{24x^2y^3 + 12x^3y^4 - 6xy^3}{2xy^3}$

44. $\dfrac{3.5ax^2 - 0.42a^2x + 14a^2x^2}{0.07ax}$

45. $\dfrac{224x^4y^2z^3 - 168x^3y^3z^4 - 112xy^4z^2}{28xy^2z^2}$

46. $\dfrac{55w^2 - 11w - 33}{11w}$ **47.** $\dfrac{24y^5 - 18y}{6y^2}$

48. $\dfrac{3a^2b + 4a^2b^2 - 6ab^2}{2ab^2}$ **49.** $\dfrac{1 - 6x^2 - }{2x^2}$

50. $\dfrac{18w^4r^4 + 27w^3r^3 - 36w^2r^2}{9w^3r^3}$

5.7 Division by a Polynomial

Dividing a polynomial by a polynomial of more than one term is very similar t
sion in arithmetic. We use the same names, as shown below.

$$
\begin{array}{r}
23 \quad \leftarrow \text{quotient} \\
\text{divisor} \rightarrow 25\overline{)593} \quad \leftarrow \text{dividend} \\
\underline{50} \\
93 \\
\underline{75} \\
18 \quad \leftarrow \text{remainder}
\end{array}
$$

$$
\begin{array}{r}
x + 3 \quad \leftarrow \text{quotient} \\
\text{divisor} \rightarrow x + 1\overline{)x^2 + 4x + 4} \quad \leftarrow \text{dividend} \\
\underline{x^2 + x} \\
3x + 4 \\
\underline{3x + 3} \\
1 \quad \leftarrow \text{remainder}
\end{array}
$$

As a check, you may use the relationship *dividend = divisor × quotient + remainder*. A similar procedure is followed in both cases. Compare the solutions of the two problems in the example that follows.

 Example 1   Divide $337 \div 16$ (arithmetic) and $(2x^2 + x - 14) \div (x + 3)$ (algebra).

Arithmetic	Line	Algebra
$21 \text{ r } 1$	**1.**	$2x - 5 \text{ r } 1$
$16\overline{)337}$	**2.**	$x + 3\overline{)2x^2 + x - 14}$
$\underline{32}$	**3.**	$\underline{2x^2 + 6x}$
17	**4.**	$-5x - 14$
$\underline{16}$	**5.**	$\underline{-5x - 15}$
1	**6.**	1

Step

1. Divide 16 into 33. It will go at most 2 times, so write the 2 above the line over the dividend (337).

2. Multiply the divisor (16) by 2. Write the result (32) under the first two digits of 337.

3. Subtract 32 from 33, leaving 1. Bring down the 7, giving 17.

4. Divide 16 into 17. It will go at most 1 time, so write the 1 to the right of the 2 above the dividend.

5. Multiply the divisor (16) by 1. Write the result (16) under 17.

6. Subtract 16 from 17, leaving 1. The remainder (1) is less than 16, so the problem is finished. The quotient is 21 with remainder 1.

Step

1. Divide the first term of the dividend, $2x^2$, by the first term of the divisor, x. Write the result, $2x$, above the dividend in line **1**.

2. Multiply the divisor, $x + 3$, by $2x$ and write the result, $2x^2 + 6x$, in line **3** as shown.

3. Subtract this result, $2x^2 + 6x$, from the first two terms of the dividend, leaving $-5x$. Bring down the last term of the dividend, -14, as shown in line **4**.

4. Divide the first term in line **4**, $-5x$, by the first term of the divisor, x. Write the result, -5, to the right of $2x$ above the dividend in line **1**.

5. Multiply the divisor, $x + 3$, by -5 and write the result, $-5x - 15$, in line **5** as shown.

6. Subtract line **5** from line **4**, leaving 1 as the remainder. Since the remainder, 1, is of a lower degree than the divisor, $x + 3$, the problem is finished. The quotient is $2x - 5$ with remainder 1. ■

> The dividend should always be arranged in decreasing order of degree. Any missing powers of x should be filled in by using zeros as coefficients. For example, $x^3 - 1 = x^3 + 0x^2 + 0x - 1$.

 Example 2   Divide $8x^3 - 22x^2 + 27x - 18$ by $2x - 3$.

$$
\begin{array}{r}
\phantom{2x - 3\overline{)}}4x^2 - 5x + 6 \quad \textbf{1.} \\
2x - 3\overline{)8x^3 - 22x^2 + 27x - 18} \quad \textbf{2.} \\
\underline{8x^3 - 12x^2} \quad \textbf{3.} \\
-10x^2 + 27x \quad \textbf{4.} \\
\underline{-10x^2 + 15x} \quad \textbf{5.} \\
12x - 18 \quad \textbf{6.} \\
\underline{12x - 18} \quad \textbf{7.} \\
0 \quad \textbf{8.}
\end{array}
$$

Step 1   Divide the first term in the dividend, $8x^3$, by the first term of the div
Write $4x^2$ above the dividend in line **1**.

Step 2   Multiply the divisor, $2x - 3$, by $4x^2$ and write the result in line **3**, as

Step 3   Subtract in line **4** and bring down the next term.

Step 4   Divide $-10x^2$ by $2x$ and write $-5x$ above the dividend in line **1**.

Step 5   Multiply $2x - 3$ by $-5x$ and write the result in line **5**.

Step 6   Subtract in line **6** and bring down the next term.

Step 7   Divide $12x$ by $2x$ and write 6 above the dividend in line **1**.

Step 8   Multiply $2x - 3$ by 6 and write the result in line **7**.

Step 9   Subtract in line **8**; the 0 indicates that there is no remainder.

Check:

$$
\begin{array}{ll}
4x^2 - 5x + 6 & \text{quotient} \\
\underline{2x - 3} & \text{divisor} \\
8x^3 - 10x^2 + 12x & \\
\underline{-12x^2 + 15x - 18} & \\
8x^3 - 22x^2 + 27x - 18 & \text{dividend}
\end{array}
$$

◆ **Example 3**   Divide: $(x^3 - 1) \div (x - 1)$.

$$
\begin{array}{r}
x^2 + x + 1 \\
x - 1 \overline{\smash{\big)}\, x^3 + 0x^2 + 0x - 1} \\
\underline{x^3 - x^2} \\
x^2 + 0x \\
\underline{x^2 - x} \\
x - 1 \\
\underline{x - 1} \\
0
\end{array}
$$

Supply missing powers.

The remainder is 0, so the quotient is $x^2 + x + 1$.

Check:

$$
\begin{array}{ll}
x^2 + x + 1 & \text{quotient} \\
\underline{x - 1} & \text{divisor} \\
x^3 + x^2 + x & \\
\underline{-x^2 - x - 1} & \\
x^3 \qquad\quad - 1 & \text{dividend}
\end{array}
$$

Note: When you subtract, be especially careful to change all of the signs of e
the expression being subtracted. Then follow the rules for addition.

Exercises 5.7

Find each quotient and check:

1. $(x^2 + 3x + 2) \div (x + 1)$

2. $(y^2 - 5y + 6) \div (y - 2)$

3. $(6a^2 - 3a + 2) \div (2a - 3)$

4. $(21y^2 + 2y - 10) \div (3y + 2)$

5. $(12x^2 - x - 9) \div (3x + 2)$

6. $(20x^2 + 57x + 30) \div (4x + 9)$

7. $\dfrac{2y^2 + 3y - 5}{2y - 1}$

8. $\dfrac{3x^2 - 5x -}{3x - 8}$

9. $\dfrac{6b^2 + 13b - 28}{2b + 7}$

10. $\dfrac{8x^2 + 13x}{x + 3}$

11. $(6x^3 + 13x^2 + x - 2) \div (x + 2)$

12. $(8x^3 - 18x^2 + 7x + 3) \div (x - 1)$

13. $(8x^3 - 14x^2 - 79x + 110) \div (2x - 7)$

14. $(3x^3 - 17x^2 + 18x + 10) \div (3x + 1)$

15. $\dfrac{2x^3 - 14x - 12}{x + 1}$

16. $\dfrac{x^3 + 7x^2 - 36}{x + 3}$

17. $\dfrac{4x^3 - 24x^2 + 128}{2x + 4}$

18. $\dfrac{72x^3 + 22x + 4}{6x - 1}$

19. $\dfrac{3x^3 + 4x^2 - 6}{x - 2}$

20. $\dfrac{2x^3 + 3x^2 - 9x + 5}{x + 3}$

21. $\dfrac{4x^3 + 2x^2 + 30x + 20}{2x - 5}$

22. $\dfrac{18x^3 + 6x^2 - 4x}{6x - 2}$

23. $\dfrac{8x^4 - 10x^3 + 16x^2 + 4x - 30}{4x - 5}$

24. $\dfrac{9x^4 + 12x^3 - 6x^2 + 10x + 24}{3x + 4}$

25. $\dfrac{8x^3 - 1}{2x + 1}$

26. $\dfrac{x^3 + 1}{x + 1}$

27. $\dfrac{x^4 - 16}{x + 2}$

28. $\dfrac{16x^4 + 1}{2x - 1}$

29. $\dfrac{3x^4 + 5x^3 - 17x^2 + 11x - 2}{x^2 + 3x - 2}$

30. $\dfrac{6x^4 + 5x^3 - 11x^2 + 9x - 5}{2x^2 - x + 1}$

Chapter 5 Group Activities

1. Look in Section 5.4 at Rules 1, 2, and 3 for exponents. Check these rules by doing three examples of each rule. For each example that you make up, do them by Method 1 and by Method 2 as in the examples of Section 2.5. Here is an example of one way to verify the following property.

Rule 3: $(xy)^3 = (xy)(xy)(xy) = (xyxyxy)$
$\qquad\qquad = (xxxyyy) = (xxx)(yyy) = x^3y^3$

There are other rules that are easily verified in a similar fashion. Try these as well.

a. $\left(\dfrac{x}{y}\right)^n = \dfrac{x^n}{y^n}$

b. $\dfrac{x^m}{x^n} = x^{m-n}$

2. Ask your instructor what type of graphing utility to use and how to type the expressions into it to graph them. Then in small groups,

a. Graph $y = x^2$, $y = x^4$, and $y = x^6$. Is there a similarity in these graphs? What do you notice about the graph of $y = x^{2n}$ for n, a natural number?

b. Next graph $y = x^3$, $y = x^5$, and $y = x^7$. Is there a similarity in these graphs? What do you notice about the graph of $y = x^{2n+1}$ for n, a natural number?

c. What do you get when you graph $y = x^2 + 1$, $y = x^2 + 2$, $y = x^2 + x$, $y = x^3 + x^2$? Graph some others. Do you see any interesting results?

Chapter 5 Summary

Glossary of Basic Terms

Binomial. The sum or difference of two unlike monomials. (p. 226)

Decreasing order. A polynomial written with each term of some degree less than the preceding term. (p. 227)

Degree of a monomial in one variable. The same as the exponent of the variable. (p. 227)

Degree of a polynomial in one variable. The same as the highest degree monomial contained in the polynomial. (p. 227)

Evaluate an expression. Replace the letters with given numbers; then do the arithmetic using the order of operations. (p. 221)

Exponent. In the expression, 2^4, the *base* is 2 and the *exponent* is 4, which indicates that the base 2 is multiplied as a factor 4 times, that is, $2^4 = 2 \cdot 2 \cdot 2 \cdot 2 = 16$. (p. 221)

Increasing order. A polynomial written with each term of some degree larger than the preceding term. (p. 227)

Like terms. Terms with the same variables with exactly the same exponents. (p. 223)

Monomial.   An algebraic expression that contains only products of numbers and variables, which have nonnegative integer exponents. (p. 226)

Numerical coefficient.   The numerical factor of a term. (p. 223)

Polynomial.   Either a monomial or the sum or difference of unlike monomials. (p. 226)

Term.   A single number or a product of a number and one or more letters raised to powers. (p. 223)

Trinomial.   The sum or difference of three unli[k]monomials. (p. 226)

Unlike terms.   Terms with different variables o[] exponents. (p. 223)

Variables.   Letters that represent unknown quar (p. 220)

5.1   Fundamental Operations

1. **Basic Mathematical Principles:**
    a. $a + b = b + a$ (Commutative Property for Addition)
    b. $ab = ba$ (Commutative Property for Multiplication)
    c. $(a + b) + c = a + (b + c)$ (Associative Property for Addition)
    d. $(ab)c = a(bc)$ (Associative Property for Multiplication)
    e. $a(b + c) = ab + ac$ or $(b + c)a = ba + ca$ (Distributive Property)
    f. $a + 0 = a$
    g. $a \cdot 0 = 0$
    h. $a + (-a) = 0$ (Additive Inverse)
    i. $a \cdot 1 = a$
    j. $a \cdot \dfrac{1}{a} = 1$ (a ≠ 0) (Multiplicative Inverse) (p. 220)

2. **Order of operations:**
    a. Perform all operations inside parentheses. If the problem contains a fraction bar, treat the numerator and the denominator separately.
    b. Evaluate all powers, if any.
    c. Perform any multiplications and divisions in order, from left to right.
    d. Do any additions and subtractions in order, from left to right. (p. 221)

5.2   Simplifying Algebraic Expressions

1. **Removing parentheses:**
    a. Parentheses preceded by a plus sign may be removed without changing the signs of the terms within.
    b. Parentheses preceded by a minus sign may be removed if the signs of *all* the terms within the parentheses are changed; then drop the minus sign that preceded the parentheses. (p. 223)

5.3   Addition and Subtraction of Polynomials

1. **Adding polynomials:** To add polynomials, add their like terms. (p. 227)

2. **Subtracting polynomials:** To subtract two polynomials, change all the signs of the ter[] second polynomial and then add the two res polynomials. (p. 228)

5.4   Multiplication of Monomial

1. **Multiplying powers:** To multiply powers w same base, add the exponents: $a^m \cdot a^n = a^{m-}$

2. **Raising a power to a power:** To raise a po power, multiply the exponents: $(a^m)^n = a^{mn}$

3. **Raising a product to a power:** To raise a [] power, raise each factor to that same power $(ab)^m = a^m b^m$. (p. 231)

5.5   Multiplication of Polynomia

1. **Multiplying a polynomial by a monomia** multiply a polynomial by a monomial, mul term of the polynomial by the monomial, a the products. (p. 232)

2. **Multiplying a polynomial by a polynomi** multiply a polynomial by a polynomial, mu term of the first polynomial by each term of polynomial, and then add the products. (p. [])

5.6   Division by a Monomial

1. **Dividing a monomial by a monomial:** To monomial by a monomial, first write the q[] fraction form. Then factor both numerator denominator into prime factors. Reduce to terms by dividing both numerator and den[] their common factors. The remaining fact[] numerator and the denominator give the q[] (p. 234)

2. **Dividing a polynomial by a monomial:** [] polynomial by a monomial, divide each te[] polynomial by the monomial. (p. 235)

5.7   Division by a Polynomial

1. **Dividing a polynomial by a polynomial:** polynomial by a polynomial, use long divi shown in Section 5.7. (p. 236)

Chapter 5 Review

1. For any number a, $a \cdot 1 = ?$
2. For any number a, $a \cdot 0 = ?$
3. For any number a except 0, $a \cdot \dfrac{1}{a} = ?$

Evaluate:

4. $10 - 4(3)$
5. $2 + 3 \cdot 4^2$
6. $(4)(12) \div 6 - 2^3 + 18 \div 3^2$

In Exercises 7–12, let $x = 3$ and $y = -2$. Evaluate each expression:

7. $x + y$
8. $x - 3y$
9. $5xy$
10. $\dfrac{x^2}{y}$
11. $y^3 - y^2$
12. $\dfrac{2x^3 - 3y}{xy^2}$

In Exercises 13–16, simplify by removing the parentheses and combining like terms:

13. $(5y - 3) - (2 - y)$
14. $(7 - 3x) - (5x + 1)$
15. $11(2x + 1) - 4(3x - 4)$

16. $(x^3 + 2x^2y) - (3y^3 - 2x^3 + x^2y + y)$
17. Is $1 - 8x^2$ a monomial, a binomial, or a trinomial?
18. Find the degree of the polynomial $x^4 + 2x^3 - 6$.

Perform the indicated operations:

19. $(3a^2 + 7a - 2) + (5a^2 - 2a + 4)$
20. $(6x^3 + 3x^2 + 1) - (-3x^3 - x^2 - x - 1)$
21. $(3x^2 + 5x + 2) + (9x^2 - 6x - 2) - (2x^2 + 6x - 4)$
22. $(6x^2)(4x^3)$
23. $(-7x^2y)(8x^3y^2)$
24. $(3x^2)^3$
25. $5a(3a + 4b)$
26. $-4x^2(8 - 2x + 3x^2)$
27. $(5x + 3)(3x - 4)$
28. $(3x^2 - 6x + 1)(2x - 4)$
29. $(49x^2) \div (7x^3)$
30. $\dfrac{15x^3y}{3xy}$
31. $\dfrac{36a^3 - 27a^2 + 9a}{9a}$
32. $\dfrac{6x^2 + x - 12}{2x + 3}$
33. $\dfrac{3x^3 + 2x^2 - 6x + 4}{x + 2}$

Chapter 5 Test

Evaluate:

1. $3 \cdot 5 - 2 \cdot 4^2$
2. $12 \div 2 \cdot 3 \div 2 + 3^3 - 16 \div 2^2$
3. Evaluate $\dfrac{3x^2y - 4x}{2y}$ when $x = 4$ and $y = -1$.

Perform the indicated operations and simplify:

4. $3a^2 - 17a + 6a^2 + 4a$
5. $5(2 + x) - 2(x + 4)$
6. Add: $\begin{array}{r} 3x^2 + 6x - 8 \\ -9x^2 + 6x \\ -3x^2 \quad\ + 15 \\ - 7x + 4 \\ \hline \end{array}$
7. $(5a - 5b + 7) - (2a - 5b - 3)$
8. $(7a^2)(-4a^4)(a)$
9. $(6x^4y^2)^3$

10. $-5x(2x - 3)$
11. $\dfrac{85x^4y^2}{17x^2y^5}$
12. $(4a + 6)(a - 5)$
13. $(x + y - 5)(x - y)$
14. $\dfrac{36a^4 - 20a^3 - 16a^2}{4a}$
15. $\dfrac{9x^4y^3 - 12x^2y + 18y^2}{3x^2y^3}$
16. $(5x^2y^3)(-7x^3y)$
17. $\dfrac{3x^2 - 13x - 10}{x - 5}$
18. $\dfrac{4x^5y^3}{-2x^3y^5}$
19. $\dfrac{6x^2 - 7x - 6}{3x - 5}$
20. $\dfrac{x^3 + 2x^2 + x + 12}{x + 3}$

6 Equations and Formulas

Mathematics at Work

Syracuse Newspapers/The Image Works

Diesel Technician
Diesel technician repairing a diesel engine.

Diesel technicians inspect, maintain, and repair diesel engines that power the agriculture, construction, trucking, locomotive, and heavy equipment industries. The diesel engine provides power for the nation's heavy vehicles and equipment because it delivers more power per unit of fuel and is more durable than the gasoline engine. Some diesel technicians specialize in farm tractors and farm equipment; large trucks; bulldozers, road graders, and construction equipment; and heavy industrial equipment, as well as automobiles and boats. These technicians work for companies or organizations that maintain their own equipment and spend their time doing preventive maintenance to keep their equipment operating dependably and safely. During routine maintenance, they inspect and repair the basic components to eliminate unnecessary wear and damage to avoid costly breakdowns.

The work of a diesel technician is becoming more complex as more electronic components are used to control engine operation. Diesel technicians use handheld computers and sophisticated equipment to diagnose problems and to adjust engine functions. They often need special equipment to handle large components. Many community colleges and trade and vocational schools offer associate degree and certificate diesel repair programs. Employers prefer graduates of formal training programs because of their basic understanding and their ability to more quickly advance to their journey mechanic level. Certifications are available within a variety of specialty areas, such as master heavy-duty truck repair and school bus repair in specific areas— for example, brakes, diesel engines, drive trains, electrical systems, and steering and suspension. For more information, go to the website listed below.

www.cengage.com/mathematics/ewen

Objectives

- Use the addition, subtraction, multiplication, and division proper[ties of] equality to solve simple equations.
- Solve equations with parentheses.
- Solve equations with fractions.
- Translate words into algebraic symbols.
- Solve application problems using equations.
- Solve a formula for a given letter.
- Substitute data into a formula and find the value of the indicate[d letter] using the rules for working with measurements.
- Substitute data into a formula involving reciprocals and find the [value of] the indicated letter using a scientific calculator.

6.1 Equations

In technical work, the ability to use equations and formulas is essential. A **varia**[ble is a sym]bol (usually a letter of the alphabet) used to represent an unknown number. A[n algebraic] **expression** is a combination of numbers, variables, symbols for operations ([plus, minus,] times, divide), and symbols for grouping (parentheses or a fraction bar). Exam[ples of alge]braic expressions are

$$4x - 9, \qquad 3x^2 + 6x + 9, \qquad 5x(6x + 4), \qquad \frac{2x + 5}{-3x}$$

An **equation** is a statement that two quantities are equal. The symbol [$=$ means] "equals" and separates an equation into two parts: the left member and the ri[ght member.] For example, in the equation

$$2x + 3 = 11$$

the left member is $2x + 3$ and the right member is 11. Other examples of equa[tions are]

$$x - 5 = 6, \qquad 3x = 12, \qquad 4m + 9 = 3m - 2, \qquad \frac{x - 2}{4} = 3(x + \blacksquare)$$

To **solve an equation** means to find what number or numbers can replace [the variable] to make the equation a true statement. In the equation $2x + 3 = 11$, the soluti[on, or root,] is, when x is replaced by 4, the resulting equation is a true statement.

$$2x + 3 = 11$$
$$\text{Let } x = 4: \quad 2(4) + 3 = 11 \qquad ?$$
$$8 + 3 = 11 \qquad \text{True}$$

A replacement number (or numbers) that produces a true statement in an equa[tion is called] a **solution** or a **root** of the equation. Note that replacing x in the equation abo[ve with a] number other than 4, such as 5, results in a false statement.

$$2x + 3 = 11$$

Let $x = 5$: $2(5) + 3 = 11$?

$$10 + 3 = 11 \quad \text{False}$$

One method of solving equations involves changing the given equation to an equivalent equation by performing the same arithmetic operation on both sides of the equation. The basic arithmetic operations used are addition, subtraction, multiplication, and division.

Equivalent equations are equations that have the same solutions or roots. For example, $3x = 6$ and $x = 2$ are equivalent equations, since 2 is the root of each. In solving an equation by this method, continue to change the given equation to another equivalent equation until you find an equation whose root is obvious.

Four Basic Rules Used to Solve Equations

1. *Addition Property of Equality:* If the same quantity is added to both sides of an equation, the resulting equation is equivalent to the original equation.

   Example: Solve $x - 2 = 8$.

   $$x - 2 + 2 = 8 + 2 \qquad \text{Add 2 to both sides.}$$
   $$x = 10 \qquad\qquad\qquad\qquad ■$$

2. *Subtraction Property of Equality:* If the same quantity is subtracted from both sides of an equation, the resulting equation is equivalent to the original equation.

   Example: Solve $x + 5 = 2$.

   $$x + 5 - 5 = 2 - 5 \qquad \text{Subtract 5 from both sides.}$$
   $$x = -3 \qquad\qquad\qquad\qquad ■$$

3. *Multiplication Property of Equality:* If both sides of an equation are multiplied by the same (nonzero) quantity, the resulting equation is equivalent to the original equation.

   Example: Solve $\dfrac{x}{9} = 4$.

   $$9\left(\frac{x}{9}\right) = (4)9 \qquad \text{Multiply both sides by 9.}$$
   $$x = 36 \qquad\qquad\qquad\qquad ■$$

4. *Division Property of Equality:* If both sides of an equation are divided by the same (nonzero) quantity, the resulting equation is equivalent to the original equation.

   Example: Solve $4x = 20$.

   $$\frac{4x}{4} = \frac{20}{4} \qquad \text{Divide both sides by 4.}$$
   $$x = 5 \qquad\qquad\qquad\qquad ■$$

Basically, to solve a simple equation, use one of the rules and use a number that will *undo* what has been done to the variable.

Example 1

Solve: $x + 3 = 8$.

Since 3 has been added to the variable, use Rule 2 and subtract 3 from both s
equation.

$$x + 3 = 8$$
$$x + 3 - 3 = 8 - 3 \qquad \text{Subtracting 3 was chosen to } undo \text{ adding 3.}$$
$$x + 0 = 5$$
$$x = 5$$

A check is recommended, since an error could have been made. To check, repla
able in the original equation by 5, the apparent root, to make sure that the res
ment is true.

Check:
$$x + 3 = 8$$
$$5 + 3 = 8 \qquad \text{True}$$

Thus, the root is 5.

Example 2

Solve: $x - 4 = 7$.

Since 4 has been subtracted from the variable, use Rule 1 and add 4 to both
equation.

$$x - 4 = 7$$
$$x - 4 + 4 = 7 + 4 \qquad \text{Adding 4 was chosen to } undo \text{ subtracting 4.}$$
$$x + 0 = 11$$
$$x = 11$$

The apparent root is 11.

Check:
$$x - 4 = 7$$
$$11 - 4 = 7 \qquad \text{True}$$

Thus, the root is 11.

Example 3

Solve: $2x = 9$.

Since the variable has been multiplied by 2, use Rule 4 and divide both sides
tion by 2.

$$2x = 9$$
$$\frac{2x}{2} = \frac{9}{2} \qquad \text{Dividing by 2 was chosen to } undo \text{ multiplying by 2.}$$
$$x = \frac{9}{2}$$

Note: Each solution should be checked by substituting it into the original equ
a check is not provided in this text, the check is left for you to do.

Example 4 Solve: $\dfrac{x}{3} = 9$.

Since the variable has been divided by 3, use Rule 3 and multiply both sides of the equation by 3.

$$\frac{x}{3} = 9$$

$$3\left(\frac{x}{3}\right) = (9)3 \qquad \text{Multiplying by 3 was chosen to } undo \text{ dividing by 3.}$$

$$x = 27$$ ■

 Example 5 Solve: $-4 = -6x$.

Since the variable has been multiplied by -6, use Rule 4 and divide both sides of the equation by -6.

$$-4 = -6x$$

$$\frac{-4}{-6} = \frac{-6x}{-6} \qquad \text{Dividing by } -6 \text{ was chosen to } undo \text{ multiplying by } -6.$$

$$\frac{2}{3} = x$$

The apparent root is $\dfrac{2}{3}$.

Check:

$$-4 = -6x$$

$$-4 = -6\left(\frac{2}{3}\right) \qquad ?$$

$$-4 = -4 \qquad \text{True}$$

Thus, the root is $\dfrac{2}{3}$. ■

Some equations have more than one operation indicated on the variable. For example, the equation $2x + 5 = 6$ has both addition of 5 and multiplication by 2 indicated on the variable. Use the following procedure to solve equations like this.

> When more than one operation is indicated on the variable, undo the additions and subtractions first, then undo the multiplications and divisions.

 Example 6 Solve: $2x + 5 = 6$.

$$2x + 5 - 5 = 6 - 5 \qquad \text{Subtract 5 from both sides.}$$

$$2x = 1$$

$$\frac{2x}{2} = \frac{1}{2} \qquad \text{Divide both sides by 2.}$$

$$x = \frac{1}{2}$$

The apparent root is $\dfrac{1}{2}$.

Check:
$$2x + 5 = 6$$

$$2\left(\frac{1}{2}\right) + 5 = 6 \qquad ?$$

$$1 + 5 = 6 \qquad \text{True}$$

Thus, the root is $\frac{1}{2}$.

Example 7

Solve: $\frac{x}{3} - 6 = 9$.

$$\frac{x}{3} - 6 + 6 = 9 + 6 \qquad \text{Add 6 to both sides.}$$

$$\frac{x}{3} = 15$$

$$3\left(\frac{x}{3}\right) = (15)3 \qquad \text{Multiply both sides by 3.}$$

$$x = 45$$

Example 8

Solve: $118 - 22m = 30$.

$$118 - 22m - 118 = 30 - 118 \qquad \text{Subtract 118 from both sides.}$$

$$-22m = -88$$

$$\frac{-22m}{-22} = \frac{-88}{-22} \qquad \text{Divide both sides by } -22.$$

$$m = 4$$

The apparent root is 4.

Check:
$$118 - 22m = 30$$

$$118 - 22(4) = 30 \qquad ?$$

$$118 - 88 = 30 \qquad ?$$

$$30 = 30 \qquad \text{True}$$

Thus, the root is 4.

Here is another approach to solving this equation:

$$118 - 22m = 30$$

$$118 - 22m + 22m = 30 + 22m \qquad \text{Add } 22m \text{ to both sides.}$$

$$118 = 30 + 22m$$

$$118 - 30 = 30 + 22m - 30 \qquad \text{Subtract 30 from both sides.}$$

$$88 = 22m$$

$$\frac{88}{22} = \frac{22m}{22} \qquad \text{Divide both sides by 22.}$$

$$4 = m$$

Exercises 6.1

Solve each equation and check:

1. $x + 2 = 8$

2. $3a = 7$

3. $y - 5 = 12$

4. $\dfrac{2}{3}n = 6$

5. $w - 7\dfrac{1}{2} = 3$

6. $2m = 28.4$

7. $\dfrac{x}{13} = 1.5$

8. $n + 12 = -5$

9. $3b = 15.6$

10. $y - 17 = 25$

11. $17x = 5117$

12. $28 + m = 3$

13. $2 = x - 5$

14. $-29 = -4y$

15. $17 = -3 + w$

16. $49 = 32 + w$

17. $14b = 57$

18. $y + 28 = 13$

19. $5m = 0$

20. $28 + m = 28$

21. $x + 5 = 5$

22. $y + 7 = -7$

23. $4x = 64$

24. $5x - 125 = 0$

25. $\dfrac{x}{7} = 56$

26. $\dfrac{y}{5} = 35$

27. $-48 = 12y$

28. $13x = -78$

29. $-x = 2$

30. $-y = 7$

31. $5y + 3 = 13$

32. $4x - 2 = 18$

33. $10 - 3x = 16$

34. $8 - 2y = 4$

35. $\dfrac{x}{4} - 5 = 3$

36. $\dfrac{x}{5} + 4 = 9$

37. $2 - x = 6$

38. $8 - y = 3$

39. $\dfrac{2}{3}y - 4 = 8$

40. $5 - \dfrac{1}{4}x = 7$

41. $3x - 5 = 12$

42. $5y + 7 = 28$

43. $\dfrac{m}{3} - 6 = 8$

44. $\dfrac{w}{5} + 7 = 13$

45. $\dfrac{2x}{3} = 7$

46. $\dfrac{4b}{5} = 15$

47. $-3y - 7 = -6$

48. $28 = -7 - 3r$

49. $5 - x = 6$

50. $17 - 5w = -68$

51. $54y - 13 = 17.8$

52. $37a - 7 = 67$

53. $28w - 56 = -8$

54. $52 - 4x = -8$

55. $29r - 13 = 57$

56. $15x - 32 = 18$

57. $31 - 3y = 41$

58. $62 = 13y - 3$

59. $-83 = 17 - 4x$

60. $58 = 5m + 52$

6.2 Equations with Variables in Both Members

To solve equations with variables in both members (both sides), such as

$$3x + 4 = 5x - 12$$

do the following:

First, add or subtract either variable term from both sides of the equation.

$$3x + 4 = 5x - 12$$
$$3x + 4 - 3x = 5x - 12 - 3x \qquad \text{Subtract } 3x \text{ from both sides.}$$
$$4 = 2x - 12$$

Then take the constant term (which now appears on the same side of the equation with the variable term) and add it to, or subtract it from, both sides. Solve the resulting equation.

$$4 + 12 = 2x - 12 + 12 \qquad \text{Add 12 to both sides.}$$
$$16 = 2x$$
$$\dfrac{16}{2} = \dfrac{2x}{2} \qquad \text{Divide both sides by 2.}$$
$$8 = x$$

This equation could also have been solved as follows:

$$3x + 4 = 5x - 12$$
$$3x + 4 - 5x = 5x - 12 - 5x \qquad \text{Subtract } 5x \text{ from both sides.}$$
$$-2x + 4 = -12$$
$$-2x + 4 - 4 = -12 - 4 \qquad \text{Subtract 4 from both sides.}$$
$$-2x = -16$$
$$\frac{-2x}{-2} = \frac{-16}{-2} \qquad \text{Divide both sides by } -2.$$
$$x = 8$$

Example 1

Solve: $5x - 4 = 8x - 13$.

$$5x - 4 - 8x = 8x - 13 - 8x \qquad \text{Subtract } 8x \text{ from both sides.}$$
$$-3x - 4 = -13$$
$$-3x - 4 + 4 = -13 + 4 \qquad \text{Add 4 to both sides.}$$
$$-3x = -9$$
$$\frac{-3x}{-3} = \frac{-9}{-3} \qquad \text{Divide both sides by } -3.$$
$$x = 3$$

Check:

$$5x - 4 = 8x - 13$$
$$5(3) - 4 = 8(3) - 13 \qquad ?$$
$$15 - 4 = 24 - 13 \qquad ?$$
$$11 = 11 \qquad \text{True}$$

Therefore, 3 is a root.

Example 2

Solve: $-2x + 5 = 6x - 11$.

$$-2x + 5 + 2x = 6x - 11 + 2x \qquad \text{Add } 2x \text{ to both sides.}$$
$$5 = 8x - 11$$
$$5 + 11 = 8x - 11 + 11 \qquad \text{Add 11 to both sides.}$$
$$16 = 8x$$
$$\frac{16}{8} = \frac{8x}{8} \qquad \text{Divide both sides by 8.}$$
$$2 = x$$

Check:

$$-2x + 5 = 6x - 11$$
$$-2(2) + 5 = 6(2) - 11 \qquad ?$$
$$-4 + 5 = 12 - 11 \qquad ?$$
$$1 = 1 \qquad \text{True}$$

Thus, 2 is a root.

Example 3

Solve: $5x + 7 = 2x - 14$.

$$5x + 7 - 2x = 2x - 14 - 2x \quad \text{Subtract } 2x \text{ from both sides.}$$
$$3x + 7 = -14$$
$$3x + 7 - 7 = -14 - 7 \quad \text{Subtract 7 from both sides.}$$
$$3x = -21$$
$$\frac{3x}{3} = \frac{-21}{3} \quad \text{Divide both sides by 3.}$$
$$x = -7$$

Check:
$$5x + 7 = 2x - 14$$
$$5(-7) + 7 = 2(-7) - 14 \quad ?$$
$$-35 + 7 = -14 - 14 \quad ?$$
$$-28 = -28 \quad \text{True}$$

So -7 is a root. ■

 Example 4

Solve: $4 - 5x = 28 + x$.

$$4 - 5x + 5x = 28 + x + 5x \quad \text{Add } 5x \text{ to both sides.}$$
$$4 = 28 + 6x$$
$$4 - 28 = 28 + 6x - 28 \quad \text{Subtract 28 from both sides.}$$
$$-24 = 6x$$
$$\frac{-24}{6} = \frac{6x}{6} \quad \text{Divide both sides by 6.}$$
$$-4 = x$$ ■

Exercises 6.2

Solve each equation and check:

1. $4y + 9 = 7y - 15$
2. $2y - 45 = -y$
3. $5x + 3 = 7x - 5$
4. $2x - 3 = 3x - 13$
5. $-2x + 7 = 5x - 21$
6. $5x + 3 = 2x - 15$
7. $3y + 5 = 5y - 1$
8. $3x - 4 = 7x - 32$
9. $-3x + 17 = 6x - 37$
10. $3x + 13 = 2x - 12$
11. $7x + 9 = 9x - 3$
12. $-5y + 12 = 12y - 5$
13. $3x - 2 = 5x + 8$
14. $13y + 2 = 20y - 5$
15. $-4x + 25 = 6x - 45$
16. $5x - 7 = 6x - 5$
17. $5x + 4 = 10x - 7$
18. $3x - 2 = 5x - 20$
19. $27 + 5x = 9 + 3x$
20. $2y + 8 = 5y - 1$
21. $-7x + 18 = 11x - 36$
22. $4x + 5 = 2x - 7$
23. $4y + 11 = 7y - 28$
24. $4x = 2x - 12$
25. $-4x + 2 = 8x - 7$
26. $6x - 1 = 9x - 9$
27. $13x + 6 = 6x - 1$
28. $6y + 7 = 18y - 1$
29. $3x + 1 = 17 - x$
30. $17 - 4y = 14 - y$

 6.3 # Equations with Parentheses

To solve an equation having parentheses in one or both members, always remove the parentheses first. Then combine like terms. Then use the previously explained methods to solve the resulting equation.

Example 1   Solve: $5 - (2x - 3) = 7$.

$$5 - 2x + 3 = 7 \qquad \text{Remove parentheses.}$$
$$8 - 2x = 7 \qquad \text{Combine like terms.}$$
$$8 - 2x - 8 = 7 - 8 \qquad \text{Subtract 8 from both sides.}$$
$$-2x = -1$$
$$\frac{-2x}{-2} = \frac{-1}{-2} \qquad \text{Divide both sides by } -2.$$
$$x = \frac{1}{2}$$

Check: $5 - (2x - 3) = 7$

$$5 - \left[2\left(\frac{1}{2} \right) - 3 \right] = 7 \qquad ?$$
$$5 - (1 - 3) = 7 \qquad ?$$
$$5 - (-2) = 7 \qquad \text{True}$$

Therefore, $\frac{1}{2}$ is the root.

 Example 2   Solve: $7x - 6(5 - x) = 9$.

$$7x - 30 + 6x = 9 \qquad \text{Remove parentheses.}$$
$$13x - 30 = 9 \qquad \text{Combine like terms.}$$
$$13x - 30 + 30 = 9 + 30 \qquad \text{Add 30 to both sides.}$$
$$13x = 39$$
$$\frac{13x}{13} = \frac{39}{13} \qquad \text{Divide both sides by 13.}$$
$$x = 3$$

In the following examples, we have parentheses as well as the variable in both

 Example 3   Solve: $3(x - 5) = 2(4 - x)$.

$$3x - 15 = 8 - 2x \qquad \text{Remove parentheses.}$$
$$3x - 15 + 2x = 8 - 2x + 2x \qquad \text{Add } 2x \text{ to both sides.}$$
$$5x - 15 = 8 \qquad \text{Combine like terms.}$$
$$5x - 15 + 15 = 8 + 15 \qquad \text{Add 15 to both sides.}$$
$$5x = 23$$
$$\frac{5x}{5} = \frac{23}{5} \qquad \text{Divide both sides by 5.}$$
$$x = \frac{23}{5}$$

Check: $3(x - 5) = 2(4 - x)$

$$3\left(\frac{23}{5} - 5\right) = 2\left(4 - \frac{23}{5}\right) \quad ?$$

$$3\left(-\frac{2}{5}\right) = 2\left(-\frac{3}{5}\right) \quad ?$$

$$-\frac{6}{5} = -\frac{6}{5} \qquad \text{True}$$

Therefore, $\frac{23}{5}$ is the root.   ■

Example 4   Solve: $8x - 4(x + 2) = 12(x + 1) - 14$.

$$8x - 4x - 8 = 12x + 12 - 14 \qquad \text{Remove parentheses.}$$
$$4x - 8 = 12x - 2 \qquad \text{Combine like terms.}$$
$$4x - 8 - 12x = 12x - 2 - 12x \qquad \text{Subtract } 12x \text{ from both sides.}$$
$$-8x - 8 = -2$$
$$-8x - 8 + 8 = -2 + 8 \qquad \text{Add 8 to both sides.}$$
$$-8x = 6$$
$$\frac{-8x}{-8} = \frac{6}{-8} \qquad \text{Divide both sides by } -8.$$
$$x = -\frac{3}{4}$$

Check: $$8x - 4(x + 2) = 12(x + 1) - 14$$

$$8\left(-\frac{3}{4}\right) - 4\left(-\frac{3}{4} + 2\right) = 12\left(-\frac{3}{4} + 1\right) - 14 \quad ?$$

$$-6 - 4\left(\frac{5}{4}\right) = 12\left(\frac{1}{4}\right) - 14 \quad ?$$

$$-6 - 5 = 3 - 14 \quad ?$$

$$-11 = -11 \qquad \text{True}$$

Therefore, $-\frac{3}{4}$ is the root.   ■

Exercises 6.3

Solve each equation and check:

1. $2(x + 3) - 6 = 10$

2. $-3x + 5(x - 6) = 32$

3. $3n + (2n + 4) = 6$

4. $5m - (2m - 7) = -5$

5. $16 = -3(x - 4)$

6. $5y + 6(y - 3) = 15$

7. $5a - (3a + 4) = 8$

8. $2(b + 4) - 3 = 15$

9. $5a - 4(a - 3) = 7$

10. $29 = 4 + (2m + 1)$

11. $5(x - 3) = 21$

12. $27 - 8(2 - y) = -13$

13. $2a - (5a - 7) = 22$

14. $2(5m - 6) - 13 = -1$

15. $2(w - 3) + 6 = 0$

16. $6r - (2r - 3) + 5 = 0$

17. $3x - 7 + 17(1 - x) = -6$

18. $4y - 6(2 - y) = 8$

19. $6b = 27 + 3b$

20. $2a + 4 = a - 3$

21. $4(25 - x) = 3x + 2$

22. $4x - 2 = 3(25 - x)$

23. $x + 3 = 4(57 - x)$

24. $2(y + 1) = y - 7$

25. $6x + 2 = 2(17 - x)$ **26.** $6(17 - x) = 2 - 4x$

27. $5(x - 8) - 3x - 4 = 0$ **28.** $5(28 - 2x) - 7x = 4$

29. $3(x + 4) + 3x = 6$

30. $8x - 4(x + 2) + 11 = 0$

31. $y - 4 = 2(y - 7)$

32. $7(w - 4) = w + 2$

33. $9m - 3(m - 5) = 7m - 3$

34. $4(x + 18) = 2(4x + 18)$

35. $3(2x + 7) = 13 + 2(4x + 2)$

36. $5y - 3(y - 2) = 6(y + 1)$

37. $8(x - 5) = 13x$

38. $4(x + 2) = 30 - (x - 3)$

39. $5 + 3(x + 7) = 26 - 6(5x + 11)$

40. $2(y - 3) = 4 + (y - 14)$

41. $5(2y - 3) = 3(7y - 6) + 19(y + 1) + 14$

42. $2(7y - 6) - 11(y + 1) = 38 - 7(9y + 4)$

43. $16(x + 3) = 7(x - 5) - 9(x + 4) - 7$

44. $31 - 2(x - 5) = -3(x + 4)$

45. $4(y + 2) = 8(y - 4) + 7$

46. $12x - 13(x + 4) = 4x - 6$

47. $4(5y - 2) + 3(2y + 6) = 25(3y + 2) - 19y$

48. $6(3x + 1) = 5x - (2x + 2)$

49. $12 + 8(2y + 3) = (y + 7) - 16$

50. $-2x + 6(2 - x) - 4 = 3(x + 1) - 6$

51. $5x - 10(3x - 6) = 3(24 - 9x)$

52. $4y + 7 - 3(2y + 3) = 4(3y - 4) - 7y + 7$

53. $6(y - 4) - 4(5y + 1) = 3(y - 2) - 4(2y +$

54. $2(5y + 1) + 16 = 4 + 3(y - 7)$

55. $-6(x - 5) + 3x = 6x - 10(-3 + x)$

56. $14x + 14(3 - 2x) + 7 = 4 - x + 5(2 - 3x$

57. $2.3x - 4.7 + 0.6(3x + 5) = 0.7(3 - x)$

58. $5.2(x + 3) + 3.7(2 - x) = 3$

59. $0.089x - 0.32 + 0.001(5 - x) = 0.231$

60. $5x - 2.5(7 - 4x) = x - 7(4 + x)$

6.4 Equations with Fractions

To Solve an Equation with Fractions

1. Find the least common denominator (LCD) of all the fractional terms on [
   sides of the equation.
2. Multiply both sides of the equation by the LCD. (If this step has been dor
   correctly, no fractions should now appear in the resulting equation.)
3. Solve the resulting equation from Step 2 using the methods introduced ea
   this chapter.

 Example 1

Solve: $\dfrac{3x}{4} = \dfrac{45}{20}$.

The LCD of 4 and 20 is 20; therefore, multiply both sides of the equation by 20

$$\frac{3x}{4} = \frac{45}{20}$$

$$20\left(\frac{3x}{4}\right) = \left(\frac{45}{20}\right)20$$

$$15x = 45$$

$$\frac{15x}{15} = \frac{45}{15} \qquad \text{Divide both sides by 15.}$$

$$x = 3$$

Example 2   Solve: $\dfrac{3}{4} + \dfrac{x}{6} = \dfrac{13}{12}$.

The LCD of 4, 6, and 12 is 12; multiply both sides of the equation by 12.

$$\frac{3}{4} + \frac{x}{6} = \frac{13}{12}$$

$$12\left(\frac{3}{4} + \frac{x}{6}\right) = \left(\frac{13}{12}\right)12$$

$$12\left(\frac{3}{4}\right) + 12\left(\frac{x}{6}\right) = \left(\frac{13}{12}\right)12 \qquad$$ Apply the Distributive Property on the left side by multiplying each term within parentheses by 12.

$$9 + 2x = 13$$

$$9 + 2x - 9 = 13 - 9 \qquad$$ Subtract 9 from both sides.

$$2x = 4$$

$$\frac{2x}{2} = \frac{4}{2} \qquad$$ Divide both sides by 2.

$$x = 2$$

Check:
$$\frac{3}{4} + \frac{x}{6} = \frac{13}{12}$$

$$\frac{3}{4} + \frac{2}{6} = \frac{13}{12} \qquad ?$$

$$\frac{9}{12} + \frac{4}{12} = \frac{13}{12} \qquad \text{True}$$

■

 Example 3   Solve: $\dfrac{2x}{9} - 4 = \dfrac{x}{6}$.

The LCD of 9 and 6 is 18; multiply both sides of the equation by 18.

$$\frac{2x}{9} - 4 = \frac{x}{6}$$

$$18\left(\frac{2x}{9} - 4\right) = \left(\frac{x}{6}\right)18$$

$$18\left(\frac{2x}{9}\right) - 18(4) = \left(\frac{x}{6}\right)18 \qquad$$ Apply the Distributive Property on the left side by multiplying each term within parentheses by 18.

$$4x - 72 = 3x$$

$$4x - 72 - 4x = 3x - 4x \qquad$$ Subtract 4x from both sides.

$$-72 = -x$$

$$\frac{-72}{-1} = \frac{-x}{-1} \qquad$$ Divide both sides by −1.

$$72 = x$$

Check:
$$\frac{2x}{9} - 4 = \frac{x}{6}$$

$$\frac{2(72)}{9} - 4 = \frac{72}{6} \qquad ?$$

$$16 - 4 = 12 \qquad \text{True}$$

■

Example 4   Solve: $\dfrac{2}{3}x + \dfrac{3}{4}(36 - 2x) = 32$.

The LCD of 3 and 4 is 12; multiply both sides of the equation by 12.

$$\frac{2}{3}x + \frac{3}{4}(36 - 2x) = 32$$

$$12\left[\frac{2}{3}x + \frac{3}{4}(36 - 2x)\right] = (32)12$$

$$12\left(\frac{2}{3}x\right) + 12\left(\frac{3}{4}\right)(36 - 2x) = (32)12 \qquad \text{Apply the Distributive Prop}$$
$$\text{left to remove brackets.}$$

$$8x + 9(36 - 2x) = 384$$

$$8x + 324 - 18x = 384 \qquad \text{Apply the Distributive Prop}$$
$$\text{remove parentheses.}$$

$$-10x + 324 = 384 \qquad \text{Combine like terms.}$$

$$-10x + 324 - 324 = 384 - 324 \qquad \text{Subtract 324 from both side}$$

$$-10x = 60$$

$$\frac{-10x}{-10} = \frac{60}{-10} \qquad \text{Divide both sides by } -10.$$

$$x = -6$$

Check: $\qquad \dfrac{2}{3}x + \dfrac{3}{4}(36 - 2x) = 32$

$$\frac{2}{3}(-6) + \frac{3}{4}[36 - 2(-6)] = 32 \qquad ?$$

$$-4 + \frac{3}{4}(36 + 12) = 32 \qquad ?$$

$$-4 + \frac{3}{4}(48) = 32 \qquad ?$$

$$-4 + 36 = 32 \qquad \text{True}$$

 Example 5   Solve: $\dfrac{2x + 1}{3} - \dfrac{x - 6}{4} = \dfrac{2x + 4}{8} + 2$.

The LCD of 3, 4, and 8 is 24; multiply both sides of the equation by 24.

$$\frac{2x + 1}{3} - \frac{x - 6}{4} = \frac{2x + 4}{8} + 2$$

$$24\left(\frac{2x + 1}{3} - \frac{x - 6}{4}\right) = \left(\frac{2x + 4}{8} + 2\right)24$$

$$24\left(\frac{2x + 1}{3}\right) - 24\left(\frac{x - 6}{4}\right) = \left(\frac{2x + 4}{8}\right)24 + 2(24)$$

$$8(2x + 1) - 6(x - 6) = 3(2x + 4) + 48$$

$$16x + 8 - 6x + 36 = 6x + 12 + 48 \qquad \text{Remove parentheses.}$$

$$10x + 44 = 6x + 60 \qquad \text{Combine like terms.}$$

$$10x + 44 - 6x = 6x + 60 - 6x \qquad \text{Subtract } 6x \text{ from both}$$

$$4x + 44 = 60$$

$$4x + 44 - 44 = 60 - 44 \qquad \text{Subtract 44 from both sides.}$$

$$4x = 16$$

$$\frac{4x}{4} = \frac{16}{4} \qquad \text{Divide both sides by 4.}$$

$$x = 4$$

Check:

$$\frac{2x + 1}{3} - \frac{x - 6}{4} = \frac{2x + 4}{8} + 2$$

$$\frac{2(4) + 1}{3} - \frac{4 - 6}{4} = \frac{2(4) + 4}{8} + 2 \qquad ?$$

$$\frac{9}{3} - \frac{-2}{4} = \frac{12}{8} + 2 \qquad ?$$

$$3 + \frac{1}{2} = \frac{3}{2} + 2 \qquad ?$$

$$\frac{7}{2} = \frac{7}{2} \qquad \text{True} \qquad ■$$

When the variable appears in the denominator of a fraction in an equation, multiply both members by the LCD. Be careful that the replacement for the variable does not make the denominator zero.

Example 6   Solve: $\dfrac{3}{x} = 2$.

$$x\left(\frac{3}{x}\right) = (2)x \qquad \text{Multiply both sides by the LCD, } x.$$

$$3 = 2x$$

$$\frac{3}{2} = \frac{2x}{2} \qquad \text{Divide both sides by 2.}$$

$$\frac{3}{2} = x$$

Check:

$$\frac{3}{x} = 2$$

$$\frac{3}{\dfrac{3}{2}} = 2 \qquad ?$$

$$3 \div \frac{3}{2} = 2 \qquad ?$$

$$3 \cdot \frac{2}{3} = 2 \qquad ?$$

$$2 = 2 \qquad \text{True}$$

Thus, the root is $\dfrac{3}{2}$.   ■

Example 7   Solve: $\dfrac{5}{x} - 2 = 3$.

$$\frac{5}{x} - 2 + 2 = 3 + 2 \qquad \text{Add 2 to both sides.}$$

$$\frac{5}{x} = 5$$

$$x\left(\frac{5}{x}\right) = (5)x \qquad \text{Multiply both sides by } x.$$

$$5 = 5x$$

$$\frac{5}{5} = \frac{5x}{5} \qquad \text{Divide both sides by 5.}$$

$$1 = x$$

Exercises 6.4

Solve each equation and check:

1. $\dfrac{2x}{3} = \dfrac{32}{6}$

2. $\dfrac{5x}{7} = \dfrac{20}{14}$

3. $\dfrac{3}{8}y = 1\dfrac{14}{16}$

4. $\dfrac{5}{3}x = -13\dfrac{1}{3}$

5. $2\dfrac{1}{2}x = 7\dfrac{1}{2}$

6. $\dfrac{1}{2} + \dfrac{x}{3} = \dfrac{5}{2}$

7. $\dfrac{2}{3} + \dfrac{x}{4} = \dfrac{28}{6}$

8. $\dfrac{x}{7} - \dfrac{1}{14} = \dfrac{70}{28}$

9. $\dfrac{3}{4} - \dfrac{x}{3} = \dfrac{5}{12}$

10. $1\dfrac{1}{4} + \dfrac{x}{3} = \dfrac{7}{12}$

11. $\dfrac{3}{5}x - 25 = \dfrac{x}{10}$

12. $\dfrac{2x}{3} - 7 = -\dfrac{x}{2}$

13. $\dfrac{y}{3} - 1 = \dfrac{y}{6}$

14. $\dfrac{1}{2}x - 3 = \dfrac{x}{5}$

15. $\dfrac{3x}{4} - \dfrac{7}{20} = \dfrac{2}{5}x$

16. $\dfrac{5x}{6} + \dfrac{1}{3}(6 + x) = 37$

17. $\dfrac{x}{2} + \dfrac{2}{3}(2x + 3) = 46$

18. $\dfrac{1}{6}x - \dfrac{1}{9}(2x - 3) = 1$

19. $0.96 = 0.06(12 + x)$

20. $\dfrac{1}{2}(x + 2) + \dfrac{3}{8}(28 - x) = 11$

21. $\dfrac{3x - 24}{16} - \dfrac{3x - 12}{12} = 3$

22. $\dfrac{4x + 3}{15} - \dfrac{2x - 3}{9} = \dfrac{6x + 4}{6} - x$

23. $5x + \dfrac{6x - 8}{14} + \dfrac{10x + 6}{6} = 43$

24. $\dfrac{3x}{5} - \dfrac{9 - 3x}{10} = \dfrac{6}{10} - \dfrac{3x + 6}{10}$

25. $\dfrac{4x}{6} - \dfrac{x + 5}{2} = \dfrac{6x - 6}{8}$

26. $\dfrac{x}{3} + \dfrac{2x + 4}{4} = \dfrac{x - 1}{6} - \dfrac{3 - 2x}{2}$

27. $\dfrac{4}{x} = 6$

28. $\dfrac{2}{x} - 8 = -7$

29. $5 - \dfrac{1}{x} = 7$

30. $\dfrac{3}{x} - 6 = 8$

31. $\dfrac{5}{y} - 1 = 4$

32. $\dfrac{17}{x} = 8$

33. $\dfrac{3}{x} - 8 = 7$

34. $\dfrac{5}{2x} + 8 = 1$

35. $7 - \dfrac{6}{x} = 5$

36. $9 + \dfrac{3}{x} = 10$

37. $\dfrac{6}{x} + 5 = 14$

38. $\dfrac{3}{x} - 3 = \dfrac{5}{2x}$

39. $1 - \dfrac{2}{x} = \dfrac{14}{3x} - \dfrac{1}{3}$

40. $\dfrac{3}{x} + 2 = \dfrac{5}{x}$

41. $\dfrac{7}{2x} + 14\dfrac{1}{2} = \dfrac{7}{x} - 10$

42. $\dfrac{8}{x} + \dfrac{1}{4} = \dfrac{5}{x}$

6.5 Translating Words into Algebraic Symbols

The ability to translate English words into algebra is very important for solving "applied" problems. To help you, we provide the following table of common English words for the common mathematical symbols:

+	−	×	÷	=
plus	minus	times	divide	equal or equals
increased by	decreased by	product	quotient	is or are
added to	subtract	multiply by	divided by	is equal to
more than	less than	double or twice		result is
sum of	difference	triple or thrice		
	subtract from			

Example 1   Translate into algebra: One number is four times another, and their sum is twenty.

Let x = first number

$4x$ = four times the number

$x + 4x$ = their sum

Sentence in algebra: $x + 4x = 20$ ■

Example 2   Translate into algebra: The sum of a number and the number decreased by six is five.

Let x = the number

$x - 6$ = number decreased by six

$x + (x - 6)$ = sum

Sentence in algebra: $x + (x - 6) = 5$ ■

Example 3   Translate into algebra: Fifteen more than twice a number is twenty-four.

Let x = the number

$2x$ = twice the number

$2x + 15$ = fifteen more than twice the number

Sentence in algebra: $2x + 15 = 24$ ■

Example 4   Translate into algebra: Twice the sum of a number and five is eighty.

Let x = the number

$x + 5$ = sum of a number and five

$2(x + 5)$ = twice the sum of a number and five

Sentence in algebra: $2(x + 5) = 80$ ■

Exercises 6.5

Translate each phrase or sentence into algebraic symbols:

1. A number decreased by twenty
2. A number increased by five
3. A number divided by six
4. A number times eighteen
5. The sum of a number and sixteen
6. Subtract twenty-six from a number
7. Subtract a number from twenty-six
8. Half a number
9. Twice a number
10. The difference between four and a number
11. The sum of six times a number and twenty-eight is forty.
12. The difference between twice a number and thirty is fifty.
13. The quotient of a number and six is five.
14. If seven is added to a number, the sum is 32.

15. If a number is increased by 28 and then the su[m] tiplied by five, the result is 150.
16. The sum of a number and the number decreas[ed] is 25.
17. Seven less than the quotient of a number and
18. The product of five and five more than a num[ber]
19. The difference between thirty and twice a [number is] four.
20. Double the difference between a number a[nd] thirty.
21. The product of a number decreased by seve[n and the] same number increased by five is thirteen.
22. Seven times a number decreased by eleven i[s]
23. The product of a number and six decreased [by] teen is seven.
24. If twelve is added to the product of a nu[mber and] twelve, the sum is 72.
25. Seventeen less than four times a number is 6

6.6 Applications Involving Equations*

An applied problem can often be expressed mathematically as a simple equation[. The prob]lem can then be solved by solving the equation. To solve such an application pr[oblem, we] suggest the following steps.

Solving Application Problems

Step 1 Read the problem carefully at least twice.

Step 2 If possible, draw a diagram. This will often help you to visualize [the] mathematical relationship needed to write the equation.

Step 3 Choose a letter to represent the unknown quantity in the problem [and] write what it represents.

Step 4 Write an equation that expresses the information given in the pr[oblem] and that involves the unknown.

Step 5 Solve the equation from Step 4.

Step 6 Check your solution both in the equation from Step 4 and in the original problem itself.

*Note: In this section, do not use the rules for calculating with measurements.

Example 1

You need to tile the floor of a rectangular room with a wooden outer border of 6 in. The floor of the room is 10 ft by 8 ft 2 in. How many rows of 4-in.-by-4-in. tiles are needed to fit across the length of the room?

The sketch shown in Figure 6.1 is helpful in solving the problem.

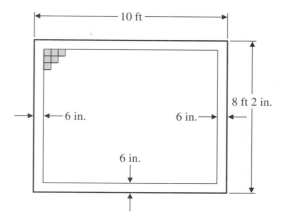

FIGURE 6.1

Let x = the number of tiles across the length of the room

$4x$ = the number of inches in x tiles

The total length of the rectangular room is then

$$4x + 6 + 6 = 120 \qquad \text{10 ft = 120 in.}$$
$$4x + 12 = 120$$
$$4x = 108 \qquad \text{Subtract 12 from both sides.}$$
$$x = 27 \qquad \text{Divide both sides by 4.}$$

So there are 27 rows of tiles.　■

Example 2

An interior wall measures 30 ft 4 in. long. It is to be divided by 10 evenly spaced posts; each post is 4 in. by 4 in. (Posts are to be located in the corners.) What is the distance between posts? Note that there are 9 spaces between posts. (See Figure 6.2.)

FIGURE 6.2

Let x = the distance between posts

$9x$ = the distance of 9 spaces

$(10)(4\text{ in.})$ = the distance used up by ten 4-in. posts

The total length of the interior wall is then

$$9x + (10)(4 \text{ in.}) = 364 \text{ in.} \qquad 30 \text{ ft } 4 \text{ in.} = 364 \text{ in.}$$
$$9x + 40 = 364$$
$$9x = 324 \qquad \text{Subtract 40 from both sides.}$$
$$x = 36 \text{ in.} \qquad \text{Divide both sides by 9.}$$

Check:
$$9x + 40 = 364$$
$$9(36) + 40 = 364 \qquad ?$$
$$324 + 40 = 364 \qquad ?$$
$$364 = 364 \qquad \text{True}$$

 Example 3   Two different automotive batteries cost a total of $270. One costs $12 more than other. Find the cost of each battery.

$$\text{Let } x = \text{the cost of one battery}$$
$$2x + 12 = \text{the cost of the other battery}$$

The total cost of both batteries is then

$$x + (2x + 12) = 270$$
$$3x + 12 = 270 \qquad \text{Combine like terms.}$$
$$3x = 258 \qquad \text{Subtract 12 from both sides.}$$
$$x = \$86, \text{ the cost of the first battery}$$
$$2x + 12 = 2(86) + 12 = \$184, \text{ the cost of the other battery}$$

 Example 4   One side of a triangle is twice another. The third side is 5 more than the shortest perimeter is 33. Find the length of each side.

First, draw and label a triangle as in Figure 6.3.

FIGURE 6.3

$$\text{Let } x = \text{the length of the first side}$$
$$2x = \text{the length of the second side}$$
$$x + 5 = \text{the length of the third side}$$

The sum of the three sides is then

$$x + 2x + (x + 5) = 33$$
$$4x + 5 = 33 \qquad \text{Combine like terms.}$$
$$4x = 28 \qquad \text{Subtract 5 from both sides.}$$
$$x = 7, \text{ the length of the first side}$$
$$2x = 2(7) = 14, \text{ the length of the second side}$$
$$x + 5 = (7) + 5 = 12, \text{ the length of the third side}$$

 Example 5   Forty acres of land were sold for $216,000. Some was sold at $6400 per acre, a was sold at $4800 per acre. How much was sold at each price?

$$\text{Let } x = \text{the amount of land sold at } \$6400/\text{acre}$$
$$40 - x = \text{the amount of land sold at } \$4800/\text{acre}$$

Then

$$6400x = \text{the value of the land sold at } \$6400/\text{acre}$$
$$4800(40 - x) = \text{the value of the land sold at } \$4800/\text{acre}$$
$$216{,}000 = \text{the total value of the land}$$

Therefore, the equation for the total value of the land is

$$6400x + 4800(40 - x) = 216{,}000$$

$$6400x + 192{,}000 - 4800x = 216{,}000 \qquad \text{Remove parentheses.}$$
$$1600x + 192{,}000 = 216{,}000 \qquad \text{Combine like terms.}$$
$$1600x = 24{,}000 \qquad \text{Subtract 192,000 from both sides.}$$
$$x = 15 \qquad \text{Divide both sides by 1600.}$$
$$40 - x = 25$$

Thus, 15 acres were sold at $6400/acre and 25 acres were sold at $4800/acre. ■

Note: When you know the total of two parts, one possible equation-solving strategy is to

$$\text{let } x = \text{one part} \quad \text{and}$$
$$\text{total} - x = \text{the other part}$$

◆ **Example 6**

How much pure alcohol must be added to 200 cm³ of a solution that is 15% alcohol to make a solution that is 40% alcohol?

Let $x =$ the amount of pure alcohol (100%) added

You may find Figure 6.4 helpful.

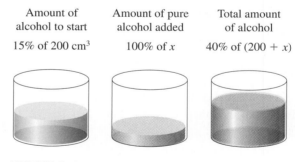

Amount of alcohol to start	Amount of pure alcohol added	Total amount of alcohol
15% of 200 cm³	100% of x	40% of (200 + x)

FIGURE 6.4

Write an equation in terms of the amount of pure alcohol; that is, the amount of pure alcohol in each separate solution equals the amount of pure alcohol in the final solution.

$$0.15(200) + 1.00x = 0.40(200 + x)$$
$$30 + x = 80 + 0.4x$$
$$x = 50 + 0.4x \qquad \text{Subtract 30 from both sides.}$$
$$0.6x = 50 \qquad \text{Subtract } 0.4x \text{ from both sides.}$$
$$x = 83.3 \qquad \text{Divide both sides by 0.6.}$$

Thus, 83.3 cm³ of pure alcohol must be added. ■

Exercises 6.6

1. A set of eight built-in bookshelves is to be constructed in a room. The floor-to-ceiling clearance is 8 ft 2 in. Each shelf is 1 in. thick. An equal space is to be left between shelves. What space should there be between each shelf and the next? (There is no shelf against the ceiling and no shelf on the floor.)

2. Saw a board 8 ft 4 in. long into nine equal pieces. If the loss per cut is $\frac{1}{8}$ in., how long will each piece be?

3. Separate an order of 256 light fixtures so that the number of fluorescent light fixtures will be 20 fewer than twice the number of incandescent light fixtures.

4. Distribute $1000 into three parts so that one part will be three times as large as the second and the third part will be as large as the sum of the other two.

5. Distribute $4950 among John, Maria, and Betsy so that Maria receives twice as much as John and Betsy receives three times as much as John.

6. Distribute $4950 among John, Maria, and Betsy so that Maria receives twice as much as John and Betsy receives three times as much as Maria.

7. A rectangle is twice as long as it is wide. Its perimeter (the sum of the lengths of its sides) is 60 cm. Find its length and width.

8. The length of a rectangle is 4 cm less than twice its width. Its perimeter is 40 cm. Find its length and width.

9. One side of a rectangular yard is bounded by the side of a house. The other three sides are to be fenced with 345 ft of fencing. The length of fence opposite the house is 15 ft less than either of the other two sides. Find the length and width of the yard.

10. A given type of concrete contains twice as much sand as cement and 1.5 times as much gravel as sand. How many cubic yards of each must be used to make 9 yd^3 of concrete? Assume no loss of volume in mixing.

11. The perimeter of a triangle is 122 ft. The lengths of two sides are the same. The length of the third side is 4 ft shorter than either of the other two sides. Find the lengths of the three sides.

12. Cut a board 20 ft long into three pieces so that the longest piece will be three times as long as each of the other two of equal lengths. Find the length of each piece.

13. Cut a 12-ft beam into two pieces so that one piece is 18 in. longer than the other. Find the length of the two pieces.

14. The total cost of three automobile batteries is $ most expensive battery is three times the c least expensive. The third is $15 more than th pensive. Find the cost of each battery.

15. The total cost of 20 boards is $166. One $6.50, and the second size costs $9.50. H boards are purchased at each price?

16. Amy and Kim earned a total of $428 by wor tal of 30 hours. If Amy earns $12/h and K $16/h, how many hours did each work?

17. Joyce invests $7500 in two savings accounts count earns interest at 4% per year; the other e per year. The total interest earned from both after one year is $132.50. How much was orig posited in each account?

18. Chuck receives loans totaling $12,000 from t One bank charges 7.5% annual interest, and bank charges 6% annual interest. He paid $8 interest in one year. How much was loaned at e

19. Regular milk has 4% butterfat. How many lit ular milk must be mixed with 40 L of milk wi terfat to have milk with 2% butterfat?

20. How much pure alcohol must be added to 75 solution that is 40% alcohol to make a solut 60% alcohol?

21. Mix a solution that is 30% alcohol with a solu 80% alcohol to make 800 mL of a solution t alcohol. How much of each solution should y

22. Mix a solution that is 50% acid with a solut 100% water to make 4 L of a solution that is How much of each solution should you use?

23. A 12-quart cooling system is checked and f filled with a solution that is 40% antifreeze. T strength of the solution is 60% antifreeze. H quarts of solution need to be drained and rep pure antifreeze to reach the desired strength?

24. In testing an engine, various mixtures of ga methanol are being tried. How much of a 90% mixture and a 75% gasoline mixture are 1200 L of an 85% gasoline mixture?

25. Assume you have sea water that weighs 64 contains 35 parts per thousand (0.0035) dis by weight. How many cubic feet of the sea w you need to place in an evaporation basin 125 lb of sea salt?

26. The carrying capacity of any rangeland area can be estimated in terms of animal equivalent units (AEUs). A white-tailed deer requires 0.25 AEU of grazing capacity and an elk requires 0.67 AEU. A wildlife manager estimates that a state park will provide grazing to support about 150 AEUs on a sustained basis. If the park has 75 elk, what is the maximum number of deer that it will support?

6.7 Formulas

A **formula** is a general rule written as an equation, usually expressed in letters, which shows the relationship between two or more quantities. For example, the formula

$$d = rt$$

states that the distance, d, that a body travels equals the product of its rate, r, of travel and the time, t, of travel. The formula

$$p = \frac{F}{A}$$

states that the pressure, p, equals the quotient of the force, F, and the area, A, over which the force is applied.

Sometimes the letters in a formula do not match the first letter of the name of the quantity. For example, Ohm's law is often written

$$E = IR$$

where E is the voltage, I is the current, and R is the resistance.

Sometimes subscripts are used to distinguish between different readings of the same quantity. For example, the final velocity of an object equals the sum of the initial velocity and the product of its acceleration and the time of the acceleration. This is written

$$v_f = v_i + at$$

where v_f is the final velocity, v_i is the initial velocity, a is the acceleration, and t is the time.

Sometimes Greek letters are used. For example, the resistance of a wire is given by the formula

$$R = \frac{\rho L}{A}$$

where R is the resistance of the wire, ρ (rho) is the resistivity constant of the wire, L is the length of the wire, and A is the cross-sectional area of the wire.

Sometimes the formula is written with the letters and symbols used by the person who discovered the relationship. The letters may have no obvious relationship with the quantity.

Solving Formulas

To solve a formula for a given letter means to isolate the given letter on one side of the equation and express it in terms of all the remaining letters.

This means that the given letter appears on one side of the equation by itself; all the other letters appear on the opposite side of the equation. We solve a formula using the same methods that we use in solving an equation.

Example 1   Solve **a.** $d = rt$ for t and **b.** $15 = 5t$ for t.

a. $d = rt$

$$\frac{d}{r} = \frac{rt}{r}$$ Divide both sides by r.

$$\frac{d}{r} = t$$

b. $15 = 5t$

$$\frac{15}{5} = \frac{5t}{5}$$ Divide both si▉

$$3 = t$$

Note that we use the same techniques for solving a formula as we learned earlier
ing equations.

 Example 2   Solve $p = \dfrac{F}{A}$ for F and then for A.

First, solve for F:

$$p = \frac{F}{A}$$

$$pA = \left(\frac{F}{A}\right)A$$ Multiply both sides by A.

$$pA = F$$

Now solve for A:

$$p = \frac{F}{A}$$

$$pA = F$$ Multiply both sides by A.

$$\frac{pA}{p} = \frac{F}{p}$$ Divide both sides by p.

$$A = \frac{F}{p}$$

Example 3   Solve $V = E - Ir$ for I.

One way:

$$V = E - Ir$$

$$V - E = E - Ir - E$$ Subtract E from both sides.

$$V - E = -Ir$$

$$\frac{V - E}{-r} = \frac{-Ir}{-r}$$ Divide both sides by $-r$.

$$\frac{V - E}{-r} = I$$

Alternative way:

$$V = E - Ir$$

$$V + Ir = E - Ir + Ir$$ Add Ir to both sides.

$$V + Ir = E$$

$$V + Ir - V = E - V$$ Subtract V from both sides.

$$Ir = E - V$$

$$\frac{Ir}{r} = \frac{E - V}{r}$$ Divide both sides by r.

$$I = \frac{E - V}{r}$$

Note that the two results are equivalent. Take the first result,

$$\frac{V - E}{-r}$$

and multiply numerator and denominator by -1:

$$\left(\frac{V - E}{-r}\right)\left(\frac{-1}{-1}\right) = \frac{-V + E}{r} = \frac{E - V}{r}$$ ■

Example 4   Solve $S = \dfrac{n(a + l)}{2}$ for n and then for l.

First, solve for n:

$$S = \frac{n(a + l)}{2}$$

$$2S = n(a + l) \qquad \text{Multiply both sides by 2.}$$

$$\frac{2S}{a + l} = n \qquad \text{Divide both sides by } (a + l).$$

Now solve for l:

$$S = \frac{n(a + l)}{2}$$

$$2S = n(a + l) \qquad \text{Multiply both sides by 2.}$$

$$2S = na + nl \qquad \text{Remove parentheses.}$$

$$2S - na = nl \qquad \text{Subtract } na \text{ from both sides.}$$

$$\frac{2S - na}{n} = l \qquad \text{Divide both sides by } n.$$

or $\qquad \dfrac{2S}{n} - a = l$ ■

Example 5   Solve $(\triangle L) = kL(T - T_0)$ for T.

$$(\triangle L) = kL(T - T_0). \qquad \text{Note: Treat } (\triangle L) \text{ as one variable.}$$

$$(\triangle L) = kLT - kLT_0 \qquad \text{Remove parentheses.}$$

$$(\triangle L) + kLT_0 = kLT \qquad \text{Add } kLT_0 \text{ to both sides.}$$

$$\frac{(\triangle L) + kLT_0}{kL} = T \qquad \text{Divide both sides by } kL.$$

Can you show that $T = \dfrac{(\triangle L)}{kL} + T_0$ is an equivalent solution?   ■

Exercises 6.7

Solve each formula for the given letter:

1. $E = Ir$ for r

2. $A = bh$ for b

3. $F = ma$ for a

4. $w = mg$ for m

5. $C = \pi d$ for d

6. $V = IR$ for R

7. $V = lwh$ for w

8. $X_L = 2\pi fL$ for f

9. $A = 2\pi rh$ for h

10. $C = 2\pi r$ for r

11. $v^2 = 2gh$ for h

12. $V = \pi r^2 h$ for h

13. $I = \dfrac{Q}{t}$ for t

14. $I = \dfrac{Q}{t}$ for Q

15. $v = \dfrac{s}{t}$ for s

16. $I = \dfrac{E}{Z}$ for Z

17. $I = \dfrac{V}{R}$ for R

18. $P = \dfrac{w}{t}$ for w

19. $E = \dfrac{I}{4\pi r^2}$ for I

20. $R = \dfrac{\pi}{2P}$ for P

21. $X_C = \dfrac{1}{2\pi fC}$ for f

22. $R = \dfrac{\rho L}{A}$ for L

23. $A = \dfrac{1}{2}bh$ for b

24. $V = \dfrac{1}{3}\pi r^2 h$ for h

25. $Q = \dfrac{I^2 Rt}{J}$ for R

26. $R = \dfrac{kl}{D^2}$ for l

27. $F = \dfrac{9}{5}C + 32$ for C

28. $C = \dfrac{5}{9}(F - 32)$ for F

29. $C_T = C_1 + C_2 + C_3 + C_4$ for C_2

30. $R_T = R_1 + R_2 + R_3 + R_4$ for R_4

31. $Ax + By + C = 0$ for x

32. $A = P + Prt$ for r

33. $Q_1 = P(Q_2 - Q_1)$ for Q_2

34. $v_f = v_i + at$ for v_i

35. $A = \left(\dfrac{a + b}{2}\right)h$ for h

36. $A = \left(\dfrac{a + b}{2}\right)h$ for b

37. $l = a + (n - 1)d$ for d

38. $A = ab + \dfrac{d}{2}(a + c)$ for d

39. $Ft = m(V_2 - V_1)$ for m

40. $l = a + (n - 1)d$ for n

41. $Q = wc(T_1 - T_2)$ for c

42. $Ft = m(V_2 - V_1)$ for V_2

43. $V = \dfrac{2\pi(3960 + h)}{P}$ for h

44. $Q = wc(T_1 - T_2)$ for T_2

6.8 Substituting Data into Formulas

Problem-solving skills are essential in all technical fields. Working with formulas
the most important tools that you can gain from this course.

> **Problem Solving**
>
> Necessary parts of problem solving include:
>
> 1. Analyzing the given data.
> 2. Finding an equation or formula that relates the given quantities with the unk
>   quantity.
> 3. Solving the formula for the unknown quantity.
> 4. Substituting the given data into this solved formula.

Actually, you may solve the formula for the unknown quantity and then subs
data. Or you may substitute the data into the formula first and then solve for the
quantity. If you use a calculator, the first method is more helpful.

Example 1

Given the formula $V = IR$, $V = 120$, and $R = 2\overline{0}0$. Find I.

First, solve for I:

$$V = IR$$

$$\dfrac{V}{R} = \dfrac{IR}{R} \qquad \text{Divide both sides by } R.$$

$$\dfrac{V}{R} = I$$

Then substitute the data:

$$I = \frac{V}{R} = \frac{120}{200} = 0.60$$

■

 Example 2

Given $v = v_0 + at$, $v = 60.0$, $v_0 = 20.0$, and $t = 5.00$. Find a.

First, solve for a:

$$v = v_0 + at$$
$$v - v_0 = at \qquad \text{Subtract } v_0 \text{ from both sides.}$$
$$\frac{v - v_0}{t} = \frac{at}{t} \qquad \text{Divide both sides by } t.$$
$$\frac{v - v_0}{t} = a$$

Then substitute the data:

$$a = \frac{v - v_0}{t} = \frac{60.0 - 20.0}{5.00} = \frac{40.0}{5.00} = 8.00$$

■

 Example 3

Given the formula $S = \frac{MC}{l}$, $S = 47.5$, $M = 190$, and $C = 8.0$. Find l.

First, solve for l:

$$S = \frac{MC}{l}$$
$$Sl = MC \qquad \text{Multiply both sides by } l.$$
$$l = \frac{MC}{S} \qquad \text{Divide both sides by } S.$$

Then substitute the data:

$$l = \frac{MC}{S} = \frac{190(8.0)}{47.5} = 32$$

■

 Example 4

Given the formula $Q = WC(T_1 - T_2)$, $Q = 15$, $W = 3.0$, $T_1 = 11\overline{0}$, and $T_2 = 6\overline{0}$. Find C.

First, solve for C:

$$Q = WC(T_1 - T_2)$$
$$\frac{Q}{W(T_1 - T_2)} = C \qquad \text{Divide both sides by } W(T_1 - T_2).$$

Then substitute the data:

$$\frac{15}{3.0(11\overline{0} - 6\overline{0})} = C$$
$$C = \frac{15}{150} = 0.10$$

■

Example 5   Given the formula $V = \frac{1}{2}lw(D + d)$, $V = 156.8$, $D = 2.00$, $l = 8.37$, and $w = 7.19$

First, solve for d:

$$V = \frac{1}{2}lw(D + d)$$

$2V = lw(D + d)$	Multiply both sides by 2.
$2V = lwD + lwd$	Remove parentheses.
$2V - lwD = lwd$	Subtract lwD from both sides.
$\dfrac{2V - lwD}{lw} = d$	Divide both sides by lw.
$\dfrac{2V}{lw} - D = d$	Simplify the left side.

Then substitute the data:

$$d = \frac{2V}{lw} - D = \frac{2(156.8)}{(8.37)(7.19)} - 2.00 = 3.21$$

Exercises 6.8

a. *Solve for the indicated letter.* **b.** *Then substitute the given values to find the value of the indicated letter (use the r working with measurements):*

	Formula	Given	Find
1.	$A = lw$	$A = 414$, $w = 18.0$	l
2.	$V = IR$	$I = 9.20$, $V = 5.52$	R
3.	$V = \dfrac{\pi r^2 h*}{3}$	$V = 753.6$, $r = 6.00$	h
4.	$I = \dfrac{V}{R}$	$R = 44$, $I = 2.5$	V
5.	$E = \dfrac{mv^2}{2}$	$E = 484{,}000$; $v = 22.0$	m
6.	$v_f = v_i + at$	$v_f = 88$, $v_i = 10.0$, $t = 12$	a
7.	$v_f = v_i + at$	$v_f = 193.1$, $v_i = 14.9$, $a = 18.0$	t
8.	$y = mx + b$	$x = 3$, $y = 2$, $b = 9$	m
9.	$v_f^2 = v_i^2 + 2gh$	$v_f = 192$, $v_i = 0$, $g = 32.0$	h
10.	$A = P + Prt$	$r = 0.07$, $P = \$1500$, $A = \$2025$	t
11.	$L = \pi(r_1 + r_2) + 2d$	$L = 37.68$, $d = 6.28$, $r_2 = 5.00$	r_1
12.	$C = \dfrac{5}{9}(F = 32)$	$C = -20$	F

Note: Use the π key on your calculator.

	Formula	Given	Find
13.	$Fgr = Wv^2$	$F = 12{,}000;\ W = 24{,}000;$ $v = 176;\ g = 32$	r
14.	$Q = WC(T_1 - T_2)$	$Q = 18.9,\ W = 3.0,\ C = 0.18,$ $T_2 = 59$	T_1
15.	$A = \frac{1}{2}h(a + b)$	$A = 1160,\ h = 22.0,\ a = 56.5$	b
16.	$A = \frac{1}{2}h(a + b)$	$A = 5502,\ h = 28.0,\ b = 183$	a
17.	$V = \frac{1}{2}lw(D + d)$	$V = 226.8,\ l = 9.00,\ w = 6.30,$ $D = 5.00$	d
18.	$S = \frac{n}{2}(a + l)$	$S = 575,\ n = 25,\ l = 15$	a
19.	$S = \frac{n}{2}(a + l)$	$S = 147.9,\ n = 14.5,\ l = 3.80$	a
20.	$S = \frac{n}{2}(a + l)$	$S = 96\frac{7}{8},\ n = 15,\ a = 8\frac{2}{3}$	l

21. A drill draws a current, I, of 4.50 A. The resistance, R, is 16.0 Ω. Find its power, P, in watts. $P = I^2R$.

22. A flashlight bulb is connected to a 1.50-V source. Its current, I, is 0.250 A. What is its resistance, R, in ohms? $V = IR$.

23. The area of a rectangle is 84.0 ft^2. Its length is 12.5 ft. Find its width. $A = lw$.

24. The volume of a box is given by $V = lwh$. Find the width if the volume is 3780 ft^3, its length is 21.0 ft, and its height is 15.0 ft.

25. The volume of a cylinder is given by the formula $V = \pi r^2 h$, where r is the radius and h is the height. Find the height in m if the volume is 8550 m^3 and the radius is 15.0 m.

26. The pressure at the bottom of a lake is found by the formula $P = hD$, where h is the depth of the water and D is the density of the water. Find the pressure in lb/in^2 at 175 ft below the surface. $D = 62.4$ lb/ft^3.

27. The equivalent resistance R of two resistances connected in parallel is given by $\frac{1}{R} = \frac{1}{R_1} + \frac{1}{R_2}$. Find the equivalent resistance for two resistances of 20.0 Ω and 60.0 Ω connected in parallel.

28. The R value of insulation is given by the formula $R = \frac{L}{K}$, where L is the thickness of the insulating material and K is the thermal conductivity of the material. Find the R value of 8.0 in. of mineral wool insulation. $K = 0.026$. *Note: L* must be in feet.

29. A steel railroad rail expands and contracts according to the formula $\triangle l = \alpha l \triangle T$, where $\triangle l$ is the change in length, α is a constant called the *coefficient of linear expansion*, l is the original length, and $\triangle T$ is the change in temperature. If a 50.0-ft steel rail is installed at 0°F, how many inches will it expand when the temperature rises to 110°F? $\alpha = 6.5 \times 10^{-6}$/°F.

30. The inductive reactance, X_L, of a coil is given by $X_L = 2\pi fL$, where f is the frequency and L is the inductance. If the inductive reactance is 245 Ω and the frequency is 60.0 cycles/s, find the inductance L in henrys, H.

6.9 Reciprocal Formulas Using a Calculat

The **reciprocal of a number** is 1 divided by that number. The product of a numbe
0) and its reciprocal is 1. Examples of numbers and their reciprocals are shown he

Number	Reciprocal
4	$\dfrac{1}{4}$
-6	$-\dfrac{1}{6}$
$\dfrac{2}{3}$	$\dfrac{3}{2}$
$-\dfrac{12}{7}$	$-\dfrac{7}{12}$
0	None

The reciprocal of a number may be found by using the $1/x$ or x^{-1} key. This ma
you to use the second function key on your calculator.

 Example 1   Find the reciprocal of 12 rounded to three significant digits.

12

0.083333333

Thus, $\dfrac{1}{12} = 0.0833$ rounded to three significant digits.

 Example 2   Find $\dfrac{1}{41.2}$ rounded to three significant digits.

41.2

0.024271845

The reciprocal of 41.2 is 0.0243 rounded to three significant digits.

Formulas involving reciprocals are often used in electronics and physics. We
sider an alternative method for substituting data into such formulas and solving f
ified letter using a calculator.

To use a calculator with formulas involving reciprocals,

1. Solve for the reciprocal of the specified letter.
2. Substitute the given data.
3. Follow the calculator steps shown in the following examples.

Example 3

Given the formula $\dfrac{1}{R} = \dfrac{1}{R_1} + \dfrac{1}{R_2}$, where $R_1 = 6.00\ \Omega$ and $R_2 = 12.0\ \Omega$, find R.

Since the formula is already solved for the reciprocal of R, substitute the data:

$$\frac{1}{R} = \frac{1}{R_1} + \frac{1}{R_2}$$

$$\frac{1}{R} = \frac{1}{6.00\ \Omega} + \frac{1}{12.0\ \Omega}$$

Then use your calculator as follows:

6   12

$$4$$

So $R = 4.00\ \Omega$.

Note: This formula relates the electrical resistances in a parallel circuit.   ■

 Example 4

Given the formula $\dfrac{1}{f} = \dfrac{1}{s_0} + \dfrac{1}{s_i}$, where $f = 8.00$ cm and $s_0 = 12.0$ cm, find s_i.

First, solve the formula for the reciprocal of s_i:

$$\frac{1}{f} = \frac{1}{s_0} + \frac{1}{s_i}$$

$$\frac{1}{s_i} = \frac{1}{f} - \frac{1}{s_0} \qquad \text{Subtract } \frac{1}{s_0} \text{ from both sides.}$$

Next, substitute the data:

$$\frac{1}{s_i} = \frac{1}{8.00\ \text{cm}} - \frac{1}{12.0\ \text{cm}}$$

Then use your calculator as follows:

8   12

$$24$$

So $s_i = 24.0$ cm.   ■

Example 5

Given the formula $\dfrac{1}{C} = \dfrac{1}{C_1} + \dfrac{1}{C_2} + \dfrac{1}{C_3}$ where $C = 2.00$ μF, $C_1 = 3.00$

$C_3 = 18.0$ μF, find C_2.

First, solve the formula for the reciprocal of C_2:

$$\frac{1}{C} = \frac{1}{C_1} + \frac{1}{C_2} + \frac{1}{C_3}$$

$$\frac{1}{C_2} = \frac{1}{C} - \frac{1}{C_1} - \frac{1}{C_3} \qquad \text{Subtract } \frac{1}{C_1} \text{ and } \frac{1}{C_3} \text{ from both sides.}$$

Next, substitute the data:

$$\frac{1}{C_2} = \frac{1}{2.00 \text{ μF}} - \frac{1}{3.00 \text{ μF}} - \frac{1}{18.0 \text{ μF}}$$

Then use your calculator as follows:

9

Therefore, $C_2 = 9.00$ μF.

Note: This formula relates the electrical capacitances of capacitors in a series ci

Exercises 6.9

Use the formula $\dfrac{1}{R} = \dfrac{1}{R_1} + \dfrac{1}{R_2}$ for Exercises 1–6:

1. Given $R_1 = 8.00 \ \Omega$ and $R_2 = 12.0 \ \Omega$, find R.
2. Given $R = 5.76 \ \Omega$ and $R_1 = 9.00 \ \Omega$, find R_2.
3. Given $R = 12.0 \ \Omega$ and $R_2 = 36.0 \ \Omega$, find R_1.
4. Given $R_1 = 24.0 \ \Omega$ and $R_2 = 18.0 \ \Omega$, find R.
5. Given $R = 15.0 \ \Omega$ and $R_2 = 24.0 \ \Omega$, find R_1.
6. Given $R = 90.0 \ \Omega$ and $R_1 = 125 \ \Omega$, find R_2.

Use the formula $\dfrac{1}{f} = \dfrac{1}{s_0} + \dfrac{1}{s_i}$ for Exercises 7–10:

7. Given $s_0 = 3.00$ cm and $s_i = 15.0$ cm, find f.
8. Given $f = 15.0$ cm and $s_i = 25.0$ cm, find s_0.
9. Given $f = 14.5$ cm and $s_0 = 21.5$ cm, find s_i.
10. Given $s_0 = 16.5$ cm and $s_i = 30.5$ cm, find f.

Use the formula $\dfrac{1}{R} = \dfrac{1}{R_1} + \dfrac{1}{R_2} + \dfrac{1}{R_3}$ for Exercises 11–16:

11. Given $R_1 = 30.0 \ \Omega$, $R_2 = 18.0 \ \Omega$, and $R_3 = 45.0 \ \Omega$, find R.

12. Given $R_1 = 75.0 \ \Omega$, $R_2 = 50.0 \ \Omega$, and $R_3 = $ find R.
13. Given $R = 80.0 \ \Omega$, $R_1 = 175 \ \Omega$, and R_2 find R_3.
14. Given $R = 145 \ \Omega$, $R_2 = 875 \ \Omega$, and R_3 find R_1.
15. Given $R = 1250 \ \Omega$, $R_1 = 3750 \ \Omega$, and $R_3 = $ find R_2.
16. Given $R = 1830 \ \Omega$, $R_1 = 4560 \ \Omega$, and $R_2 = $ find R_3.

Use the formula $\dfrac{1}{C} = \dfrac{1}{C_1} + \dfrac{1}{C_2} + \dfrac{1}{C_3}$ for Exercis

17. Given $C_1 = 12.0$ μF, $C_2 = 24.0$ μF, and $C_3 = $ find C.
18. Given $C = 45.0$ μF, $C_1 = 85.0$ μF, and C_3 find C_2.
19. Given $C = 1.25 \times 10^{-6}$ F, $C_1 = 8.75 \times 1($ $C_2 = 6.15 \times 10^{-6}$ F, find C_3.

20. Given $C = 1.75 \times 10^{-12}$ F, $C_2 = 7.25 \times 10^{-12}$ F, and $C_3 = 5.75 \times 10^{-12}$ F, find C_1.

21. Given $C_1 = 6.56 \times 10^{-7}$ F, $C_2 = 5.05 \times 10^{-6}$ F, and $C_3 = 1.79 \times 10^{-8}$ F, find C.

22. Given $C = 4.45 \times 10^{-9}$ F, $C_1 = 5.08 \times 10^{-8}$ F, and $C_3 = 7.79 \times 10^{-9}$ F, find C_2.

Use the formula $\dfrac{1}{R} = \dfrac{1}{R_1} + \dfrac{1}{R_2} + \dfrac{1}{R_3} + \dfrac{1}{R_4}$ *for Exercises 23–24:*

23. Given $R_1 = 655\ \Omega$, $R_2 = 775\ \Omega$, $R_3 = 1050\ \Omega$, and $R_4 = 1250\ \Omega$, find R.

24. Given $R = 155\ \Omega$, $R_1 = 625\ \Omega$, $R_3 = 775\ \Omega$, and $R_4 = 1150\ \Omega$, find R_2.

Chapter 6 Group Activities

1. There are all sorts of math tricks—some performed by magicians and others performed by friends. Here is an example of a short trick.

Joel: "Think of a positive natural number but do not tell me the number."

Wayne: "Okay, I have a number." (Wayne's number is five.)

Joel: "Now add two to your number."

Wayne: "Okay, I have done that." (Now Wayne's number is 7.)

Joel: "Multiply by three."

Wayne: "Okay, that is done." (Wayne's number is now 21.)

Joel: "Subtract 6 from this number you have."

Wayne: "All right." (Wayne's number is now 15.)

Joel: "Okay, Wayne, tell me the number you have now and I'll tell you the original number you chose."

Wayne: "Sure. My number is now 15."

Joel: "Hmm, 15. Well, I guess your original number was 5."

Wayne: "Wow, that was neat. How did you do it?"

What did Joel do to figure out Wayne's number? Use algrebra to figure out this trick. To help, let the natural number chosen be represented by n.

Once you see how this trick works algebraically, you can easily make up your own similar trick. Each person in the group should try to make up a trick like this and test it on the members. The model above and the algebraic expression that represents the steps will help you.

Chapter 6 Summary

Glossary of Basic Terms

Algebraic expression. A combination of numbers, variables, symbols for operations, and symbols for grouping. (p. 244)

Equation. A statement that two quantities are equal. (p. 244)

Equivalent equations. Equations with the same solutions or roots. (p. 245)

Formula. A general rule written as an equation, usually expressed in letters, which shows the relationship between two or more quantities. (p. 265)

Reciprocal of a number. 1 divided by that number. The product of a number (except 0) and its reciprocal is 1. (p. 272)

Solution or **root.** A replacement number (or numbers) that produces a true statement in an equation. (p. 244)

Solve an equation. Find what number or numbers can replace the variable to make the equation a true statement. (p. 244)

Variable. A symbol (usually a letter) used to represent an unknown number. (p. 244)

6.1   Equations

1. **Four basic rules used to solve equations:**
   a. If the same quantity is added to both sides of an equation, the resulting equation is equivalent to the original equation.
   b. If the same quantity is subtracted from both sides of an equation, the resulting equation is equivalent to the original equation.
   c. If both sides of an equation are multiplied by the same (nonzero) quantity, the resulting equation is equivalent to the original equation.
   d. If both sides of an equation are divided by the same (nonzero) quantity, the resulting equation is equivalent to the original equation. (p. 245)

2. **When more than one operation is indicated on the variable, undo the additions and subtractions first, then undo the multiplications and divisions. (p. 247)**

6.3   Equations with Parentheses

1. **Solving an equation with parentheses:** To solve an equation having parentheses, remove the parentheses first. Then solve the resulting equation using other methods explained in this chapter. (p. 251)

6.4   Equations with Fractions

1. **Solving an equation with fractions:** To solve an equation with fractions:
   a. Find the least common denominator (LCD) of all the fractional terms on both sides of the equation.
   b. Multiply both sides of the equation by the LCD. (If this step has been done correctly, no fractions should now appear in the resulting equation.)
   c. Then solve the resulting equation using other methods explained in this chapter. (p. 254)

6.6   Applications Involving Equations

1. **Solving application problems:** To solve application problems:
   a. Read the problem carefully at least twice.
   b. If possible, draw a diagram. This will ofter you to visualize the mathematical relations needed to write the equation.
   c. Choose a letter to represent the unknown q in the problem, and write what it represent:
   d. Write an equation that expresses the inforn given in the problem and that involves the unknown.
   e. Solve the equation.
   f. Check your solution in the equation and in original problem itself. (p. 260)

6.7   Formulas

1. **Solving formulas:** To solve a formula for a g letter, isolate the given letter on one side of th equation and express it in terms of all the rem letters. (p. 265)

6.8   Substituting Data into Form

1. **Problem Solving:** To solve problems:
   a. Analyze the given data.
   b. Find an equation or formula that relates th quantities with the unknown quantity.
   c. Solve the formula for the unknown quanti
   d. Substitute the given data into this solved f (p. 268)

6.9   Reciprocal Formulas Using a Calculator

1. **To use a calculator with formulas involvin; reciprocals:**
   a. Solve for the reciprocal of the specified le
   b. Substitute the given data.
   c. Follow the calculator steps shown in Sect (p. 272)

Chapter 6 Review

Solve each equation and check:

1. $2x + 4 = 7$

2. $11 - 3x = 23$

3. $\dfrac{x}{3} - 7 = 12$

4. $5 - \dfrac{x}{6} = 1$

5. $78 - 16y = 190$

6. $25 = 3x - 2$

7. $2x + 9 = 5x - 15$

8. $-6x + 5 = 2$

9. $3 - 2x = 9 - 3x$

10. $4x + 1 = 4$

11. $7 - (x - 5) = 11$

12. $4x + 2(x + 3)$

13. $3y - 5(2 - y) = 22$

14. $6(x + 7) - 5(x + 8) = 0$

15. $3x - 4(x - 3) = 3(x - 4)$

16. $4(x + 3) - 9(x - 2) = x + 27$

17. $\dfrac{2x}{3} = \dfrac{16}{9}$

18. $\dfrac{x}{3} - 2 = \dfrac{3x}{5}$

19. $\dfrac{3x}{4} - \dfrac{x - 1}{5} = \dfrac{3 + x}{2}$

20. $\dfrac{7}{x} - 3 = \dfrac{1}{x}$

21. $5 - \dfrac{7}{x} = 3\dfrac{3}{5}$

22. The length of a rectangle is 6 more than twice its width. Its perimeter is 48 in. Find its length and width.

23. Mix a solution that is 60% acid with a solution that is 100% acid to make 12 L of a solution that is 75% acid. How much of each solution should you use?

Solve each formula for the given letter:

24. $F = Wg$ for g

25. $P = \dfrac{W}{A}$ for A

26. $L = A + B + \dfrac{1}{2}t$ for t

27. $k = \dfrac{1}{2}mv^2$ for m

28. $P_2 = \dfrac{P_1 T_2}{T_1}$ for T_1

29. $v = \dfrac{v_f + v_0}{2}$ for v_0

30. $K = \dfrac{5}{9}(F - 32) + 273$; find F if $K = 175$.

31. $P = 2(l + w)$; find w if $P = 112.8$ and $l = 36.9$.

32. $k = \dfrac{1}{2}mv^2$; find m if $k = 460$ and $v = 5.0$.

33. Given $\dfrac{1}{R} = \dfrac{1}{R_1} + \dfrac{1}{R_2}$, $R_1 = 50.0\ \Omega$, and $R_2 = 75.0\ \Omega$, find R.

34. Given $\dfrac{1}{C} = \dfrac{1}{C_1} + \dfrac{1}{C_2} + \dfrac{1}{C_3}$, $C = 25.0\ \mu\text{F}$, $C_1 = 75.0\ \mu\text{F}$, and $C_3 = 80.0\ \mu\text{F}$, find C_2.

◆ Chapter 6 Test

Solve each equation:

1. $x - 8 = -6$

2. $4x = 60$

3. $10 - 2x = 42$

4. $3x + 14 = 29$

5. $7x - 20 = 5x + 4$

6. $-2(x + 10) = 3(5 - 2x) + 5$

7. $\dfrac{1}{2}(3x - 6) = 3(x - 2)$

8. $\dfrac{8x}{9} = \dfrac{5}{6}$

9. $\dfrac{3x}{5} - 2 = \dfrac{x}{5} - \dfrac{x}{10}$

10. $\dfrac{8}{x} + 6 = 2$

11. $\dfrac{x}{2} - \dfrac{2}{5} = \dfrac{2x}{5} - \dfrac{3}{4}$

12. Distribute $2700 among Jose, Maria, and George so that Maria receives $200 more than Jose and George receives half of what Jose receives.

13. How much pure antifreeze must be added to 20 L of a solution that is 60% antifreeze to make a solution that is 80% antifreeze?

14. Solve $P = 2(l + w)$ for l.

15. Solve $C_T = C_1 + C_2 + C_3$ for C_2.

16. Solve $V = lwh$ for w.

17. Given $P = I^2R$, $P = 480$, and $I = 5.0$, find R.

18. Given $A = \dfrac{h}{2}(a + b)$, $A = 260$, $h = 13$, and $a = 15$, find b.

19. Given $\dfrac{1}{C} = \dfrac{1}{C_1} + \dfrac{1}{C_2}$, $C = 20.0\ \mu\text{F}$ and $C_2 = 30.0\ \mu\text{F}$, find C_1.

20. Given $\dfrac{1}{R} = \dfrac{1}{R_1} + \dfrac{1}{R_2} + \dfrac{1}{R_3}$, $R_1 = 225\ \Omega$, $R_2 = 475\ \Omega$, and $R_3 = 925\ \Omega$, find R.

Cumulative Review

1. Find the prime factorization of 696.
2. Change 0.081 to a percent.
3. Write 3.015×10^{-4} in decimal form.
4. Write 28,500 in scientific notation.
5. 5 ha = _____ m^2 6. 101°F = _____ °C
7. 6250 in^2 = _____ ft^2
8. Give the number of significant digits (accuracy) of each measurement:
   a. 110 cm   **b.** 6000 mi   **c.** 24.005 s
9. Read the measurement shown on the vernier caliper in Illustration 1 **a.** in metric units and **b.** in U.S. units.

ILLUSTRATION 1

10. Read the measurement shown on the U.S. micrometer in Illustration 2.

ILLUSTRATION 2

11. Use the rules for addition of measurements to sum of 25,000 W; 17,900 W; 13,962 W; 875 428,000 W.

Simplify:

12. $(2x - 5y) + (3y - 4x) - 2(3x - 5y)$
13. $(4y^3 + 3y - 5) - (2y^3 - 4y^2 - 2y + 6)$
14. $(3y^3)^3$
15. $-2x(x^2 - 3x + 4)$
16. $(6y^3 - 5y^2 - y + 2)(2y - 1)$
17. $(4x - 3y)(5x + 2y)$ 18. $\dfrac{215\, x^2 y^3}{45 x^3 y^5}$
19. $(16x^2 y^3)(-5x^4 y^5)$
20. $\dfrac{x^3 + 2x^2 - 11x - 20}{x + 5}$
21. $3x^2 - 4xy + 5y^2 - (-3x^2) + (-7xy) + 10y^2$

Solve:

22. $4x - 2 = 12$ 23. $\dfrac{x}{4} - 5 = 9$
24. $4x - 3 = 7x + 15$ 25. $\dfrac{5x}{8} = \dfrac{3}{2}$
26. $5 - (x - 3) = (2 + x) - 5$
27. $C = \dfrac{1}{2}(a + b + c)$ for a
28. $A = lw$; find w if $l = 8.20$ m and $A = 91.3$ m
29. Translate into algebraic symbols: The pro number and 7 is 250.
30. The perimeter of a rectangle is 30 ft. The wid half of the length. What are the dimensio rectangle?

7 Ratio and Proportion

Mathematics at Work

Heating, ventilation, air-conditioning, and refrigeration (HVAC/R) technicians install and repair such systems. Duties include installation and repair of oil burners, hot-air furnaces, heating stoves, and similar equipment in residential, commercial, and industrial buildings using hand and pipe threading. Heating and air-conditioning systems control the temperature, humidity, and total air quality in such locations. Refrigeration systems allow for the storing and transport of food, medicine, and other perishable items.

Heating, ventilation, air-conditioning, and refrigeration systems consist of many mechanical, electrical, and electronic components such as motors, compressors, pumps, fans, ducts, pipes, thermostats, and switches. These technicians must be able to maintain, diagnose, and correct problems throughout the entire system by adjusting system controls to recommended settings and test the performance of the system using special tools and test equipment. Although trained to do installation or repair and maintenance, technicians often specialize in one or the other. Some specialize in one type of equipment, such as commercial refrigerators, oil burners, or solar panels. Technicians work for large or small contractors or directly for

Heating, Ventilation, Air-Conditioning, and Refrigeration Technician
HVAC technician checking and explaining a furnace problem.

manufacturers or wholesalers. Employers prefer to employ those with technical school or apprenticeship training. Many community colleges and postsecondary and trade schools offer associate degree and certificate programs in which students study theory, design, equipment construction, and electronics as well as the basics of installation, maintenance, and repair. All technicians who work with refrigerants must be certified in their proper handling. North American Technician Excellence, Inc., offers one standard for certification of experienced technicians. For more information, go to the website listed below.

www.cengage.com/mathematics/ewen

Objectives

■ Express a ratio and a rate in lowest terms.

■ Solve a proportion.

■ Solve application problems using ratios, rates, and proportions.

■ Solve application problems involving direct variation.

■ Solve application problems involving inverse variation.

7.1 Ratio*

The comparison of two numbers is a very important concept, and one of the mo[st impor]tant of all comparisons is the ratio. The **ratio of two numbers**, a and b, is the fir[st] divided by the second number. Ratios may be written in several different ways. [For exam]ple, the ratio of 3 to 4 may be written as $\frac{3}{4}$, 3/4, 3 : 4, or 3 ÷ 4. Each form is rea[d "the ra]tio of 3 to 4."

If the quantities to be compared include units, the units should be the same [if] possible. To find the ratio of 1 ft to 15 in., first express both quantities in inche[s. Then] find the ratio:

$$\frac{1 \text{ ft}}{15 \text{ in.}} = \frac{12 \text{ in.}}{15 \text{ in.}} = \frac{12}{15} = \frac{4}{5}$$

Ratios are usually given in lowest terms.

Example 1

Express the ratio 18 : 45 in lowest terms.

$$18 : 45 = \frac{18}{45} = \frac{\cancel{9} \cdot 2}{\cancel{9} \cdot 5} = \frac{2}{5}$$

Example 2

Express the ratio of 3 ft to 18 in. in lowest terms.

$$\frac{3 \text{ ft}}{18 \text{ in.}} = \frac{36 \text{ in.}}{18 \text{ in.}} = \frac{\cancel{18} \times 2 \text{ in.}}{\cancel{18} \times 1 \text{ in.}} = \frac{2}{1} \text{ or } 2$$

Note: $\frac{2}{1}$ and 2 indicate the same ratio, "the ratio of 2 to 1."

Example 3

Express the ratio of 50 cm to 2 m in lowest terms.

First express the measurements in the same units.
1 m = 100 cm, so 2 m = 200 cm.

$$\frac{50 \text{ cm}}{2 \text{ m}} = \frac{50 \text{ cm}}{200 \text{ cm}} = \frac{1}{4}$$

To find the ratio of two fractions, use the technique for dividing fractions.

Note: In this chapter, do not use rules for calculating with measurements.

Example 4   Express the ratio $\dfrac{2}{3} : \dfrac{8}{9}$ in lowest terms.

$$\frac{2}{3} : \frac{8}{9} = \frac{2}{3} \div \frac{8}{9} = \frac{2}{3} \times \frac{9}{8} = \frac{18}{24} = \frac{\cancel{6} \cdot 3}{\cancel{6} \cdot 4} = \frac{3}{4}$$

 Example 5   Express the ratio of $2\frac{1}{2}$ to 10 in lowest terms.

$$2\frac{1}{2} \text{ to } 10 = \frac{5}{2} \div 10 = \frac{5}{2} \times \frac{1}{10} = \frac{5}{20} = \frac{\cancel{5} \cdot 1}{\cancel{5} \cdot 4} = \frac{1}{4}$$

 Example 6   Steel can be worked in a lathe at a cutting speed of 25 ft/min. Stainless steel can be worked in a lathe at a cutting speed of 15 ft/min. What is the ratio of the cutting speed of steel to the cutting speed of stainless steel?

$$\frac{\text{cutting speed of steel}}{\text{cutting speed of stainless steel}} = \frac{25 \text{ ft/min}}{15 \text{ ft/min}} = \frac{5}{3}$$

 Example 7   A construction crew uses 4 buckets of cement and 12 buckets of sand to mix a supply of concrete. What is the ratio of cement to sand?

$$\frac{\text{amount of cement}}{\text{amount of sand}} = \frac{4 \text{ buckets}}{12 \text{ buckets}} = \frac{1}{3}$$

In a ratio, we compare like or related quantities; for example,

$$\frac{18 \text{ ft}}{12 \text{ ft}} = \frac{3}{2} \quad \text{or} \quad \frac{50 \text{ cm}}{2 \text{ m}} = \frac{50 \text{ cm}}{200 \text{ cm}} = \frac{1}{4}$$

A ratio simplified into its lowest terms is a pair of unitless numbers.

Suppose you drive 75 miles and use 3 gallons of gasoline. Your mileage would be found as follows:

$$\frac{75 \text{ mi}}{3 \text{ gal}} = \frac{25 \text{ mi}}{1 \text{ gal}}$$

We say that your mileage is 25 miles per gallon. Note that each of these two fractions compares unlike quantities: miles and gallons. A *rate* is the comparison of two unlike quantities whose units do not cancel.

 Example 8   Express the rate of $\dfrac{250 \text{ gal}}{50 \text{ acres}}$ in lowest terms.

$$\frac{250 \text{ gal}}{50 \text{ acres}} = \frac{5 \text{ gal}}{1 \text{ acre}} \text{ or } 5 \text{ gal/acre}$$

The symbol "/" is read "per." The rate is read "5 gallons per acre."

A common medical practice is to give nourishment and/or medication to a patient by IV (intravenously). The number of drops per minute is related to the type of equipment being used. The number of drops per mL is called the *drop factor*. Common drop factors are 10 drops/mL, 12 drops/mL, and 15 drops/mL.

Example 9

A doctor orders 500 mL of glucose to be given to an adult patient by IV in 6 h. The [...] tor of the equipment is 15. Determine the number of drops per minute in order to set [...]

First, change 6 h to minutes:

$$6 \text{ h} \times \frac{60 \text{ min}}{1 \text{ h}} = 360 \text{ min (time for IV)}$$

Thus, 500 mL of glucose is to be given during a 360-min time period, which give [...] of $\frac{500 \text{ mL}}{360 \text{ min}}$. Since the equipment has a drop factor of 15 drops/mL, the flow rate is [...]

$$\frac{500 \text{ mL}}{360 \text{ min}} \times \frac{15 \text{ drops}}{\text{mL}} = 21 \text{ drops/min (rounded to the nearest whole numbe[...]}$$

Sometimes the doctor orders an IV as a rate of flow, and the nurse must fin[...] needed to administer the IV.

Example 10

Give 1500 mL of saline solution IV with a drop factor of 10 at a rate of 50 drops[...] adult patient. Find how long the IV should be administered.

First, determine the total number of drops to be administered:

$$1500 \text{ mL} \times \frac{10 \text{ drops}}{\text{mL}} = 15,000 \text{ drops}$$

Then, divide the total number of drops by the flow rate to find the time:

$$\frac{15,000 \text{ drops}}{50 \text{ drops/min}} = 300 \text{ min} \qquad \frac{\text{drops}}{\dfrac{\text{drops}}{\text{min}}} = \text{drops} \div \frac{\text{drops}}{\text{min}}$$

$$= \text{drops} \times \frac{\text{min}}{\text{drops}} = \text{min}$$

Exercises 7.1

Express each ratio in lowest terms:

1. 3 to 15

2. 6 : 12

3. 7 : 21

4. $\dfrac{4}{22}$

5. $\dfrac{80}{48}$

6. 28 to 20

7. 3 in. to 15 in.

8. 3 ft to 15 in.

9. 3 cm to 15 mm

10. 1 in. to 8 ft

11. 9 in^2 : 2 ft^2

12. 4 m : 30 cm

13. $\dfrac{3}{4}$ to $\dfrac{7}{6}$

14. $\dfrac{2}{3} : \dfrac{22}{9}$

15. $2\dfrac{3}{4} : 4$

16. 6 to $4\dfrac{2}{3}$

17. $\dfrac{5\frac{1}{3}}{2\frac{2}{3}}$

18. $\dfrac{18\frac{1}{2}}{2\frac{1}{4}}$

19. 10 to $2\dfrac{1}{2}$

20. $\dfrac{7}{8} : \dfrac{9}{16}$

21. [...]

22. $2\dfrac{2}{3} : 3\dfrac{3}{4}$

23. $1\dfrac{3}{4}$ to 7

24. [...]

Express each rate in lowest terms:

25. $\dfrac{240 \text{ mi}}{8 \text{ gal}}$

26. $\dfrac{360 \text{ gal}}{18 \text{ acres}}$

27. [...]

28. $\dfrac{\$36}{3 \text{ h}}$

29. $\dfrac{625 \text{ mi}}{12\frac{1}{2} \text{ h}}$

30. [...]

31. $\dfrac{2\frac{1}{4} \text{ lb}}{6 \text{ gal}}$

32. $\dfrac{\$64,800}{1800 \text{ ft}^2}$

33. A bearing bronze mix includes 96 lb of copper and 15 lb of lead. Find the ratio of copper to lead.

34. What is the alternator-to-engine drive ratio if the alternator turns at 1125 rpm when the engine is idling at 500 rpm?

35. Suppose 165 gal of oil flow through a feeder pipe in 5 min. Find the flow rate in gallons per minute.

36. A flywheel has 72 teeth, and a starter drive-gear has 15 teeth. Find the ratio of flywheel teeth to drive-gear teeth.

37. A transformer has a voltage of 18 V in the primary circuit and 4950 V in the secondary circuit. Find the ratio of the primary voltage to the secondary voltage.

38. The ratio of the voltage drops across two resistors wired in series equals the ratio of their resistances. Find the ratio of a 720-Ω resistor to a 400-Ω resistor.

39. A transformer has 45 turns in the primary coil and 540 turns in the secondary coil. Find the ratio of secondary turns to primary turns.

40. The resistance in ohms of a resistor is the ratio of the voltage drop across the resistor, in volts, to the current through the resistor, in amperes. A resistor has a voltage drop across it of 117 V and a current through it of 2.6 A. What is the resistance in ohms of the resistor?

41. A 150-bu wagon holds 2.7 tons of grain. Express the weight of grain in pounds per bushel.

42. The total yield from a 55-acre field is 7425 bu. Express the yield in bushels per acre.

43. A 350-gal spray tank covers 14 acres. Find the rate of application in gallons per acre.

44. Suppose 12 gal of herbicide concentrate are used for 28 acres. Find the ratio of gallons of concentrate to acres.

45. Suppose 16 ft of copper tubing costs $27.04. Find its cost per foot.

46. A structure has 3290 ft^2 of wall area (excluding windows) and 1880 ft^2 of window area. Find the ratio of wall area to window area.

47. A 2150-ft^2 home sells for $225,750. Find the ratio of cost to area (price per ft^2).

48. You need 15 ft^3 of cement to make 80 ft^3 of concrete. Find the ratio of volume of concrete to volume of cement.

49. A welder has 9 pieces of 4-ft steel angle and 12 pieces of 2-ft steel angle. What is the ratio of pieces of 4-ft steel angle to 2-ft steel angle?

50. A welder grabs a handful of 6011 welding rods and another handful of Super Strength 100 welding rods. When the welder sees how many of each she has, she has 32 of the 6011 welding rods and 60 of the Super Strength 100 welding rods. What is the ratio of the 6011 welding rods to the Super Strength 100 welding rods?

51. The total number of hours required for a private-pilot, single-engine land rating is 40 h of flight time. The total number of hours of flight time required for a commercial rating is 250 h. What is the ratio of the number of hours required for a private rating to those required for a commercial rating?

52. Two small window air conditioner units were purchased and put into opposite sides of a house. One air conditioner was 5000 Btu and the other, 7500 Btu. What is the ratio of the 5000-Btu to the 7500-Btu air conditioner?

53. Suppose 2.8 cm^3 of medication are drawn from a vial of hydrocortisone that contains 140 mg of medication. How many milligrams of medication per cubic centimetre are in the vial?

54. A 250-cm^3 bottle contains 4000 mg of aminophylline. Find the ratio of milligrams of aminophylline to each cubic centimetre.

55. A 45-cm^3 vial contains 180 mg of Demerol. Find the ratio of milligrams of Demerol to each cubic centimetre.

56. Over a period of 5 h, 1200 cm^3 of a solution will be administered intravenously. How many cubic centimetres per minute is this?

57. Adult male cougars commonly require 50 to 150 mi^2 of territory to live. If 280 male cougars live in a forest area that covers 45,000 mi^2, of which 3000 mi^2 is covered by water, what is the ratio of male cougars per living area?

58. In tilapia fish farming, 5 tons of high-quality feed resulted in a weight gain of 5,000 lb. What is the amount of feed-to-weight gain ratio?

Find the flow rate for each given IV (assume a drop factor of 15 drops/mL):

59. 1200 mL in 6 h   60. 900 mL in 3 h

61. 1 L in 5.5 h   62. 2 L in 5 h

Find the length of time each IV should be administered (assume a drop factor of 10 drops/mL):

63. 1000 mL at a rate of 50 drops/min

64. 1600 mL at a rate of 40 drops/min

65. 2 L at a rate of 40 drops/min

66. 1.4 L at a rate of 35 drops/min

7.2 Proportion

A **proportion** states that two ratios or two rates are equal. Thus,

$$\frac{3}{4} = \frac{9}{12}, \qquad 2 : 3 = 4 : 6, \qquad \text{and} \qquad \frac{a}{b} = \frac{c}{d}$$

are proportions. A proportion has four terms. In the proportion $\frac{2}{5} = \frac{4}{10}$, the first
the second term is 5, the third term is 4, and the fourth term is 10.

The first and fourth terms of a proportion are called the **extremes**, and the se
third terms are called the **means** of the proportion. This is more easily seen whe
portion $\frac{a}{b} = \frac{c}{d}$ is written in the form

$$
\overset{\textstyle \text{means}}{\overbrace{a : b = c : d}}
$$

extremes

Example 1

Given the proportion $\dfrac{2}{3} = \dfrac{4}{6}$.

   a. The first term is 2.

   b. The second term is 3.

   c. The third term is 4.

   d. The fourth term is 6.

   e. The means are 3 and 4.

   f. The extremes are 2 and 6.

   g. The product of the means $= 3 \cdot 4 = 12$.

   h. The product of the extremes $= 2 \cdot 6 = 12$.

We see in **g** and **h** that the product of the means (that is, 12) equals the pro
extremes (also 12). Let us look at another proportion and see if this is true agair

Example 2

Given the proportion $\dfrac{5}{13} = \dfrac{10}{26}$, find the product of the means and the product of the

The extremes are 5 and 26, and the means are 13 and 10. The product of the e
130, and the product of the means is 130. Here again, the product of the means
product of the extremes. As a matter of fact, this will always be the case.

> ### Proportion
>
> In any proportion, the *product of the means* equals the *product of the extrem*
> is, if $\frac{a}{b} = \frac{c}{d}$, then $bc = ad$.

To determine whether two ratios are equal, put the two ratios in the form c
tion. If the product of the means equals the product of the extremes, the ratios a

Example 3 Determine whether or not the ratios $\dfrac{13}{36}$ and $\dfrac{29}{84}$ are equal.

If $36 \times 29 = 13 \times 84$, then $\dfrac{13}{36} = \dfrac{29}{84}$.

However, $36 \times 29 = 1044$ and $13 \times 84 = 1092$. Therefore, $\dfrac{13}{36} \neq \dfrac{29}{84}$. ■

To *solve* a proportion means to find the missing term. To do this, form an equation by setting the product of the means equal to the product of the extremes. Then solve the resulting equation.

 Example 4 Solve the proportion $\dfrac{x}{3} = \dfrac{8}{12}$.

$$\frac{x}{3} = \frac{8}{12}$$

$12x = 24 \qquad$ The product of the means equals the product of the extremes.

$x = 2$ ■

 Example 5 Solve the proportion $\dfrac{5}{x} = \dfrac{10}{3}$.

$$\frac{5}{x} = \frac{10}{3}$$

$10x = 15 \qquad$ The product of the means equals the product of the extremes.

$x = \dfrac{3}{2} \quad$ or $\quad 1.5$ ■

A calculator is helpful in solving a proportion with decimal fractions.

 Example 6 Solve $\dfrac{32.3}{x} = \dfrac{17.9}{25.1}$.

$17.9x = (32.3)(25.1) \qquad$ The product of the means equals the product of the extremes.

$x = \dfrac{(32.3)(25.1)}{17.9} = 45.3$, rounded to three significant digits

32.3 ☒ 25.1 ÷ 17.9 =

$\boxed{45.292179}$ ■

 Example 7 If 125 bolts cost \$7.50, how much do 75 bolts cost?

First, let's find the rate of dollars/bolts in each case.

$$\frac{\$7.50}{125 \text{ bolts}} \quad \text{and} \quad \frac{x}{75 \text{ bolts}} \quad \text{where } x = \text{the cost of 75 bolts}$$

Since these two rates are equal, we have the proportion

$$\frac{7.5}{125} = \frac{x}{75}$$

$125x = (7.5)(75)$ The product of the means equals the product of the extrem

$$x = \frac{(7.5)(75)}{125}$$ Divide both sides by 125.

$$x = 4.5$$

That is, the cost of 75 bolts is $4.50.

Note: A key to solving proportions like the one in Example 7 is to set up the
with the same units in each ratio—in this case

$$\frac{\$}{\text{bolts}} = \frac{\$}{\text{bolts}}$$

    Example 8   The pitch of a roof is the ratio of the rise to the run of a rafter. (See Figure 7.1.)
of the roof shown is 2 : 7. Find the rise if the run is 21 ft.

FIGURE 7.1

$$\text{pitch} = \frac{\text{rise}}{\text{run}}$$

$$\frac{2}{7} = \frac{x}{21 \text{ ft}}$$

$$7x = (2)(21 \text{ ft})$$

$$x = \frac{(2)(21 \text{ ft})}{7}$$

$$x = 6 \text{ ft, which is the rise}$$

In Section 1.14, you studied percent, using the formula $P = BR$, where R
written as a decimal. Knowing this formula and knowing the fact that *percent*
hundred," we can write the proportion

$$\frac{P}{B} = \frac{R}{100}$$

where R is the rate written as a percent. We can use this proportion to solve percen

Note: You may find it helpful to review the meanings of P (part), B (base), an
Section 1.14.

Example 9

A student answered 27 out of 30 questions correctly. What percent of the answers were correct?

$$P \text{ (part)} = 27$$
$$B \text{ (base)} = 30$$
$$R \text{ (rate)} = x$$
$$\frac{27}{30} = \frac{x}{100}$$
$$30x = 2700 \qquad \text{The product of the means equals the product of the extremes.}$$
$$x = 90$$

Therefore, the student answered 90% of the questions correctly.   ■

 Example 10

A factory produces bearings used in automobiles. After inspecting 4500 bearings, the inspectors find that 127 are defective. What percent are defective?

$$P \text{ (part)} = 127$$
$$B \text{ (base)} = 4500$$
$$R \text{ (rate)} = x$$
$$\frac{127}{4500} = \frac{x}{100}$$
$$4500x = 12{,}700$$
$$x = 2.8$$

Therefore, 2.8% of the bearings are defective.   ■

 Example 11

A nurse must prepare 300 mL of 10% glucose solution from pure crystalline glucose. How much pure crystalline glucose is needed?

$$B \text{ (base)} = 300 \text{ mL}$$
$$R \text{ (rate)} = 10\%$$
$$P \text{ (part)} = x$$
$$\frac{x}{300} = \frac{10}{100}$$
$$100x = 3000$$
$$x = 30 \text{ mL}$$ ■

 Example 12

Prepare 2000 mL of a Lysol solution containing 1 part Lysol and 19 parts water from pure Lysol. How much pure Lysol is needed?

$$R = 1 : (1 + 19) = 1 : 20 = 1/20 = 0.05 = 5\%$$
$$B = 2000 \text{ mL}$$
$$P = x$$
$$\frac{x}{2000} = \frac{5}{100}$$
$$100x = 10{,}000$$
$$x = 100 \text{ mL}$$ ■

Exercises 7.2

In each proportion, find **a.** *the means,* **b.** *the extremes,* **c.** *the product of the means, and* **d.** *the product of the extremes:*

1. $\dfrac{1}{2} = \dfrac{3}{6}$ **2.** $\dfrac{3}{4} = \dfrac{6}{8}$ **3.** $\dfrac{7}{9} = \dfrac{28}{36}$

4. $\dfrac{x}{3} = \dfrac{6}{9}$ **5.** $\dfrac{x}{7} = \dfrac{w}{z}$ **6.** $\dfrac{a}{b} = \dfrac{4}{5}$

Determine whether or not each pair of ratios is equal:

7. $\dfrac{2}{3}, \dfrac{10}{15}$ **8.** $\dfrac{2}{3}, \dfrac{9}{6}$ **9.** $\dfrac{3}{5}, \dfrac{18}{20}$

10. $\dfrac{3}{7}, \dfrac{9}{21}$ **11.** $\dfrac{1}{3}, \dfrac{4}{12}$ **12.** $\dfrac{125}{45}, \dfrac{25}{9}$

Solve each proportion (round each result to three significant digits when necessary):

13. $\dfrac{x}{4} = \dfrac{9}{12}$ **14.** $\dfrac{1}{a} = \dfrac{4}{16}$ **15.** $\dfrac{5}{7} = \dfrac{4}{y}$

16. $\dfrac{12}{5} = \dfrac{x}{10}$ **17.** $\dfrac{2}{x} = \dfrac{4}{28}$ **18.** $\dfrac{10}{15} = \dfrac{y}{75}$

19. $\dfrac{5}{7} = \dfrac{3x}{14}$ **20.** $\dfrac{x}{18} = \dfrac{7}{9}$ **21.** $\dfrac{5}{7} = \dfrac{25}{y}$

22. $\dfrac{1.1}{6} = \dfrac{x}{12}$ **23.** $\dfrac{-5}{x} = \dfrac{2}{3}$ **24.** $\dfrac{4x}{9} = \dfrac{12}{7}$

25. $\dfrac{8}{0.04} = \dfrac{700}{x}$ **26.** $\dfrac{x}{9} = \dfrac{2}{0.6}$ **27.** $\dfrac{3x}{27} = \dfrac{0.5}{9}$

28. $\dfrac{0.25}{2x} = \dfrac{8}{48}$ **29.** $\dfrac{17}{28} = \dfrac{153}{2x}$ **30.** $\dfrac{3x}{10} = \dfrac{7}{50}$

31. $\dfrac{12}{y} = \dfrac{84}{144}$ **32.** $\dfrac{13}{169} = \dfrac{27}{x}$ **33.** $\dfrac{x}{48} = \dfrac{56}{72}$

34. $\dfrac{124}{67} = \dfrac{149}{x}$ **35.** $\dfrac{472}{x} = \dfrac{793}{64.2}$ **36.** $\dfrac{94.7}{6.72} = \dfrac{x}{19.3}$

37. $\dfrac{30.1}{442} = \dfrac{55.7}{x}$ **38.** $\dfrac{9.4}{291} = \dfrac{44.1}{x}$ **39.** $\dfrac{36.9}{104} = \dfrac{3210}{x}$

40. $\dfrac{0.0417}{0.355} = \dfrac{26.9}{x}$ **41.** $\dfrac{x}{4.2} = \dfrac{19.6}{3.87}$

42. $\dfrac{0.120}{3x} = \dfrac{0.575}{277}$

43. You need $2\frac{3}{4}$ ft^3 of sand to make 8 ft^3 of concrete. How much sand would you need to make 128 ft^3 of concrete?

44. The pitch of a roof is $\frac{1}{3}$. If the run is 15 ft, find the rise. (See Example 8.)

45. A builder sells an 1800-ft^2 home for $171,000. What would be the price of a 2400-ft^2 home of similar quality? Assume that the price per square foot remains constant.

46. Suppose 826 bricks are used in constructing a wall 14 ft long. How many bricks will be needed for a similar wall 35 ft long?

47. A buyer purchases 75 yd of material for $120. Then an additional 90 yd are ordered. What is the additional cost?

48. A salesperson is paid a commission of $75 for selling $300 worth of goods. What is the commission on $760 of sales at the same rate of commission?

49. A plane flies for 3 h and uses 25 gal of 100*LL* aviation fuel. How much will be used if the plane flies for only 1.2 h?

50. Metal duct that is 6 in. in diameter costs $7.50 for 5 ft. If 16.5 ft are needed for an order, what is the cost?

51. Suppose 20 gal of water and 3 lb of pesticide are applied per acre. How much pesticide should you put in a 350-gal spray tank? Assume that the pesticide dissolves in the water and has no volume.

52. A farmer uses 150 lb of a chemical on a 40-acre field. How many pounds will he need for a 220-acre field? Assume the same rate of application.

53. Suppose a yield of 100 bu of corn per acre removes 90 lb of nitrogen, phosphorus, and potash (or potassium) (N, P, and K). How many pounds of N, P, and K would be removed by a yield of 120 bu per acre?

54. A farmer has a total yield of 42,000 bu of corn from a 350-acre farm. What total yield should he expect from a similar 560-acre farm?

55. A copper wire 750 ft long has a resistance of 1.563 Ω. How long is a copper wire of the same size whose resistance is 2.605 Ω? (The resistance of these wires is proportional to their length.)

56. The voltage drop across a 28-Ω resistor is 52 V. What is the voltage drop across a 63-Ω resistor that is in series with the first one? (Resistors in series have voltage drops proportional to their resistances.)

57. The ratio of secondary turns to primary turns in a transformer is 35 to 4. How many secondary turns are there if the primary coil has 68 turns?

58. If welding rods cost $75 per 50 lb, how much would 75 lb cost?

59. An engine with displacement of 380 in^3 develops 212 hp. How many horsepower would be developed by a 318-in^3 engine of the same design?

60. An 8-V automotive coil has 250 turns of wire in the primary circuit. The secondary voltage is 15,000 V. How many secondary turns are in the coil? (The ratio of secondary voltage to primary voltage equals the ratio of secondary turns to primary turns.)

61. A car uses 18 gal of gas to go 560 mi. How many gallons are required to go 820 mi?

62. A fuel pump delivers 35 mL of fuel in 420 strokes. How many strokes are needed to pump 50 mL of fuel?

63. A label reads: "4 cm³ of solution contains gr X [10 gr] of potassium chloride." How many cubic centimetres are needed to give gr XXXV [35 gr]?

64. A multiple-dose vial has been mixed and labeled "200,000 units in 1 cm³." How many cubic centimetres are needed to give 900,000 units?

65. You are to administer 150 mg of aminophylline from a bottle marked 250 mg/10 cm³. How many cubic centimetres should you draw?

66. A label reads: "Gantrisin, 1.5 g in 20 cm³." How many cubic centimetres are needed to give 10.5 g of Gantrisin?

When finding the percent, round to the nearest tenth of a percent when necessary:

67. Carla bought a used car for $15,000. She paid $3500 down. What percent of the price was her down payment?

68. A baseball team last year won 18 games and lost 12. What percent did they win? What percent did they lose?

69. A car is listed to sell at $20,400. The salesperson offers to sell it to you for $19,200. What percent of the list price is the reduction?

70. A live hog weighs 254 lb, and its carcass weighs 198 lb. What percent of the live hog is carcass? What percent is waste?

71. In a 100-g sample of beef, there are 18 g of fat.
   a. What is the percent of fat in the beef?
   b. How many pounds of fat would there be in a 650-lb beef carcass? Assume the same percent of fat.

72. At the beginning of a trip, a tire on a car has a pressure of 32 psi (lb/in²). At the end of the trip, the pressure is 38 psi. What is the percent increase in pressure?

73. A gasoline tank contains 5.7 hectolitres (hL) when it is 30% full. What is the capacity of the tank?

74. Jamal had a pay raise from $1975 to $2370 per month. Find the percent increase.

75. A concrete mix is composed of 1 part cement, 2.5 parts sand, and 4 parts gravel by volume. What is the percent by volume in the dry mix of **a.** cement, **b.** sand, and **c.** gravel?

76. You are to put 4 qt of pure antifreeze in a tractor radiator and then fill the radiator with 5 gal of water. What percent of antifreeze will be in the radiator?

77. In developing a paint line process for an appliance manufacturer, a given component on a conveyor belt passes a given point at a rate of 25 ft in 3 min into a drying booth. If an object must be in the drying booth for 10 min, how long should this booth be?

78. A barn with dimensions 32 ft × 14 ft × 8 ft (wall height) with a roof peak of 19 ft is being reproduced at the scale 1 in. = 4 ft to show the board of directors what it will look like when finished. Find the overall dimensions of the model.

79. A pound of sea water contains 35 parts per thousand of dissolved salt by weight. How much salt is contained in 1 ton of sea water?

80. The content of common fertilizer is listed using three numbers that represent the percentages of nitrogen, phosphorus, and potassium, commonly abbreviated as N-P-K. For example, a bag of 5-10-15 fertilizer contains 5% nitrogen, 10% phosphorus, and 15% potassium. If you apply enough 5-10-15 fertilizer to your lawn to apply a total of 24 lb of nitrogen, how much phosphorus and how much potassium are applied?

81. The gear ratio of a fishing reel refers to the number of times the spool rotates for each turn of the handle. If a reel requires 20 turns of the handle to retrieve 100 ft of line, how many turns would it take to retrieve 175 ft of line?

82. An acre inch is the amount of water needed to provide 1 inch of water on an acre of surface. There are approximately 27,150 gal of water in an acre inch. According to the USDA, only about $\frac{1}{4}$ of the initial water that is used in irrigation ever reaches the crop for which it is intended with the rest evaporating or soaking into the ground before it reaches the crop. How many gallons of water would be required to apply an actual inch of water in a given field of one acre?

Find the amount of pure ingredient needed to prepare each solution as indicated:

83. 150 mL of 3% cresol solution from pure cresol

84. 1000 mL of 5% Lysol solution from pure Lysol

85. 500 mL of 1% sodium bicarbonate solution from pure powdered sodium bicarbonate

86. 600 mL of 10% glucose solution from pure crystalline glucose

87. 1.5 L of 1 : 1000 epinephrine solution from pure epinephrine

88. 20 mL of 1 : 200 silver nitrate solution from pure silver nitrate

89. 300 mL of 1 : 10 glucose solution from pure glucose

90. 400 mL of 1 : 50 sodium bicarbonate solution from pure sodium bicarbonate

7.3 Direct Variation

When two quantities, x and y, change so that their *ratios are constant*—that is,

$$\frac{y_1}{x_1} = \frac{y_2}{x_2}$$

they are said to *vary directly*. This relationship between the two quantities is called **direct variation**. If one quantity increases, the other increases by the same factor. Likewise, if one decreases, the other decreases by the same factor.

Consider the following data:

y	6	24	15	18	9	30
x	2	8	5	6	3	10

Note that y varies directly with x because the ratio $\frac{y}{x}$ is always 3, a constant. This x relationship may also be written $y = 3x$.

> **Direct Variation**
>
> $$\frac{y_1}{x_1} = \frac{y_2}{x_2}$$

Examples of direct variation are scale drawings such as maps and blueprints where

$$\frac{\text{scale measurement 1}}{\text{scale measurement 2}} = \frac{\text{actual measurement 1}}{\text{actual measurement 2}}$$

or

$$\frac{\text{scale measurement 1}}{\text{actual measurement 1}} = \frac{\text{scale measurement 2}}{\text{actual measurement 2}}$$

A *scale drawing* of an object has the same shape as the actual object, but the scale drawing may be smaller than, equal to, or larger than the actual object. The scale used in a drawing indicates what the ratio is between the size of the scale drawing and the size of the object drawn.

A portion of a map of the state of Illinois is shown in Figure 7.2. The scale is 1 in. = 32 mi.

Example 1

Find the approximate distance between Champaign and Kankakee using the map in Figure 7.2.

The distance on the map measures $2\frac{3}{8}$ in. Set up a proportion that has as its first ratio the scale drawing ratio and as its second ratio the length measured on the map to the actual distance.

$$\frac{1}{32} = \frac{2\frac{3}{8}}{d}$$

$$1d = 32\left(2\frac{3}{8}\right) \qquad \text{The product of the means equals the product of the extremes.}$$

$$d = 32\left(\frac{19}{8}\right) = 76 \text{ mi}$$

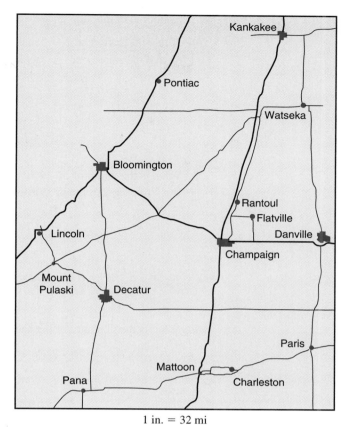

1 in. = 32 mi

FIGURE 7.2 ■

Square-ruled paper may also be used to represent scale drawings. Each square represents a unit of length according to some scale.

◆ Example 2 The scale drawing in Figure 7.3 represents a metal plate a machinist is to make.

a. How long is the plate?

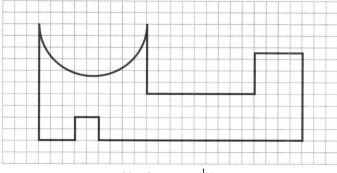

side of square = $\frac{1}{4}$ in.

FIGURE 7.3

Count the number of spaces, then set up the proportion:

$$\frac{\text{scale measurement 1}}{\text{actual measurement 1}} = \frac{\text{scale measurement 2}}{\text{actual measurement 2}}$$

$$\frac{1 \text{ space}}{\frac{1}{4} \text{ in.}} = \frac{22 \text{ spaces}}{x \text{ in.}}$$

$$x = 22\left(\frac{1}{4}\right) = 5\frac{1}{2} \text{ in.}$$

b. What is the width of the plate at its right end?
It is $7\frac{1}{2}$ spaces, so the proportion is

$$\frac{1 \text{ space}}{\frac{1}{4} \text{ in.}} = \frac{7\frac{1}{2} \text{ spaces}}{x \text{ in.}}$$

$$x = \left(7\frac{1}{2}\right)\left(\frac{1}{4}\right) = 1\frac{7}{8} \text{ in.}$$

c. What is the diameter of the semicircle?

$$\frac{1 \text{ space}}{\frac{1}{4} \text{ in.}} = \frac{9 \text{ spaces}}{x \text{ in.}}$$

$$x = 9\left(\frac{1}{4}\right) = 2\frac{1}{4} \text{ in.} \qquad\qquad ■$$

Another example of direct variation is the hydraulic press or hydraulic pump, which allows one to exert a small force to move or raise a large object, such as a car. Other uses of hydraulics include compressing junk cars, stamping metal sheets to form car parts, and lifting truck beds.

A hydraulic press is shown in Figure 7.4. When someone presses a force of 50 lb on the small piston, a force of 5000 lb is exerted by the large piston. The *mechanical advantage* (MA) *of a hydraulic press* is the ratio of the force from the large piston (F_l) to the force on the small piston (F_s). The formula is

$$MA = \frac{F_l}{F_s}$$

FIGURE 7.4
Hydraulic press

Example 3   Find the mechanical advantage of the press shown in Figure 7.4.

$$MA = \frac{F_l}{F_s}$$

$$= \frac{5000 \text{ lb}}{50 \text{ lb}} = \frac{100}{1}$$

Thus, for every pound exerted on the small piston, 100 lb is exerted by the large piston. ■

The *mechanical advantage of a hydraulic press* can also be calculated when the radii of the pistons are known.

$$MA = \frac{r_l^2}{r_s^2}$$

 Example 4   The radius of the large piston of a hydraulic press is 12 in. The radius of the small piston is 2 in. Find the MA.

$$MA = \frac{r_l^2}{r_s^2}$$

$$= \frac{(12 \text{ in.})^2}{(2 \text{ in.})^2}$$

$$= \frac{144 \text{ in}^2}{4 \text{ in}^2}$$

$$= \frac{36}{1}$$

That is, for every pound exerted on the small piston, 36 lb is exerted by the large piston.
■

You now have two ways of finding mechanical advantage: when F_l and F_s are known and when r_l and r_s are known. From this knowledge, you can find a relationship among F_l, F_s, r_l, and r_s.

Since $MA = \dfrac{F_l}{F_s}$ and $MA = \dfrac{r_l^2}{r_s^2}$ then $\dfrac{F_l}{F_s} = \dfrac{r_l^2}{r_s^2}$

 Example 5   Given $F_s = 240$ lb, $r_l = 16$ in., and $r_s = 2$ in., find F_l.

$$\frac{F_l}{F_s} = \frac{r_l^2}{r_s^2}$$

$$\frac{F_l}{240 \text{ lb}} = \frac{(16 \text{ in.})^2}{(2 \text{ in.})^2}$$

$$F_l = \frac{(240 \text{ lb})(16 \text{ in.})^2}{(2 \text{ in.})^2}$$

$$= \frac{(240 \text{ lb})(256 \text{ in}^2)}{4 \text{ in}^2}$$

$$= 15{,}360 \text{ lb}$$

■

Exercises 7.3

Use the map in Figure 7.2 (page 291) to find the approximate distance between each pair of cities (find straight-line [air] distances only):

1. Champaign and Bloomington
2. Bloomington and Decatur
3. Rantoul and Kankakee
4. Rantoul and Bloomington
5. Pana and Rantoul
6. Champaign and Mattoon
7. Charleston and Pontiac
8. Paris and Bloomington
9. Lincoln and Danville
10. Flatville and Mt. Pulaski

Use the map in Illustration 1 to find the approximate air distance between each pair of cities:

11. St. Louis and Kansas City
12. Memphis and St. Louis
13. Memphis and Little Rock
14. Sedalia, MO, and Tulsa, OK
15. Fort Smith, AR, and Springfield, MO
16. Pine Bluff, AR, and Jefferson City, MO

Use the scale drawing of a metal plate cover in Illustration 2 in Exercises 17–26:

17. What is the length of the plate cover?
18. What is the width of the plate cover?
19. What is the area of the plate cover?
20. What is the diameter of the circular holes?
21. What are the dimensions of the square hole?
22. What are the dimensions of the rectangular holes?
23. What is the distance between the rectangular holes, center to center?
24. What is the distance between the centers of the upper pair of circular holes?
25. What is the distance between the centers of the right pair of circular holes?
26. Answer each of the questions in Exercises 17–25 if the scale were changed so that the side of the square = $2\frac{1}{16}$ in.

With pencil and ruler make line drawings on square-ruled paper to fit each description in Exercises 27–30:

27. A rectangle 8 ft by 6 ft. Use this scale: Side of a square = 1 ft.
28. A square 16 cm on a side. Use this scale: Side of a square = 2 cm.

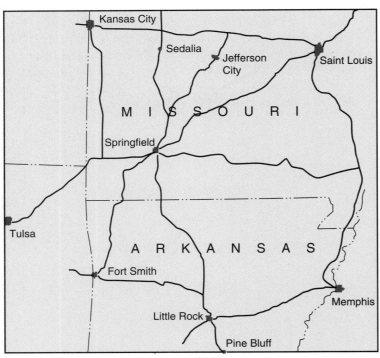

1 cm = 85 km

ILLUSTRATION 1

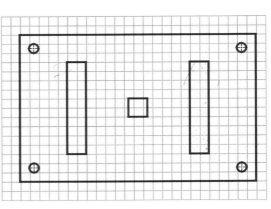

Side of square = 0.5 cm

ILLUSTRATION 2

29. A circle 36 mm in diameter. Use this scale: Side of a square = 3 mm.

30. A rectangle 12 in. by 8 in. with a circle in its center 4 in. in diameter. Use this scale: Side of a square = 2 in.

31. a. Can the actual circle in Exercise 29 be placed within the actual rectangle in Exercise 27?
    b. Can the scale drawing of the circle be placed within the scale drawing of the rectangle?

32. a. Can the actual circle in Exercise 29 be placed within the actual square in Exercise 28?
    b. Can the scale drawing of the circle be placed within the scale drawing of the square?

Use the formulas for the hydraulic press to find each value in Exercises 33–50:

33. F_l = 4000 lb and F_s = 200 lb. Find MA.

34. When a force of 160 lb is applied to the small piston of a hydraulic press, a force of 4800 lb is exerted by the large piston. Find its mechanical advantage.

35. A 400-lb force applied to the small piston of a hydraulic press produces a 3600-lb force by the large piston. Find its mechanical advantage.

36. F_l = 2400 lb and MA = $\frac{50}{1}$. Find F_s.

37. F_l = 5100 lb and MA = $\frac{75}{1}$. Find F_s.

38. A hydraulic press has a mechanical advantage of 36 : 1. If a force of 2750 lb is applied to the small piston, what force is produced by the large piston?

39. A hydraulic press with an MA of 90 : 1 has a force of 2650 lb applied to its small piston. What force is produced by its large piston?

40. A hydraulic system with an MA of 125 : 1 has a force of 2450 lb exerted by its large piston. What force is applied to its small piston?

41. r_l = 27 in. and r_s = 3 in. Find MA.

42. r_l = 36 in. and r_s = 4 in. Find MA.

43. The radii of the pistons of a hydraulic press are 3 in. and 15 in. Find its mechanical advantage.

44. The radius of the small piston of a hydraulic press is 2 in., and the radius of its large piston is 18 in. What is its mechanical advantage?

45. F_s = 25 lb, r_l = 8 in., and r_s = 2 in. Find F_l.

46. F_s = 81 lb, r_l = 9 in., and r_s = 1 in. Find F_l.

47. F_l = 6400 lb, r_l = 16 in., and r_s = 4 in. Find F_s.

48. F_l = 7500 lb, r_l = 15 in., and r_s = 3 in. Find F_s.

49. A force of 40 lb is applied to a piston of radius 7 in. of a hydraulic press. The large piston has a radius of 28 in. What force is exerted by the large piston?

50. A force of 8100 lb is exerted by a piston of radius 30 in. of a hydraulic press. What force was applied to its piston of radius 3 in.?

51. Distance and time vary directly when driving at a constant speed. Mackenzie driving at a constant speed of 60 mi/h travels 90.0 mi in $1\frac{1}{2}$ h. How far does she travel at the same speed in $3\frac{3}{4}$ h?

52. Distance and the amount of gasoline used vary directly when driving at a constant speed. Zachary driving at a constant speed of 65 mi/h travels 495 mi and uses 14.1 gal of gasoline. How much gasoline does he use driving 912 mi traveling at the same speed?

53. The number of patio blocks varies directly with the area covered. If 93 patio blocks each having an area of $\frac{3}{4}$ ft^2 covers 124 ft^2, how many of these same size patio blocks would it take to cover an area of 188 ft^2?

54. The number of people working on a project and the amount of work completed vary directly assuming all work at the same rate. If three painters can paint 12 motel rooms per day, how many painters are needed to paint 20 motel rooms per day?

7.4 Inverse Variation

If two quantities, y and x, change so that their *product is constant* (that is, $y_1x_1 = y_2x_2$), they are said to *vary inversely*. This relationship between the two quantities is called **inverse variation**. This means that if one quantity increases, the other decreases and vice versa so that their product is always the same. Compare this with direct variation, where the ratio of the two quantities is always the same.

Consider the following data:

y	8	24	12	3	6	48
x	6	2	4	16	8	1

FIGURE 7.5
Pulley system

Note that y varies inversely with x because the product is always 48, a constant. This relationship may also be written $xy = 48$ or $y = \frac{48}{x}$.

Inverse Variation

$$x_1 y_1 = x_2 y_2$$

or

$$\frac{y_1}{y_2} = \frac{x_2}{x_1}$$

One example of inverse variation is the relationship between two rotating pulleys connected by a belt (see Figure 7.5). This relationship is given by the following formula.

Pulley System Relationship

(diameter of A)(rpm of A) = (diameter of B)(rpm of B)

 Example 1

FIGURE 7.6
Gears

A small pulley is 11 in. in diameter, and a larger one is 20 in. in diameter. How many rpm does the smaller pulley make if the larger one rotates at 44 rpm (revolutions per minute)?

(diameter of A)(rpm of A) = (diameter of B)(rpm of B)

(11   )(x) = (20   )(44   )

$$x = \frac{(20)(44)}{11} = 80 \text{ rpm} \qquad ■$$

Another example of inverse variation is the relationship between the number of teeth and the number of rpm of two rotating gears, as shown in Figure 7.6.

Gear System Relationship

(no. of teeth in A)(rpm of A) = (no. of teeth in B)(rpm of B)

Example 2

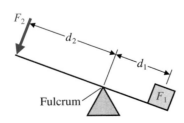

FIGURE 7.7
Lever

A large gear with 14 teeth rotates at 40 rpm. It turns a small gear with 8 teeth. How fast does the small gear rotate?

(no. of teeth in A)(rpm of A) = (no. of teeth in B)(rpm of B)

(14   )(40 rpm) = (8   )(x)

$$\frac{(14)(40 \text{ rpm})}{8} = x$$

$$70 \text{ rpm} = x \qquad ■$$

Figure 7.7 shows a lever, which is a rigid bar, pivoted to turn on a point (or edge) called a *fulcrum*. The parts of the lever on either side of the fulcrum are called *lever arms*.

To lift the box requires a force F_1. This is produced by pushing down on the other end of the lever with a force F_2. The distance from F_2 to the fulcrum is d_2. The distance from F_1 to the fulcrum is d_1. The principle of the lever is another example of inverse variation. It can be expressed by the following formula.

Lever Principle Relationship

$$F_1 d_1 = F_2 d_2$$

When the products are equal, the lever is balanced.

◆ Example 3

A man places one end of a lever under a large rock, as in Figure 7.8. He places a second rock under the lever, 2 ft from the first rock, to act as a fulcrum. He exerts a force of 180 pounds at a distance of 6 ft from the fulcrum. Find F_1, the maximum weight of rock that could be lifted.

$$F_1 d_1 = F_2 d_2$$
$$(F_1)(2 \text{ ft}) = (180 \text{ lb})(6 \text{ ft})$$

$$F_1 = \frac{(180 \text{ lb})(\overset{3}{\cancel{6}} \cancel{\text{ft}})}{\cancel{2 \text{ ft}}}$$

$$F_1 = 540 \text{ lb}$$

■

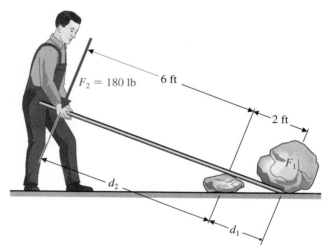

FIGURE 7.8

Exercises 7.4

Fill in the blanks:

	Pulley A		Pulley B	
	Diameter	rpm	Diameter	rpm
1.	25 cm	72	50 cm	
2.	18 cm		12 cm	96
3.	10 cm	120	15 cm	
4.		84	8 in.	48
5.	34 cm	440		680
6.	25 cm	600	48 cm	
7.		225	15 in.	465
8.	98 cm	240		360

9. A small pulley is 13 in. in diameter, and a larger one is 18 in. in diameter. How many rpm does the larger pulley make if the smaller one rotates at 720 rpm?

10. A 21-in. pulley, rotating at 65 rpm, turns a smaller pulley at 210 rpm. What is the diameter of the smaller pulley?

11. A large pulley turns at 48 rpm. A smaller pulley 8 in. in diameter turns at 300 rpm. What is the diameter of the larger pulley?

12. A pulley 32 in. in diameter turns at 825 rpm. At how many rpm will a pulley 25 in. in diameter turn?

13. A motor turning at 1870 rpm has a 4.0-in.-diameter pulley driving a fan that must turn at 680 rpm. What diameter pulley must be put on the fan?

14. A hydraulic pump is driven with an electric motor (see Illustration 1). The pump must rotate at 1200 rpm. The pump is equipped with a 6.0-in.-diameter belt pulley. The motor runs at 1800 rpm. What diameter pulley is required on the motor?

ILLUSTRATION 1

15. One pulley is 7 cm larger in diameter than a second pulley. The larger pulley turns at 80 rpm, and the smaller pulley turns at 136 rpm. What is the diameter of each pulley?

16. One pulley is twice as large in diameter as a second pulley. If the larger pulley turns at 256 rpm, what is the rpm of the smaller?

Fill in the blanks:

	Gear A Number of teeth	rpm	Gear B Number of teeth	rpm
17.	50	400		125
18.	220	?	45	440
19.	42	600	25	
20.	50	64		80
21.		$6\frac{1}{4}$	120	30
22.	80	$1\frac{1}{4}$		$3\frac{1}{3}$

23. A small gear with 25 teeth turns a large gear with 75 teeth at 32 rpm. How many rpm does the small gear make?

24. A large gear with 180 teeth running at 600 rpm turns a small gear at 900 rpm. How many teeth does the small gear have?

25. A large gear with 60 teeth turning at 72 rpm turns a small gear with 30 teeth. At how many rpm does the small gear turn?

26. A large gear with 80 teeth turning at 150 rpm turns a small gear with 12 teeth. At how many rpm does the small gear turn?

27. A large gear with 120 teeth turning at 30 rpm turns a small gear at 90 rpm. How many teeth does the small gear have?

28. A large gear with 200 teeth turning at 17 rpm turns a small gear at 100 rpm. How many teeth does the small gear have?

Complete the table:

	F_1	d_1	F_2	d_2
29.	18 lb	5 in.	9 lb	
30.	30 lb		70 lb	8 in.
31.	40 lb	9 in.		3 in.
32.		.6.3 ft	458.2 lb	8.7 ft

In Exercises 33–37, draw a sketch for each and solve:

33. An object is 6 ft from the fulcrum and balances a second object 8 ft from the fulcrum. The first object weighs 180 lb. How much does the second object weigh?

34. A block of steel weighing 1800 lb is to be raised by a lever extending under the block 9 in. from the fulcrum. How far from the fulcrum must a 150-lb man apply his weight to the bar to balance the steel?

35. A rocker arm raises oil from an oil well. On each stroke, it lifts a weight of 1 ton on a weight arm 4 ft long. What force is needed to lift the oil on a force arm 8 ft long?

36. A carpenter needs to raise one side of a building with a lever 3.65 m in length. The lever, with one end under the building, is placed on a fulcrum 0.45 m from the building. A mass of 90 kg pulls down on the other end. What mass is being lifted when the building begins to rise?

37. A 1200-g mass is placed 72 cm from the fulcrum of a lever. How far from the fulcrum is a 1350-g mass that balances it?

38. A lever is in balance when a mass of 2000 g is placed 28 cm from a fulcrum. An unknown mass is placed 20 cm from the fulcrum on the other side. What is the amount of the unknown mass?

39. A 210-lb object is placed on a lever. It balances a 190-lb weight that is 28 in. from the fulcrum. How far from the fulcrum should the 210-lb weight be placed?

40. A piece of machinery weighs 3 tons. It is to be balanced by two men whose combined weight is 330 lb. The piece of machinery is placed 11 in. from the fulcrum. How far from the fulcrum must the two men exert their weight in order to balance it?

41. The length and width vary inversely in a rectangle of a given area. A given rectangle has sides of length 18 cm and width 12 cm. Find the length of a rectangle of the same area whose width is 8 cm.

42. The speed and time of travel vary inversely when driving a given distance. Juan drives 45 mi/h for 6 h to visit his grandmother on a snowy day. He drives at 55 mi/h for his return trip home. How long did the return trip take?

43. The current and resistance vary inversely in a circuit with a given voltage. In a 240-volt circuit, the current is 8 amperes and the resistance is 30 ohms. Find the current if the resistance is 80 ohms in a circuit with the same voltage.

44. The number of people working on a project and the time it takes to complete a project vary inversely, assuming everyone works at the same rate. If it takes three painters 24 days to paint all of the rooms in a motel, how many painters are needed to paint all of the rooms in 8 working days?

Chapter 7 Group Activities

1. When people build a house or a car or make a floor plan, they often start with a model. If you look at the side of a box of a model car or plane, you find it has a scale such as $\frac{1}{24}$ or $\frac{1}{48}$. For this activity, your group is to scale something. You may choose a room, a book, a desk, etc. Measure the object you choose and scale the measurements using a ratio. For example, suppose a car has a length of 180 in. We want to use the ratio of 6 to 1. That is, 6 in. of the large car is to be 1 in. in the model. So we take $\frac{180}{6} = 30$ in. for every 1 in. You can see that $\frac{180}{30}$ reduces to a ratio of 6 to 1. Either draw the dimensions of the object you choose to scale or make an actual scale model.

2. Estimate the height of a building or some other tall object on a sunny day as follows: First measure the height and the length of the shadow of a classmate. Measure the length of the shadow of the object. Find the height of the building by using similar triangles.

Chapter 7 Summary

Glossary of Basic Terms

Direct variation. When two quantities change so that their ratios are constant. When one quantity increases, the other quantity increases (or when one quantity decreases, the other quantity decreases) so that their ratio is always the same. (p. 290)

Extremes. The first and fourth terms of a proportion. (p. 284)

Inverse variation. When two quantities change so that their products are constant. When one quantity increases, the other quantity decreases (or when one quantity decreases, the other quantity increases) so that their product is always the same. (p. 295)

Means. The second and third terms of a proportion. (p. 284)

Proportion. An equation with two equal ratios. (p. 284)

Ratio of two numbers. The first number divided by the second number. (p. 280)

7.2   Proportion

1. **In any proportion, the product of the means equals the product of the extremes. That is, if**

$$\frac{a}{b} = \frac{c}{d}, \text{ then } bc = ad. \text{ (p. 284)}$$

7.3   Direct Variation

1. **Direct variation:** When two quantities vary directly, their ratios are constant in the form $\dfrac{y_1}{x_1} = \dfrac{y_2}{x_2}$. (p. 290)

7.4   Inverse Variation

1. **Inverse variation:** When two quantities vary inversely, their ratios are constant in the form

$$\frac{y_1}{y_2} = \frac{x_2}{x_1} \text{ or } x_1 y_1 = x_2 y_2. \text{ (p. 296)}$$

◆ Chapter 7   Review

Write each ratio in lowest terms:

1. 7 to 28

2. 60 : 40

3. 1 g to 500 mg

4. $\dfrac{5 \text{ ft 6 in.}}{9 \text{ ft}}$

Determine whether or not each pair of ratios is equal:

5. $\dfrac{7}{2}, \dfrac{35}{10}$

6. $\dfrac{5}{18}, \dfrac{30}{115}$

Solve each proportion (round each to three significant digits when necessary):

7. $\dfrac{x}{4} = \dfrac{5}{20}$

8. $\dfrac{10}{25} = \dfrac{x}{75}$

9. $\dfrac{3}{x} = \dfrac{8}{64}$

10. $\dfrac{72}{96} = \dfrac{30}{x}$

11. $\dfrac{73.4}{x} = \dfrac{25.9}{37.4}$

12. $\dfrac{x}{19.7} = \dfrac{144}{68.7}$

13. $\dfrac{61.1}{81.3} = \dfrac{592}{x}$

14. $\dfrac{243}{58.3} = \dfrac{x}{127}$

15. A piece of cable 180 ft long costs $67.50. How much will 500 ft cost at the same unit price?

16. A copper wire 750 ft long has a resistance of 1.89 Ω. How long is a copper wire of the same size whose resistance is 3.15 Ω?

17. A crew of electricians can wire 6 houses in 144 h. How many hours will it take them to wire 9 houses?

18. An automobile braking system has a 12-to-1 lever advantage on the master cylinder. A 25-lb force is applied to the pedal. What force is applied to the master cylinder?

19. Jones invests $6380 and Hernandez invests $4620 in a partnership business. What percent of the total investment does each have?

20. One gallon of a pesticide mixture weighs 7 lb 13 oz. It contains 11 oz of pesticide. What percent of the mixture is pesticide?

21. Indicate what kind of variation is shown by each equation.

     (a) $\dfrac{y_1}{x_1} = \dfrac{y_2}{x_2}$

     (b) $y_1 x_1 = y_2 x_2$

22. What kind of variation is indicated when one quantity increases while the other increases?

23. Suppose $\frac{1}{4}$ in. on a map represents 25 mi. What distance is represented by $3\frac{5}{8}$ in.?

24. The scale on a map is 1 in. = 600 ft. Two places are known to be 2 mi apart. What distance will show between them on the map?

25. Two pulleys are connected by a belt. The numbers of rpm of the two pulleys vary inversely as their diameters. A pulley having a diameter of 25 cm is turning at 900 rpm. What is the number of rpm of the second pulley, which has a diameter of 40 cm?

26. A large gear with 42 teeth rotates at 25 rpm. It turns a small gear with 14 teeth. How fast does the small gear rotate?

27. In hydraulics, the formula relating the forces and the radii of the pistons is

$$\frac{F_l}{F_s} = \frac{r_l^2}{r_s^2}$$

Given $F_l = 6050$ kg, $r_l = 22$ cm, and $r_s = 2$ cm, find F_s.

28. An object 9 ft from the fulcrum of a lever balances a second object 12 ft from the fulcrum. The first object weighs 240 lb. How much does the second object weigh?

29. The current I varies directly as the voltage E. Suppose $I = 0.6$ A when $E = 30$ V. Find the value of I when $E = 100$ V.

30. The number of workers needed to complete a particular job is inversely proportional to the number of hours that they work. If 12 electricians can complete a job in 72 h, how long will it take 8 electricians to complete the same job? Assume that each person works at the same rate, no matter how many people are assigned to the job.

 ## Chapter 7   Test

Write each ratio in lowest terms:

1. 16 m to 64 m   **2.** 3 ft to 6 in.   **3.** 400 mL to 5 L

Solve each proportion:

4. $\dfrac{x}{8} = \dfrac{18}{48}$ **5.** $\dfrac{8}{x} = \dfrac{24}{5}$ **6.** $\dfrac{7200}{84} = \dfrac{x}{252}$

7. If 60 ft of fencing costs \$115, how much does 80 ft cost?

8. Five quarts of pure antifreeze are added to 10 quarts of water to fill a radiator. What percent of antifreeze is in the mixture?

9. The scale on a map is 1 cm = 10 km. If two cities are 4.8 cm apart on the map, what is the actual distance between them?

10. A used car sells for \$7500. The down payment is \$900. What percent of the selling price is the down payment?

11. A small gear with 36 teeth turns a large gear with 48 teeth at 150 rpm. What is the speed (in rpm) of the small gear?

12. A pulley is 20 cm in diameter and rotating at 150 rpm. Find the diameter of a smaller pulley that must rotate at 200 rpm.

13. Given the lever formula, $F_1 d_1 = F_2 d_2$, and $F_1 = 800$ kg, $d_1 = 9$ m, $d_2 = 3.6$ m, find F_2.

14. A man who weighs 200 lb is to be raised by a lever extending under the man 15 in. from the fulcrum. How much force must be applied at a distance of 24 in. from the fulcrum in order to lift him?

8 Graphing Linear Equations

Mathematics at Work

Drafters prepare working plans and detailed technical drawings used by construction and production workers to build a wide variety of products ranging from manufactured products to industrial machinery to buildings to oil and gas pipelines. Their drawings provide visual and technical details of the products and structures as well as specifying dimensions, materials to be used, and procedures and processes to be followed. Drafters also provide rough sketches, specifications, codes, and any calculations provided by engineers, architects, scientists, or surveyors. Most use computer-assisted drafting (CAD) equipment and software to prepare drawings. These systems use computer workstations to create a drawing on a video screen and store the drawings electronically so that revisions, copies, or variations can be made easily and quickly. While CAD is a useful tool, drafters need the basic drafting skills and standards as well as the CAD skills and knowledge.

Drafter
Draftsperson using CAD to design industrial components.

Drafting work has many specializations due to special design and applications. Architectural drafters draw structures and buildings. Aeronautical drafters prepare engineering drawings detailing plans and specifications used for manufacturing aircraft, missiles, and related parts. Electrical drafters prepare wiring and layout diagrams used by workers to erect, install, and repair electrical equipment and wiring in a wide variety of settings. Civil drafters prepare drawings and topographical and relief maps used in construction projects such as highways, bridges, pipelines, and water and sewage systems. Mechanical drafters prepare detail and assembly drawings of a wide variety of machinery and mechanical devices showing dimensions, fastening methods, and other requirements.

Many community colleges and postsecondary and trade schools offer associate degree and certificate programs in which students develop drafting and mechanical skills; a basic knowledge of drafting standards, mathematics, science, and engineering technology; computer-aided drafting and design techniques; and communication and problem-solving skills. For more information, go to the website listed below.

www.cengage.com/mathematics/ewen

Objectives

■ Find ordered pairs of numbers that are solutions to a linear equation with two variables.

■ Plot points in the number plane.

■ Graph a linear equation by plotting points.

■ Find the slope of a line.

■ Determine when two lines are parallel, perpendicular, or neither by finding the slope of the lines.

■ Graph a linear equation given its slope and y intercept and through a given point with a given slope.

■ Find the equation of a line with a given slope and y intercept, with a given slope through a given point, and through two given points.

8.1   Linear Equations with Two Variables

In Chapter 6, we studied linear equations with one variable, such as $2x + 4 = 10$ and $3x - 7 = 5$. We found that most linear equations in one variable have only one root. In this chapter, we study equations with *two* variables, such as

$$3x + 4y = 12 \quad \text{or} \quad x + y = 7$$

How many solutions does the equation $x + y = 7$ have? Any two numbers whose sum is 7 is a solution—for example, 1 for x and 6 for y, 2 for x and 5 for y, -2 for x and 9 for y, $5\frac{1}{2}$ for x and $1\frac{1}{2}$ for y, and so on. Most linear equations with *two variables* have *many possible solutions.*

Since it is very time consuming to write pairs of replacements in this manner, we use **ordered pairs** in the form (x, y) to write solutions of equations with two variables. Therefore, instead of writing the solutions of the equation $x + y = 7$ as above, we write them as $(1, 6), (2, 5), (-2, 9), \left(5\frac{1}{2}, 1\frac{1}{2}\right)$, and so on.

Linear Equation with Two Variables

A **linear equation with two variables** can be written in the form

$$ax + by = c$$

where the numbers a, b, and c are all real numbers such that a and b are not both 0.

    Example 1

Determine whether the given ordered pair is a solution of the given equation.

a. $(5, 2)$; $3x + 4y = 23$

To determine whether the ordered pair $(5, 2)$ is a solution to $3x + 4y = 23$, substitute 5 for x and 2 for y as follows:

$$3x + 4y = 23$$
$$3(5) + 4(2) = 23 \qquad \text{Substitute } x = 5 \text{ and } y = 2.$$
$$15 + 8 = 23$$
$$23 = 23 \qquad \text{True}$$

The result is true, so (5, 2) is a solution of $3x + 4y = 23$.

b. $(-5, 6)$; $2x - 4y = -32$

$$2(-5) - 4(6) = -32 \qquad \text{Substitute } x = -5 \text{ and } y = 6.$$
$$-10 - 24 = -32$$
$$-34 = -32 \qquad \text{False}$$

The result is false, so $(-5, 6)$ is not a solution of $2x - 4y = -32$. ■

To find solutions of a linear equation with two variables, replace one variable with a number you have chosen and then solve the resulting linear equation for the remaining variable.

◆ Example 2

Complete the three ordered-pair solutions of $2x + y = 5$.

a. $(4, \quad)$

Replace x with 4. Any number could be used, but for this example, we will use 4. The resulting equation is

$$2(4) + y = 5$$
$$8 + y = 5$$
$$8 + y - 8 = 5 - 8 \qquad \text{Subtract 8 from both sides.}$$
$$y = -3$$

Check: Replace x with 4 and y with -3.

$$2(4) + (-3) = 5 \qquad ?$$
$$8 - 3 = 5 \qquad \text{True}$$

Therefore, $(4, -3)$ is a solution.

b. $(-2, \quad)$

Replace x with -2. The resulting equation is

$$2(-2) + y = 5$$
$$-4 + y = 5$$
$$-4 + y + 4 = 5 + 4 \qquad \text{Add 4 to both sides.}$$
$$y = 9$$

Check: Replace x with -2 and y with 9.

$$2(-2) + 9 = 5 \qquad ?$$
$$-4 + 9 = 5 \qquad \text{True}$$

Thus, $(-2, 9)$ is a solution.

c. $(0, \quad)$

Replace x with 0. The resulting equation is

$$2(0) + y = 5$$
$$0 + y = 5$$
$$y = 5$$

Check: Replace x with 0 and y with 5.

$$2(0) + (5) = 5 \qquad ?$$
$$0 + 5 = 5 \qquad \text{True}$$

Therefore, $(0, 5)$ is a solution. ■

You may find it easier first to solve the equation for y and then make each replacement for x.

Example 3   Complete the three ordered-pair solutions of $3x - y = 4$ by first solving the equation for y.

    a. $(5, \quad)$ **b.** $(-2, \quad)$ **c.** $(0, \quad)$

$$3x - y = 4$$
$$3x - y - 3x = 4 - 3x \qquad \text{Subtract } 3x \text{ from both sides.}$$
$$-y = 4 - 3x$$
$$y = -4 + 3x \qquad \text{Divide both sides by } -1.$$

You may make a table to keep your work in order.

	x	$3x - 4$	y	Ordered pairs
a.	5	$3(5) - 4 = 15 - 4$	11	$(5, 11)$
b.	-2	$3(-2) - 4 = -6 - 4$	-10	$(-2, -10)$
c.	0	$3(0) - 4 = 0 - 4$	-4	$(0, -4)$

Therefore, the three solutions are $(5, 11)$, $(-2, -10)$, and $(0, -4)$. Choosing a positive, a negative, and 0 value for x is often a good approach. ■

Example 4   Complete the three ordered-pair solutions of $5x + 3y = 7$.

    a. $(2, \quad)$ **b.** $(0, \quad)$ **c.** $(-1, \quad)$

We will first solve for y.

$$5x + 3y = 7$$
$$5x + 3y - 5x = 7 - 5x \qquad \text{Subtract } 5x \text{ from both sides.}$$
$$3y = 7 - 5x$$
$$\frac{3y}{3} = \frac{7 - 5x}{3} \qquad \text{Divide both sides by 3.}$$
$$y = \frac{7 - 5x}{3}$$

	x	$\dfrac{7 - 5x}{3}$	y	Ordered pairs
a.	2	$\dfrac{7 - 5(2)}{3} = \dfrac{-3}{3}$	-1	$(2, -1)$
b.	0	$\dfrac{7 - 5(0)}{3} = \dfrac{7}{3}$	$\dfrac{7}{3}$	$\left(0, \dfrac{7}{3}\right)$
c.	-1	$\dfrac{7 - 5(-1)}{3} = \dfrac{12}{3}$	4	$(-1, 4)$

The three solutions are $(2, -1)$, $\left(0, \dfrac{7}{3}\right)$, and $(-1, 4)$. ■

    Solutions to linear equations with two variables may be shown visually by graphing them in a number plane. To construct a number plane, draw a horizontal number line, which is called the **x axis,** as in Figure 8.1. Then draw a second number line intersecting the first line at right angles so that both number lines have the same zero point, called the **origin.** The vertical number line is called the **y axis.**

Each number line, or axis, has a scale. The numbers on the *x* axis are *positive to the right* of the origin and *negative to the left* of the origin. Similarly, the numbers on the *y* axis are *positive above* the origin and *negative below* the origin.

All the points in the plane determined by these two intersecting axes make up the **number plane.** The axes divide the number plane into four regions, called **quadrants.** The quadrants are numbered as shown in Figure 8.1.

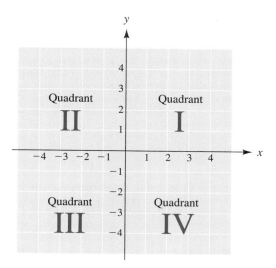

FIGURE 8.1
Rectangular coordinate system

Points in the number plane are usually indicated by an **ordered pair** of numbers written in the form (x, y), where x is the first number in the ordered pair and y is the second number in the ordered pair. The numbers x and y are also called the **coordinates** of a point in the number plane. Figure 8.1 is often called the **rectangular coordinate system** or the *Cartesian coordinate system.*

Plotting Points in the Number Plane

To locate the point in the number plane which corresponds to an ordered pair (x, y):

Step 1   Count right or left, from 0 (the origin) along the *x* axis, the number of spaces corresponding to the first number of the ordered pair (right if positive, left if negative).

Step 2   Count up or down, from the point reached on the *x* axis in Step 1, the number of spaces corresponding to the second number of the ordered pair (up if positive, down if negative).

Step 3   Mark the last point reached with a dot.

Example 5   Plot the point corresponding to the ordered pair $(3, -4)$ in Figure 8.2.

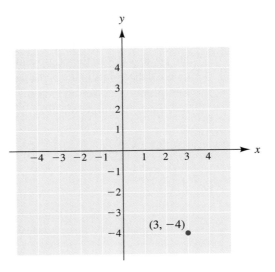

FIGURE 8.2

First, count three spaces to the right along the x axis. Then count down four spaces from that point. Mark the final point with a dot.   ■

 Example 6   Plot the points corresponding to the ordered pairs in the number plane in Figure 8.3: $A(1, 2)$, $B(3, -2)$, $C(-4, 7)$, $D(5, 0)$, $E(-2, -3)$, $F(-5, -1)$, $G(-2, 4)$.

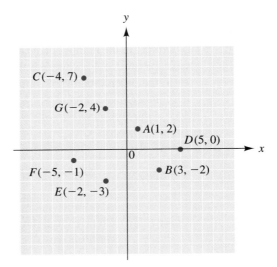

FIGURE 8.3   ■

Exercises 8.1 _____

Complete the three ordered-pair solutions of each equation:

Equation	Ordered Pairs		
1. $x + y = 5$	(3,)	(8,)	(−2,)
2. $-2x + y = 8$	(2,)	(7,)	(−4,)
3. $6x + 2y = 10$	(2,)	(0,)	(−2,)
4. $6x - y = 0$	(3,)	(5,)	(−2,)
5. $3x - 4y = 8$	(0,)	(2,)	(−4,)
6. $5x - 3y = 8$	(1,)	(0,)	(−2,)
7. $-2x + 5y = 10$	(5,)	(0,)	(−3,)
8. $-4x - 7y = -3$	(−1,)	(0,)	(−8,)
9. $9x - 2y = 10$	(2,)	(0,)	(−4,)
10. $2x + 3y = 6$	(3,)	(0,)	(−6,)
11. $y = 3x + 4$	(2,)	(0,)	(−3,)
12. $y = 4x - 8$	(3,)	(0,)	(−4,)
13. $5x + y = 7$	(2,)	(0,)	(−4,)
14. $4x - y = 8$	(1,)	(0,)	(−3,)
15. $2x = y - 4$	(3,)	(0,)	(−1,)
16. $3y - x = 5$	(1,)	(0,)	(−4,)
17. $5x - 2y = -8$	(4,)	(0,)	(−2,)
18. $2x - 3y = 1$	(2,)	(0,)	(−4,)
19. $9x - 2y = 5$	(1,)	(0,)	(−3,)
20. $2x + 7y = -12$	(1,)	(0,)	(−8,)
21. $y = 3$	(2,)	(0,)	(−4,)

(Think: $0x + 1y = 3$)

22. $y + 4 = 0$	(3,)	(0,)	(−7,)
23. $x = 5$	(, 4)	(, 0)	(, −2)

(Think: $1x + 0y = 5$)

24. $x + 7 = 0$	(, 5)	(, 0)	(, −6)

Solve for y in terms of x:

25. $2x + 3y = 6$ **26.** $4x + 5y = 10$

27. $x + 2y = 7$ **28.** $2x + 2y = 5$

29. $x - 2y = 6$ **30.** $x - 3y = 9$

31. $2x - 3y = 9$ **32.** $4x - 5y = 10$

33. $-2x + 3y = 6$ **34.** $-3x + 5y = 25$

35. $-2x - 3y = -15$ **36.** $-3x - 4y = -8$

Write the ordered pair corresponding to each point in Illustration 1:

37. A **38.** B **39.** C **40.** D **41.** E

42. F **43.** G **44.** H **45.** I **46.** J

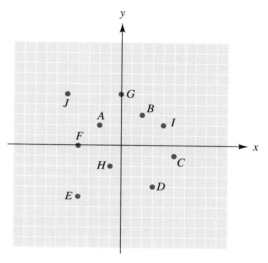

ILLUSTRATION 1

Plot each point in the number plane. Label each point by writing its ordered pair and letter:

47. $A(1, 3)$ **48.** $B(4, 0)$ **49.** $C(-6, -2)$

50. $D(-2, 4)$ **51.** $E(5, -4)$ **52.** $F(-4, -4)$

53. $G(0, 9)$ **54.** $H(3, 7)$ **55.** $I(5, -5)$

56. $J(-5, 5)$ **57.** $K(-6, -3)$ **58.** $L(3, -7)$

59. $M(-4, 5)$ **60.** $N(-2, -6)$ **61.** $O(1, -3)$

62. $P(5, 2)$ **63.** $Q\left(3, \dfrac{1}{2}\right)$ **64.** $R\left(4\dfrac{1}{2}, -3\dfrac{1}{2}\right)$

65. $S\left(-6, 2\dfrac{1}{2}\right)$ **66.** $T\left(-4\dfrac{1}{2}, 6\dfrac{1}{2}\right)$

8.2 Graphing Linear Equations

In Section 8.1, you learned that a linear equation with two variables has many solutions. In Example 2, you found that three of the solutions of $2x + y = 5$ were $(4, -3)$, $(-2, 9)$, and $(0, 5)$. Now plot the points corresponding to these ordered pairs and connect the points, as shown in Figure 8.4.

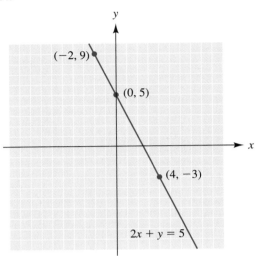

FIGURE 8.4

You can see from the figure that the three points lie on the same straight line. If you find another solution of $2x + y = 5$—say, $(1, 3)$—the point corresponding to this ordered pair also lies on the same straight line. The solutions of a linear equation with two variables always correspond to points lying on a straight line. Therefore, *the graph of the solutions of a linear equation with two variables is always a straight line*. Only part of this line can be shown on the graph; the line actually extends without limit in both directions.

Graphing Linear Equations

To draw the graph of a linear equation with two variables:

Step 1   Find any three solutions of the equation.

(*Note:* Two solutions would be enough, since two points determine a straight line. However, a third solution gives a third point as a check. If the three points do not lie on the same straight line, you have made an error.)

Step 2   Plot the three points corresponding to the three ordered pairs that you found in Step 1.

Step 3   Draw a line through the three points. If the line is not straight, check your solutions.

We will show two methods for finding the three solutions of the equation. The first method involves solving the equation for y and then substituting three different values of x to find each corresponding y value. The second method involves substituting three different values of x and then solving each resulting equation for y. You may use either method. Depending on the equation, one method may be easier to use than the other.

◆ Example 1

Draw the graph of $3x + 4y = 12$.

Step 1 Find any three solutions of $3x + 4y = 12$. First, solve for y:

$$3x + 4y - 3x = 12 - 3x \qquad \text{Subtract } 3x \text{ from both sides.}$$
$$4y = 12 - 3x$$
$$\frac{4y}{4} = \frac{12 - 3x}{4} \qquad \text{Divide both sides by 4.}$$
$$y = \frac{12 - 3x}{4}$$

Choose any three values of x and solve for y. Here we have chosen $x = 4$, $x = 0$, and $x = -2$.

Note: We often choose a positive number, 0, and a negative number for x to obtain a range of points in the graph. Although finding and plotting any two points will allow you to graph the straight line, the third point provides an excellent check.

x	$\dfrac{12 - 3x}{4}$	y	Ordered pairs
4	$\dfrac{12 - 3(4)}{4} = \dfrac{0}{4}$	0	$(4, 0)$
0	$\dfrac{12 - 3(0)}{4} = \dfrac{12}{4}$	3	$(0, 3)$
-2	$\dfrac{12 - 3(-2)}{4} = \dfrac{18}{4}$	$\dfrac{9}{2}$	$\left(-2, \dfrac{9}{2}\right)$

Three solutions are $(4, 0)$, $(0, 3)$, and $\left(-2, \dfrac{9}{2}\right)$.

Step 2 Plot the points corresponding to $(4, 0)$, $(0, 3)$, and $\left(-2, \dfrac{9}{2}\right)$.

Step 3 Draw a straight line through these three points. (See Figure 8.5.)

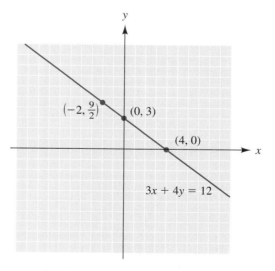

FIGURE 8.5

An alternative method is shown in Example 2. ■

Example 2   Draw the graph of $2x - 3y = 6$.

Step 1   Set up a table and write the values you choose for x—say, 3, 0, and -3.

	a.	b.	c.
x	3	0	-3
y			

Step 2   Substitute the chosen values of x in the given equation and solve for y.

a.
$$2x - 3y = 6$$
$$2(3) - 3y = 6$$
$$6 - 3y = 6$$
$$-3y = 0$$
$$y = 0$$

b.
$$2x - 3y = 6$$
$$2(0) - 3y = 6$$
$$0 - 3y = 6$$
$$-3y = 6$$
$$y = -2$$

c.
$$2x - 3y = 6$$
$$2(-3) - 3y = 6$$
$$-6 - 3y = 6$$
$$-3y = 12$$
$$y = -4$$

Step 3   Write the values for y that correspond to the chosen values for x in the table, thus:

	a.	b.	c.
x	3	0	-3
y	0	-2	-4

That is, three solutions of $2x - 3y = 6$ are the ordered pairs $(3, 0)$, $(0, -2)$, and $(-3, -4)$.

Step 4   Plot the points from Step 3 and draw a straight line through them, as in Figure 8.6.

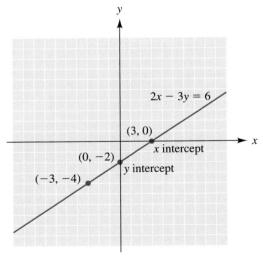

FIGURE 8.6

The *x* Intercept and the *y* Intercept of a Line

The line in Figure 8.6 crosses the x axis at the point $(3, 0)$. The number 3 is called the
***x* intercept**—the x coordinate of the point where the graph crosses the x axis.

To find the x intercept of a line, replace y in the equation by 0 and solve for x.

The line in Figure 8.6 crosses the y axis at the point $(0, -2)$. The number -2 is called the **y intercept**—the y coordinate of the point where the graph crosses the y axis.

To find the y intercept of a line, replace x in the equation by 0 and solve for y.

 Example 3

Find the x and y intercepts of the graph of $3x - 5y = 30$ and then graph the equation.

To find the x intercept, replace y in the equation by 0 and solve for x as follows:

$$3x - 5y = 30$$
$$3x - 5(0) = 30$$
$$3x = 30 \qquad \text{Divide both sides by 3.}$$
$$x = 10$$

So, the x intercept is $(10, 0)$.

To find the y intercept, replace x in the equation by 0 and solve for y as follows:

$$3x - 5y = 30$$
$$3(0) - 5y = 30$$
$$-5y = 30 \qquad \text{Divide both sides by } -5.$$
$$y = -6$$

So, the y intercept is $(0, -6)$.

Let's find a third point by letting $x = 5$ and solve for y as follows:

$$3x - 5y = 30$$
$$3(5) - 5y = 30$$
$$15 - 5y = 30$$
$$-5y = 15$$
$$y = -3$$

The third point is $(5, -3)$.

Plot these three points and draw a line through them. (See Figure 8.7.)

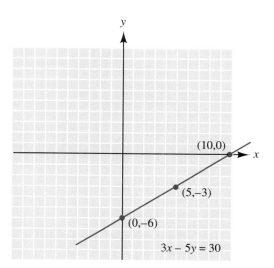

FIGURE 8.7

Finding the x and y intercepts is an excellent method for graphing a linear equation or for checking the graph of a linear equation.

Two special cases of linear equations have a graph of a horizontal line or a vertical line. The equation $y = 5$ is a linear equation with an x coefficient of 0. (This equation may also be written as $0x + 1y = 5$.) Similarly, $x = -7$ is a linear equation with a y coefficient of 0. (This equation may also be written as $1x + 0y = -7$.) These equations have graphs that are horizontal or vertical straight lines, as shown in the next two examples.

 Example 4   Draw the graph of $y = 5$.

Set up a table and write the values you choose for x—say, 3, 0, and -4. As the equation states, y is always 5 for any value of x that you choose.

x	3	0	-4
y	5	5	5

Plot the points from the table: $(3, 5)$, $(0, 5)$, and $(-4, 5)$. Then draw a straight line through them, as in Figure 8.8.

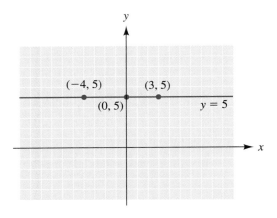

FIGURE 8.8   ■

Horizontal Line

The graph of the linear equation $y = k$, where k is a constant, is the horizontal line through the point $(0, k)$. That is, $y = k$ is a horizontal line with a y intercept of k.

Example 5   Draw the graph of $x = -7$.

All ordered pairs that are solutions of $x = -7$ have an x value of -7. You can choose any number for y. Three ordered pairs that satisfy $x = -7$ are $(-7, 3)$, $(-7, 1)$, and $(-7, -4)$. Plot these three points and draw a straight line through them, as in Figure 8.9.

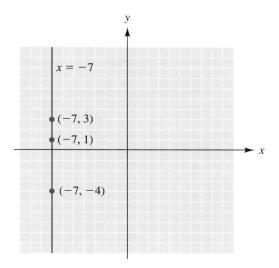

FIGURE 8.9 ■

Vertical Line

The graph of the linear equation $x = k$, where k is a constant, is the vertical line through the point $(k, 0)$. That is, $x = k$ is a vertical line with an x intercept of k.

Solve the equation $2x - y = 8$ for y. The solution is $y = 2x - 8$. To graph this equation, assign values for x and find the corresponding y values. We call x the independent variable, because we may choose any value for x that we wish. The **independent variable** is the first element of an ordered pair, usually x. Since the value of y depends on the value of x, we call y the dependent variable. The **dependent variable** is the second element of an ordered pair, usually y.

In many technical classes, variables other than x and y are often used. These other variables are usually related to formulas. Recall that a formula can be solved for one variable in terms of another. For example, Ohm's law can be expressed as $V = IR$, or as $V = 10I$ when $R = 10$. For the equation $V = 10I$, I is called the *independent variable* and V is called the *dependent variable*. The dependent variable is the variable for which the formula is solved.

When graphing an equation, the horizontal axis corresponds to the independent variable; the vertical axis corresponds to the dependent variable. Think of graphing the ordered pairs

(independent variable, dependent variable)

 Example 6

Draw the graph of $V = 10I$.

Since I is the independent variable, graph ordered pairs in the form (I, V). Set up a table and write the values you choose for I—say, 0, 5, and 8. For this example, limit your values of I to nonnegative numbers.

I	0	5	8
V	0	50	80

Next, choose a suitable scale for the vertical axis and graph the ordered pairs (0, 0), (5, 50), and (8, 80), as shown in Figure 8.10.

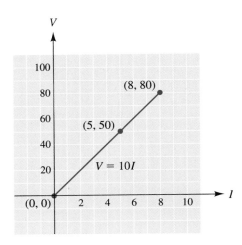

FIGURE 8.10   ■

Exercises 8.2

Draw the graph of each equation:

1. $x + y = 7$

2. $x + 3y = 9$

3. $y = 2x + 3$

4. $y = 4x - 5$

5. $4y = x$

6. $2x + y = 6$

7. $6x - 2y = 10$

8. $2x + 3y = 9$

9. $3x - 4y - 12$

10. $3x - 5y = 15$

11. $5x + 4y = 20$

12. $2x - 3y = 18$

13. $2x + 7y = 14$

14. $2x - 5y = 20$

15. $y = 2x$

16. $y = -3x$

17. $3x + 5y = 11$

18. $4x - 3y = 15$

19. $y = -\dfrac{1}{2}x + 4$

20. $y = \dfrac{2}{3}x - 6$

21. $y = 3$

22. $y = -2$

23. $x = -4$

24. $x = 5$

25. $y - 6 = 0$

26. $y + 10 = 0$

27. $x + 3\dfrac{1}{2} = 0$

28. $x - 4 = 0$

29. $y = 0$

30. $x = 0$

Identify the independent and dependent variables for each equation:

31. $s = 4t + 7$

32. $V = 5t - 2$

33. $R = 0.5V$

34. $s = 65t$

35. $i = 30t - 10$

36. $E = 4V + 2$

37. $v = 50 - 6t$

38. $i = 18 - 3t$

39. $s = 3t^2 + 5t - 1$

40. $v = 2i^2 - 3i + 10$

41. The distance, s (in feet), that a body travels in t seconds is given by the equation $s = 5t + 10$. Graph the equation for nonnegative values of t.

42. The voltage, v (in mV), in an electrical circuit varies according to the equation $v = 10t - 5$, where t is in seconds. Graph the equation for nonnegative values of v.

43. The resistance, R, in an electrical circuit varies according to the equation $R = 1.5V$, where V is in volts. Graph the equation for nonnegative values of V.

44. The current, I (in amps), in an electrical circuit varies according to the equation $I = 0.05V$, where V is in volts. Choose a suitable scale and graph the equation for nonnegative values of V.

45. The voltage, v, in an electrical circuit is given by the equation $v = 60 - 5t$, where t is in μs. Graph the equation for nonnegative values of v and t.

46. The distance, s (in metres), that a point travels in t milliseconds is given by the equation $s = 24 - 2t$. Graph the equation for nonnegative values of s and t.

8.3 The Slope of a Line

The slope of a line or the "steepness" of a roof (see Figure 8.11) can be measured by the following ratio:

FIGURE 8.11
Slope

$$\text{slope} = \frac{\text{vertical change}}{\text{horizontal change}} = \frac{\text{rise}}{\text{run}}$$

A straight line can also be graphed by using its slope and knowing one point on the line.

If two points on a line (x_1, y_1) (read "x-sub-one, y-sub-one") and (x_2, y_2) (read "x-sub-two, y-sub-two") are known (see Figure 8.12), the slope of the line is defined as follows.

Slope of a Line

$$\text{slope} = m = \frac{\text{vertical change}}{\text{horizontal change}} = \frac{\text{rise}}{\text{run}} = \frac{\text{difference in } y \text{ values}}{\text{difference in } x \text{ values}} = \frac{y_2 - y_1}{x_2 - x_1}$$

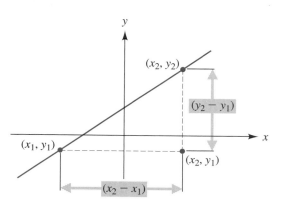

FIGURE 8.12
Slope of a line through two points

Example 1   Find the slope of the line passing through the points $(-2, 3)$ and $(4, 7)$. (See Figure 8.13.)

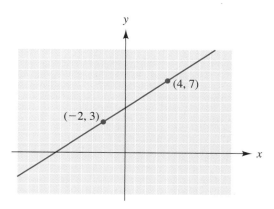

FIGURE 8.13

If we let $x_1 = -2$, $y_1 = 3$, $x_2 = 4$, and $y_2 = 7$, then

$$m = \frac{y_2 - y_1}{x_2 - x_1} = \frac{7 - 3}{4 - (-2)} = \frac{4}{6} = \frac{2}{3}$$

Note that if we reverse the order of taking the differences of the coordinates, the result is the same:

$$m = \frac{y_1 - y_2}{x_1 - x_2} = \frac{3 - 7}{-2 - 4} = \frac{-4}{-6} = \frac{2}{3}$$ ■

 Example 2   Find the slope of the line passing through $(-3, 2)$ and $(3, -6)$. (See Figure 8.14.)

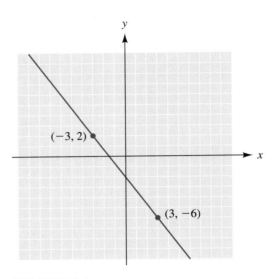

FIGURE 8.14

If we let $x_1 = -3$, $y_1 = 2$, $x_2 = 3$, and $y_2 = -6$, then

$$m = \frac{y_2 - y_1}{x_2 - x_1} = \frac{-6 - 2}{3 - (-3)} = \frac{-8}{6} = -\frac{4}{3}$$ ■

 Example 3   Find the slope of the line through $(-5, 2)$ and $(3, 2)$. (See Figure 8.15.)

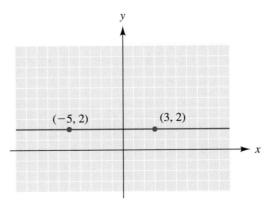

FIGURE 8.15

$$m = \frac{y_2 - y_1}{x_2 - x_1} = \frac{2 - 2}{3 - (-5)} = \frac{0}{8} = 0$$ ■

Note that all points on any horizontal line have the same y value. Therefore, *the slope of any horizontal line is 0.*

Example 4   Find the slope of the line through $(4, 2)$ and $(4, -5)$. (See Figure 8.16.)

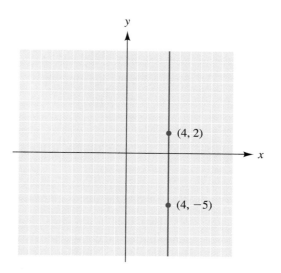

FIGURE 8.16

$$m = \frac{y_2 - y_1}{x_2 - x_1} = \frac{-5 - 2}{4 - 4} = \frac{-7}{0} \quad \text{(undefined)}$$

Division by zero is not possible, so the slope is undefined. Note that all points on any vertical line have the same x value. Therefore, *the slope of any vertical line is undefined.*   ■

   Note that in Example 1, the line slopes upward from left to right, whereas in Example 2, the line slopes downward. In general, the following is true.

General Statements About the Slope of a Line

1. If a line has positive slope, then the line slopes upward from left to right.
2. If a line has negative slope, then the line slopes downward from left to right.
3. If the slope of a line is zero, then the line is *horizontal.*
4. If the slope of a line is undefined, then the line is *vertical.*

The slope of a straight line can be found directly from its equation as follows:

1. Solve the equation for y.
2. The slope of the line is given by the coefficient of x.

Example 5

Find the slope of the line $4x + 6y = 15$.

First, solve the equation for y.

$$4x + 6y = 15$$
$$6y = -4x + 15 \qquad \text{Subtract } 4x \text{ from both sides.}$$
$$y = -\frac{2}{3}x + \frac{5}{2} \qquad \text{Divide both sides by 6.}$$
$$\underset{\text{slope}}{\uparrow}$$

The slope of the line is given by the coefficient of x, or $m = -\frac{2}{3}$.   ■

Example 6

Find the slope of the line $9x - 3y = 10$.

First, solve the equation for y.

$$9x - 3y = 10$$
$$-3y = -9x + 10 \qquad \text{Subtract } 9x \text{ from both sides.}$$
$$y = 3x - \frac{10}{3} \qquad \text{Divide both sides by } -3.$$
$$\underset{\text{slope}}{\uparrow}$$

The slope of the line is given by the coefficient of x, or $m = 3$.   ■

Two lines in the same plane are parallel if they do not intersect even if they are extended. (See Figure 8.17a.) Two lines in the same plane are perpendicular if they intersect at right angles, as in Figure 8.17(b).

Since parallel lines have the same steepness, they have the same slope.

(a) Parallel lines

(b) Perpendicular lines

FIGURE 8.17

Parallel Lines

Two lines are parallel if either one of the following conditions holds:
1. Both lines are perpendicular to the x axis (Figure 8.18a), or
2. Both lines have the same slope (Figure 8.18b)—that is, if the equations of the two lines are $L_1: y = m_1x + b_1$ and $L_2: y = m_2x + b_2$; then

$$m_1 = m_2$$

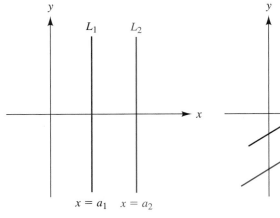

(a) Both lines are perpendicular to the x axis

 (b) $m_1 = m_2$

FIGURE 8.18
Parallel lines

> ## Perpendicular Lines
>
> Two lines are perpendicular if either one of the following conditions holds:
> 1. One line is vertical with equation $x = a$ and the other line is horizontal with equation $y = b$, or
> 2. Neither is vertical and the product of the slopes of the two lines is -1; that is, if the equations of the lines are
>
> $$L_1: y = m_1 x + b_1 \quad \text{and} \quad L_2: y = m_2 x + b_2$$
>
> then
>
> $$m_1 \cdot m_2 = -1$$

Note: The slopes of perpendicular lines are also negative reciprocals of each other; that is, $m_1 = -\dfrac{1}{m_2}$.

Example 7

Determine whether the lines given by the equations $2x + 3y = 6$ and $6x - 4y = 9$ are parallel, perpendicular, or neither.

First, find the slope of each line by solving its equation for y.

$$2x + 3y = 6 \qquad\qquad 6x - 4y = 9$$
$$3y = -2x + 6 \qquad\qquad -4y = -6x + 9$$
$$y = -\frac{2}{3}x + 2 \qquad\qquad y = \frac{3}{2}x - \frac{9}{4}$$
$$m_1 = -\frac{2}{3} \qquad\qquad m_2 = \frac{3}{2}$$

Since the slopes are not equal, the lines are not parallel. Next, find the product of the slopes.

$$m_1 \cdot m_2 = \left(-\frac{2}{3}\right)\left(\frac{3}{2}\right) = -1$$

Thus, the lines are perpendicular. ■

Example 8

Determine whether the lines given by the equations $5x + y = 7$ and $15x + 3y = -10$ are parallel, perpendicular, or neither.

First, find the slope of each line by solving its equation for y.

$$5x + y = 7 \qquad\qquad 15x + 3y = -10$$
$$y = -5x + 7 \qquad\qquad 3y = -15x - 10$$
$$\qquad\qquad\qquad\qquad y = -5x - \frac{10}{3}$$
$$m_1 = -5 \qquad\qquad m_2 = -5$$

Since both lines have the same slope, -5, and different y intercepts, they are parallel. ■

Example 9   Determine whether the lines given by the equations $4x + 5y = 15$ and $-3x - 2y = 12$ are parallel, perpendicular, or neither.

First, find the slope of each line by solving its equation for y.

$$4x + 5y = 15 \qquad\qquad -3x - 2y = 12$$
$$5y = -4x + 15 \qquad\qquad -2y = 3x + 12$$
$$y = -\frac{4}{5}x + 3 \qquad\qquad y = -\frac{3}{2}x - 6$$
$$m_1 = -\frac{4}{5} \qquad\qquad m_2 = -\frac{3}{2}$$

Since the slopes are not equal and do not have a product of -1, the lines are neither parallel nor perpendicular. That is, the lines intersect but not at right angles. ■

Exercises 8.3

Find the slope of the line passing through each pair of points:

1. $(-3, 1), (2, 6)$

2. $(4, 7), (6, 2)$

3. $(-2, 1), (-3, -5)$

4. $(5, -3), (-4, -9)$

5. $(4, 0), (0, 5)$

6. $(1, -6), (-2, 0)$

7. $(-2, 4), (5, 4)$

8. $(-3, -7), (-2, -7)$

9. $(6, 1), (6, -4)$

10. $(-8, 5), (-8, -3)$

11. $(4, -2), (-6, -8)$

12. $(-9, -1), (-3, -5)$

Find the slope of each line:

13.

14.

15.

16.

17.

18.

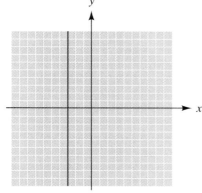

Find the slope of each line:

19. $y = 6x + 2$

20. $y = -4x + 3$

21. $y = -5x - 7$

22. $y = 9x - 13$

23. $3x + 5y = 6$

24. $9x + 12y = 8$

25. $-2x + 8y = 3$

26. $4x - 6y = 9$

27. $5x - 2y = 16$

28. $-4x - 2y = 7$

29. $x - 3 = 0$

30. $y + 15 = 0$

Determine whether the lines given by the equations are parallel, perpendicular, or neither:

31. $y = 4x - 5$
  $y = 4x + 5$

32. $y = \dfrac{2}{3}x + 4$
  $y = -\dfrac{3}{2}x - 5$

33. $y = \dfrac{3}{4}x - 2$
  $y = -\dfrac{4}{3}x + \dfrac{5}{3}$

34. $y = \dfrac{4}{5}x - 2$
  $y = -\dfrac{4}{5}x + \dfrac{2}{3}$

35. $x + 3y = 9$
  $3x - y = 14$

36. $x + 2y = 11$
  $2x + 4y = -5$

37. $x - 4y = 12$
  $x + 4y = 16$

38. $2x + 7y = 6$
  $14x - 4y = 18$

39. $y - 5x = 12$
  $5x - y = -6$

40. $-3x + 9y = 20$
  $x = 3y$

8.4 The Equation of a Line

We have learned to graph the equation of a straight line and to find the slope of a straight line given its equation or any two points on it. In this section, we will use the slope to graph the equation and to write its equation.

A fast and easy way to draw the graph of a straight line when the *slope* and *y intercept* are known is to first plot the *y*-intercept point on the graph and consider this to be the starting point. Then from this starting point and using the slope (*rise* over *run*), move up or down the number of units indicated by the rise and then move right or left the number of units indicated by the run to a second point. Mark this second point. Then draw a straight line through these two points.

Note: We suggest that if the slope is a fraction and negative, you include the negative sign with the numerator (rise). If the slope is an integer, write the slope with 1 as its denominator (run). Also note that the *y* intercept does not have to be the starting point; any point of the line can be the starting point, if it is known.

Example 1

Draw the graph of the line with slope $\dfrac{2}{3}$ and *y* intercept 4.

The slope $\dfrac{2}{3}$ corresponds to $\dfrac{\text{difference in } y \text{ values}}{\text{difference in } x \text{ values}} = \dfrac{2}{3}$. From the *y* intercept 4 [the point (0, 4)], move 2 units up and then 3 units to the right, as shown in Figure 8.19. Then draw a straight line through (0, 4) and (3, 6).

FIGURE 8.19   ■

In Section 8.3, we learned how to find the slope of a line, given its equation, by solving for y. For example, for a line whose equation is $y = 3x + 5$ the slope is 3, the coefficient of x. What does the number 5 have to do with its graph? If we let $x = 0$, the equation is

$$y = 3x + 5$$
$$y = 3(0) + 5$$
$$= 0 + 5 = 5$$

The line crosses the y axis when $x = 0$. Therefore, the y intercept of the graph is 5.

Slope-Intercept Form

When the equation of a straight line is written in the form

$$y = mx + b$$

the slope of the line is m and the y intercept is b.

Example 2

Draw the graph of the equation $8x + 2y = -10$ using its slope and y intercept.

First, find the slope and y intercept by solving the equation for y as follows.

$$8x + 2y = -10$$
$$2y = -8x - 10 \qquad \text{Subtract } 8x \text{ from both sides.}$$
$$y = -4x - 5 \qquad \text{Divide both sides by 2.}$$

$$\qquad\qquad \text{slope} \qquad y \text{ intercept}$$

The slope is -4, and the y intercept is -5. The slope -4 corresponds to

$$\frac{\text{difference in } y \text{ values}}{\text{difference in } x \text{ values}} = \frac{-4}{1}$$

When the slope is an integer, write it as a ratio with 1 in the denominator. From the y intercept -5, move 4 units down and 1 unit to the right, as shown in Figure 8.20. Draw a straight line through the points $(0, -5)$ and $(1, -9)$.

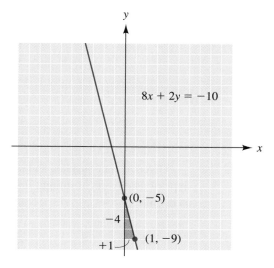

$8x + 2y = -10$

$(0, -5)$

-4

$+1$ $(1, -9)$

FIGURE 8.20   ■

 Example 3

Find the equation of the line with slope $\frac{3}{4}$ and y intercept -2.

Use the slope-intercept form with $m = \frac{3}{4}$ and $b = -2$.

$$y = mx + b$$

$$y = \frac{3}{4}x + (-2)$$

$$y = \frac{3}{4}x - 2$$

$4y = 3x - 8$	Multiply both sides by 4.
$0 = 3x - 4y - 8$	Subtract $4y$ from both sides.
$8 = 3x - 4y$	Add 8 to both sides.

$$3x - 4y = 8 \qquad ■$$

Note: Any of the last five equations in Example 3 is correct. The most common ways of writing equations are $y = mx + b$ and $cx + dy = f$.

Example 4

Draw the graph of the straight line through the point $(-3, 6)$ with slope $-\frac{2}{5}$.

The slope $-\frac{2}{5}$ corresponds to $\frac{\text{difference in } y \text{ values}}{\text{difference in } x \text{ values}} = -\frac{2}{5}$. From the point $(-3, 6)$, move 2 units down and 5 units to the right, as shown in Figure 8.21. Draw a straight line through the points $(-3, 6)$ and $(2, 4)$.

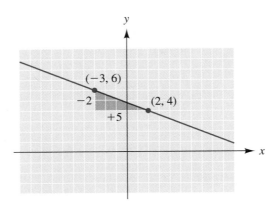

FIGURE 8.21   ■

Given a point on a line and its slope, we can find its equation. To show this, let m be the slope of a nonvertical straight line; let (x_1, y_1) be the coordinates of a known or given point on the line; and let (x, y) be the coordinates of any other point on the line. (See Figure 8.22.)

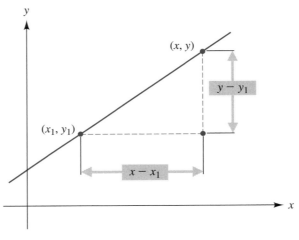

FIGURE 8.22

Then, by the definition of slope, we have

$$\frac{\text{difference in } y \text{ values}}{\text{difference in } x \text{ values}} = m$$

$$\frac{y - y_1}{x - x_1} = m$$

$$y - y_1 = m(x - x_1) \qquad \text{Multiply both sides by } (x - x_1).$$

The result is the point-slope form of the equation of a straight line.

Point-Slope Form

If m is the slope and (x_1, y_1) is any point on a nonvertical straight line, its equation is

$$y - y_1 = m(x - x_1)$$

Example 5 Find the equation of the line with slope -3 that passes through the point $(-1, 4)$.

Here, $m = -3$, $x_1 = -1$, and $y_1 = 4$. Using the point-slope form, we have

$$y - y_1 = m(x - x_1)$$
$$y - 4 = -3[x - (-1)]$$
$$y - 4 = -3(x + 1)$$
$$y - 4 = -3x - 3 \qquad \text{Remove parentheses.}$$
$$y = -3x + 1 \qquad \text{Add 4 to both sides.} \qquad ■$$

The point-slope form also can be used to find the equation of a line when two points on the line are known.

 Example 6 Find the equation of the line through the points $(5, -4)$ and $(-1, 8)$.

First, find the slope.

$$m = \frac{y_2 - y_1}{x_2 - x_1} = \frac{8 - (-4)}{-1 - 5} = \frac{12}{-6} = -2$$

Substitute $m = -2$, $x_1 = 5$, and $y_1 = -4$ in the point-slope form.

$$y - y_1 = m(x - x_1)$$
$$y - (-4) = -2(x - 5)$$
$$y + 4 = -2x + 10 \qquad \text{Remove parentheses.}$$
$$y = -2x + 6 \qquad \text{Subtract 4 from both sides.}$$

We could have used the other point $(-1, 8)$, as follows:

$$y - y_1 = m(x - x_1)$$
$$y - 8 = -2[x - (-1)]$$
$$y - 8 = -2(x + 1)$$
$$y - 8 = -2x - 2 \qquad \text{Remove parentheses.}$$
$$y = -2x + 6 \qquad \text{Add 8 to both sides.} \qquad ■$$

Exercises 8.4

Draw the graph of each line with the given slope and y intercept:

1. $m = 2, b = 5$
2. $m = 4, b = -3$
3. $m = -5, b = 4$
4. $m = -1, b = 0$
5. $m = \dfrac{2}{3}, b = -4$
6. $m = \dfrac{3}{4}, b = 2$
7. $m = -\dfrac{2}{5}, b = -4$
8. $m = \dfrac{5}{6}, b = 3$
9. $m = \dfrac{4}{3}, b = -1$
10. $m = -\dfrac{9}{4}, b = 2$

Draw the graph of each equation using the slope and y intercept:

11. $2x + y = 6$
12. $4x + y = -3$
13. $3x + 5y = 10$
14. $4x + 3y = 9$
15. $3x - y = 7$
16. $5x - y = -2$
17. $3x - 2y = 12$
18. $2x - 5y = 20$
19. $2x - 6y = 0$
20. $4x + 7y = 0$

Find the equation of the line with the given slope and y intercept:

21. $m = 2, b = 5$
22. $m = 4, b = -3$
23. $m = -5, b = 4$
24. $m = -1, b = 0$

25. $m = \dfrac{2}{3}, b = -4$

26. $m = \dfrac{3}{4}, b = 2$

27. $m = -\dfrac{6}{5}, b = 3$

28. $m = -\dfrac{12}{5}, b = -1$

29. $m = -\dfrac{3}{5}, b = 0$

30. $m = 0, b = 0$

Draw the graph of the line through the given point with the given slope:

31. $(3, 5), m = 2$

32. $(-1, 4), m = -3$

33. $(-5, 0), m = \dfrac{3}{4}$

34. $(0, 0), m = -\dfrac{1}{3}$

35. $(6, -2), m = \dfrac{3}{2}$

36. $(-3, -3), m = \dfrac{1}{2}$

37. $(-1, -1), m = -1$

38. $(0, -4), m = 0$

39. $(2, -7), m = 1$

40. $(-8, 6), m = \dfrac{1}{8}$

Find the equation of the line through the given point with the given slope:

41. $(3, 5), m = 2$

42. $(-1, 4), m = -3$

43. $(-5, 0), m = \dfrac{3}{4}$

44. $(0, 0), m = -\dfrac{1}{3}$

45. $(6, -2), m = \dfrac{3}{2}$

46. $(-3, -3), m = \dfrac{1}{2}$

47. $(-3, -1), m = -\dfrac{10}{3}$

48. $(4, 5), m = -\dfrac{3}{7}$

49. $(12, -10), m = -1$

50. $(15, 20), m = 3$

Find the equation of the line through the given points:

51. $(2, 3), (5, 1)$

52. $(8, 5), (2, 1)$

53. $(-1, 4), (2, -2)$

54. $(-1, -2), (-4, 0)$

55. $(0, 1), (-3, 0)$

56. $(5, 5), (-2, -2)$

57. $(7, 4), (-1, 0)$

58. $(6, -3), (-3, 6)$

59. $(3, 3), (1, 5)$

60. $(16, 12), (4, 8)$

Chapter 8 Group Activities

1. Stack sugar cubes in a stair-stepping pattern or lay them so that they resemble stairs, as shown in Illustration 1. Notice how these steps have a ratio of $\frac{1}{1}$, or 1 to 1. That is, one step up and one step right. This is rise 1 and run 1. Is this familiar to you?

   a. Make other different stepping patterns that have a ratio, expressed as rise over run. Keep a record of your patterns.

   b. Write the equation of a line with the slope that you recorded in part **a**. Assume that the steps start at $(0, 0)$. Also, let each cube be one unit of length.

   For example, using the pattern illustrated with rise 1 and run 1, the slope is 1. So the equation of the line passing through the origin $(0, 0)$ would be $y - 0 = 1(x - 0)$ or just $y = x$.

2. Draw a sketch of how stacked sugar cubes would look for the equations

   a. $y = 3x$ and **b.** $y = -2x$.

ILLUSTRATION 1

Chapter 8 Summary

Glossary of Basic Terms

Coordinates. The numbers x and y written in the form (x, y). (p. 307)

Dependent variable. The second element of an ordered pair, usually y. (p. 315)

Independent variable. The first element of an ordered pair, usually x. (p. 315)

Linear equation with two variables. An equation that may be written in the form $ax + by = c$, where the numbers a, b, and c are all real numbers such that a and b are not both 0. (p. 304)

Number plane. All points in the plane determined by the intersecting, perpendicular x axis (horizontal axis) and y axis (vertical axis). (p. 307)

Ordered pair. Numbers in the form (x, y) that correspond to points in the number plane; also used to write solutions of systems of equations with two variables. (pp. 304, 307)

Origin. The zero point where the x and y axes intersect in the number plane. (p. 307)

Quadrant. Each of the four regions of the number plane formed by the intersection of the x and y axes. (p. 307)

Rectangular coordinate system. The number plane formed by the intersection of the x and y axes. Also called the Cartesian coordinate system. (p. 307)

x axis. The horizontal axis in the number plane. (p. 307)

x intercept. The x coordinate of the point where the graph crosses the x axis. (p. 312)

y axis. The vertical axis in the number plane. (p. 307)

y intercept. The y coordinate of the point where the graph crosses the y axis. (p. 313)

8.1 Linear Equations with Two Variables

1. **Plotting points in the number plane:** To locate or plot the point in the number plane which corresponds to an ordered pair (x, y):
   a. Count right or left, from 0 (the origin) along the x axis, the number of spaces corresponding to the first number of the ordered pair (right if positive, left if negative).
   b. Count up or down, from the point reached on the x axis in the step above, the number of spaces corresponding to the second number of the ordered pair (up if positive, down if negative).
   c. Mark the last point reached with a dot. (p. 307)

8.2 Graphing Linear Equations

1. **The graph of the solutions of a linear equation in two variables is always a straight line. (p. 310)**

2. **Graphing linear equations:** To graph a linear equation with two variables:
   a. Find any three solutions. Note: Two solutions would be enough, since two points determine a straight line. However, a third solution gives a third point as a check.
   b. Plot the three points corresponding to the three ordered pairs that you found above.
   c. Draw a line through the three points. If it is not a straight line, check your solutions. (p. 310)

3. To find the x intercept of a line, replace y in the equation by 0 and solve for x. (p. 313)

4. To find the y intercept of a line, replace x in the equation by 0 and solve for y. (p. 313)

5. **Horizontal line:** The graph of the linear equation $y = k$, where k is a constant, is the horizontal line through the point $(0, k)$. That is, $y = k$ is a horizontal line with a y intercept of k. (p. 314)

6. **Vertical line:** The graph of the linear equation $x = k$, where k is a constant, is the vertical line through the point $(k, 0)$. That is, $x = k$ is a vertical line with an x intercept of k. (p. 315)

7. **Independent/dependent variables and ordered pairs:** In graphing an equation, the horizontal axis corresponds to the independent variable; the vertical axis corresponds to the dependent variable. In general, think of graphing the ordered pairs:

   (independent variable, dependent variable) (p. 315)

8.3 The Slope of a Line

1. **Slope of a line:** If two points (x_1, y_1) and (x_2, y_2) on a line are known, the slope of the line is $m = \dfrac{y_2 - y_1}{x_2 - x_1}$.
   (p. 317)

2. General statements about the slope of a line:
   a. If a line has a positive slope, the line slopes upward from left to right.
   b. If a line has a negative slope, the line slopes downward from left to right.
   c. If the slope of a line is zero, the line is *horizontal*.
   d. If the slope of a line is undefined, the line is *vertical*. (p. 319)

3. Parallel lines: Two lines are parallel if either one of the following conditions holds:
   a. Both lines are perpendicular to the x axis, or
   b. Both lines have the same slope; that is, $m_1 = m_2$. (p. 320)

4. Perpendicular lines: Two lines are perpendicular if either one of the following conditions holds:
   a. One is a vertical line with equation $x = a$ and the other line is horizontal with equation $y = b$, or
   b. Neither is vertical and the product of the slopes of the two lines is -1; that is, $m_1 \cdot m_2 = -1$. (p. 321)

8.4   The Equation of a Line

1. Slope-Intercept Form of a Line: When the equation of a straight line is written in the form $y = mx + b$, the slope of the line is m and the y intercept is b. (p. 324)

2. Point-Slope Form of a Line: If m is the slope and (x_1, y_1) is any point on a nonvertical straight line, its equation is $y - y_1 = m(x - x_1)$. (p. 326)

Chapter 8   Review

Complete the ordered-pair solutions of each equation:

1. $x + 2y = 8$ (3,) (0,) (−4,)

2. $2x − 3y = 12$ (3,) (0,) (−3,)

3. Solve for y: $6x + y = 15$.

4. Solve for y: $3x − 5y = −10$.

Write the ordered pair corresponding to each point in Illustration 1:

5. A

6. B

7. C

8. D

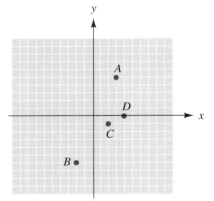

ILLUSTRATION 1

Plot each point in the number plane. Label each point by writing its ordered pair:

9. $E (3, −2)$ **10.** $F (−7, −4)$

11. $G (−1, 5)$ **12.** $H (0, −5)$

Draw the graph of each equation:

13. $x + y = 8$ **14.** $x − 2y = 5$

15. $3x + 6y = 12$ **16.** $4x − 5y = 15$

17. $4x = 9y$ **18.** $y = \dfrac{1}{3}x − 4$

19. $x = −6$ **20.** $y = 7$

Find the slope of the line passing through each pair of points:

21. $(3, −4), (10, 5)$ **22.** $(−4, 0), (2, 6)$

Find the slope of each line:

23. $y = 4x − 7$ **24.** $2x + 5y = 8$

25. $5x − 9y = −2$

Determine whether the lines given by the equations are parallel, perpendicular, or neither:

26. $y = 3x − 5$ **27.** $3x − 4y = 12$

   $y = −\dfrac{1}{3}x − 5$ $8x − 6y = 15$

28. $2x + 5y = 8$ **29.** $x = 4$

   $4x + 10y = 25$ $y = −6$

Draw the graph of each line with the given slope and y intercept:

30. $m = −2, b = 9$ **31.** $m = −\dfrac{2}{3}, b = −5$

Draw the graph of each equation using the slope and y intercept:

32. $3x + 5y = 20$ **33.** $5x - 8y = 32$

Find the equation of the line with the given slope and y intercept:

34. $m = -\dfrac{1}{2}, b = 3$ **35.** $m = \dfrac{8}{3}, b = 0$

36. $m = 0, b = 0$

Draw the graph of the line through the given point with the given slope:

37. $(6, -1), m = -3$ **38.** $(-5, -2), m = \dfrac{7}{2}$

Find the equation of the line through the given point with the given slope:

39. $(-2, 8), m = -1$ **40.** $(0, -5), m = -\dfrac{1}{4}$

Find the equation of the line through the given points:

41. $(2, -3), (10, 5)$ **42.** $(12, 0), (2, -5)$

Chapter 8   Test

Given the equation $3x - 4y = 24$, complete each ordered pair:

1. $(-4, \ )$ **2.** $(0, \ )$ **3.** $(4, \ )$

4–5. Write the ordered pair corresponding to each point in Illustration 1.

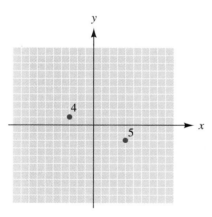

ILLUSTRATION 1

6. Draw the graph of $3x + y = 3$.

7. Draw the graph of $-2x + y = 4$.

8. Draw the graph of $s = 5 + 2t$ for nonnegative values of t.

9. Find the slope of the line containing the points $(-2, 4)$ and $(5, 6)$.

Find the slope of each line:

10. $y = 3x - 2$ **11.** $2x - 5y = 10$

12. Determine whether the graphs of the following pair of equations are parallel, perpendicular, or neither.

$$2x - y = 10$$
$$y = 2x - 3$$

13. Find the equation of the line having y intercept -3 and slope $\frac{1}{2}$.

14. Find the equation of the line containing the point $(-2, 3)$ and slope -2.

15. Draw the graph of the line containing the point $(3, 4)$ and having slope -3.

16. Draw the graph of the line $y = \frac{1}{2}x + 4$, using its slope and y intercept.

Cumulative Review

1. Evaluate: $2(6 - 5) + 3$

2. Subtract: $+6 - (-9)$

3. Find missing dimensions a and b in the figure in Illustration 1.

ILLUSTRATION 1

4. Change 250 cm to inches.

5. Multiply: $(6.2 \times 10^{-3})(1.8 \times 10^5)$

6. 61 mm = _____ m

7. Give the number of significant digits: 306,760 kg

8. Read the measurement shown on the metric micrometer in Illustration 2.

ILLUSTRATION 2

9. Use the rules for multiplication of measurements to evaluate: 1.8 m $\times$ 61.2 m $\times$ 3.2 m

10. Simplify: $(4x - 5) - (6 - 3x)$

11. Simplify: $(-5xy^2)(8x^3y^2)$

12. Simplify: $2x(4x - 3y)$

13. Solve: $3(x - 2) + 4(3 - 2x) = 9$

14. Solve: $\dfrac{2x}{3} + \dfrac{1}{5} = \dfrac{x}{4} - \dfrac{2}{3}$

15. Solve: $s = \dfrac{2V + t}{3}$ for V

Write each ratio in lowest terms:

16. 5 to 65

17. 32 in. : 3 yd

Solve each proportion (round each result to three significant digits):

18. $\dfrac{5}{13} = \dfrac{x}{156}$

19. $\dfrac{29.1}{73.8} = \dfrac{x}{104}$

20. $\dfrac{286}{x} = \dfrac{11.8}{59.7}$

21. If it costs \$28.50 to repair 5 ft^2 of sidewalk, how much would it cost to repair 18 ft^2?

22. A map shows a scale of 1 in. = 40 mi. What distance is represented by $4\frac{1}{4}$ in. on the map?

23. A large gear with 16 teeth rotates at 40 rpm. It turns a small gear at 64 rpm. How many teeth does the smaller gear have?

24. Complete the ordered pair solutions of the equation:

$2x + 3y = 12$ $(3,\ )$, $(0,\ )$, $(-3,\ )$

25. Solve for y: $4x + 2y = 7$

26. Draw the graph of $y = -2x - 5$.

27. Draw the graph of $3x - 2y = 12$.

28. Find the slope of the line containing the points $(-1, 3)$ and $(2, 6)$.

29. Find the equation of the line with slope 1/2 and containing the point $(2, -4)$.

30. Determine whether the graphs of $2x - 3y = 6$ and $3x + 5y = 7$ are parallel, perpendicular, or neither.

9 Systems of Linear Equations

Mathematics at Work

Computer support specialists provide technical assistance, support, and training to computer system users; investigate and resolve computer hardware and software problems; and answer user questions and concerns in person or via a telephone help desk using automated diagnostic programs. Such concerns may include word processing, printing, e-mail, programming languages, and operating systems. Computer support specialists may work within an organization, work directly for a computer or software vendor, or work for help-desk or computer support service companies that provide customer support on an outsourced contract basis.

Other related computer specialists include database administrators who work with database management systems software and determine ways to organize and store data. Network administrators design, install, and support an organization's local area network (LAN), wide area network (WAN), and Internet or intranet system. Computer security specialists plan, coordinate, and implement an organization's information security system. A computer scientist title applies widely to computer professionals who design computers and computer software, develop information technologies, and develop ways to use computers for new tasks. Many of these computer-related jobs require a two-year associate degree; a bachelor's degree is usually required for the computer scientist. For more information, go to the website listed below.

Comstock Premium/Alamy

Computer Support Specialist
Computer technician repairing a computer.

www.cengage.com/mathematics/ewen

Objectives

■ Solve a pair of linear equations by graphing.

■ Solve a pair of linear equations by addition.

■ Solve a pair of linear equations by substitution.

■ Solve application problems involving pairs of linear equations.

9.1 # Solving Pairs of Linear Equations by Graphing

Many problems can be solved by using two equations with two variables and solving them simultaneously. **To solve a pair of linear equations** *with two variables simultaneously, you must find an ordered pair that will make both equations true at the same time.*

As you know, the graph of a linear equation with two variables is a straight line. As shown in Figure 9.1, two straight lines (the graphs of *two* linear equations with two variables) in the same plane may be arranged as follows:

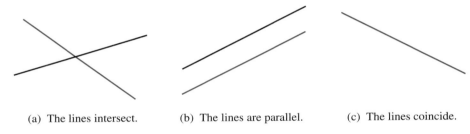

(a) The lines intersect.   (b) The lines are parallel.   (c) The lines coincide.

FIGURE 9.1
Possible relationships of two straight lines in the same plane

   a. The lines may intersect. If so, they have one point in common. The equations have one common solution. The coordinates of the point of intersection define the common solution.

   b. The lines may be parallel. If so, they have no point in common. The equations have no common solution.

   c. The lines may coincide. That is, one line lies on top of the other. If so, any solution of one equation is also a solution of the other. With infinitely many points of intersection, there are infinitely many common solutions. The solution of the pair of equations consists of all points on the common line.

 Example 1

Draw the graphs of $x - y = 2$ and $x + 3y = 6$ in the same number plane. Find the common solution of the equations.

Step 1   Draw the graph of $x - y = 2$. First, solve for y.

$$x - y = 2$$
$$x - y - x = 2 - x \qquad \text{Subtract } x \text{ from both sides.}$$
$$-y = 2 - x$$
$$\frac{-y}{-1} = \frac{2 - x}{-1} \qquad \text{Divide both sides by } -1.$$
$$y = -2 + x$$

Then find three solutions.

x	−2 + x	y
2	−2 + 2	0
0	−2 + 0	−2
−2	−2 + (−2)	−4

Three solutions are $(2, 0)$, $(0, -2)$, and $(-2, -4)$. Plot the three points that correspond to these three ordered pairs. Then draw a straight line through these three points, as in Figure 9.2.

Step 2   Draw the graph of $x + 3y = 6$. First, solve for y.

$$x + 3y = 6$$
$$x + 3y - x = 6 - x \qquad \text{Subtract } x \text{ from both sides.}$$
$$3y = 6 - x$$
$$\frac{3y}{3} = \frac{6 - x}{3} \qquad \text{Divide both sides by 3.}$$
$$y = \frac{6 - x}{3}$$

x	$\frac{6 - x}{3}$	y
6	$\frac{6 - 6}{3} = \frac{0}{3}$	0
0	$\frac{6 - 0}{3} = \frac{6}{3}$	2
−3	$\frac{6 - (-3)}{3} = \frac{9}{3}$	3

Three solutions are $(6, 0)$, $(0, 2)$, and $(-3, 3)$. Plot the three points that correspond to these three ordered pairs. Then draw a straight line through these three points, as in Figure 9.2.

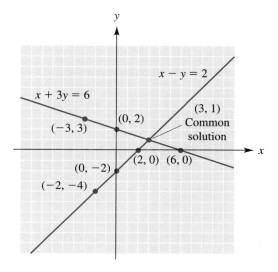

FIGURE 9.2

The point that corresponds to (3, 1) is the point of intersection. Therefore, (3, 1) is the common solution. That is, $x = 3$ and $y = 1$ are the only values that satisfy both equations. The solution (3, 1) should be checked in both equations.   ■

    Example 2

Draw the graphs of $2x + 3y = 6$ and $4x + 6y = 30$ in the same number plane. Find the common solution of the equations.

Step 1   Draw the graph of $2x + 3y = 6$. First, solve for y.

$$2x + 3y = 6$$
$$2x + 3y - 2x = 6 - 2x \qquad \text{Subtract } 2x \text{ from both sides.}$$
$$3y = 6 - 2x$$
$$\frac{3y}{3} = \frac{6 - 2x}{3} \qquad \text{Divide both sides by 3.}$$
$$y = \frac{6 - 2x}{3}$$

Then find three solutions.

x	$\frac{6 - 2x}{3}$	y
3	$\frac{6 - 2(3)}{3} = \frac{0}{3}$	0
0	$\frac{6 - 2(0)}{3} = \frac{6}{3}$	2
−3	$\frac{6 - 2(-3)}{3} = \frac{12}{3}$	4

Three solutions are (3, 0), (0, 2), and (−3, 4). Plot the three points that correspond to these three ordered pairs. Then draw a straight line through these three points, as in Figure 9.3.

Step 2   Draw the graph of $4x + 6y = 30$. First, solve for y.

$$4x + 6y = 30$$
$$4x + 6y - 4x = 30 - 4x \qquad \text{Subtract } 4x \text{ from both sides.}$$
$$6y = 30 - 4x$$
$$\frac{6y}{6} = \frac{30 - 4x}{6} \qquad \text{Divide both sides by 6.}$$
$$y = \frac{30 - 4x}{6}$$

Then find three solutions.

x	$\frac{30 - 4x}{6}$	y
3	$\frac{30 - 4(3)}{6} = \frac{18}{6}$	3
0	$\frac{30 - 4(0)}{6} = \frac{30}{6}$	5
−6	$\frac{30 - 4(-6)}{6} = \frac{54}{6}$	9

Three solutions are (3, 3), (0, 5), and (−6, 9). Plot the three points that correspond to these ordered pairs in the same number plane as the points found in Step 1. Then draw a straight line through these three points. As you see in Figure 9.3, the lines are parallel. The lines have no points in common; therefore, there is no common solution.

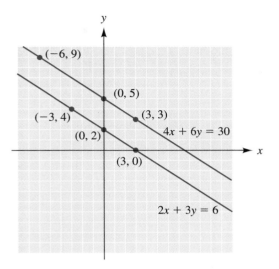

FIGURE 9.3

You may verify that the lines are parallel by showing that the slopes are equal. ■

 Example 3   Draw the graphs of $-2x + 8y = 24$ and $-3x + 12y = 36$ in the same number plane. Find the common solution.

Step 1   Draw the graph of $-2x + 8y = 24$. First, solve for y.

$$-2x + 8y = 24$$

$$-2x + 8y + 2x = 24 + 2x \qquad \text{Add } 2x \text{ to both sides.}$$

$$8y = 24 + 2x$$

$$\frac{8y}{8} = \frac{24 + 2x}{8} \qquad \text{Divide both sides by 8.}$$

$$y = \frac{24 + 2x}{8}$$

Then find three solutions.

x	$\dfrac{24 + 2x}{8}$	y
−4	$\dfrac{24 + 2(-4)}{8} = \dfrac{16}{8}$	2
0	$\dfrac{24 + 2(0)}{8} = \dfrac{24}{8}$	3
8	$\dfrac{24 + 2(8)}{8} = \dfrac{40}{8}$	5

Three solutions are $(-4, 2)$, $(0, 3)$, and $(8, 5)$. Plot the three points that correspond to these three ordered pairs. Then draw a straight line through these three points.

Step 2   Draw the graph of $-3x + 12y = 36$. First, solve for y.

$$-3x + 12y = 36$$
$$-3x + 12y + 3x = 36 + 3x \qquad \text{Add } 3x \text{ to both sides.}$$
$$12y = 36 + 3x$$
$$\frac{12y}{12} = \frac{36 + 3x}{12} \qquad \text{Divide both sides by 12.}$$
$$y = \frac{36 + 3x}{12}$$

Then find three solutions.

x	$\dfrac{36 + 3x}{12}$	y
4	$\dfrac{36 + 3(4)}{12} = \dfrac{48}{12}$	4
-8	$\dfrac{36 + 3(-8)}{12} = \dfrac{12}{12}$	1
-5	$\dfrac{36 + 3(-5)}{12} = \dfrac{21}{12}$	$\frac{7}{4}$ or $1\frac{3}{4}$

Three solutions are $(4, 4)$, $(-8, 1)$, $\left(-5, 1\frac{3}{4}\right)$. Plot the three points that correspond to these ordered pairs in the same number plane as the points found in Step 1. Then draw a straight line through these points.

   Note that the lines coincide. Any solution of one equation is also a solution of the other. Hence, there are infinitely many points of intersection and infinitely many common solutions (see Figure 9.4). The solutions are the coordinates of the points on either line.

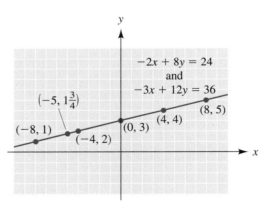

FIGURE 9.4   ■

◆ **Example 4**   The sum of two electric currents is 12 A. One current is three times the other. Find the two currents graphically.

   Let x = first current
   y = second current

The equations are then: $x + y = 12$

$$y = 3x$$

Draw the graph of each equation, as shown in Figure 9.5.

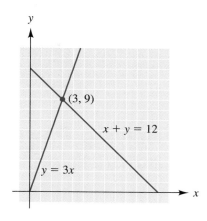

FIGURE 9.5

As you can see from Figure 9.5, the point of intersection is (3, 9). Thus, $x = 3$ and $y = 9$ is the common solution; the currents are 3 A and 9 A. ◼

Exercises 9.1

Draw the graphs of each pair of linear equations. Find the point of intersection. If the lines do not intersect, tell whether the lines are parallel or coincide:

1. $y = 3x$
$y = x + 4$

2. $x - y = 2$
$x + 3y = 6$

3. $y = -x$
$y - x = 2$

4. $x + y = 3$
$2x + 2y = 6$

5. $x - 3y = 6$
$2x - 6y = 18$

6. $x + y = 4$
$2x + y = 5$

7. $2x - 4y = 8$
$3x - 6y = 12$

8. $4x - 3y = 11$
$6x + 5y = -12$

9. $3x - 6y = 12$
$4x - 8y = 12$

10. $2x + y = 6$
$2x - y = 6$

11. $3x + 2y = 10$
$2x - 3y = 11$

12. $5x - y = 10$
$x - 3y = -12$

13. $5x + 8y = -58$
$2x + 2y = -18$

14. $6x + 2y = 24$
$3x - 4y = 12$

15. $3x + 2y = 17$
$x = 3$

16. $5x - 4y = 28$
$y = -2$

17. $y = 2x$
$y = -x + 2$

18. $y = -5$
$y = x + 3$

19. $2x + y = 6$
$y = -2x + 1$

20. $3x + y = -5$
$2x + 5y = 1$

21. $4x + 3y = 2$
$5x - y = 12$

22. $4x - 6y = 10$
$2x - 3y = 5$

23. $2x - y = 9$
$-2x + 3y = -11$

24. $x - y = 5$
$2x - 3y = 5$

25. $8x - 3y = 0$
$4x + 3y = 3$

26. $2x + 8y = 9$
$4x + 4y = 3$

Solve Exercises 27–30 graphically:

27. The sum of two resistances is 14 Ω. Their difference is 6 Ω. Find the two resistances. If we let R_1 and R_2 be the two resistances, the equations are

$$R_1 + R_2 = 14$$
$$R_1 - R_2 = 6$$

28. A board 36 in. long is cut into two pieces so that one piece is 8 in. longer than the other. Find the length of each piece. If we let x and y be the two lengths, the equations are

$$x + y = 36$$
$$y = x + 8$$

29. In a concrete mix, there is four times as much gravel as concrete. The total volume is 20 ft^3. How much of each is in the mix? If

x = the amount of concrete

y = the amount of gravel,

the equations are

$$y = 4x$$
$$x + y = 20$$

30. An electric circuit containing two currents may be expressed by the equations

$$3i_1 + 4i_2 = 15$$
$$5i_1 - 2i_2 = -1$$

where i_1 and i_2 are the currents in microamperes (μA). Find the two currents.

9.2 Solving Pairs of Linear Equations by Addition

Often, solving a pair of linear equations by graphing results in only an approximate solution when an exact solution is needed.

 Example 1

Solve the following pair of linear equations by graphing.

$$3x + 4y = -2$$
$$6x - 8y = 32$$

First, find any three solutions to the first equation by any method. Our solutions are $(2, -2)$, $(6, -5)$, and $\left(-3, 1\frac{3}{4}\right)$. Plot these three points and connect them with a straight line as shown in Figure 9.6.

Then find any three solutions to the second equation by any method. Our solutions are $(0, -4)$, $(4, -1)$, and $\left(-2, -5\frac{1}{2}\right)$. Plot these three points and connect them with a straight line as shown in Figure 9.6.

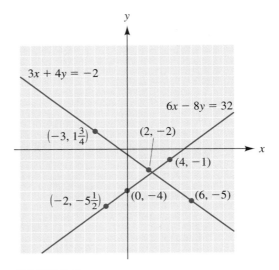

FIGURE 9.6

As you can see, only an approximate solution (point of intersection) can be read from this graph. The exact solution is found in Example 6.   ■

The *addition method* outlined below provides exact solutions to pairs of equations.

Solving a Pair of Linear Equations by the Addition Method

Step 1 If necessary, multiply both sides of one or both equations by a number (or numbers) so that the numerical coefficients of one of the variables are negatives of each other.

Step 2 Add the two equations from Step 1 to obtain an equation containing only one variable.

Step 3 Solve the equation from Step 2 for the one remaining variable.

Step 4 Solve for the second variable by substituting the solution from Step 3 in either of the original equations.

Step 5 Check your solution by substituting the ordered pair in both original equations.

 Example 2 Solve the following pair of linear equations by addition. Check your solution.

$$2x - y = 6$$
$$x + y = 9$$

Step 1 of the preceding rules is unnecessary, since you can eliminate the y variable by adding the two equations as they are.

$$
\begin{array}{ll}
2x - y = 6 & \\
\underline{x + y = 9} & \\
3x + 0 = 15 & \text{Add the equations.} \\
3x \quad = 15 & \\
x \quad = 5 & \text{Divide both sides by 3.}
\end{array}
$$

Now substitute 5 for x in either of the original equations to solve for y. (Choose the simpler equation to make the arithmetic easier.)

$$
\begin{array}{ll}
x + y = 9 & \\
5 + y = 9 & \text{Substitute 5 for } x. \\
y = 4 & \text{Subtract 5 from both sides.}
\end{array}
$$

The apparent solution is $(5, 4)$.

Check: Substitute 5 for x and 4 for y in both original equations.

$$
\begin{array}{lll}
2x - y = 6 & \qquad & x + y = 9 \\
2(5) - 4 = 6 \quad ? & \qquad & (5) + (4) = 9 \quad \text{True} \\
10 - 4 = 6 \quad \text{True} & &
\end{array}
$$

The solution checks. Thus, the solution is $(5, 4)$. ■

 Example 3 Solve the following pair of linear equations by addition. Check your solution.

$$2x + y = 5$$
$$x + y = 4$$

First, multiply both sides of the second equation by -1 to eliminate y by addition.

$$2x + y = 5$$
$$(-1)(x + y) = (-1)(4)$$

or

$$2x + y = 5$$
$$\underline{-x - y = -4}$$
$$x \quad = 1$$

Now substitute 1 for x in the equation $x + y = 4$ to solve for y.

$$x + y = 4$$
$$1 + y = 4$$
$$y = 3$$

The solution is $(1, 3)$. The check is left to the student.   ■

 Example 4   Solve the following pair of linear equations by addition. Check your solution.

$$4x + 2y = 2$$
$$3x - 4y = 18$$

Multiply both sides of the first equation by 2 to eliminate y by addition.

$$2(4x + 2y) = 2(2)$$
$$\underline{3x - 4y = 18}$$

or

$$8x + 4y = 4$$
$$\underline{3x - 4y = 18}$$
$$11x \quad = 22$$
$$x \quad = 2$$

Now substitute 2 for x in the equation $4x + 2y = 2$ to solve for y.

$$4x + 2y = 2$$
$$4(2) + 2y = 2$$
$$8 + 2y = 2$$
$$2y = -6$$
$$y = -3$$

The apparent solution is $(2, -3)$.

Check:

$$4x + 2y = 2$$
$$4(2) + 2(-3) = 2 \qquad ?$$
$$8 - 6 = 2 \qquad \text{True}$$

$$3x - 4y = 18$$
$$3(2) - 4(-3) = 18 \qquad ?$$
$$6 + 12 = 18 \qquad \text{True}$$

The solution checks. Thus, the solution is $(2, -3)$.   ■

 Example 5   Solve the following pair of linear equations by addition. Check your solution.

$$3x - 4y = 11$$
$$4x - 5y = 14$$

Multiply both sides of the first equation by 4. Then multiply both sides of the second equation by -3 to eliminate x by addition.

$$4(3x - 4y) = 4(11)$$
$$\underline{-3(4x - 5y) = -3(14)}$$

or

$$12x - 16y = 44$$
$$\underline{-12x + 15y = -42}$$
$$0 - 1y = 2$$
$$y = -2$$

Now substitute -2 for y in the equation $3x - 4y = 11$ to solve for x.

$$3x - 4y = 11$$
$$3x - 4(-2) = 11$$
$$3x + 8 = 11$$
$$3x = 3$$
$$x = 1$$

The solution is $(1, -2)$.

Check: Left to the student.

Example 6

Solve the following pair of linear equations from Example 1 by addition. Check your solution.

$$3x + 4y = -2$$
$$6x - 8y = 32$$

Multiply both sides of the first equation by 2 and eliminate y by addition.

$$2(3x + 4y) = (2)(-2)$$
$$6x - 8y = 32$$

or

$$6x + 8y = -4$$
$$6x - 8y = 32$$
$$12x \quad = 28$$
$$x \quad = \tfrac{7}{3} \text{ or } 2\tfrac{1}{3} \qquad \text{Divide both sides by 12.}$$

Substitute $\tfrac{7}{3}$ for x in the equation $3x + 4y = -2$ to solve for y.

$$3x + 4y = -2$$
$$3\left(\tfrac{7}{3}\right) + 4y = -2$$
$$7 + 4y = -2$$
$$4y = -9 \qquad \text{Subtract 7 from both sides.}$$
$$y = -\tfrac{9}{4} \text{ or } -2\tfrac{1}{4} \qquad \text{Divide both sides by 4.}$$

The apparent solution is $\left(2\tfrac{1}{3}, -2\tfrac{1}{4}\right)$.

Check: Substitute $2\tfrac{1}{3}$ for x and $-2\tfrac{1}{4}$ for y in both equations.

$$3x + 4y = -2 \qquad\qquad 6x - 8y = 32$$
$$3\left(2\tfrac{1}{3}\right) + 4\left(-2\tfrac{1}{4}\right) = -2 \quad ? \qquad 6\left(2\tfrac{1}{3}\right) - 8\left(-2\tfrac{1}{4}\right) = 32 \quad ?$$
$$3\left(\tfrac{7}{3}\right) + 4\left(-\tfrac{9}{4}\right) = -2 \quad ? \qquad 6\left(\tfrac{7}{3}\right) - 8\left(-\tfrac{9}{4}\right) = 32 \quad ?$$
$$7 - 9 = -2 \quad \text{True} \qquad 14 + 18 = 32 \quad \text{True}$$

The solution checks. Compare the result in Figure 9.6 with this exact result. ■

In the preceding examples, we considered only pairs of linear equations with one common solution. Thus, the graphs of these equations intersect at a point, and the ordered pair that names this point is the common solution for the pair of equations. Sometimes in solving a pair of linear equations by addition, the final statement is that two unequal numbers are equal, such as $0 = -2$. If so, the pair of equations does not have a common solution. The graphs of these equations are parallel lines.

Example 7   Solve the following pair of linear equations by addition.

$$2x + 3y = 7$$
$$4x + 6y = 12$$

Multiply both sides of the first equation by -2 to eliminate x by addition.

$$\begin{array}{r} -2(2x + 3y) = -2(7) \\ 4x + 6y = 12 \end{array} \quad \text{or} \quad \begin{array}{r} -4x - 6y = -14 \\ 4x + 6y = 12 \\ \hline 0 + \ \ 0 = -2 \\ 0 = -2 \end{array}$$

Since $0 \neq -2$, there is no common solution, and the graphs of these two equations are parallel lines.   ■

If addition is used to solve a pair of linear equations and the resulting statement is $0 = 0$, then there are many common solutions. In fact, any solution of one equation is also a solution of the other. In this case, the graphs of the two equations coincide.

 Example 8   Solve the following pair of linear equations by addition.

$$2x + \ 5y = 7$$
$$4x + 10y = 14$$

Multiply both sides of the first equation by -2 to eliminate x by addition.

$$\begin{array}{r} -2(2x + 5y) = -2(7) \\ 4x + 10y = 14 \end{array} \quad \text{or} \quad \begin{array}{r} -4x - 10y = -14 \\ 4x + 10y = 14 \\ \hline 0 + \ \ 0 = 0 \\ 0 = 0 \end{array}$$

Since $0 = 0$, there are many common solutions, and the graphs of the two equations coincide.

Note: If you multiply both sides of the first equation by 2, you obtain the second equation. Thus, the two equations are equivalent. If two equations are equivalent, they should have the same graph.   ■

In Section 9.1, we saw that the graphs of two straight lines in the same plane may (a) intersect, (b) be parallel, or (c) coincide as shown in Figure 9.1. When using the addition method to solve a pair of linear equations, one of the same three possibilities occurs, as follows.

> ◆ **Addition Method Possible Cases**
>
> 1. The steps of the addition method result in exactly one ordered pair, such as $x = 2$ and $y = -5$. This ordered pair is the point at which the graphs of the two linear equations intersect.
> 2. The steps of the addition method result in a false statement, such as $0 = 7$ or $0 = -2$. This means that there is no common solution and that the graphs of the two linear equations are parallel.
> 3. The steps of the addition method result in a true statement, such as $0 = 0$. This means that there are many common solutions and that the graphs of the two linear equations coincide.

Exercises 9.2

Solve each pair of linear equations by addition. If there is one common solution, give the ordered pair of the point of intersection. If there is no one common solution, tell whether the lines are parallel or coincide.

1. $3x + y = 7$
$x - y = 1$

2. $x + y = 8$
$x - y = 4$

3. $2x + 5y = 18$
$4x - 5y = 6$

4. $3x - y = 9$
$2x + y = 6$

5. $-2x + 5y = 39$
$2x - 3y = -25$

6. $-4x + 2y = 12$
$-3x - 2y = 9$

7. $x + 3y = 6$
$x - y = 2$

8. $3x - 2y = 10$
$3x + 4y = 20$

9. $2x + 5y = 15$
$7x + 5y = -10$

10. $4x + 5y = -17$
$4x + y = 13$

11. $5x + 6y = 31$
$2x + 6y = 16$

12. $6x + 7y = 0$
$2x - 3y = 32$

13. $4x - 5y = 14$
$2x + 3y = -4$

14. $6x - 4y = 10$
$2x + y = 4$

15. $3x - 2y = -11$
$7x - 10y = -47$

16. $3x + 2y = 10$
$x + 5y = -27$

17. $x + 2y = -3$
$2x + y = 9$

18. $5x - 2y = 6$
$3x - 4y = 12$

19. $3x + 5y = 7$
$2x - 7y = 15$

20. $12x + 5y = 21$
$13x + 6y = 21$

21. $8x - 7y = -51$
$12x + 13y = 41$

22. $5x - 7y = -20$
$3x - 19y = -12$

23. $5x - 12y = -5$
$9x - 16y = -2$

24. $2x + 3y = 2$
$3x - 2y = 3$

25. $2x + 3y = 8$
$x + y = 2$

26. $4x + 7y = 9$
$12x + 21y = 12$

27. $3x - 5y = 7$
$9x - 15y = 21$

28. $2x - 3y = 8$
$4x - 3y = 0$

29. $2x + 5y = -1$
$3x - 2y = 8$

30. $3x - 7y = -9$
$2x + 14y = -6$

31. $16x - 36y = 70$
$4x - 9y = 17$

32. $8x + 12y = 36$
$16x + 15y = 45$

33. $4x + 3y = 17$
$2x - y = -4$

34. $12x + 15y = 36$
$7x - 12y = 187$

35. $2x - 5y = 8$
$4x - 10y = 16$

36. $3x - 2y = 5$
$7x + 3y = 4$

37. $5x - 8y = 10$
$-10x + 16y = 8$

38. $-3x + 2y = 5$
$-30x + y = 12$

39. $8x - 5y = 426$
$7x - 2y = 444$

40. $3x - 10y = -21$
$5x + 4y = 27$

41. $16x + 5y = 6$
$7x + \dfrac{5}{8}y = 2$

42. $\dfrac{1}{4}x - \dfrac{2}{5}y = 1$
$5x - 8y = 20$

43. $7x + 8y = 47$
$5x - 3y = 51$

44. $2x - 5y = 13$
$5x + 7y = 13$

9.3 Solving Pairs of Linear Equations by Substitution

For many problems, the *substitution* method is easier than the addition method for finding exact solutions. Use the substitution method when one equation has been solved or is easily solved for one of the variables.

> ### Solving a Pair of Linear Equations by the Substitution Method
>
> **1.** From either of the two given equations, solve for one variable in terms of the other.
> **2.** Substitute the result from Step 1 into the remaining equation. Note that this step should eliminate one variable.
>
> *continued*

3. Solve the equation from Step 2 for the remaining variable.

4. Solve for the second variable by substituting the solution from Step 3 into the equation resulting from Step 1.

5. Check by substituting the solution in both original equations.

 Example 1 Solve the following pair of linear equations by substitution. Check your solution.

$$x + 3y = 15$$
$$x = 2y$$

First, substitute $2y$ for x in the first equation.

$$x + 3y = 15$$
$$2y + 3y = 15$$
$$5y = 15$$
$$y = 3$$

Now substitute 3 for y in the equation $x = 2y$ to solve for x.

$$x = 2y$$
$$x = 2(3)$$
$$x = 6$$

The apparent solution is (6, 3).

Check: Substitute 6 for x and 3 for y.

$x + 3y = 15$	$x = 2y$
$6 + 3(3) = 15$?	$6 = 2(3)$ True
$6 + 9 = 15$ True	

Thus, the solution is (6, 3).   ■

The addition method is often preferred if the pair of linear equations has no numerical coefficients equal to 1. For example,

$$3x + 4y = 7$$
$$5x + 7y = 12$$

Example 2 Solve the following pair of linear equations by substitution.

$$2x + 5y = 22$$
$$-3x + \ y = 18$$

First, solve the second equation for y and substitute in the first equation.

$$-3x + y = 18$$
$$y = 3x + 18 \qquad \text{Add } 3x \text{ to both sides.}$$
$$2x + 5y = 22$$
$$2x + 5(3x + 18) = 22 \qquad \text{Substitute } y = 3x + 18.$$
$$2x + 15x + 90 = 22 \qquad \text{Remove parentheses.}$$
$$17x + 90 = 22 \qquad \text{Combine like terms.}$$
$$17x = -68 \qquad \text{Subtract 90 from both sides.}$$
$$x = -4 \qquad \text{Divide both sides by 17.}$$

Now substitute -4 for x in the equation $-3x + y = 18$ to solve for y.

$$-3x + y = 18$$
$$(-3)(-4) + y = 18 \qquad \text{Substitute } x = -4.$$
$$12 + y = 18$$
$$y = 6 \qquad \text{Subtract 12 from both sides.}$$

The solution is $(-4, 6)$. The check is left to the student.   ■

Knowing both the addition and substitution methods, you may choose the one that seems easier to you for each problem.

Exercises 9.3

Solve, using the substitution method, and check:

1. $2x + y = 12$
$\quad y = 3x$

2. $3x + 4y = -8$
$\quad x = 2y$

3. $5x - 2y = 46$
$\quad x = 5y$

4. $2x - y = 4$
$\quad y = -x$

5. $3x + 2y = 30$
$\quad x = y$

6. $3x - 2y = 49$
$\quad y = -2x$

7. $5x - y = 18$
$\quad y = \dfrac{1}{2}x$

8. $15x + 3y = 9$
$\quad y = -2x$

9. $x - 6y = 3$
$\quad 3y = x$

10. $4x + 5y = 10$
$\quad 4x = -10y$

11. $3x + y = 7$
$\quad 4x - y = 0$

12. $5x + 2y = 1$
$\quad y = -3x$

13. $4x + 3y = -2$
$\quad x + y = 0$

14. $7x + 8y + 93 = 0$
$\quad y = 3x$

15. $6x - 8y = 115$
$\quad x = -\dfrac{y}{5}$

16. $2x + 8y = 12$
$\quad x = -4y$

17. $3x + 8y = 27$
$\quad y = 2x + 1$

18. $4x - 5y = -40$
$\quad x = 3 - 2y$

19. $8y - 2x = -34$
$\quad x = 1 - 4y$

20. $2y + 7x = 48$
$\quad y = 3x - 2$

21. $3x + 4y = 25$
$\quad x - 5y = -17$

22. $5x - 4y = 29$
$\quad 2x + y = 9$

23. $4x + y = 30$
$\quad -2x + 5y = 18$

24. $x + 6y = -20$
$\quad 5x - 8y = 14$

25. $3x + 4y = 22$
$\quad -5x + 2y = -2$

26. $2x + 3y = 12$
$\quad 5x - 6y = -51$

9.4 Applications Involving Pairs of Linear Equations*

Often, a technical application can be expressed mathematically as a system of linear equations. The procedure is similar to that outlined in Section 6.6, except that here you need to write two equations that express the information given in the problem and that involve both unknowns.

Solving Applications Involving Equations with Two Variables

1. Choose a different variable for each of the two unknowns that you need to find. Write what each variable represents.

continued

*In this chapter, do not use the rules for calculating with measurements.

2. Write the problem as two equations using both variables. To obtain these two equations, look for two different relationships that express the two unknown quantities in equation form.

3. Solve this resulting system of equations using the methods given in this chapter.

4. Answer the question or questions asked in the problem. Check your solution using the original problem to make certain that it makes sense.

5. Check your solution in both original equations.

 **Example 1**

The sum of two voltages is 120 V. The difference between them is 24 V. Find each voltage.

Let x = large voltage

    y = small voltage

The sum of two voltages is 120 V; that is,

$$x + y = 120$$

The difference between them is 24 V; that is,

$$x - y = 24$$

The system of equations is

$$x + y = 120$$
$$\underline{x - y = 24}$$
$$2x = 144 \qquad \text{Add the equations.}$$
$$x = 72$$

Substitute $x = 72$ in the equation $x + y = 120$ and solve for y.

$$x + y = 120$$
$$72 + y = 120$$
$$y = 48 \qquad \text{Subtract 72 from both sides.}$$

Thus, the voltages are 72 V and 48 V.

Check: The sum of the voltages, 72 V + 48 V, is 120 V. The difference between them, 72 V − 48 V, is 24 V.   ■

 Example 2

How many pounds of feed mix A that is 75% corn and how many pounds of feed mix B that is 50% corn will need to be mixed to make a 400-lb mixture that is 65% corn?

Let x = number of pounds of mix A (75% corn)

    y = number of pounds of mix B (50% corn)

The sum of the two mixtures is 400 lb; that is,

$$x + y = 400$$

Thus, 75% of x is corn and 50% of y is corn. Adding these amounts together results in a 400-lb final mixture that is 65% corn. Write an equation in terms of the amount of corn; that is, the amount of corn in each mix equals the amount of corn in the final mixture.

$$0.75x + 0.50y = (0.65)(400)$$
$$\text{or} \quad 0.75x + 0.50y = 260$$

The system of equations is

$$x + y = 400$$
$$0.75x + 0.50y = 260$$

First, let's multiply both sides of the second equation by 100 to eliminate the decimals.

$$x + y = 400$$
$$75x + 50y = 26,000$$

Then multiply both sides of the first equation by -50 to eliminate y by addition.

$$-50x - 50y = -20,000$$
$$\underline{75x + 50y = 26,000}$$
$$25x = 6,000$$
$$\frac{25x}{25} = \frac{6000}{25} \qquad \text{Divide both sides by 25.}$$
$$x = 240$$

Now substitute 240 for x in the equation $x + y = 400$ to solve for y.

$$x + y = 400$$
$$240 + y = 400$$
$$y = 160$$

Therefore, we need 240 lb of mix A and 160 lb of mix B.

Check: Substitute 240 for x and 160 for y in both original equations.

$x + y = 400$	$0.75x + 0.50y = 260$
$240 + 160 = 400$ True	$0.75(240) + 0.50(160) = 260$?
	$180 + 80 = 260$ True ■

Example 3

A company sells two grades of sand. One grade sells for 15¢/lb, and the other sells for 25¢/lb. How much of each grade needs to be mixed to obtain 1000 lb of a mixture worth 18¢/lb?

Let x = amount of sand selling at 15¢/lb

y = amount of sand selling at 25¢/lb

The total amount of sand is 1000 lb; that is,

$$x + y = 1000$$

One grade sells at 15¢/lb, and the other sells at 25¢/lb. The two grades are mixed to obtain 1000 lb of a mixture worth 18¢/lb. Here, we need to write an equation that relates the cost of the sand; that is, the cost of the sand separately equals the cost of the sand mixed.

$$15x + 25y = 18(1000)$$

That is, the cost of x pounds of sand at 15¢/lb is $15x$ cents. The cost of y pounds of sand at 25¢/lb is $25y$ cents. The cost of 1000 pounds of sand at 18¢/lb is 18(1000) cents. Therefore, the system of equations is

$$x + y = 1000$$
$$15x + 25y = 18,000$$

Multiply the first equation by -15 to eliminate x by addition.

$$-15x - 15y = -15{,}000$$
$$\underline{15x + 25y = \quad 18{,}000}$$
$$10y = \quad 3000$$
$$y = 300$$

Substitute $y = 300$ in the equation $x + y = 1000$ and solve for x.

$$x + y = 1000$$
$$x + 300 = 1000$$
$$x = 700$$

That is, 700 lb of sand selling at 15¢/lb and 300 lb of sand selling at 25¢/lb are needed to obtain 1000 lb of a mixture worth 18¢/lb.

Check: Left to the student.   ■

Example 4

Enclose a rectangular yard (Figure 9.7) with a fence so that the length is twice the width. The length of the 80-ft house is used to enclose part of one side of the yard. If 580 ft of fencing are used, what are the dimensions of the yard?

Let x = length of the yard
y = width of the yard

The amount of fencing used (two lengths plus two widths minus the length of the house) is 580 ft; that is,

$$2x + 2y - 80 = 580$$
$$2x + 2y = 660$$

The length of the yard is twice the width; that is,

$$x = 2y$$

The system of equations is

$$2x + 2y = 660$$
$$x = 2y$$

Substitute $2y$ for x in the equation $2x + 2y = 660$ and solve for y.

$$2(2y) + 2y = 660$$
$$4y + 2y = 660$$
$$6y = 660$$
$$y = 110$$

Now substitute 110 for y in the equation $x = 2y$ and solve for x.

$$x = 2y$$
$$x = 2(110)$$
$$x = 220$$

Therefore, the length is 220 ft and the width is 110 ft. (The check is left to the student.)

FIGURE 9.7

Exercises 9.4 _____

1. A board 96 cm long is cut into two pieces so that one piece is 12 cm longer than the other. Find the length of each piece.

2. Find the capacity of two trucks if 6 trips of the smaller and 4 trips of the larger make a total haul of 36 tons, and 8 trips of the larger and 4 trips of the smaller make a total haul of 48 tons.

3. A plumbing contractor decides to field-test two new pumps. One is rated at 180 gal/h and the other at 250 gal/h. She tests one, then the other. Over a period of 6 h, she pumps a total of 1325 gal. Assume that both pumps operate as rated. How long is each in operation?

4. A bricklayer lays an average of 150 bricks per hour. During the job, he is called away and replaced by a less experienced man, who averages 120 bricks an hour. The two men laid 930 bricks in 7 h. How long did each work?

5. Two welders worked a total of 48 h on a project. One welder made $32/h, while the other made $41/h. If the gross earnings of the two welders was $1734 for the job, how many hours did each welder work?

6. A contractor finds a bill for $1110 for 720 ceiling tiles. She knows that there were two types of tiles used; one selling at $1.25 a tile and the other at $1.75 a tile. How many of each type were used?

7. A farmer has two types of feed. One has 5% digestible protein and the other 15% digestible protein. How much of each type will she need to mix 100 lb of 12% digestible-protein feed?

8. A dairyman wants to make 125 lb of 12% butterfat cream. How many pounds of 40% butterfat cream and how many pounds of 2% butterfat milk must he mix?

9. A farmer sells corn for $4.60/bu and soybeans for $11.60/bu. The entire 3150 bu sells for $17,640. How much of each is sold?

10. A farmer has a 1.4% solution and a 2.9% solution of a pesticide. How much of each would he mix to get 2000 gal of 2% solution for his sprayer?

11. A farmer has a 6% solution and a 12% solution of pesticide. How much of each must she mix to have 300 gal of an 8% solution for her sprayer?

12. The sum of two capacitors is 85 microfarads (μF). The difference between them is 25 μF. Find the size of each capacitor.

13. Nine batteries are hooked in series to provide a 33-V power source. Some of the batteries are 3 V and some are 4.5 V. How many of each type are used?

14. In a parallel circuit, the total current is 1.25 A through the two branches. One branch has a resistance of 50 Ω, and the other has a resistance of 200 Ω. What current is flowing through each branch? *Note:* In a parallel circuit, the products of the current in amperes and the resistance in ohms are equal in all branches.

15. How much of an 8% solution and a 12% solution would you use to make 140 mL of a 9% electrolyte solution?

16. The total current in a parallel circuit with seven branches is 1.95 A. Some of the branches have currents

of 0.25 A and others 0.35 A. How many of each type of branch are in the circuit? *Note:* The total current in a parallel circuit equals the sum of the currents in each branch.

17. A small single-cylinder engine was operated on a test stand for 14 min. It was run first at 850 rpm and was then increased to 1250 rpm. A total of 15,500 revolutions was counted during the test. How long was the engine operated at each speed?

18. In testing a hybrid engine, various mixtures of gasoline and methanol are being tried. How much of a 95% gasoline mixture and how much of an 80% gasoline mixture would be needed to make 240 gal of a 90% gasoline mixture?

19. An engine on a test stand was operated at two fixed settings, each with an appropriate load. At the first setting, fuel consumption was 1 gal every 12 min. At the second setting, it was 1 gal every 15 min. The test took 5 h, and 22 gal of fuel were used. How long did the engine run at each setting?

20. A mechanic stores a parts cleaner as a 65% solution, which is to be diluted to a 25% solution for use. Someone accidentally prepares a 15% solution. How much of the 65% solution and the 15% solution should be mixed to make 100 gal of the 25% solution?

21. Amy has a 3% solution and an 8% solution of a pesticide. How much of each must she mix to have 200 L of 4% solution?

22. Two rental airplanes were flown a total of 54 h in one month. One plane rents for $105/h, and the other rents for $130/h. The total income from the two planes was $6570. Find the number of hours each plane was flown for the month.

23. When three identical compressors and five air-handling units are in operation, a total of 26.4 A are needed. When only two compressors and three air-handling units are being used, the current requirement is 17.2 A. How many amps are required
   a. by each compressor and
   b. by each air-handling unit?

24. A hospital has 35% saline solution on hand. How much water and how much of this solution should be used to prepare 140 mL of a 20% saline solution?

25. A nurse gives 1000 cm^3 of an intravenous (IV) solution over a period of 8 h. It is given first at a rate of 140 cm^3/h, then at a reduced rate of 100 cm^3/h. How long should it be given at each rate?

26. A hospital has a 4% saline solution and an 8% saline solution on hand. How much of each should be used to prepare 1000 cm^3 of 5% saline solution?

27. A medication is available in 2-cm^3 vials and in 5-cm^3 vials. In a certain month, 42 vials were used, totaling 117 cm^3 of medication. How many of each type of vial were used?

28. A salesman turns in a ticket on two carpets for $2360. He sold a total of 75 yd^2 of carpet. One type was worth $27.50/$yd^2$, and the second type was worth $36/$yd^2$. He neglects to note, however, how much of each type he sold. How much did he sell of each type?

29. An apartment owner rents one-bedroom apartments for $575 and two-bedroom apartments for $650. A total of 13 apartments rent for $8150 a month. How many of each type does she have?

30. A sporting goods store carries two types of snorkels. One sells for $14.95, and the other sells for $21.75. Records for July show that 23 snorkels were sold, for $357.45. How many of each type were sold?

31. The sum of two resistors is 550 Ω. One is 4.5 times the other. Find the size of each resistor.

32. One concrete mix contains four times as much gravel as cement. The total volume is 15 yd^3. How much of each ingredient is used?

33. A wire 120 cm long is to be cut into two pieces so that one piece is three times as long as the other. Find the length of each piece.

34. The sum of two resistances is 1500 Ω. The larger is four times the smaller. Find the size of each resistance.

35. A rectangle is twice as long as it is wide. Its perimeter is 240 cm. Find the length and the width of the rectangle.

36. A duct from a furnace forks into two ducts. The air coming into the two ducts was 2300 cubic feet per minute. If the one duct took 800 cubic feet per minute more than the other duct, what was the flow in cubic feet per minute in each duct?

37. The sum of three currents is 210 mA. Two currents are the same. The third is five times either of the other two. Find the third current.

38. An experienced welder makes $\frac{5}{3}$ as many welds as a beginning welder. If an experienced welder and a beginning welder complete 240 welds, how many were from the experienced welder and how many from the beginning welder?

39. On a survey plat, the perimeter is noted as 2800 ft. If the length is 40 ft more than the width, find the dimensions of the plot.

40. A supply duct is to run the perimeter of a building having a length twice the width. The total duct length is 696 ft. What are the dimensions of the building?

41. In a drawing, the perimeter of a room is 40 ft. If the length is decreased by 6 ft and the width is doubled, the room would have the same perimeter.
   a. Find the original dimensions.
   b. Which room would have the greater square footage, or would they be equal?

42. A rectangular walkway in front of an office building has a perimeter of 150 ft. If the length is three times the width, find the dimensions of the walkway.

43. If the length of a building is $2\frac{1}{2}$ times the width and each dimension is increased by 5 ft, then the perimeter is 230 ft. Find the dimensions of the original building.

44. The center-to-center distance between a fan and a motor shaft is 30.0 in. See Illustration 1. Pulleys with a 4.5 : 1 ratio are installed. The distance between the pulleys is 19.0 in. Find the diameter of each pulley.

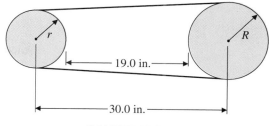

ILLUSTRATION 1

45. A carpenter starts with a board that is 12 ft long and cuts it into two pieces that differ in length by 8 in. Assuming no loss in length from the saw blade kerf, how long are the two pieces?

46. Taughannock Falls near Ithaca, New York, is about 39 ft taller than Niagara Falls. The two falls together would be 391 ft. How tall is each fall?

47. Ground corn and soybean meal are blended to produce 100 lb of animal feed that is 15% crude protein (CP). The ground corn is rated at 10% CP and supplies 8.5 lb of protein to the mix. How much of each grain is required for the mixture?

Chapter 9 Group Activities

1. In this chapter, we used the addition and substitution methods to solve a system of two linear equations with two unknowns. Investigate the following to see if you can discover a new way to solve a system with two equations and two unknowns. First, solve the following any way you know how.
   a. $x + y = 5$
   $\quad\ x - y = 1$
   b. $2x + 3y = 7$
   $\quad\ 3x + 4y = 10$
   c. $4x - y = 11$
   $\quad\ x + 5y = 8$

You need to know that a determinant is a square array of numbers. The value of the determinant is calculated by the following formula for an array of numbers with 2 rows and 2 columns.

$$\begin{vmatrix} a & b \\ c & d \end{vmatrix} = ad - cb$$

For example,

$$\begin{vmatrix} 2 & 1 \\ 3 & 4 \end{vmatrix} = 2 \cdot 4 - 3 \cdot 1 = 8 - 3 = 5$$

The value of this determinant is 5. With this in mind, part **a** is

$$x + y = 5$$
$$x - y = 1$$

Evaluate the numbers given by the ratio of the following determinants:

$$\frac{\begin{vmatrix} 5 & 1 \\ 1 & -1 \end{vmatrix}}{\begin{vmatrix} 1 & 1 \\ 1 & -1 \end{vmatrix}} =$$

Also evaluate

$$\frac{\begin{vmatrix} 1 & 5 \\ 1 & 1 \end{vmatrix}}{\begin{vmatrix} 1 & 1 \\ 1 & -1 \end{vmatrix}} =$$

What are these two numbers when compared to the solutions of part **a**? Similarly for part **b**, evaluate

$$\frac{\begin{vmatrix} 7 & 3 \\ 10 & 4 \end{vmatrix}}{\begin{vmatrix} 2 & 3 \\ 3 & 4 \end{vmatrix}} =$$

$$\frac{\begin{vmatrix} 2 & 7 \\ 3 & 10 \end{vmatrix}}{\begin{vmatrix} 2 & 3 \\ 3 & 4 \end{vmatrix}} =$$

What are these two numbers when compared to the solutions of part **b**?

From where did the numbers in the array come? Use what you have discovered to solve part **c**.

Note: This method is called Cramer's Rule. You can use this as long as the denominators are not zero. You can use this method to solve a system of two equations with two unknowns.

Chapter 9   Summary

Glossary of Basic Terms

Solving a pair of linear equations.   To solve a pair of linear equations with two variables simultaneously, you must find an ordered pair that will make both equations true at the same time. (p. 334)

9.1   Solving Pairs of Linear Equations by Graphing

1. **Graphs of linear equations with two variables:** The graphs of a pair of linear equations with two variables consist of two straight lines and may be arranged as follows:
   a. The lines intersect. If so, they have one point in common whose coordinates define the point of intersection, which is the common solution.
   b. The lines are parallel. If so, they have no point in common, and the equations have no common solution.
   c. The lines coincide; that is, one line lies on top of the other. If so, any solution of one equation is also a solution of the other. With infinitely many points of intersection, there are infinitely many common solutions. The solution of the pair of equations consists of all points on the common line. (See Figure 9.1 on p. 334.)

9.2   Solving Pairs of Linear Equations by Addition

1. **Solving a pair of linear equations by the addition method:** The addition method is outlined as follows:
   a. If necessary, multiply both sides of one or both equations by a number (or numbers) so that the numerical coefficients of one of the variables are negatives of each other.
   b. Add the two equations from Step **a** to obtain an equation containing only one variable.
   c. Solve the equation from Step **b** for the one remaining variable.
   d. Solve for the second variable by substituting the solution from Step **c** in either of the original equations.
   e. Check your solutions by substituting the ordered pair in both original equations. (p. 341)

2. **Addition method possible cases:** When solving a pair of linear equations by graphing, we found three possible cases; that is, the lines may intersect, be parallel, or coincide. In using the addition method, one of these same three possibilities occurs as follows:
   a. The steps of the addition method result in exactly one ordered pair, which is the point of intersection, whose coordinates are the common solution.

b. The steps of the addition method result in a false statement, such as $0 = 7$ or $0 = -2$. This means that there is no common solution and that the graphs of the two linear equations are parallel.

c. The steps of the addition method result in a true statement, such as $0 = 0$. This means that there are many common solutions and that the graphs of the two linear equations coincide. (p. 344)

9.3 Solving Pairs of Linear Equations by Substitution

1. Solving a pair of linear equations by the substitution method: The substitution method is outlined as follows:

a. From either of the two given equations, solve for one variable in terms of the other.

b. Substitute the result from Step **a** into the remaining equation. Note that this step should eliminate one variable.

c. Solve the equation from Step **b** for the remaining variable.

d. Solve for the second variable by substituting the solution from Step **c** into the equation resulting from Step **a**.

e. Check by substituting the solution in both original equations. (pp. 345–346)

9.4 Applications Involving Pairs of Linear Equations

1. Solving applications involving equations with two variables: Follow these steps:

a. Choose a different variable for each of the two unknowns you need to find. Write what each variable represents.

b. Write the problem as two equations using both variables. To obtain these two equations, look for two different relationships that express the two unknown quantities in equation form.

c. Solve this resulting system of equations using the methods given in this chapter.

d. Answer the question or questions asked in the problem. Check your solution using the original problem to make certain that it makes sense.

e. Check your solution in both original equations. (pp. 347–348)

 ## Chapter 9 Review

Draw the graphs of each pair of linear equations on the same set of coordinate axes. Find the point of intersection. If the lines do not intersect, tell whether the lines are parallel or coincide:

1. $x + y = 6$
$2x - y = 3$

2. $y = 2x + 5$
$y = x + 2$

3. $4x + 6y = 12$
$6x + 9y = 18$

4. $5x - 2y = 10$
$10x - 4y = -20$

5. $3x + 4y = -1$
$x = -3$

6. $y = 2x$
$y = -5$

Solve each system of equations:

7. $x + y = 7$
$2x - y = 2$

8. $3x + 2y = 11$
$x + 2y = 5$

9. $3x - 5y = -3$
$2x - 3y = -1$

10. $2x - 3y = 1$
$4x - 6y = 5$

11. $3x + 5y = 8$
$6x - 4y = 44$

12. $5x + 7y = 22$
$4x + 8y = 20$

13. $x + 2y = 3$
$3x + 6y = 9$

14. $3x + 5y = 52$
$y = 2x$

15. $5y - 4x = -6$
$x = \dfrac{1}{2}y$

16. $3x - 7y = -69$
$y = 4x + 5$

17. You can buy twenty 20-amp electrical switches and eight 15-amp 4-way electrical switches for $162 or sixty 20-amp electrical switches and forty 15-amp 4-way electrical switches for $670. Find the price of each type of switch.

18. The sum of the length and width of a rectangular lot is 190 ft. The lot is 75 ft longer than it is wide. Find the length and width of the lot.

19. The sum of two inductors is 90 millihenrys (mH). The larger is 3.5 times the smaller. What is the size of each inductor?

20. The sum of two lengths is 90 ft, and their difference is 20 ft. Find the two lengths.

Chapter 9   Test

Solve each system of equations by the method indicated.

1. $3x - y = 5$
   $2x - y = 0$ by graphing

2. $y = 3x - 5$
   $y = 2x - 1$ by graphing

3. $2x + 7y = -1$
   $x + 2y = 1$ by addition

4. $x - 3y = 8$
   $x + 4y = -6$ by addition

5. $\quad\quad y = -3x$
   $2x + 3y = 13$ by substitution

6. $\quad\quad x = 7y$
   $2x - 8y = 12$ by substitution

Solve each pair of linear equations. If there is one common solution, give the ordered pair of the point of intersection. If there is no one common solution, tell whether the lines are parallel or coincide:

7. $4x - 5y = 10$
   $-8x + 10y = 6$

8. $3x - y = 8$
   $12x - 4y = 32$

9. $x - 3y = -8$
   $2x + y = 5$

10. The perimeter of a rectangular lot is 600 m. The length is twice the width. Find its length and width.

11. The sum of two resistances is 550 Ω. The difference between them is 250 Ω. Find the size of each resistance.

10 Factoring Algebraic Expressions

Mathematics at Work

Agriculture requires a wide variety of support specialists. Those working in agricultural business management often specialize in agrimarketing, management, animal science, soils and fertilizers, grain merchandising, and crop production. Other areas include agricultural equipment management and marketing, equine management, and landscape design, construction, and management, as well as turf management. Precision farming requires specialist support in satellite-controlled soil sampling, fertilizer application, crop scouting, and yield mapping using geographic information systems (GIS) and precision farming technology. Farmers, soil testing labs, fertilizer and chemical companies, banks, and other agribusinesses need employees trained in these new techniques. Agricultural research colleges and companies need agricultural research technicians to assist in research development projects for seed, chemical production, and other agricultural products. Training and education for these careers are available at many community colleges and trade schools. For more information, go to the website listed below.

Maximiliam Stock Ltd./Photo Researchers Inc.

Agriculture Support Specialists
Agricultural research technician investigating the effects of herbicides and fertilizer on crop sizes.

www.cengage.com/mathematics/ewen

Objectives

■ Find the greatest common monomial factor in an algebraic expression.

■ Find the product of two binomials mentally.

■ Factor trinomials.

■ Find the square of a binomial.

■ Identify and find the product of the sum and difference of two terms.

■ Identify and factor perfect square trinomials.

■ Identify and factor the difference of two squares.

10.1 Finding Monomial Factors

Factoring an algebraic expression, like finding the prime factors of a number, means *writing the expression as a product of factors*. The prime factorization of 12 is $2 \cdot 2 \cdot 3$. Other factorizations of 12 are $2 \cdot 6$ and $4 \cdot 3$. Since factorization means writing a number or an algebraic expression as a product, then a number or an algebraic expression divided by one factor generates another factor. Thus, 12 divided by 2 gives 6, so 2 times 6 is a factorization of 12.

To factor the expression $2x + 2y$, notice that 2 is a factor common to both terms of the expression. In other words, 2 is a factor of $2x + 2y$. To find the other factor, divide by 2.

$$\frac{2x + 2y}{2} = \frac{2x}{2} + \frac{2y}{2} = x + y$$

Therefore, a factorization of $2x + 2y$ is $2(x + y)$.

A **monomial factor** is a one-term factor that divides each term of an algebraic expression. Here, 2 divides each term of the algebraic expression and is called a monomial factor. When factoring any algebraic expression, *always look first for monomial factors* that are common to all terms.

 Example 1

Factor: $3a + 6b$.

First, look for a common monomial factor. Since 3 divides both $3a$ and $6b$, 3 is a common monomial factor of $3a + 6b$. Divide $3a + 6b$ by 3.

$$\frac{3a + 6b}{3} = \frac{3a}{3} + \frac{6b}{3}$$
$$= a + 2b$$

Thus, $3a + 6b = 3(a + 2b)$.

Check this result by multiplication: $3(a + 2b) = 3a + 6b$.   ■

Example 2

Factor: $4x^2 + 8x + 12$.

Since 4 divides each term of the expression, divide $4x^2 + 8x + 12$ by 4 to obtain the other factor.

$$\frac{4x^2 + 8x + 12}{4} = \frac{4x^2}{4} + \frac{8x}{4} + \frac{12}{4}$$
$$= x^2 + 2x + 3$$

Thus, $4x^2 + 8x + 12 = 4(x^2 + 2x + 3)$.

Check: $4(x^2 + 2x + 3) = 4x^2 + 8x + 12$

Your product should be the original expression.

Note: In this example, 2 is also a common factor of each term of the expression. However, 4 is the greatest common factor. The **greatest common factor of a polynomial** is the largest common factor that divides all terms in the expression. *When factoring, always choose the monomial factor that is the greatest common factor.* ■

 Example 3

Factor: $15ax - 6ay$.

Note that 3 divides both $15ax$ and $6ay$, so 3 is a common factor. However, a also divides $15ax$ and $6ay$, so a is also a common factor. We are looking for the greatest common factor (GCF), which in this case is $3a$. Then we divide $15ax - 6ay$ by $3a$ to obtain the other factor.

$$\frac{15ax - 6ay}{3a} = \frac{15ax}{3a} - \frac{6ay}{3a}$$
$$= 5x - 2y$$

Thus, $15ax - 6ay = 3a(5x - 2y)$.

Note that $3(5ax - 2ay)$ or $a(15x - 6y)$ are also factored forms of $15ax - 6ay$. However, we use the monomial factor that is the greatest common factor. ■

 Example 4

Factor: $15xy^2 - 25x^2y + 10xy$.

The greatest common factor is $5xy$. Dividing each term by $5xy$, we have
$15xy^2 - 25x^2y + 10xy = 5xy(3y - 5x + 2)$. ■

Exercises 10.1

Factor:

1. $4a + 4$

2. $3x - 6$

3. $bx + by$

4. $9 - 18y$

5. $15b - 20$

6. $12ab + 30ac$

7. $x^2 - 7x$

8. $3x^2 - 6x$

9. $a^2 - 4a$

10. $7xy - 21y$

11. $4n^2 - 8n$

12. $10x^2 + 5x$

13. $10x^2 + 25x$

14. $y^2 - 8y$

15. $3r^2 - 6r$

16. $x^3 + 13x^2 + 25x$

17. $4x^4 + 8x^3 + 12x^2$

18. $9x^4 - 15x^2 - 18x$

19. $9a^2 - 9ax^2$

20. $a - a^3$

21. $10x + 10y - 10z$

22. $2x^2 - 2x$

23. $3y - 6$

24. $y - 3y^2$

25. $14xy - 7x^2y^2$

26. $25a^2 - 25b^2$

27. $12x^2m - 7m$

28. $90r^2 - 10R^2$

29. $60ax - 12a$

30. $2x^2 - 100x^3$

31. $52m^2n^2 - 13mn$

32. $40x - 8x^3 + 4x^4$

33. $52m^2 - 14m + 2$

34. $27x^3 - 54x$

35. $36y^2 - 18y^3 + 54y^4$

36. $20y^3 - 10y^2 + 5y$

37. $6m^4 - 12m^2 + 3m$

38. $-16x^3 - 32x^2 - 16x$

39. $-4x^2y^3 - 6x^2y^4 - 10x^2y^5$

40. $18x^3y - 30x^4y + 48xy$

41. $3a^2b^2c^2 + 27a^3b^3c^2 - 81abc$

42. $15x^2yz^4 - 20x^3y^2z^2 + 25x^2y^3z^2$

43. $4x^3z^4 - 8x^2y^2z^3 + 12xyz^2$

44. $18a^2b^2c^2 + 24ab^2c^2 - 30a^2c^2$

10.2 Finding the Product of Two Binomials Mentally

In Section 5.5, you learned how to multiply two binomials such as $(2x + 3)(4x - 5)$ by the following method:

$$
\begin{array}{r}
2x + 3 \\
4x - 5 \\
\hline
-10x - 15 \\
8x^2 + 12x \\
\hline
8x^2 + 2x - 15
\end{array}
$$

This process of multiplying two binomials can be shortened as follows.

Finding the Product of Two Binomials Mentally

1. The *first term* of the product is the product of the first terms of the binomials.
2. The *middle term* of the product is the sum of the outer product and the inner product of the binomials.
3. The *last term* of the product is the product of the last terms of the binomials.

Let's use the above steps to find the product $(2x + 3)(4x - 5)$.

Step 1   Product of the first terms: $(2x)(4x) = 8x^2$

Step 2   Outer product $= (2x)(-5) = -10x$

$$(2x + 3)(4x - 5)$$

   Inner product $= (3)(4x)$ $=$ $\underline{12x}$

                  Sum: $2x$

Step 3   Product of the last terms: $(3)(-5) = \underline{-15}$

   Therefore, $(2x + 3)(4x - 5)$ $= 8x^2 + 2x - 15$

Note that in each method, we found the exact same terms. The second method is much quicker, especially when you become more familiar and successful with it. The second method is used to factor polynomials. Factoring polynomials is the content of the rest of this chapter and a necessary part of the next chapter. Therefore, it is very important that you learn to find the product of two binomials, mentally, before proceeding with the next section.

This method is often called the *FOIL method*, where *F* refers to the product of the *first* terms, *O* refers to the *outer* product, *I* refers to the *inner* product, and *L* refers to the product of the *last* terms.

Example 1   Find the product $(2x - 7)(3x - 4)$ mentally.

Step 1   Product of the first terms: $(2x)(3x) = 6x^2$

Step 2   Outer product $= (2x)(-4) = -8x$

$$(2x - 7)(3x - 4)$$

Inner product $= (-7)(3x) = \underline{-21x}$

Sum: $-29x$

Step 3   Product of the last terms: $(-7)(-4) = \underline{\hspace{1cm} 28}$

Therefore, $(2x - 7)(3x - 4)$ $= 6x^2 - 29x + 28$

                                                                                    ■

 Example 2   Find the product $(x + 4)(3x + 5)$ mentally.

Step 1   Product of the first terms: $(x)(3x) = 3x^2$

Step 2   Outer product $= (x)(5) = 5x$

$$(x + 4)(3x + 5)$$

Inner product $= (4)(3x) = \underline{12x}$

Sum: $17x$

Step 3   Product of the last terms: $(4)(5) = \underline{\hspace{1cm} 20}$

Therefore, $(x + 4)(3x + 5)$ $= 3x^2 + 17x + 20$

                                                                                    ■

By now, you should be writing only the final result of each product. If you need some help, refer to the three steps at the beginning of this section and the outline shown in Examples 1 and 2.

 Example 3   Find the product $(x + 8)(x + 5)$ mentally.

$(x + 8)(x + 5) = x^2 + (5x + 8x) + 40$ mental step

$= x^2 + 13x + 40$ product   ■

 Example 4   Find the product $(x - 6)(x - 9)$ mentally.

$(x - 6)(x - 9) = x^2 + (-9x - 6x) + 54$ mental step

$= x^2 - 15x + 54$ product   ■

 Example 5   Find the product $(x + 2)(x - 5)$ mentally.

$(x + 2)(x - 5) = x^2 + (-5x + 2x) - 10$

$= x^2 - 3x - 10$ ■

 Example 6   Find the product $(4x + 1)(5x + 8)$ mentally.

$(4x + 1)(5x + 8) = 20x^2 + (32x + 5x) + 8$

$= 20x^2 + 37x + 8$ ■

Example 7   Find the product $(6x + 5)(2x - 3)$ mentally.

$$(6x + 5)(2x - 3) = 12x^2 + (-18x + 10x) - 15$$
$$= 12x^2 - 8x - 15 \qquad ■$$

 Example 8   Find the product $(4x - 5)(4x - 5)$ mentally.

$$(4x - 5)(4x - 5) = 16x^2 + (-20x - 20x) + 25$$
$$= 16x^2 - 40x + 25 \qquad ■$$

Exercises 10.2

Find each product mentally:

1. $(x + 5)(x + 2)$
2. $(x + 3)(2x + 7)$
3. $(2x + 3)(3x + 4)$
4. $(x + 3)(x + 18)$
5. $(x - 5)(x - 6)$
6. $(x - 9)(x - 8)$
7. $(x - 12)(x - 2)$
8. $(x - 9)(x - 4)$
9. $(x + 8)(2x + 3)$
10. $(3x - 7)(2x - 5)$
11. $(x + 6)(x - 2)$
12. $(x - 7)(x - 3)$
13. $(x - 9)(x - 10)$
14. $(x - 9)(x + 10)$
15. $(x - 12)(x + 6)$
16. $(2x + 7)(4x - 5)$
17. $(2x - 7)(4x + 5)$
18. $(2x - 5)(4x + 7)$
19. $(2x + 5)(4x - 7)$
20. $(6x + 5)(5x - 1)$
21. $(7x + 3)(2x + 5)$
22. $(5x - 7)(2x + 1)$
23. $(x - 9)(3x + 8)$
24. $(x - 8)(2x + 9)$
25. $(6x + 5)(x + 7)$
26. $(16x + 3)(x - 1)$

27. $(13x - 4)(13x - 4)$
28. $(12x + 1)(12x + 5)$
29. $(10x + 7)(12x - 3)$
30. $(10x - 7)(12x + 3)$
31. $(10x - 7)(10x - 3)$
32. $(10x + 7)(10x + 3)$
33. $(2x - 3)(2x - 5)$
34. $(2x + 3)(2x + 5)$
35. $(2x - 3)(2x + 5)$
36. $(2x + 3)(2x - 5)$
37. $(3x - 8)(2x + 7)$
38. $(3x + 8)(2x - 7)$
39. $(3x + 8)(2x + 7)$
40. $(3x - 8)(2x - 7)$
41. $(8x - 5)(2x + 3)$
42. $(x - 7)(x + 5)$
43. $(y - 7)(2y + 3)$
44. $(m - 9)(m + 2)$
45. $(3n - 6y)(2n + 5y)$
46. $(6a - b)(2a + 3b)$
47. $(4x - y)(2x + 7y)$
48. $(8x - 12)(2x + 3)$
49. $\left(\dfrac{1}{2}x - 8\right)\left(\dfrac{1}{4}x - 6\right)$
50. $\left(\dfrac{2}{3}x - 6\right)\left(\dfrac{1}{3}x + 9\right)$

 # 10.3 Finding Binomial Factors

A **binomial factor** is a two-term factor of an algebraic expression. The factors of a trinomial are often binomial factors. To find these binomial factors, you must "undo" the process of multiplication as presented in Section 10.2. The following steps will enable you to undo the multiplication in a trinomial such as $x^2 + 7x + 10$.

Step 1   Factor any common monomial. In the expression $x^2 + 7x + 10$, there is no common factor.

Step 2   If $x^2 + 7x + 10$ can be factored, the two factors will probably be binomials. Write parentheses for the binomials.
$$x^2 + 7x + 10 = (\qquad)(\qquad)$$

Step 3   The product of the first two terms of the binomials is the first term of the trinomial. So the first term in each binomial must be x.
$$x^2 + 7x + 10 = (x\qquad)(x\qquad)$$

Step 4   Here, all the signs in the trinomial are positive, so the signs in the binomials are also positive.
$$x^2 + 7x + 10 = (x + \qquad)(x + \qquad)$$

Step 5 Find the last terms of the binomials by finding two numbers that have a product of $+10$ and a sum of $+7$. The only possible factorizations of 10 are $1 \cdot 10$ and $2 \cdot 5$. The sums of the pairs of factors are $1 + 10 = 11$ and $2 + 5 = 7$. Thus, the numbers that you want are 2 and 5.

$$x^2 + 7x + 10 = (x + 2)(x + 5)$$

Step 6 Multiply the two binomials as a check, to see whether their product is the same as the original trinomial.

 Example 1   Factor the trinomial $x^2 + 15x + 56$.

Step 1 $x^2 + 15x + 56$ has no common monomial factor.
Step 2 $x^2 + 15x + 56 = (\quad)(\quad)$
Step 3 $x^2 + 15x + 56 = (x \quad)(x \quad)$
Step 4 $x^2 + 15x + 56 = (x + \quad)(x + \quad)$ All signs in the trinomial are positive.

To determine which factors of 56 to use, list all possible pairs.

$$1 \cdot 56 = 56 \qquad 1 + 56 = 57$$
$$2 \cdot 28 = 56 \qquad 2 + 28 = 30$$
$$4 \cdot 14 = 56 \qquad 4 + 14 = 18$$
$$7 \cdot 8 = 56 \qquad 7 + 8 = 15$$

Since the coefficient of x in the trinomial is 15, choose 7 and 8 for the second term of the binomial factors. There are no other pairs of positive whole numbers with a product of 56 and a sum of 15.

Step 5 $x^2 + 15x + 56 = (x + 7)(x + 8)$

In actual work, all five steps are completed in one or two lines, depending on whether or not there is a common monomial.

Step 6   *Check:* $(x + 7)(x + 8) = x^2 + 15x + 56$ ■

Example 2   Factor the trinomial $x^2 - 13x + 36$.

Note that the only difference between this trinomial and the ones we have considered previously is the sign of the second term. Here, the sign of the second term is negative instead of positive. Thus, the steps for factoring will be the same except for Step 4.

Step 1 $x^2 - 13x + 36$ has no common monomial factor.
Step 2 $x^2 - 13x + 36 = x^2 + (-13)x + 36$
$$= (\quad)(\quad)$$
Step 3 $x^2 + (-13)x + 36 = (x \quad)(x \quad)$

Note that the sign of the third term ($+36$) is positive and the coefficient of the second term ($-13x$) is negative. Since 36 is positive, the two factors of $+36$ must have like signs; and since the coefficient of $-13x$ is negative, the signs in the two factors must both be negative.

Step 4 $x^2 + (-13)x + 36 = (x - \quad)(x - \quad)$

Find two integers whose product is 36 and whose sum is -13. Since $(-4)(-9) = 36$ and $(-4) + (-9) = -13$, these are the factors of 36 to be used.

Step 5 $x^2 - 13x + 36 = (x - 4)(x - 9)$
Step 6   *Check:* $(x - 4)(x - 9) = x^2 - 13x + 36$ ■

Example 3   Factor the trinomial $3x^2 + 12x - 36$.

Step 1 $3x^2 + 12x - 36 = 3[x^2 + 4x - 12]$ 3 is a common factor.

Step 2 $3[x^2 + 4x + (-12)] = 3[(\quad)(\quad)]$

Step 3 $3[x^2 + 4x + (-12)] = 3[(x\quad)(x\quad)]$

Note that the last term of the trinomial (-12) is negative. This means that the two factors of -12 must have unlike signs, since a positive number times a negative number gives a negative product.

Step 4 $3[x^2 + 4x - 12] = 3[(x +\quad)(x -\quad)]$

Find two integers with a product of -12 and a sum of $+4$. All possible pairs of factors are shown below.

$$(-12)(+1) = -12 \qquad (-12) + (+1) = -11$$
$$(+12)(-1) = -12 \qquad (+12) + (-1) = 11$$
$$(+6)(-2) = -12 \qquad (+6) + (-2) = 4$$
$$(-6)(+2) = -12 \qquad (-6) + (+2) = -4$$
$$(-4)(+3) = -12 \qquad (-4) + (+3) = -1$$
$$(+4)(-3) = -12 \qquad (+4) + (-3) = 1$$

From these possibilities, you see that the two integers with a product of -12 and a sum of $+4$ are $+6$ and -2. Write these numbers as the last terms of the binomials.

Step 5 $3[x^2 + 4x - 12] = 3(x + 6)(x - 2)$

Step 6   *Check:* $3(x + 6)(x - 2) = 3[x^2 + 4x - 12]$
$$= 3x^2 + 12x - 36 \qquad ■$$

Example 4   Factor the trinomial $x^2 - 11x - 12$.

The signs of the factors of -12 must be different. From the list in Example 3, choose the two factors with a sum of -11.

$$x^2 - 11x - 12 = (x - 12)(x + 1)$$

Check: $(x - 12)(x + 1) = x^2 - 11x - 12 \qquad ■$

Factoring Trinomials

To factor a trinomial $x^2 + bx + c$, use the following steps. Assume that b and c are both positive numbers.

Step 1   First, look for any common monomial factors.

Step 2   **a.** For the trinomial $x^2 + bx + c$, use the form
$$x^2 + bx + c = (x +\quad)(x +\quad)$$

b. For the trinomial $x^2 - bx + c$, use the form
$$x^2 - bx + c = (x -\quad)(x -\quad)$$

c. For the trinomials $x^2 - bx - c$ and $x^2 + bx - c$, use the forms
$$x^2 - bx - c = (x +\quad)(x -\quad)$$
$$x^2 + bx - c = (x +\quad)(x -\quad)$$

Exercises 10.3

Factor each trinomial completely:

1. $x^2 + 6x + 8$
2. $x^2 + 8x + 15$
3. $y^2 + 9y + 20$
4. $2w^2 + 20w + 32$
5. $3r^2 + 30r + 75$
6. $a^2 + 14a + 24$
7. $b^2 + 11b + 30$
8. $c^2 + 21c + 54$
9. $x^2 + 17x + 72$
10. $y^2 + 18y + 81$
11. $5a^2 + 35a + 60$
12. $r^2 + 12r + 27$
13. $x^2 - 7x + 12$
14. $y^2 - 6y + 9$
15. $2a^2 - 18a + 28$
16. $c^2 - 9c + 18$
17. $3x^2 - 30x + 63$
18. $r^2 - 12r + 35$
19. $w^2 - 13w + 42$
20. $x^2 - 14x + 49$
21. $x^2 - 19x + 90$
22. $4x^2 - 84x + 80$
23. $t^2 - 12t + 20$
24. $b^2 - 15b + 54$
25. $x^2 + 2x - 8$
26. $x^2 - 2x - 15$
27. $y^2 + y - 20$
28. $2w^2 - 12w - 32$
29. $a^2 + 5a - 24$
30. $b^2 + b - 30$
31. $c^2 - 15c - 54$
32. $b^2 - 6b - 72$
33. $3x^2 - 3x - 36$
34. $a^2 + 5a - 14$
35. $c^2 + 3c - 18$
36. $x^2 - 4x - 21$
37. $y^2 + 17y + 42$
38. $m^2 - 18m + 72$
39. $r^2 - 2r - 35$
40. $x^2 + 11x - 42$
41. $m^2 - 22m + 40$
42. $y^2 + 17y + 70$
43. $x^2 - 9x - 90$
44. $x^2 - 8x + 15$
45. $a^2 + 27a + 92$
46. $x^2 + 17x - 110$
47. $2a^2 - 12a - 110$
48. $y^2 - 14y + 40$
49. $a^2 + 29a + 100$
50. $y^2 + 14y - 120$
51. $y^2 - 14y - 95$
52. $b^2 + 20b + 36$
53. $y^2 - 18y + 32$
54. $x^2 - 8x - 128$
55. $7x^2 + 7x - 14$
56. $2x^2 - 6x - 36$
57. $6x^2 + 12x - 6$
58. $4x^2 + 16x + 16$
59. $y^2 - 12y + 35$
60. $a^2 + 16a + 63$
61. $a^2 + 2a - 63$
62. $y^2 - y - 42$
63. $x^2 + 18x + 56$
64. $x^2 + 11x - 26$
65. $2y^2 - 36y + 90$
66. $ax^2 + 2ax + a$
67. $3xy^2 - 18xy + 27x$
68. $x^3 - x^2 - 156x$
69. $x^2 + 30x + 225$
70. $x^2 - 2x - 360$
71. $x^2 - 26x + 153$
72. $x^2 + 8x - 384$
73. $x^2 + 28x + 192$
74. $x^2 + 3x - 154$
75. $x^2 + 14x - 176$
76. $x^2 - 59x + 798$
77. $2a^2b + 4ab - 48b$
78. $ax^2 - 15ax + 44a$
79. $y^2 - y - 72$
80. $x^2 + 19x + 60$

10.4 Special Products

The square of a is $a \cdot a$ or a^2 (read "a squared"). The square of the binomial $x + y$ is $(x + y)(x + y)$ or $(x + y)^2$, which is read "the quantity $x + y$ squared." When the multiplication is performed, the product is

$$(x + y)^2 = (x + y)(x + y) = x^2 + 2xy + y^2$$

A **perfect square trinomial** is a trinomial with the same two binomial factors.

The Square of a Binomial

The square of the *sum* of two terms of a binomial equals the square of the first term *plus* twice the product of the two terms plus the square of the second term.

$$(a + b)(a + b) = (a + b)^2 = a^2 + 2ab + b^2$$

Similarly, the square of the *difference* of two terms of a binomial equals the square of the first term *minus* twice the product of the two terms plus the square of the second term.

$$(a - b)(a - b) = (a - b)^2 = a^2 - 2ab + b^2$$

Example 1   Find $(x + 12)^2$.

The square of the first term is x^2. Twice the product of the terms is $2(x \cdot 12)$, or $24x$. The square of the second term is 144. Thus,

$$(x + 12)^2 = x^2 + 24x + 144$$ ■

Example 2   Find $(5xy - 3)^2$.

The square of the first term is $25x^2y^2$. Twice the product of the terms is $2(5xy \cdot 3)$, or $30xy$. The square of the second term is 9. Thus,

$$(5xy - 3)^2 = 25x^2y^2 - 30xy + 9$$ ■

Finding the *product of the sum and difference of two terms*, $(a + b)(a - b)$, is another special case in which the product is a binomial.

The Product of the Sum and Difference of Two Terms

This product is the difference of two squares: the square of the first term *minus* the square of the second term.

$$(a + b)(a - b) = a^2 - b^2$$

Example 3   Find the product $(x + 3)(x - 3)$.

The square of the first term is x^2. The square of the second term is 9. Thus,

$$(x + 3)(x - 3) = x^2 - 9$$

Note that the sum of the outer and inner products is zero.   ■

Example 4   Find the product $(4y + 7)(4y - 7)$.

The square of the first term is $16y^2$. The square of the second term is 49. Thus,

$$(4y + 7)(4y - 7) = 16y^2 - 49$$ ■

Example 5   Find the product $(3x - 8y)(3x + 8y)$.

The square of the first term is $9x^2$. The square of the second term is $64y^2$. Thus,

$$(3x - 8y)(3x + 8y) = 9x^2 - 64y^2$$ ■

Exercises 10.4

Find each product:

1. $(x + 3)(x - 3)$
2. $(x + 3)^2$
3. $(a + 5)(a - 5)$
4. $(y^2 + 9)(y^2 - 9)$
5. $(2b + 11)(2b - 11)$
6. $(x - 6)^2$
7. $(100 + 3)(100 - 3)$
8. $(90 + 2)(90 - 2)$
9. $(3y^2 + 14)(3y^2 - 14)$
10. $(y + 8)^2$
11. $(r - 12)^2$
12. $(t + 10)^2$
13. $(4y + 5)(4y - 5)$
14. $(200 + 5)(200 - 5)$

15. $(xy - 4)^2$

16. $(x^2 + y)(x^2 - y)$

29. $(x - 3)^2$

30. $(x + 4)^2$

17. $(ab + d)^2$

18. $(ab + c)(ab - c)$

31. $(ab + 2)(ab - 2)$

32. $(m - 3)(m + 3)$

19. $(z - 11)^2$

20. $(x^3 + 8)(x^3 - 8)$

33. $(x^2 + 2)(x^2 - 2)$

34. $(m + 15)(m - 15)$

21. $(st - 7)^2$

22. $(w + 14)(w - 14)$

35. $(r - 15)^2$

36. $(t + 7a)^2$

23. $(x + y^2)(x - y^2)$

24. $(1 - x)^2$

37. $(y^3 - 5)^2$

38. $(4 - x^2)^2$

25. $(x + 5)^2$

26. $(x - 6)^2$

39. $(10 - x)(10 + x)$

40. $(ay^2 - 3)(ay^2 + 3)$

27. $(x + 7)(x - 7)$

28. $(y - 12)(y + 12)$

10.5 Finding Factors of Special Products

To find the square root of a variable raised to a power, divide the exponent by 2 and use the result as the exponent of the given variable.

 Example 1

Assuming that x and y are positive, find the square roots of **a.** x^2, **b.** x^4, **c.** x^6, and **d.** x^8y^{10}.

a. $\sqrt{x^2} = x$

c. $\sqrt{x^6} = x^3$

b. $\sqrt{x^4} = x^2$

d. $\sqrt{x^8y^{10}} = x^4y^5$ ■

To factor a trinomial, first look for a common monomial factor. Then inspect the remaining trinomial to see if it is one of the special products. If it is not a perfect square trinomial and if it can be factored, use the methods shown in Section 10.3. If it is a perfect square trinomial, it may be factored using the reverse of the rule in Section 10.4.

Factoring Perfect Square Trinomials

Each of the two factors of a perfect square trinomial with a *positive* middle term is the square root of the first term *plus* the square root of the third term. That is,

$$a^2 + 2ab + b^2 = (a + b)(a + b)$$

Similarly, each of the two factors of the perfect square trinomial with a *negative* middle term is the square root of the first term *minus* the square root of the third term. That is,

$$a^2 - 2ab + b^2 = (a - b)(a - b)$$

 Example 2

Factor: $9x^2 + 30x + 25$.

This perfect square trinomial has no common monomial factor. Since the middle term is positive, each of its two factors is the square root of the first term plus the square root of the third term. The square root of the first term is $3x$; the square root of the third term is 5. The sum is $3x + 5$. Therefore,

$$9x^2 + 30x + 25 = (3x + 5)(3x + 5)$$ ■

Example 3

Factor: $x^2 - 12x + 36$.

This perfect square trinomial has no common monomial factor. Since the middle term is negative, each of its two factors is the square root of the first term minus the square root of the third term. The square root of the first term is x; the square root of the third term is 6. The difference is $x - 6$. Therefore,

$$x^2 - 12x + 36 = (x - 6)(x - 6) \qquad ■$$

Note: If you do not recognize $x^2 - 12x + 36$ as a perfect square trinomial, you can factor it by trial and error as you would any trinomial (see Section 10.3). Your result should be the same.

Example 4

Factor: $4x^2 + 24xy + 36y^2$.

First, find the common monomial factor, 4.

$$4x^2 + 24xy + 36y^2 = 4(x^2 + 6xy + 9y^2)$$

This perfect square trinomial has a positive middle term. Each of its two factors is the square root of the first term plus the square root of the third term. The square root of the first term is x; the square root of the third term is $3y$. The sum is $x + 3y$. Therefore,

$$4x^2 + 24xy + 36y^2 = 4(x + 3y)(x + 3y) \qquad ■$$

To factor a binomial that is the difference of two squares, use the reverse of the rule in Section 10.4. The factors of the difference of two squares are the square root of the first term *plus* the square root of the second term times the square root of the first term *minus* the square root of the second term as shown below.

> **Factoring the Difference of Two Squares**
> $$a^2 - b^2 = (a + b)(a - b)$$

Note that a is the square root of a^2 and b is the square root of b^2.

Example 5

Factor: $x^2 - 4$.

First, find the square root of each term of the expression. The square root of x^2 is x, and the square root of 4 is 2. Thus, $x + 2$ is the sum of the square roots, and $x - 2$ is the difference of the square roots.

$$x^2 - 4 = (x + 2)(x - 2)$$

Check: $\qquad (x + 2)(x - 2) = x^2 - 4 \qquad ■$

Example 6

Factor: $1 - 36y^4$.

The square root of 1 is 1, and the square root of $36y^4$ is $6y^2$. Thus, the sum, $1 + 6y^2$, and the difference, $1 - 6y^2$, of the square roots are the factors.

$$1 - 36y^4 = (1 + 6y^2)(1 - 6y^2)$$

Check: $\qquad (1 + 6y^2)(1 - 6y^2) = 1 - 36y^4 \qquad ■$

Example 7   Factor: $81y^4 - 1$.

The square root of $81y^4$ is $9y^2$, and the square root of 1 is 1. The factors are the sum of the square roots, $9y^2 + 1$, and the difference of the square roots, $9y^2 - 1$.

$$81y^4 - 1 = (9y^2 + 1)(9y^2 - 1)$$

However, $9y^2 - 1$ is also the difference of two squares. Its factors are $9y^2 - 1 = (3y + 1)(3y - 1)$. Therefore,

$$81y^4 - 1 = (9y^2 + 1)(3y + 1)(3y - 1)$$ ■

 Example 8   Factor: $2x^2 - 18$.

First, find the common monomial factor, 2.

$$2x^2 - 18 = 2(x^2 - 9)$$

Then $x^2 - 9$ is the difference of two squares whose factors are $x + 3$ and $x - 3$. Therefore,

$$2x^2 - 18 = 2(x + 3)(x - 3)$$ ■

Exercises 10.5

Factor completely. Check by multiplying the factors:

1. $a^2 + 8a + 16$
2. $b^2 - 2b + 1$
3. $b^2 - c^2$
4. $m^2 - 1$
5. $x^2 - 4x + 4$
6. $2c^2 - 4c + 2$
7. $4 - x^2$
8. $4x^2 - 1$
9. $y^2 - 36$
10. $a^2 - 64$
11. $5a^2 + 10a + 5$
12. $9x^2 - 25$
13. $1 - 81y^2$
14. $16x^2 - 100$
15. $49 - a^4$
16. $m^2 - 2mn + n^2$
17. $49x^2 - 64y^2$
18. $x^2y^2 - 1$
19. $1 - x^2y^2$
20. $c^2d^2 - 16$
21. $4x^2 - 12x + 9$
22. $16x^2 - 1$
23. $R^2 - r^2$
24. $36x^2 - 12x + 1$
25. $49x^2 - 25$
26. $1 - 100y^2$
27. $y^2 - 10y + 25$
28. $x^2 + 6x + 9$
29. $b^2 - 9$
30. $16 - c^2d^2$
31. $m^2 + 22m + 121$
32. $n^2 - 30n + 225$
33. $4m^2 - 9$
34. $16b^2 - 81$
35. $4x^2 + 24x + 36$
36. $-2y^2 + 12y - 18$
37. $27x^2 - 3$
38. $-9x^2 + 225x^4$
39. $am^2 - 14am + 49a$
40. $-bx^2 - 12bx - 36b$

 ## 10.6 Factoring General Trinomials

In previous examples, such as $x^2 + 7x + 10 = (x + 2)(x + 5)$, there was only one possible choice for the first terms of the binomials: x and x. When the coefficient of x^2 is greater than 1, however, there may be more than one possible choice. The idea is still the same: Find two binomial factors whose product equals the trinomial. Use the relationships outlined in Section 10.3 for finding the signs.

Example 1

Factor: $6x^2 - x - 2$.

The first terms of the binomial factors are either $6x$ and x or $3x$ and $2x$. The minus sign of the constant term of the trinomial tells you that one sign will be plus and the other will be minus in the last terms of the binomials. The last terms are either $+2$ and -1 or -2 and $+1$. The eight possibilities are

1. $(6x + 2)(x - 1) = 6x^2 - 4x - 2$
2. $(6x - 1)(x + 2) = 6x^2 + 11x - 2$
3. $(6x - 2)(x + 1) = 6x^2 + 4x - 2$
4. $(6x + 1)(x - 2) = 6x^2 - 11x - 2$
5. $(3x + 2)(2x - 1) = 6x^2 + x - 2$
6. $(3x - 1)(2x + 2) = 6x^2 + 4x - 2$
7. $(3x - 2)(2x + 1) = 6x^2 - x - 2$
8. $(3x + 1)(2x - 2) = 6x^2 - 4x - 2$

Only Equation 7 gives the desired middle term, $-x$. Therefore,

$$6x^2 - x - 2 = (3x - 2)(2x + 1)$$ ■

When factoring trinomials of this type, sometimes you may have to make several guesses or look at several combinations until you find the correct one. It is a *trial-and-error process*. The rules for the signs, outlined in Section 10.3, simply reduce the number of possibilities you need to try.

Another way to reduce the possibilities is to eliminate any combination in which either binomial contains a common factor. In the above list, the factors $6x + 2$, $6x - 2$, $2x + 2$, and $2x - 2$ all have the common factor 2 and cannot be correct, so Equations 1, 3, 6, and 8 can be eliminated. It is important to look for common monomial factors as the first step in factoring any trinomial.

Example 2

Factor: $12x^2 - 23x + 10$.

First terms of the binomial: $12x$ and x, $6x$ and $2x$, or $4x$ and $3x$.
Signs of the last term of the binomial: both $(-)$.
Last terms of the binomial: -1 and -10 or -2 and -5.

You cannot use $12x$, $6x$, $4x$, or $2x$ with -10 or -2, since they contain the common factor 2, so the list of possible combinations is narrowed to

1. $(12x - 1)(x - 10) = 12x^2 - 121x + 10$
2. $(12x - 5)(x - 2) = 12x^2 - 29x + 10$
3. $(4x - 5)(3x - 2) = 12x^2 - 23x + 10$
4. $(4x - 1)(3x - 10) = 12x^2 - 43x + 10$

As you can see, Equation 3 is the correct one, since it gives the desired middle term, $-23x$. Therefore,

$$12x^2 - 23x + 10 = (4x - 5)(3x - 2)$$ ■

Example 3

Factor: $12x^2 - 2x - 4$.

First look for a common factor. In this case, it is 2, so we write

$$12x^2 - 2x - 4 = 2(6x^2 - x - 2)$$

Next we try to factor the trinomial $6x^2 - x - 2$ into two binomial factors, as we did in Examples 1 and 2. Since the third term is negative (-2), we know that the signs of the second terms of the binomials are different. For the first terms of the binomials, we can try $6x$ and x or $3x$ and $2x$. The second terms can be $+2$ and -1 or -2 and $+1$. After eliminating all combinations with common factors, we have the following possibilities:

1. $(6x + 1)(x - 2) = 6x^2 - 11x - 2$
2. $(6x - 1)(x + 2) = 6x^2 + 11x - 2$
3. $(3x + 2)(2x - 1) = 6x^2 + x - 2$
4. $(3x - 2)(2x + 1) = 6x^2 - x - 2$

The correct one is Equation 4, since the middle term, $-x$, is the one we want. Therefore,

$$12x^2 - 2x - 4 = 2(3x - 2)(2x + 1)$$

　■

Another method for factoring binomials in the form $ax^2 + bx + c$ is to list the pairs of the factors of ac and find the pair whose sum is b.

 Example 4

Factor: $6x^2 - 19x + 10$.

First, list all factors of $ac = (6)(10) = 60$.

$1 \cdot 60$
$2 \cdot 30$
$3 \cdot 20$
$4 \cdot 15$
$5 \cdot 12$
$6 \cdot 10$

Note that only the factors $4 \cdot 15$ have a sum of 19. That is, the factors in both binomials must be some combination of 4 and 15.

Since the sign of the last term in both binomials is $(-)$, we have

$$6x^2 - 19x + 10 = (3x - 2)(2x - 5)$$

　■

This method is especially helpful if you are not successful in guessing by trial and error.

Example 5

Factor: $20x^2 + x - 12$.

First, list all the factors of $ac = (20)(12) = 240$:

$1 \cdot 240$	$6 \cdot 40$
$2 \cdot 120$	$8 \cdot 30$
$3 \cdot 80$	$10 \cdot 24$
$4 \cdot 60$	$12 \cdot 20$
$5 \cdot 48$	$15 \cdot 16$

Note that only the factors $15 \cdot 16$ can result in the middle term being 1. That is, the factors in both binomials must be some combination of 15 and 16. Since the sign of the last term of the trinomial is negative, we have

$$20x^2 + x - 12 = (5x + 4)(4x - 3)$$

　■

Exercises 10.6

Factor completely:

1. $5x^2 - 28x - 12$
2. $4x^2 - 4x - 3$
3. $10x^2 - 29x + 21$
4. $4x^2 + 4x + 1$
5. $12x^2 - 28x + 15$
6. $9x^2 - 36x + 32$
7. $8x^2 + 26x - 45$
8. $4x^2 + 15x - 4$
9. $16x^2 - 11x - 5$
10. $6x^2 + 3x - 3$

11. $12x^2 - 16x - 16$ **12.** $10x^2 - 35x + 15$ **27.** $90 + 17c - 3c^2$ **28.** $10x^2 - x - 2$
13. $15y^2 - y - 6$ **14.** $6y^2 + y - 2$ **29.** $6x^2 - 13x + 5$ **30.** $56x^2 - 29x + 3$
15. $8m^2 - 10m - 3$ **16.** $2m^2 - 7m - 30$ **31.** $2y^4 + 9y^2 - 35$ **32.** $2y^2 + 7y - 99$
17. $35a^2 - 2a - 1$ **18.** $12a^2 - 28a + 15$ **33.** $4b^2 + 52b + 169$ **34.** $6x^2 - 19x + 15$
19. $16y^2 - 8y + 1$ **20.** $25y^2 + 20y + 4$ **35.** $14x^2 - 51x + 40$ **36.** $42x^4 - 13x^2 - 40$
21. $3x^2 + 20x - 63$ **22.** $4x^2 + 7x - 15$ **37.** $28x^3 + 140x^2 + 175x$ **38.** $-24x^3 - 54x^2 - 21x$
23. $12b^2 + 5b - 2$ **24.** $10b^2 - 7b - 12$ **39.** $10ab^2 - 15ab - 175a$ **40.** $40bx^2 - 72bx - 70b$
25. $15y^2 - 14y - 8$ **26.** $5y^2 + 11y + 2$

Chapter 10 Group Activities

1. In small groups, discuss how to find the product $(2x + 3y)^3$. Then, find the result individually. Then as a group, compare the results and resolve any differences before checking the result in Appendix C (the answer section).

Chapter 10 Summary

Glossary of Basic Terms

Binomial factor.   A two-term factor of an algebraic expression. (p. 362)
Factoring an algebraic expression.   Writing the algebraic expression as a product of factors. (p. 358)
Greatest common factor of a polynomial.   The largest common factor that divides all terms in the expression. (p. 359)

Monomial factor.   A one-term factor that divides each term of an algebraic expression. (p. 358)
Perfect square trinomial.   A trinomial with the same two binomial factors. (p. 365)

10.2 Finding the Product of Two Binomials Mentally

1. **Finding the product of two binomials mentally:** The mental process is outlined as follows:
   a. The *first term* of the product is the product of the first terms of the binomials.
   b. The *middle term* of the product is the sum of the outer product and the inner product of the binomials.
   c. The *last term* of the product is the product of the last terms of the binomials.

This method is often called the *FOIL method,* where *F* refers to the product of the *first* terms, *O* refers to the *outer* product, *I* refers to the *inner* product, and *L* refers to the product of the *last* terms. (p. 360)

10.3 Finding Binomial Factors

1. **Factoring trinomials:** To factor a trinomial $x^2 + bx + c$, use the following steps.

Assume that b and c are both positive numbers.

First, look for any common monomial factors. Then,
   a. For the trinomial $x^2 + bx + c$, use the form
   $$x^2 + bx + c = (x +\ \ )(x +\ \ )$$
   b. For the trinomial $x^2 - bx + c$, use the form
   $$x^2 - bx + c = (x -\ \ )(x -\ \ )$$
   c. For the trinomials $x^2 - bx - c$ and $x^2 + bx - c$, use the forms
   $$x^2 - bx - c = (x +\ \ )(x -\ \ )$$
   $$x^2 + bx - c = (x +\ \ )(x -\ \ )\ \text{(p. 364)}$$

10.4 Special Products

1. **The square of a binomial:** There are two forms.
   a. The square of the *sum* of two terms of a binomial equals the square of the first term *plus* twice the product of the two terms plus the square of the second term; that is, $(a + b)(a + b) = (a + b)^2 = a^2 + 2ab + b^2$.
   b. The square of the *difference* of two terms of a binomial equals the square of the first term *minus* twice the product of the two terms plus the square of the second term; that is, $(a - b)(a - b) = (a - b)^2 = a^2 - 2ab + b^2$. (p. 365)

2. **The product of the sum and difference of two terms:** This product is the difference of two squares; that is, the square of the first term *minus* the square of the second term, $(a + b)(a - b) = a^2 - b^2$. (p. 366)

10.5 Finding Factors of Special Products

1. **Factoring Perfect Square Trinomials:** There are two forms.
   a. Each of the two factors of a perfect square trinomial with a *positive* middle term is the square root of the first term *plus* the square root of the third term; that is, $a^2 + 2ab + b^2 = (a + b)(a + b)$.
   b. Each of the two factors of a perfect square trinomial with a *negative* middle term is the square root of the first term *minus* the square root of the third term; that is, $a^2 - 2ab + b^2 = (a - b)(a - b)$. (p. 367)

2. **Factoring the difference of two squares:** The factors of the difference of two squares are the square root of the first term *plus* the square root of the second term times the square root of the first term *minus* the square root of the second term; that is, $a^2 - b^2 = (a + b)(a - b)$. (p. 368)

 ## Chapter 10 Review

Find each product mentally:

1. $(c + d)(c - d)$
2. $(x - 6)(x + 6)$
3. $(y + 7)(y - 4)$
4. $(2x + 5)(2x - 9)$
5. $(x + 8)(x - 3)$
6. $(x - 4)(x - 9)$
7. $(x - 3)^2$
8. $(2x - 6)^2$
9. $(1 - 5x^2)^2$

Factor each expression completely:

10. $6a + 6$
11. $5x - 15$
12. $xy + 2xz$
13. $y^4 + 17y^3 - 18y^2$
14. $y^2 - 6y - 7$
15. $z^2 + 18z + 81$
16. $x^2 + 10x + 16$
17. $4a^2 + 4x^2$

18. $x^2 - 17x + 72$
19. $x^2 - 18x + 81$
20. $x^2 + 19x + 60$
21. $y^2 - 2y + 1$
22. $x^2 - 3x - 28$
23. $x^2 - 4x - 96$
24. $x^2 + x - 110$
25. $x^2 - 49$
26. $16y^2 - 9x^2$
27. $x^2 - 144$
28. $25x^2 - 81y^2$
29. $4x^2 - 24x - 364$
30. $5x^2 - 5x - 780$
31. $2x^2 + 11x + 14$
32. $12x^2 - 19x + 4$
33. $30x^2 + 7x - 15$
34. $12x^2 + 143x - 12$
35. $4x^2 - 6x + 2$
36. $36x^2 - 49y^2$
37. $28x^2 + 82x + 30$
38. $30x^2 - 27x - 21$
39. $4x^3 - 4x$
40. $25y^2 - 100$

 ## Chapter 10 Test

Find each product mentally:

1. $(x + 8)(x - 3)$
2. $(2x - 8)(5x - 6)$
3. $(2x - 8)(2x + 8)$
4. $(3x - 5)^2$
5. $(4x - 7)(2x + 3)$
6. $(9x - 7)(5x + 4)$

Factor each expression completely:

7. $x^2 + 4x + 3$
8. $x^2 - 12x + 35$
9. $6x^2 - 7x - 90$
10. $9x^2 + 24x + 16$

11. $x^2 + 7x - 18$
12. $4x^2 - 25$
13. $6x^2 + 13x + 6$
14. $3x^2y^2 - 18x^2y + 27x^2$
15. $3x^2 - 11x - 4$
16. $15x^2 - 19x - 10$
17. $5x^2 + 7x - 6$
18. $3x^2 - 3x - 6$
19. $9x^2 - 121$
20. $9x^2 - 30x + 25$

Cumulative Review

1. Perform the indicated operations and simplify:
   $2 + 6^2 - 24 \div 3(4)$

2. Round 746.83 to the **a.** nearest tenth and **b.** nearest ten.

3. Do as indicated and simplify: $-\dfrac{2}{3} \div \dfrac{1}{5} + \dfrac{2}{3}$

4. Write 0.000318 in **a.** scientific notation and **b.** engineering notation.

5. Change 625 g to kg.

6. Change 6 m² to ft².

7. Read the voltmeter scale in Illustration 1.

ILLUSTRATION 1

8. Use the rules of measurement to multiply:
   (5.0 cm)(148 cm)(0.128 cm)

Combine like terms and simplify:

9. $3(x - 2) - 4(2 - 3x)$

10. $(6a - 3b + 2c) - (-2a - 3b + c)$

11. Solve: $\dfrac{x}{3} - 4 = \dfrac{2x}{5}$

12. A rectangle is 5 m longer than it is wide. Its perimeter is 58 m. Find the length and the width.

13. Solve the proportion and round the result to three significant digits:

    $\dfrac{15.7}{8.2} = \dfrac{x}{10}$

14. A pulley is 18 in. in diameter, is rotating at 125 rpm, and is connected to a smaller pulley rotating at 225 rpm. Find the diameter of the smaller pulley.

15. Complete the ordered pair solutions of the equation:
    $2x + 3y = 12$

    $(3, \ \ ), (0, \ \ ), (-3, \ \ )$

16. Solve for y: $3x - y = 5$

17. Draw the graph of $3x + 4y = 24$.

18. Draw the graphs of $2x - y = 4$ and $x + 3y = -5$. Find the point of intersection.

Solve each pair of linear equations:

19. $\dfrac{1}{2}x - \dfrac{2}{3}y = \dfrac{1}{12}$
    $6x - 8y = 12$

20. $y = 3x - 5$
    $x + 3y = 8$

21. $x - y = 6$
    $3x + y = 2$

22. $3x - 5y = 7$
    $-6x + 10y = 5$

23. $135x + 40y = 29$
    $60x - 45y = 38$

24. Two rental cars were leased for a total of 16 days. One car rents for $53.95 per day, and the other car rents for $42.95 per day. The total cost for leasing the two cars was $753.20. Find the number of days each car was rented.

Find each product mentally:

25. $(2x - 5)(3x + 8)$ 26. $(5x - 7y)^2$

27. $(3x - 5)(5x - 7)$

Factor each expression completely:

28. $7x^3 - 63x$ 29. $4x^3 + 12x^2$

30. $2x^2 - 7x - 4$

11 ◆ Quadratic Equations

Mathematics at Work

Aircraft mechanics and service technicians perform scheduled maintenance, make repairs, and make inspections to keep aircraft in peak operating condition. Many specialize in preventive maintenance by inspecting engines, landing gears, instruments, pressurized sections, and various accessories such as brakes, valves, pumps, and air-conditioning systems. Such inspections occur following a schedule based on the number of hours the aircraft has flown, calendar days, cycles of operation, or a combination of these factors. Powerplant mechanics are authorized to work on engines and do limited work on propellers. Airframe mechanics are authorized to work on any part of the aircraft except the instruments, power plants, and propellers. Combination airframe-and-powerplant mechanics (A&P mechanics) work on all parts of the plane except instruments. The majority of mechanics working on civilian aircraft are A&P mechanics. Avionics technicians repair and maintain components used for aircraft navigation and radio communications, weather radar systems, and other instruments and computers that control flight, engine, and primary functions. The Federal Aviation Administration (FAA) regulates certification of aircraft mechanics and service technicians as well as training programs. Mathematics, physics, chemistry, electronics, computer science, mechanical drawing, and communications skills are key to training programs and success on the job. For more information, go to the website listed below.

Michelle D. Bridwell/PhotoEdit, Inc.

Aircraft Mechanics and Service Technicians

Airline mechanic doing routine engine maintenance.

www.cengage.com/mathematics/ewen

Objectives

- Solve quadratic equations by factoring.
- Solve quadratic equations by using the quadratic formula.
- Solve application problems involving quadratic equations.
- Graph quadratic equations.
- Find the vertex of a parabola.
- Express the square root of a negative number as an imaginary number in terms of j.
- Simplify powers of j.
- Solve quadratic equations with imaginary roots.

11.1 Solving Quadratic Equations by Factoring

A **quadratic equation** in one variable is an equation in the form $ax^2 + bx + c = 0$, where $a \neq 0$.

Recall that linear equations, such as $2x + 3 = 0$, have at most *one* solution. Quadratic equations have at most *two* solutions. One way to solve quadratic equations is by factoring and using the following:

$$\text{If } ab = 0, \text{ then either } a = 0 \text{ or } b = 0.$$

That is, if you multiply two factors and the product is 0, then one or both factors are 0.

 Example 1

Solve: $4(x - 2) = 0$.

If $4(x - 2) = 0$, then $4 = 0$ or $x - 2 = 0$. However, the first statement, $4 = 0$, is false; thus, the solution is $x - 2 = 0$, or $x = 2$.   ■

 Example 2

Solve: $(x - 2)(x + 3) = 0$.

If $(x - 2)(x + 3) = 0$, then either

$$x - 2 = 0 \qquad \text{or} \qquad x + 3 = 0$$

Therefore,

$$x = 2 \qquad \text{or} \qquad x = -3$$ ■

Solving Quadratic Equations by Factoring

1. If necessary, write an equivalent equation in the form $ax^2 + bx + c = 0$.
2. Factor the polynomial.
3. Write equations by setting each factor containing a variable equal to zero.
4. Solve the two resulting first-degree equations.
5. Check.

Example 3

Solve $x^2 + 6x + 5 = 0$ for x.

Step 1 Not needed.

Step 2 $(x + 5)(x + 1) = 0$

Step 3 $x + 5 = 0$ or $x + 1 = 0$

Step 4 $x = -5$ or $x = -1$

Step 5 *Check:*

Replace x with -5.   Replace x with -1.

$$x^2 + 6x + 5 = 0$$ $$x^2 + 6x + 5 = 0$$
$$(-5)^2 + 6(-5) + 5 = 0 \quad ?$$ $$(-1)^2 + 6(-1) + 5 = 0 \quad ?$$
$$25 - 30 + 5 = 0$$ $$1 - 6 + 5 = 0$$
$$0 = 0 \quad \text{True}$$ $$0 = 0 \quad \text{True}$$

Thus, the roots are -5 and -1. ■

 Example 4

Solve $x^2 + 5x = 36$ for x.

Step 1 $x^2 + 5x - 36 = 0$

Step 2 $(x + 9)(x - 4) = 0$

Step 3 $x + 9 = 0$ or $x - 4 = 0$

Step 4 $x = -9$ or $x = 4$

Step 5 *Check:*

Replace x with -9.   Replace x with 4.

$$x^2 + 5x = 36$$ $$x^2 + 5x = 36$$
$$(-9)^2 + 5(-9) = 36 \quad ?$$ $$4^2 + 5(4) = 36 \quad ?$$
$$81 - 45 = 36$$ $$16 + 20 = 36$$
$$36 = 36 \quad \text{True}$$ $$36 = 36 \quad \text{True}$$

Thus, the roots are -9 and 4. ■

 Example 5

Solve $3x^2 + 9x = 0$ for x.

Step 1 Not needed.

Step 2 $3x(x + 3) = 0$

Step 3 $3x = 0$ or $x + 3 = 0$

Step 4 $x = 0$ or $x = -3$

Step 5 *Check:* Left to the student. ■

 Example 6

Solve $x^2 = 4$ for x.

Step 1 $x^2 - 4 = 0$

Step 2 $(x + 2)(x - 2) = 0$

Step 3 $x + 2 = 0$ or $x - 2 = 0$

Step 4 $x = -2$ or $x = 2$

Step 5 *Check:* Left to the student. ■

Example 7   Solve: $6x^2 = 7x + 20$.

Step 1 $6x^2 - 7x - 20 = 0$

Step 2 $(3x + 4)(2x - 5) = 0$

Step 3 $3x + 4 = 0$ or $2x - 5 = 0$

Step 4 $3x = -4$ $2x = 5$

$$x = -\frac{4}{3} \qquad\qquad x = \frac{5}{2}$$

So the possible roots are $-\frac{4}{3}$ and $\frac{5}{2}$.

Step 5   *Check:* Replace x with $-\frac{4}{3}$ and with $\frac{5}{2}$ in the original equation.

$$6x^2 = 7x + 20 \qquad\qquad\qquad 6x^2 = 7x + 20$$

$$6\left(-\frac{4}{3}\right)^2 = 7\left(-\frac{4}{3}\right) + 20 \qquad 6\left(\frac{5}{2}\right)^2 = 7\left(\frac{5}{2}\right) + 20$$

$$6\left(\frac{16}{9}\right) = -\frac{28}{3} + 20 \qquad\qquad 6\left(\frac{25}{4}\right) = \frac{35}{2} + 20$$

$$\frac{32}{3} = -\frac{28}{3} + \frac{60}{3} \qquad\qquad \frac{75}{2} = \frac{35}{2} + \frac{40}{2}$$

$$\frac{32}{3} = \frac{32}{3} \quad\text{True} \qquad\qquad \frac{75}{2} = \frac{75}{2} \quad\text{True}$$

So the roots are $-\frac{4}{3}$ and $\frac{5}{2}$.   ■

Exercises 11.1 _____

Solve each equation:

1. $x^2 + x = 12$

2. $x^2 - 3x + 2 = 0$

3. $x^2 + x - 20 = 0$

4. $d^2 + 2d - 15 = 0$

5. $x^2 - 2 = x$

6. $x^2 - 15x = -54$

7. $x^2 - 1 = 0$

8. $16n^2 = 49$

9. $x^2 - 49 = 0$

10. $4n^2 = 64$

11. $w^2 + 5w + 6 = 0$

12. $x^2 - 6x = 0$

13. $y^2 - 4y = 21$

14. $c^2 + 2 = 3c$

15. $n^2 - 6n - 40 = 0$

16. $x^2 - 17x + 16 = 0$

17. $9m = m^2$

18. $6n^2 - 15n = 0$

19. $x^2 = 108 + 3x$

20. $x^2 - x = 42$

21. $c^2 + 6c = 16$

22. $4x^2 + 4x - 3 = 0$

23. $10x^2 + 29x + 10 = 0$

24. $2x^2 = 17x - 8$

25. $4x^2 = 25$

26. $25x = x^2$

27. $9x^2 + 16 = 24x$

28. $24x^2 + 10 = 31x$

29. $3x^2 + 9x = 0$

30. A rectangle is 5 ft longer than it is wide. (See Illustration 1.) The area of the rectangle is 84 ft². Use a quadratic equation to find the dimensions of the rectangle.

ILLUSTRATION 1

31. The area of a triangle is 66 m², and its base is 1 m more than the height. (See Illustration 2.) Find the base and height of the triangle. (Use a quadratic equation.)

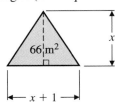

ILLUSTRATION 2

32. A rectangle is 9 ft longer than it is wide, and its area is 360 ft². Use a quadratic equation to find its length and width.

33. A heating duct has a rectangular cross section whose area is 40 in². If it is 3 in. longer than it is wide, find its length and width.

11.2 The Quadratic Formula

Many quadratic equations cannot be solved by factoring, so let's study a method by which any quadratic equation can be solved.

The roots of a quadratic equation in the form

$$ax^2 + bx + c = 0$$

may be found by using the following formula:

Quadratic Formula

$$x = \frac{-b \pm \sqrt{b^2 - 4ac}}{2a}$$

where a is the coefficient of the x^2 term,

         b is the coefficient of the x term, and

         c is the constant term.

The symbol ($\pm$) is used to combine two expressions into one. For example, "$a \pm 4$" means "$a + 4$ or $a - 4$." Similarly,

$$x = \frac{-b \pm \sqrt{b^2 - 4ac}}{2a} \quad \text{means}$$

$$x = \frac{-b + \sqrt{b^2 - 4ac}}{2a} \quad \text{or} \quad x = \frac{-b - \sqrt{b^2 - 4ac}}{2a}$$

 Example 1

In the quadratic equation $3x^2 - x - 7 = 0$, find the values of a, b, and c.

$$a = 3, \qquad b = -1, \qquad \text{and} \qquad c = -7 \qquad\qquad ■$$

 Example 2

Solve $x^2 + 5x - 14 = 0$ using the quadratic formula.

$$x = \frac{-b \pm \sqrt{b^2 - 4ac}}{2a}, \quad a = 1, \quad b = 5, \quad c = -14$$

So $x = \dfrac{-5 \pm \sqrt{5^2 - 4(1)(-14)}}{2(1)}$ Substitute.

$$= \frac{-5 \pm \sqrt{25 + 56}}{2} \qquad \text{Simplify.}$$

$$= \frac{-5 \pm \sqrt{81}}{2}$$

$$= \frac{-5 \pm 9}{2} \qquad\qquad \sqrt{81} = 9$$

$$= \frac{-5 + 9}{2} \quad \text{or} \quad \frac{-5 - 9}{2}$$

$$= 2 \quad \text{or} \quad -7$$

Check: Replace x with 2.   Replace x with -7.

$$x^2 + 5x - 14 = 0 \qquad\qquad\qquad\qquad x^2 + 5x - 14 = 0$$
$$2^2 + 5(2) - 14 = 0 \qquad ? \qquad\qquad (-7)^2 + 5(-7) - 14 = 0 \qquad ?$$
$$4 + 10 - 14 = 0 \qquad\qquad\qquad\qquad 49 - 35 - 14 = 0$$
$$0 = 0 \qquad\qquad\qquad\qquad\qquad\qquad 0 = 0$$

The roots are 2 and -7.   ■

> Before using the quadratic formula, make certain the equation is written in the form $ax^2 + bx + c = 0$, such that one member is zero.

Example 3

Solve $2x^2 = x + 21$ by using the quadratic formula.
First, write the equation in the form $2x^2 - x - 21 = 0$.

$$x = \frac{-b \pm \sqrt{b^2 - 4ac}}{2a}, \qquad a = 2, \qquad b = -1, \qquad c = -21$$

So $\quad x = \dfrac{-(-1) \pm \sqrt{(-1)^2 - 4(2)(-21)}}{2(2)} \qquad$ Substitute.

$$= \frac{1 \pm \sqrt{1 + 168}}{4} \qquad\qquad\qquad \text{Simplify.}$$

$$= \frac{1 \pm \sqrt{169}}{4}$$

$$= \frac{1 \pm 13}{4} \qquad\qquad\qquad\qquad \sqrt{169} = 13$$

$$= \frac{1 \pm 13}{4} \quad \text{or} \quad \frac{1 - 13}{4}$$

$$= \frac{7}{2} \quad \text{or} \quad -3$$

Check: Left to the student.   ■

The quantity under the radical sign, $b^2 - 4ac$, is called the **discriminant**. If the discriminant is not a perfect square, find the square root of the number by using a calculator and proceed as before. Round each final result to three significant digits.

Example 4

Solve $3x^2 + x - 5 = 0$ using the quadratic formula.

$$x = \frac{-b \pm \sqrt{b^2 - 4ac}}{2a}, \qquad a = 3, \qquad b = 1, \qquad c = -5$$

So $\quad x = \dfrac{-1 \pm \sqrt{1^2 - 4(3)(-5)}}{2(3)} \qquad$ Substitute.

$$= \frac{-1 \pm \sqrt{1 + 60}}{6} \qquad\qquad \text{Simplify.}$$

$$= \frac{-1 \pm \sqrt{61}}{6}$$

$$= \frac{-1 \pm 7.81}{6}$$ $\sqrt{61} = 7.81$

$$= \frac{-1 + 7.81}{6} \quad \text{or} \quad \frac{-1 - 7.81}{6}$$

$$= \frac{6.81}{6} \quad \text{or} \quad \frac{-8.81}{6}$$

$$= 1.14 \quad \text{or} \quad -1.47$$

The roots are 1.14 and -1.47.   ■

The check will not work out *exactly* when the number under the radical is not a perfect square.

Exercises 11.2

Find the values of a, b, and c in each equation:

1. $x^2 - 7x + 4 = 0$
2. $2x^2 + x - 3 = 0$
3. $3x^2 + 4x + 9 = 0$
4. $2x^2 - 14x + 37 = 0$
5. $-3x^2 + 4x + 7 = 0$
6. $17x^2 - x + 34 = 0$
7. $3x^2 - 14 = 0$
8. $2x^2 + 7x = 0$

Solve each equation using the quadratic formula. Check your solutions:

9. $x^2 + x - 6 = 0$
10. $x^2 - 4x - 21 = 0$
11. $x^2 + 8x - 9 = 0$
12. $2x^2 + 5x - 12 = 0$
13. $5x^2 + 2x = 0$
14. $3x^2 - 75 = 0$
15. $48x^2 - 32x - 35 = 0$
16. $13x^2 + 178x - 56 = 0$

Solve each equation using the quadratic formula (when necessary, round results to three significant digits):

17. $2x^2 + x - 5 = 0$
18. $-3x^2 + 2x + 5 = 0$
19. $3x^2 - 5x = 0$
20. $7x^2 + 9x + 2 = 0$
21. $-2x^2 + x + 3 = 0$
22. $5x^2 - 7x + 2 = 0$
23. $6x^2 + 9x + 1 = 0$
24. $16x^2 - 25 = 0$
25. $-4x^2 = 5x + 1$
26. $9x^2 = 21x - 10$
27. $3x^2 = 17$
28. $8x^2 = 11x - 3$
29. $x^2 = 15x + 7$
30. $x^2 + x = 1$
31. $3x^2 - 31 = 5x$
32. $-3x^2 - 5 = -7x^2$
33. $52.3x = -23.8x^2 + 11.8$
34. $18.9x^2 - 44.2x = 21.5$

11.3 Applications Involving Quadratic Equations

We now present some applications that involve quadratic equations. For consistency, all final results are rounded to three significant digits when necessary.

Example 1

A variable voltage in an electric circuit is given by the equation $V = 8t^2 - 28t + 20$, where t is in milliseconds (ms). Find the values of t when the voltage V equals **a.** 8 V and **b.** 15 V.

a. Substitute $V = 8$ into the equation

$$V = 8t^2 - 28t + 20$$
$$8 = 8t^2 - 28t + 20$$
$$0 = 8t^2 - 28t + 12 \qquad \text{Subtract 8 from both sides.}$$
$$0 = 2t^2 - 7t + 3 \qquad \text{Divide both sides by 4 to make the work easier.}$$
$$0 = (2t - 1)(t - 3) \qquad \text{Factor.}$$
$$2t - 1 = 0 \quad \text{or} \quad t - 3 = 0 \qquad \text{Set each factor equal to 0 and solve for } t.$$
$$t = \frac{1}{2} \text{ ms} \qquad\qquad t = 3 \text{ ms}$$

b. Substitute $V = 15$ into the equation

$$V = 8t^2 - 28t + 20$$
$$15 = 8t^2 - 28t + 20$$
$$0 = 8t^2 - 28t + 5 \qquad \text{Subtract 15 from both sides.}$$

Note that the right side of the equation does not factor, so we use the quadratic formula with $a = 8$, $b = -28$, and $c = 5$.

$$t = \frac{-b \pm \sqrt{b^2 - 4ac}}{2a}$$

$$t = \frac{-(-28) \pm \sqrt{(-28)^2 - 4(8)(5)}}{2(8)}$$

$$t = \frac{28 \pm \sqrt{784 - 160}}{16}$$

$$t = \frac{28 \pm \sqrt{624}}{16}$$

$$t = \frac{28 \pm 25.0}{16}$$

$$t = \frac{28 + 25.0}{16} \qquad \text{or} \qquad t = \frac{28 - 25.0}{16}$$

$$t = 3.31 \text{ ms} \qquad\qquad t = 0.188 \text{ ms}$$
■

◆ Example 2

Design a rectangular metal plate so that its length is 6 cm more than twice its width and its area is 360 cm^2.

First, draw a diagram as in Figure 11.1 and let

$$x = \text{the width}$$
$$2x + 6 = \text{the length}$$

Then use the formula for the area of a rectangle and substitute as follows:

$$A = lw$$
$$360 = (2x + 6)x$$
$$360 = 2x^2 + 6x \qquad\qquad \text{Remove parentheses.}$$
$$0 = 2x^2 + 6x - 360 \qquad \text{Subtract 360 from both sides.}$$
$$0 = x^2 + 3x - 180 \qquad\quad \text{Divide both sides by 2 to make the work easier.}$$
$$0 = (x + 15)(x - 12) \qquad \text{Factor.}$$
$$x + 15 = 0 \qquad \text{or} \qquad x - 12 = 0 \qquad \text{Set each factor equal to 0 and solve for } x.$$
$$x = -15 \qquad\qquad\qquad x = 12$$

FIGURE 11.1

Note that the solution $x = -15$ is not meaningful, as x refers to a width measurement, which must be a positive quantity. Therefore,

$$x = \text{the width} = 12 \text{ cm}$$
$$2x + 6 = \text{the length} = 2(12) + 6 = 30 \text{ cm}$$

Check: $A = lw = (30 \text{ cm})(12 \text{ cm}) = 360 \text{ cm}^2$, which is the given area.
■

Example 3

The perimeter of a rectangle is 20 cm, and its area is 16 cm². Find its dimensions (the length and the width).

First, note that the perimeter of a rectangle is the sum of the lengths of all four sides. Thus, if the perimeter is 20 cm, one width plus one length is 10 cm. So if the width is x, then the length must be $10 - x$.

Then draw a diagram as in Figure 11.2 and let

$$x = \text{the width}$$
$$10 - x = \text{the length}$$

Then use the formula for the area of a rectangle and substitute as follows:

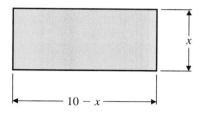

FIGURE 11.2

$$A = lw$$
$$16 = (10 - x)x$$
$$16 = 10x - x^2 \qquad \text{Remove parentheses.}$$
$$x^2 - 10x + 16 = 0 \qquad \text{Set the equation equal to 0.}$$
$$(x - 2)(x - 8) = 0 \qquad \text{Factor.}$$
$$x - 2 = 0 \quad \text{or} \quad x - 8 = 0 \qquad \text{Set each factor equal to 0 and solve for } x.$$
$$x = 2 \qquad\qquad x = 8$$

If $x = 2$,   If $x = 8$,
$$x = \text{the width} = 2 \text{ cm} \qquad\qquad x = \text{the width} = 8 \text{ cm}$$
$$10 - x = \text{the length} = 8 \text{ cm} \qquad 10 - x = \text{the length} = 2 \text{ cm}$$

Note that the dimensions are the same 2 cm by 8 cm in both cases. Since the length is greater than the width, the length is 8 cm and the width is 2 cm. ■

Example 4

A square is cut out of each corner of a rectangular sheet of metal 42 cm × 52 cm. The sides are then folded up to form a rectangular container with no top. What are the dimensions of the square if the area of the bottom of the container is 1200 cm²? Find the volume of the container.

First, draw a diagram as in Figure 11.3 and let

$$x = \text{the side of each square}$$
$$\text{cutout}$$
$$42 - 2x = \text{the width of the container}$$
$$52 - 2x = \text{the length of the container}$$

Use the formula for the area of the rectangular bottom of the container and substitute as follows:

FIGURE 11.3

$$A = lw$$
$$1200 = (52 - 2x)(42 - 2x)$$
$$1200 = 2184 - 188x + 4x^2 \qquad \text{Remove parentheses.}$$
$$0 = 4x^2 - 188x + 984 \qquad \text{Subtract 1200 from both sides.}$$
$$0 = x^2 - 47x + 246 \qquad \text{Divide both sides by 4 to make the work easier.}$$
$$0 = (x - 41)(x - 6) \qquad \text{Factor.}$$
$$x - 41 = 0 \quad \text{or} \quad x - 6 = 0 \qquad \text{Set each factor equal to 0 and solve for } x.$$
$$x = 41 \qquad\qquad x = 6$$

Note that $x = 41$ cm is not physically possible. So the side of each square is 6 cm.

The length of the container is $52 - 2x = 52 - 2(6) = 52 - 12 = 40$ cm.

The width of the container is $42 - 2x = 42 - 2(6) = 42 - 12 = 30$ cm.

The volume of the container is

$$V = lwh$$
$$V = (40 \text{ cm})(30 \text{ cm})(6 \text{ cm}) = 7200 \text{ cm}^3$$ ■

Exercises 11.3

1. A variable voltage in an electrical circuit is given by $V = t^2 - 12t + 40$, where t is in seconds. Find the values of t when the voltage V equals **a.** 8 V, **b.** 25 V, **c.** 104 V.

2. A variable electric current is given by $i = t^2 - 7t + 12$, where t is in seconds. At what times is the current i equal to **a.** 2 A? **b.** 0 A? **c.** 4 A?

3. A rectangular piece of sheet metal is 4 ft longer than it is wide. (See Illustration 1.) The area of the piece of sheet metal is 21 ft². Find its length and width.

21 ft²

x

$x + 4$

ILLUSTRATION 1

4. A hole in the side of a large metal tank needs to be repaired. A piece of rectangular sheet metal of area 16 ft² will patch the hole. If the length of the sheet metal must be 8 ft longer than its width, what will the dimensions of the sheet metal be?

5. The area of the wings of a small Cessna is 175 ft². If the length is 30 ft longer than the width, what are the dimensions of the wings? (This wing is one piece and goes along the top of the aircraft.)

6. The perimeter of a rectangle is 46 cm, and its area is 120 cm². Find its dimensions.

7. The perimeter of a rectangle is 160 m, and its area is 1200 m². Find its dimensions.

8. A rectangular field is fenced in by using a river as one side. If 1800 m of fencing are used for the 385,000-m² field, find its dimensions.

9. The dimensions of a door are 3 ft by 7 ft 6 in. If the same amount is added to each dimension of the door, the area is increased by 18 ft². Find the dimensions of the new door.

10. A square, 4 in. on a side, is cut out of each corner of a square sheet of aluminum. (See Illustration 2.) The sides are folded up to form a rectangular container with no top. The volume of the resulting container is 400 in³. What was the size of the original sheet of aluminum?

x 4

4

x

ILLUSTRATION 2

11. A square is cut out of each corner of a rectangular sheet of aluminum that is 40 cm by 60 cm. (See Illustration 3.) The sides are folded up to form a rectangular container with no top. The area of the bottom of the container is 1500 cm². **a.** What are the dimensions of each cut-out square? **b.** Find the volume of the container. ($V = lwh$)

ILLUSTRATION 3

12. The area of a rectangular lot 80 m by 100 m is to be increased by 4000 m². (See Illustration 4.) The length and the width will be increased by the same amount. What are the dimensions of the larger lot?

ILLUSTRATION 4

13. A border of uniform width is built around a rectangular garden that measures 16 ft by 20 ft. (See Illustration 5.) The area of the border is 160 ft². Find the width of the border.

ILLUSTRATION 5

14. A border of uniform width is printed on a page measuring 11 in. by 14 in. (See Illustration 6.) The area of the border is 66 in². Find the width of the border.

ILLUSTRATION 6

15. A company needs to build a warehouse with perimeter 300 ft. Find the dimensions to give maximum floor space.
  a. If the length is 10 ft, what is the area?
  b. If the length is 20 ft, what is the area?
  c. Write a formula (model) for the area in terms of the length.
  d. Complete the following table:

Length (ft)	30	40	50	60	70	80	90	100	110	120	130	140
Area (ft²)												

  e. Does one of these values give a maximum area? Explain.
  f. Graph the equation.
  g. Is there a different maximum?

16. A 2000-ft² storage building 9 ft high is needed to store yard maintenance equipment. What dimensions should be used to minimize the outside walls?

17. A landscaper is laying sod in a rectangular front lawn that is 76 ft longer than it is wide. Its area is 9165 ft². Find its dimensions.

18. A rectangular forest plot contains 120 acres and is three times as long as it is wide. Find its dimensions.

11.4 Graphs of Quadratic Equations

The graph of a quadratic equation written in the form $y = ax^2 + bx + c = 0$, where a, b, and c are real numbers and $a \neq 0$, is called a **parabola**. The quadratic equation written in the form $x = ay^2 + by + c = 0$ also represents a parabola, but we will not work with this equation in this book. Many physical phenomena follow a curved path of a parabola. The trajectory of a rocket (Figure 11.4) or any projectile is one such example.

To draw the graph of a parabola, find points whose ordered pairs satisfy the equation by choosing various values of x and solving for y. Since this graph is not a straight line, you will need to find and plot many points to get an accurate graph of the curve. A table is helpful for listing these ordered pairs.

FIGURE 11.4
The trajectory of a rocket follows a path in the shape of a parabola.

  Example 1

Graph the equation $y = x^2$.

First, set up a table as follows.

x	-4	-3	-2	-1	0	1	2	3	4
$y = x^2$	16	9	4	1	0	1	4	9	16

Then, using a rectangular coordinate system, plot these points. Notice that with only the points shown in Figure 11.5(a), there isn't a definite outline of the curve. So let's look more closely at values of x between 0 and 1.

x	$\frac{1}{6}$	$\frac{1}{5}$	$\frac{1}{4}$	$\frac{1}{3}$	$\frac{2}{5}$	$\frac{1}{2}$	$\frac{3}{5}$	$\frac{3}{4}$	$\frac{2}{3}$	$\frac{4}{5}$	$\frac{5}{6}$
$y = x^2$	$\frac{1}{36}$	$\frac{1}{25}$	$\frac{1}{16}$	$\frac{1}{9}$	$\frac{4}{25}$	$\frac{1}{4}$	$\frac{9}{25}$	$\frac{9}{16}$	$\frac{4}{9}$	$\frac{16}{25}$	$\frac{25}{36}$

Plotting these additional ordered pairs, we get a better graph (Figure 11.5b). If we were to continue to choose more and more values, the graph would appear as a solid line.

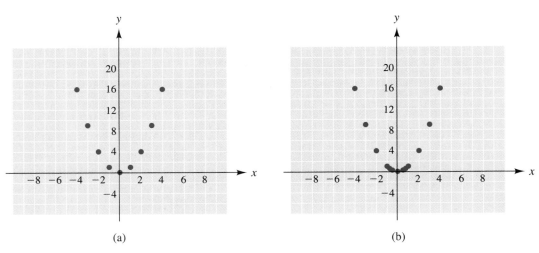

(a)   (b)

FIGURE 11.5
Plotting points that satisfy the equation $y = x^2$.

Since it is impossible to find *all* ordered pairs that satisfy the equation, we will assume that all the points between any two of the ordered pairs already located could be found and that they do lie on the graph. Thus, assume that the graph of $y = x^2$ looks like the graph in Figure 11.6.

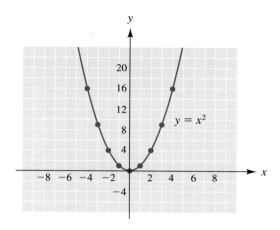

FIGURE 11.6   ■

In summary, to draw the graph of a quadratic equation, form a table to find many ordered pairs that satisfy the equation by choosing various values of x and solving for y. Then plot these ordered pairs and connect them with a smooth curved line.

◆ Example 2   Graph the equation $y = 2x^2 - 4x + 5$.

Let $x = -7$; then $y = 2(-7)^2 - 4(-7) + 5 = 131$.
Let $x = -6$; then $y = 2(-6)^2 - 4(-6) + 5 = 101$.
Let $x = -5$; then $y = 2(-5)^2 - 4(-5) + 5 = 75$.
Let $x = -4$; then $y = 2(-4)^2 - 4(-4) + 5 = 53$.
Let $x = -3$; then $y = 2(-3)^2 - 4(-3) + 5 = 35$.
Let $x = -2$; then $y = 2(-2)^2 - 4(-2) + 5 = 21$.
Let $x = -1$; then $y = 2(-1)^2 - 4(-1) + 5 = 11$.
Let $x = 0$; then $y = 2(0)^2 - 4(0) + 5 = 5$.
Let $x = 1$; then $y = 2(1)^2 - 4(1) + 5 = 3$.
Let $x = 2$; then $y = 2(2)^2 - 4(2) + 5 = 5$.
Let $x = 3$; then $y = 2(3)^2 - 4(3) + 5 = 11$.
Let $x = 4$; then $y = 2(4)^2 - 4(4) + 5 = 21$.
Let $x = 5$; then $y = 2(5)^2 - 4(5) + 5 = 35$.
Let $x = 6$; then $y = 2(6)^2 - 4(6) + 5 = 53$.
Let $x = 7$; then $y = 2(7)^2 - 4(7) + 5 = 75$.
Let $x = 8$; then $y = 2(8)^2 - 4(8) + 5 = 101$.
Let $x = 9$; then $y = 2(9)^2 - 4(9) + 5 = 131$.

x	-7	-6	-5	-4	-3	-2	-1	0	1	2	3	4	5	6	7	8	9
y	131	101	75	53	35	21	11	5	3	5	11	21	35	53	75	101	131

Then plot the points from the table. Disregard those coordinates that cannot be plotted. Connect the points with a smooth curved line, as shown in Figure 11.7.

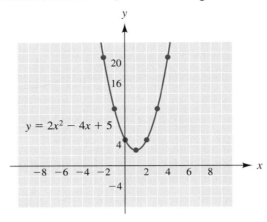

FIGURE 11.7   ■

All parabolas have a property called **symmetry**, which means that a line can be drawn through a parabola dividing it into two parts that are mirror images of each other. (See Figure 11.8.)

The point of intersection of this line of symmetry and the graph of the parabola is called the **vertex**. In this section, the vertex is either the highest or lowest point of the parabola. Locating the vertex is most helpful in drawing the graph of a parabola.

The vertex of a parabola whose equation is in the form $y = ax^2 + bx + c$ may be found as follows.

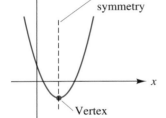

FIGURE 11.8
The line of symmetry of a parabola divides the parabola into two parts that are mirror images of each other.

Vertex of a Parabola

1. The x coordinate is the value of $-\dfrac{b}{2a}$.

2. The y coordinate is found by substituting the x coordinate from Step 1 into the parabola equation and solving for y.

  Example 3

Find the vertex of the parabola $y = 2x^2 - 4x + 5$ in Example 2.

Note that $a = 2$ and $b = -4$.

1. The x coordinate is $-\dfrac{b}{2a} = -\dfrac{-4}{2(2)} = 1$.

2. Substitute $x = 1$ into $y = 2x^2 - 4x + 5$ and solve for y.
$$y = 2(1)^2 - 4(1) + 5$$
$$= 2 - 4 + 5 = 3$$

Thus, the vertex is (1, 3).   ■

  Example 4

Graph the equation $y = -x^2 + 6x$.

First, find the vertex. The x coordinate is $-\dfrac{b}{2a} = -\dfrac{6}{2(-1)} = 3$. Then substitute $x = 3$ into $y = -x^2 + 6x$.

$$y = -(3)^2 + 6(3)$$
$$y = -9 + 18 = 9$$

The vertex is (3, 9). Graph the vertex in Figure 11.9.

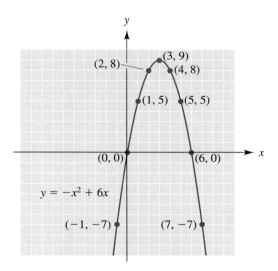

FIGURE 11.9

To find other points, let

$x = 4$, then $y = -(4)^2 + 6(4) = -16 + 24 = 8$; graph (4, 8).
$x = 5$, then $y = -(5)^2 + 6(5) = -25 + 30 = 5$; graph (5, 5).
$x = 6$, then $y = -(6)^2 + 6(6) = -36 + 36 = 0$; graph (6, 0).
$x = 7$, then $y = -(7)^2 + 6(7) = -49 + 42 = -7$; graph (7, −7).

From symmetry, do you see that you can graph the points (2, 8), (1, 5), (0, 0), and (−1, −7) in Figure 11.9 without calculation? If not, let $x = 2$ and solve for y, etc.   ■

You may also note that the parabola in Figure 11.9 opens down. In general, the graph of $y = ax^2 + bx + c$

opens up with $a > 0$ and
opens down when $a < 0$.

Exercises 11.4

Draw the graph of each equation and label each vertex:

1. $y = 2x^2$

2. $y = -2x^2$

3. $y = \frac{1}{2}x^2$

4. $y = -\frac{1}{2}x^2$

5. $y = x^2 + 3$

6. $y = x^2 - 4$

7. $y = 2(x - 3)^2$

8. $y = -(x + 2)^2$

9. $y = x^2 - 2x + 1$

10. $y = 2(x + 1)^2 - 3$

11. $y = 2x^2 - 5$

12. $y = -3x^2 - 2x$

13. $y = x^2 - 2x - 5$

14. $y = -3x^2 + 6x + 15$

15. $y = x^2 - 2x - 15$

16. $y = 2x^2 - x - 15$

17. $y = -4x^2 - 5x + 9$

18. $y = 4x^2 - 12x + 9$

19. $y = \frac{1}{5}x^2 - \frac{2}{5}x + 4$

20. $y = -0.4x^2 + 2.4x + 0.7$

11.5 Imaginary Numbers

What is the meaning of $\sqrt{-4}$? What number squared is -4? Try to find its value on your calculator.

As you can see, this is a different kind of number. Up to now, we have considered only real numbers. The number $\sqrt{-4}$ is not a real number. The square root of a negative number is called an **imaginary number**. The imaginary unit is defined as $\sqrt{-1}$ and in many mathematics texts is given by the symbol i. However, in technical work, i is commonly used for current. To avoid confusion, many technical books use j for $\sqrt{-1}$, which is what we use in this book.

> **Imaginary Unit**
>
> $$\sqrt{-1} = j$$

Then $\sqrt{-4} = \sqrt{(-1)(4)} = (\sqrt{-1})(\sqrt{4}) = (j)(2)$ or $2j$.

 Example 1

Express each number in terms of j: **a.** $\sqrt{-25}$, **b.** $\sqrt{-45}$, **c.** $\sqrt{-183}$.

a. $\sqrt{-25}$
$= \sqrt{(-1)(25)}$
$= (\sqrt{-1})(\sqrt{25})$
$= (j)(5)$ or $5j$

b. $\sqrt{-45}$
$= \sqrt{(-1)(45)}$
$= (\sqrt{-1})(\sqrt{45})$
$= (j)(6.71)$ or $6.71j$

c. $\sqrt{-183}$
$= \sqrt{(-1)(183)}$
$= (\sqrt{-1})(\sqrt{183})$
$= (j)(13.5)$ or $13.5j$ ■

Now let's consider powers of j, or $\sqrt{-1}$. Using the rules of exponents and the definition of j, carefully study the following powers of j:

$$j = j$$
$$j^2 = (\sqrt{-1})^2 = -1$$
$$j^3 = (j^2)(j) = (-1)(j) = -j$$
$$j^4 = (j^2)(j^2) = (-1)(-1) = 1$$
$$j^5 = (j^4)(j) = (1)(j) = j$$
$$j^6 = (j^4)(j^2) = (1)(-1) = -1$$
$$j^7 = (j^4)(j^3) = (1)(-j) = -j$$
$$j^8 = (j^4)^2 = 1^2 = 1$$
$$j^9 = (j^8)(j) = (1)(j) = j$$
$$j^{10} = (j^8)(j^2) = (1)(-1) = -1$$

As you can see, the values of j to a power repeat in the order of $j, -1, -j, 1, j, -1, -j, 1, \ldots$ Also, j to any power divisible by 4 equals 1.

 Example 2

Simplify **a.** j^{15}, **b.** j^{21}, **c.** j^{72}.

a. $j^{15} = (j^{12})(j^3)$
$= (1)(-j) = -j$

b. $j^{21} = (j^{20})(j)$
$= (1)(j) = j$

c. $j^{72} = 1$ ■

In general, we define an **imaginary number** as any number in the form bj, where b is a real number. We define a **complex number** as any number in the form $a + bj$, where a and b are real numbers. Note that in the general complex number $a + bj$, when $a = 0$, we have an imaginary number, and when $b = 0$, we have a real number. The following table contains examples of complex numbers, imaginary numbers, and real numbers.

Complex numbers	Imaginary numbers	Real numbers
$3 + 5j$	$8j$	9
$-2 - 6j$	$-12.5j$	$-5\frac{1}{2}$
$-1.6 + 4.44j$	$\frac{3}{4}j$	27.5
$\frac{1}{2} - \frac{1}{3}j$	$-0.322j$	-0.75

The solutions of the quadratic equation $ax^2 + bx + c = 0$ are given by the quadratic formula

$$x = \frac{-b \pm \sqrt{b^2 - 4ac}}{2a}$$

The part under the radical sign, $b^2 - 4ac$, is called the **discriminant**. The value of $b^2 - 4ac$ determines what kind of solutions (or roots) the quadratic equation has and how many solutions it has when a, b, and c are integers.

If $b^2 - 4ac$ is	Roots
positive and a perfect square,	both roots are rational.
positive and not a perfect square,	both roots are irrational.
zero,	there is only one rational root.
negative,	both roots are imaginary.

The relationship between the graph of $y = ax^2 + bx + c$ and the value of the discriminant is shown in Figure 11.10.

(a)
$b^2 - 4ac > 0$
Two solutions, as indicated
by the two points
of intersection on the x axis

(b)
$b^2 - 4ac = 0$
One solution, as indicated
by the one point
of intersection on the x axis

(c)
$b^2 - 4ac < 0$
The graph does not cross
the x axis; both roots
are imaginary.

FIGURE 11.10
The value of the discriminant, $b^2 - 4ac$, determines the kinds and the number of solutions of the equation $y = ax^2 + bx + c$, where a, b, and c are integers.

Example 3

Determine the nature of the roots of $3x^2 + 5x - 2 = 0$ without solving the equation.

$$a = 3, \qquad b = 5, \qquad c = -2$$

The value of the discriminant is

$$b^2 - 4ac = (5)^2 - 4(3)(-2)$$
$$= 25 + 24 = 49$$

Since 49 is a perfect square, both roots are rational.   ■

Example 4

Determine the nature of the roots of $4x^2 - 12x + 9 = 0$ without solving the equation.

$$a = 4, \qquad b = -12, \qquad c = 9$$

The value of the discriminant is

$$b^2 - 4ac = (-12)^2 - 4(4)(9)$$
$$= 144 - 144 = 0$$

Therefore, there is only one rational root.   ■

Example 5

Determine the nature of the roots of $x^2 - 3x + 8 = 0$ without solving the equation.

$$a = 1, \qquad b = -3, \qquad c = 8$$

The value of the discriminant is

$$b^2 - 4ac = (-3)^2 - 4(1)(8)$$
$$= 9 - 32 = -23$$

Since -23 is negative, both roots are imaginary.   ■

Example 6

Solve $4x^2 - 6x + 5 = 0$ using the quadratic formula.

$$x = \frac{-b \pm \sqrt{b^2 - 4ac}}{2a}, \qquad a = 4, \qquad b = -6, \qquad c = 5$$

$$x = \frac{-(-6) \pm \sqrt{(-6)^2 - 4(4)(5)}}{2(4)}$$

$$= \frac{6 \pm \sqrt{36 - 80}}{8}$$

$$= \frac{6 \pm \sqrt{-44}}{8}$$

$$= \frac{6 \pm 6.63j}{8} \qquad \sqrt{-44} = (\sqrt{-1})(\sqrt{44}) = 6.63j$$

$$= \frac{6 + 6.63j}{8} \quad \text{or} \quad \frac{6 - 6.63j}{8}$$

$$= 0.75 + 0.829j \quad \text{or} \quad 0.75 - 0.829j$$

The roots are $0.75 + 0.829j$ and $0.75 - 0.829j$.   ■

Exercises 11.5

Express each number in terms of j (when necessary, round the result to three significant digits):

1. $\sqrt{-49}$ **2.** $\sqrt{-64}$ **3.** $\sqrt{-14}$

4. $\sqrt{-5}$ **5.** $\sqrt{-2}$ **6.** $\sqrt{-3}$

7. $\sqrt{-56}$ **8.** $\sqrt{-121}$ **9.** $\sqrt{-169}$

10. $\sqrt{-60}$ **11.** $\sqrt{-27}$ **12.** $\sqrt{-40}$

Simplify:

13. j^3 **14.** j^6 **15.** j^{13} **16.** j^{16}

17. j^{19} **18.** j^{31} **19.** j^{24} **20.** j^{26}

21. j^{38} **22.** j^{81} **23.** $\dfrac{1}{j}$ **24.** $\dfrac{1}{j^6}$

Determine the nature of the roots of each quadratic equation without solving it:

25. $x^2 + 3x - 10 = 0$ **26.** $2x^2 - 7x + 3 = 0$

27. $5x^2 + 4x + 1 = 0$ **28.** $9x^2 + 12x + 4 = 0$

29. $3x + 1 = 2x^2$ **30.** $3x^2 = 4x - 8$

31. $2x^2 + 6 = x$ **32.** $2x^2 + 7x = 4$

33. $x^2 + 25 = 0$ **34.** $x^2 - 4 = 0$

Solve each quadratic equation using the quadratic formula (when necessary, round results to three significant digits):

35. $x^2 - 6x + 10 = 0$ **36.** $x^2 - x + 2 = 0$

37. $x^2 - 14x + 53 = 0$ **38.** $x^2 + 10x + 34 = 0$

39. $x^2 + 8x + 41 = 0$ **40.** $x^2 - 6x + 13 = 0$

41. $6x^2 + 5x + 8 = 0$ **42.** $4x^2 + 3x - 1 = 0$

43. $3x^2 = 6x - 7$ **44.** $5x^2 + 2x = -3$

45. $5x^2 + 8x + 4 = 0$ **46.** $2x^2 + x + 3 = 0$

47. $5x^2 + 14x = 3$ **48.** $2x^2 + 1 = x$

49. $x^2 + x + 1 = 0$ **50.** $12x^2 + 23x + 10 = 0$

Chapter 11 Group Activities

1. Graph each of the following:

a. $y = x^2$ **b.** $y = (x + 2)^2$

c. $y = (x - 2)^2$ **d.** $y = x^2 + 2$

e. $y = x^2 - 2$

Compare the graph of $y = x^2$ with the graphs of parts **b** through **e**.

Now graph the following:

f. $y = x^3$ **g.** $y = (x + 2)^3$

h. $y = (x - 2)^3$ **i.** $y = x^3 + 2$

j. $y = x^3 - 2$

Compare the graphs of parts **g** through **j** with the graph of part **f**. In your group, summarize your findings. Does this help us graph more quickly? If so, how?

Chapter 11 Summary

Glossary of Basic Terms

Complex number. Any number in the form $a + bj$, where a and b are real numbers. (p. 390)

Discriminant. The quantity under the radical sign, $b^2 - 4ac$. (p. 380)

Imaginary number. Any number in the form bj, where b is a real number; the square root of a negative number is an imaginary number. The imaginary unit is defined as $\sqrt{-1} = j$. (p. 390)

Parabola. The graph of the quadratic equation $y = ax^2 + bx + c$, where a, b, and c are real numbers and $a \neq 0$. (p. 386)

Quadratic equation. A quadratic equation in one variable is in the form $ax^2 + bx + c = 0$, where $a \neq 0$. A quadratic equation has at most two solutions. (p. 376)

Symmetry with respect to a parabola. A line, called a *line of symmetry*, can be drawn through a parabola, dividing it into two parts that are mirror images of each other. (p. 388)

Vertex of a parabola. The point of intersection of the line of symmetry and the graph of the parabola. (p. 388)

11.1 Solving Quadratic Equations by Factoring

1. **Solving quadratic equations by factoring:**
   a. If necessary, write an equivalent equation in the form $ax^2 + bx + c = 0$.
   b. Factor the polynomial.
   c. Write equations by setting each factor containing a variable equal to zero.
   d. Solve the two resulting first-degree equations.
   e. Check. (p. 376)

11.2 The Quadratic Formula

1. **Quadratic formula:** The roots of a quadratic equation in the form $ax^2 + bx + c = 0$ may be found by using the formula $x = \dfrac{-b \pm \sqrt{b^2 - 4ac}}{2a}$. (p. 379)

11.4 Graphs of Quadratic Equations

1. **Graphing quadratic equations:** To graph a quadratic equation with two variables, find and plot points whose ordered pairs satisfy the equation by choosing various values of x and solving for y. Then connect the points with a smooth curved line. Finding additional points may be necessary.

2. **Vertex of a parabola:** The coordinates of the vertex of a parabola may be found as follows:
   a. The x coordinate is the value of $\dfrac{-b}{2a}$.
   b. The y coordinate is found by substituting the x coordinate from Step **a** into the parabola equation and solving for y. (p. 388)

11.5 Imaginary Numbers

1. The roots of a quadratic equation in the form $ax^2 + bx + c = 0$, where a, b, and c are integers, may be described by using the value of the discriminant as follows. (p. 391)

If $b^2 - 4ac$ is	Roots
positive and a perfect square,	both roots are rational.
positive and not a perfect square,	both roots are irrational.
zero,	there is only one rational root.
negative,	both roots are imaginary.

Chapter 11 Review

1. If $ab = 0$, what is known about either a or b?
2. Solve for x: $3x(x - 2) = 0$.

Solve each equation by factoring:

3. $x^2 - 4 = 0$
4. $x^2 - x = 6$
5. $5x^2 - 6x = 0$
6. $x^2 - 3x - 28 = 0$
7. $x^2 - 14x = -45$
8. $x^2 - 18 - 3x = 0$
9. $3x^2 + 20x + 32 = 0$

Solve each equation using the quadratic formula (when necessary, round results to three significant digits):

10. $3x^2 - 16x - 12 = 0$
11. $x^2 + 7x - 5 = 0$
12. $2x^2 + x = 15$
13. $x^2 - 4x = 2$
14. $3x^2 - 4x = 5$
15. The area of a piece of plywood is 36 ft². Its length is 5 ft more than its width. Find its length and width.

16. A variable electric current is given by the formula $i = t^2 - 12t + 36$, where t is in μs. At what times is the current i equal to **a.** 4 A? **b.** 0 A? **c.** 10 A?

Draw the graph of each equation and label each vertex:

17. $y = x^2 - x - 6$
18. $y = -3x^2 + 2$

Express each number in terms of j:

19. $\sqrt{-36}$
20. $\sqrt{-73}$

Simplify:

21. j^{12}
22. j^{27}

Determine the nature of the roots of each quadratic equation without solving it:

23. $9x^2 + 30x + 25 = 0$
24. $3x^2 - 2x + 4 = 0$

Solve each equation using the quadratic formula (when necessary, round results to three significant digits):

25. $x^2 - 4x + 5 = 0$ **26.** $5x^2 - 6x + 4 = 0$

27. A solar-heated house has a rectangular heat collector with a length 1 ft more than three times its width. The area of the collector is 21.25 ft^2. Find its length and width.

28. A rectangular opening is 15 in. wide and 26 in. long. (See Illustration 1.) A strip of constant width is to be removed from around the opening to increase the area to 672 in^2. How wide must the strip be?

ILLUSTRATION 1

 ## Chapter 11 Test

Solve each equation:

1. $x^2 = 64$ **2.** $x^2 - 8x = 0$

3. $x^2 + 9x - 36 = 0$ **4.** $12x^2 + 4x = 1$

Solve each equation using the quadratic formula (when necessary, round results to three significant digits):

5. $5x^2 + 6x - 10 = 0$ **6.** $3x^2 = 4x + 9$

Solve each equation (when necessary, round results to three significant digits):

7. $21x^2 - 29x - 10 = 0$ **8.** $5x^2 - 7x = 2$

9. $3x^2 - 39x + 90 = 0$ **10.** $6x^2 = 8x + 5$

11. Draw the graph of $y = -x^2 - 8x - 15$ and label the vertex.

12. Draw the graph of $y = 2x^2 + 8x + 11$ and label the vertex.

Express each number in terms of j:

13. $\sqrt{-16}$ **14.** $\sqrt{-29}$

Simplify:

15. j^9 **16.** j^{28}

17. Determine the nature of the roots of $3x^2 - x + 4 = 0$ without solving it.

18. One side of a rectangle is 5 cm more than another. Its area is 204 cm^2. Find its length and width.

12 Geometry

Mathematics at Work

Modern manufacturing companies require a wide variety of technology specialists for their operations. Manufacturing technology specialists set up, operate, and maintain industrial and manufacturing equipment as well as computer-numeric-controlled (CNC) and other automated equipment that make a large variety of products according to controlled specifications. Some focus on systematic equipment maintenance and repair. Others specialize in materials transportation and distribution; that is, they are responsible for moving and distributing the products to the sales locations and/or consumers after they are manufactured. Other key team members include designers, engineers, draftspersons, and quality control specialists. Training and education for these careers are available at many community colleges and trade schools. Some require a bachelor's degree. For more information, go to the website listed below.

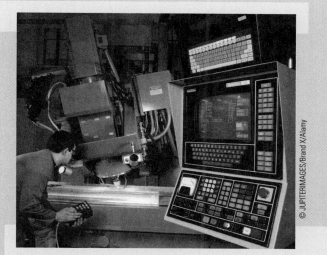

Manufacturing Technology Specialist
Technician working with numerically controlled industrial equipment.

www.cengage.com/mathematics/ewen

Objectives

■ Use a protractor to measure an angle.

■ Apply the basic definitions and relationships for angles, lines, and geometric figures to solve application problems.

■ Find the area and perimeter of quadrilaterals and triangles.

■ Use the Pythagorean theorem to find the side of a right triangle when two sides are known.

■ Use the relationships of similar polygons to solve application problems.

■ Find the area and circumference of circles.

■ Use the relationships of chords, secants, and tangent lines of a circle, arcs of a circle, and inscribed and central angles to solve application problems.

■ Use radian measure to solve application problems.

■ Find the volume, the lateral surface area, and the total surface area of prisms, cylinders, pyramids, cones, and spheres.

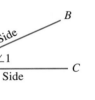

FIGURE 12.1
Basic parts of an angle

12.1 Angles and Polygons

Some fundamentals of geometry must be understood in order to solve many technical applications. Geometry is also needed to follow some of the mathematical developments in technical mathematics courses, in technical support courses, and in on-the-job training programs. Here, we will cover the most basic and most often used geometric terms and relationships. **Plane geometry** is the study of the properties, measurement, and relationships of points, angles, lines, and curves in two dimensions: length and width.

An **angle** is formed by two lines that have a common point. The common point is called the **vertex** of the angle. The parts of the lines are called the **sides of the angle**. An angle is designated by a number, by a single letter, or by three letters. For example, the angle in Figure 12.1 is referred to as $\angle A$ or $\angle BAC$. The middle letter of the three letters is always the one at the vertex.

The measure of an angle is the amount of rotation needed to make one side coincide with the other side. The measure can be expressed in any one of many different units. The standard metric unit of plane angles is the radian (rad). While the radian is the metric unit of angle measurement, many ordinary measurements continue to be made in degrees (°). Although some trades subdivide the degree into the traditional minutes and seconds, most others use tenths and hundredths of degrees. Radian measure is developed in Section 12.6.

One degree is $\frac{1}{360}$ of one complete revolution; that is, $360° =$ one revolution. The protractor in Figure 12.2 is an instrument, marked in degrees, used to measure angles.

FIGURE 12.2
Protractor

Using a Protractor

Step 1 Place the protractor so that the center mark on its base coincides with the vertex of the angle and so that the 0° mark is on one side of the angle.

Step 2 Read the mark on the protractor that is on the other side of the angle (extended, if necessary).

    a. If the side of the angle under the 0° mark extends to the *right* from the vertex, read the inner scale to find the degree measure.

    b. If the side of the angle under the 0° mark extends to the *left* from the vertex, read the outer scale to find the degree measure.

Example 1   Find the degree measure of the angle in Figure 12.3.

FIGURE 12.3

The measure of the angle is 38°, using Step 2(a).   ■

Example 2   Find the degree measure of the angle in Figure 12.4.

FIGURE 12.4

The measure of the angle is 120°, using Step 2(b).   ■

Angles are often classified by degree measure. A **right angle** is an angle with a measure of 90°. In a sketch or diagram, a 90° angle is often noted by placing ∟ or ⌐ in the angle, as shown in Figure 12.5. An **acute angle** is an angle with a measure less than 90°. An **obtuse angle** is an angle with a measure greater than 90° but less than 180°.

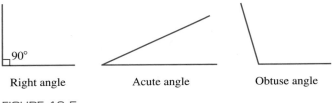

Right angle   Acute angle   Obtuse angle

FIGURE 12.5
Angles classified by degree measure.

We will first study some geometric relationships of angles and lines in the same plane. Two lines **intersect** if they have only one point in common. (See Figure 12.6.)

Two lines in the same plane are **parallel** (∥) if they do not intersect even when extended. (See Figure 12.7.)

FIGURE 12.6
Lines *l* and *m* intersect at point *A*.

FIGURE 12.7
Lines *l* and *m* are parallel.

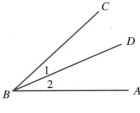

FIGURE 12.8
Adjacent angles

Two angles are **adjacent** if they have a common vertex and a common side lying between them. See Figure 12.8, where ∠1 and ∠2 are adjacent angles because they have a common vertex *B* and a common side *BD* between them. Angles *CBA* and *CBD* are not adjacent because although they have a common vertex, point *B*, their common side, *BA*, is not between the angles.

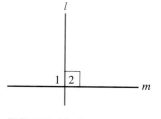

FIGURE 12.9
Perpendicular lines

Two lines in the same plane are **perpendicular** (⊥) if they intersect and form equal adjacent angles. Each of these equal adjacent angles is a right angle. See Figure 12.9, where $l \perp m$ because $\angle 1 = \angle 2$. Angles 1 and 2 are right angles.

Two angles are **complementary** if the sum of their measures is 90°. (See Figure 12.10.) Angles A and B in Figure 12.10(a) are complementary:

$$52° + 38° = 90°$$

Angles LMN and NMP in Figure 12.10(b) are complementary:

$$25° + 65° = 90°$$

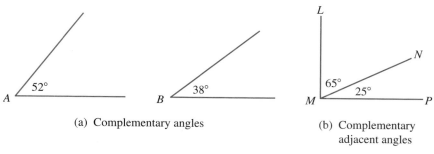

(a) Complementary angles   (b) Complementary
                                 adjacent angles

FIGURE 12.10

Two angles are **supplementary** if the sum of their measures is 180°. (See Figure 12.11.) Angles C and D in Figure 12.11(a) are supplementary: $70° + 110° = 180°$.

Two adjacent angles with their exterior sides in a straight line are *supplementary*. (See Figure 12.11b.) Angles 1 and 2 have their exterior sides in a straight line, so they are supplementary: $\angle 1 + \angle 2 = 180°$.

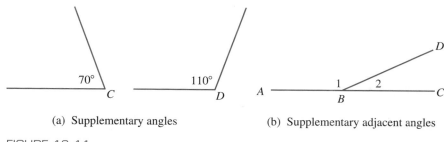

(a) Supplementary angles   (b) Supplementary adjacent angles

FIGURE 12.11

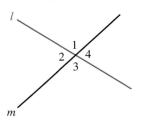

FIGURE 12.12
Vertical angles

When two lines intersect, the angles opposite each other are called **vertical angles**. (See Figure 12.12.) If two straight lines intersect, the vertical angles that are formed are equal. In Figure 12.12, angles 1 and 3 are vertical angles, so $\angle 1 = \angle 3$. Angles 2 and 4 are vertical angles, so $\angle 2 = \angle 4$.

A **transversal** is a line that intersects two or more lines in different points in the same plane (Figure 12.13). **Interior angles** are angles formed inside the lines by the transversal. Angles formed between lines l and m are called interior angles. **Exterior angles** are angles

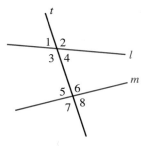

FIGURE 12.13
Line *t* is a transversal of
lines *l* and *m*.

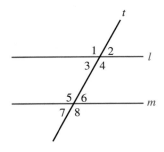

FIGURE 12.14
Line *t* is a transversal for
parallel lines *l* and *m*.

formed outside the lines by the transversal. Those angles outside lines *l* and *m* are called exterior angles.

Interior angles: $\angle 3$, $\angle 4$, $\angle 5$, $\angle 6$
Exterior angles: $\angle 1$, $\angle 2$, $\angle 7$, $\angle 8$

Exterior-interior angles on the same side of the transversal are **corresponding angles**. For example, $\angle 3$ and $\angle 7$ in Figure 12.13 are corresponding angles. Angles 2 and 6 are also corresponding angles.

Angles on opposite sides of the transversal with different vertices are **alternate angles**. Angles 1 and 6 in Figure 12.13 are alternate angles. Angles 1 and 8 are also alternate angles.

If two *parallel* lines are cut by a transversal, then

* *corresponding* angles are *equal.*
* *alternate-interior* angles are *equal.*
* *alternate-exterior* angles are *equal.*
* *interior* angles on the same side of the transversal are *supplementary.*

In Figure 12.14, lines *l* and *m* are parallel and *t* is a transversal. The corresponding angles are equal. That is, $\angle 1 = \angle 5$, $\angle 2 = \angle 6$, $\angle 3 = \angle 7$, and $\angle 4 = \angle 8$.

The alternate-interior angles are equal. That is, $\angle 3 = \angle 6$ and $\angle 4 = \angle 5$.
The alternate-exterior angles are equal. That is, $\angle 1 = \angle 8$ and $\angle 2 = \angle 7$.
The interior angles on the same side of the transversal are supplementary. That is,
$\angle 3 + \angle 5 = 180°$ and $\angle 4 + \angle 6 = 180°$.

◆   Example 3

In Figure 12.15, lines *l* and *m* are parallel and line *t* is a transversal. The measure of $\angle 2$ is 65°. Find the measure of $\angle 5$.

There are many ways of finding $\angle 5$. We show two ways.

Method 1:
Angles 2 and 4 are supplementary, so

$$\angle 4 = 180° - 65° = 115°$$

Angles 4 and 5 are alternate-interior angles, so
$$\angle 4 = \angle 5 = 115°$$

Method 2:
Angles 2 and 3 are vertical angles, so

$$\angle 2 = \angle 3 = 65°$$

Angles 3 and 5 are interior angles on the same side of the transversal; therefore they are supplementary. $\angle 5 = 180° - 65° = 115°$ ■

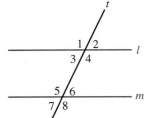

FIGURE 12.15

◆   Example 4

In Figure 12.16, lines *l* and *m* are parallel and line *t* is a transversal. Given $\angle 2 = 2x + 10$ and $\angle 3 = 3x - 5$, find the measure of $\angle 1$.

Angles 2 and 3 are alternate-interior angles, so they are equal. That is,

$$2x + 10 = 3x - 5$$
$$15 = x$$

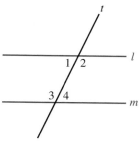

FIGURE 12.16

Then $\angle 2 = 2x + 10 = 2(15) + 10 = 40°$. Since $\angle 1$ and $\angle 2$ are supplementary, $\angle 1 = 180° - 40° = 140°$.

■

Note: We let $\overline{AB}$ be the line segment with endpoints at A and B:

A   B

Let $\overleftrightarrow{AB}$ be the line containing A and B:

A   B

And let AB be the length of $\overline{AB}$.

A **polygon** is a closed figure whose sides are straight line segments. A polygon is shown in Figure 12.17. Polygons are named according to the number of sides they have (see Figure 12.18). A **triangle** is a polygon with three sides. A **quadrilateral** is a polygon with four sides. A **pentagon** is a polygon with five sides. A **regular polygon** has all its sides and interior angles equal.

Polygon

FIGURE 12.17

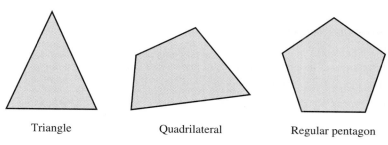

Triangle   Quadrilateral   Regular pentagon

FIGURE 12.18
A polygon is named according to the number of its sides.

Some polygons with more than five sides are named as follows.

Number of sides	Name of polygon
6	**Hexagon**
7	**Heptagon**
8	**Octagon**
9	**Nonagon**

Exercises 12.1 _____

Classify each angle as right, acute, or obtuse:

1.

2.

5.

6.

3.

Right

4.

obtuse

7.

8.

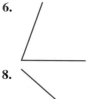

9. In Illustration 1, line *l* intersects line *m* and forms a right angle. Then ∠1 is a(n) ___?___ angle. Lines *l* and *m* are ___?___.

10. Suppose *l* ∥ *m* and *t* ⊥ *l*. Is *t* ⊥ *m*? Why or why not?

11. In Illustration 2, **a.** name the pairs of adjacent angles; **b.** name the pairs of vertical angles.

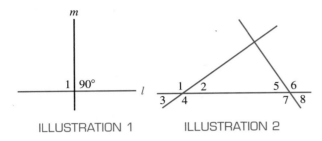

ILLUSTRATION 1   ILLUSTRATION 2

12. In Illustration 2, suppose ∠3 = 40° and ∠7 = 97°. Find the measures of the other angles.

13. In Illustration 3, suppose *l* ∥ *m* and ∠1 = 57°. What are the measures of the other angles?

14. In Illustration 4, suppose *a* ∥ *b*, *a* ⊥ *c*, and ∠1 = 37°. Find the measures of angles 2 and 3.

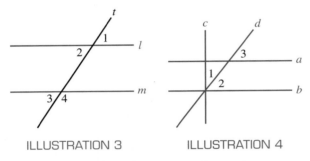

ILLUSTRATION 3   ILLUSTRATION 4

15. In Illustration 5, suppose $\overleftrightarrow{AOB}$ is a straight line and ∠*AOC* = 119°. What is the measure of ∠*COB*?

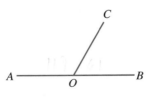

ILLUSTRATION 5

16. Suppose angles 1 and 2 are supplementary and ∠1 = 63°. Then ∠2 = ?

17. Suppose angles 3 and 4 are complementary and ∠3 = 38°. Then ∠4 = ?

18. In Illustration 6, suppose *l* ∥ *m*, $\overleftrightarrow{AOB}$ is a straight line, and ∠3 = ∠6 = 68°. Find the measure of each of the other angles.

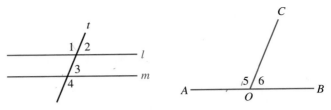

ILLUSTRATION 6

19. In Illustration 7, suppose *a* ∥ *b*, *t* ∥ *x*, ∠3 = 38°, and ∠1 = 52°.
   a. Is *x* ⊥ *a*?   **b.** Is *x* ⊥ *b*?

20. Suppose angles 1 and 2 are complementary and ∠1 = ∠2. Find the measure of each angle.

21. In Illustration 8, suppose ∠1 and ∠3 are supplementary. Find the measure of each angle.

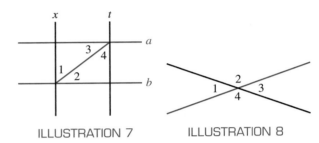

ILLUSTRATION 7   ILLUSTRATION 8

22. In Illustration 9, suppose *l* ∥ *m*, ∠1 = 3x − 50, and ∠2 = x + 60. Find the value of *x*.

23. In Illustration 9, suppose *l* ∥ *m*, ∠1 = 4x + 55, and ∠3 = 10x − 85. Find the value of *x*.

24. In Illustration 9, suppose *l* ∥ *m*, ∠1 = 8x + 60 and ∠4 = 3x + 10. Find the value of *x*.

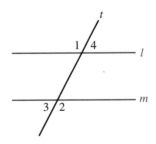

ILLUSTRATION 9

25. A plumber wishes to add a pipe parallel to an existing pipe as shown in Illustration 10. Find angle *x*.

ILLUSTRATION 10

26. A machinist needs to weld a piece of iron parallel to an existing piece of iron as shown in Illustration 11. What is angle *y*?

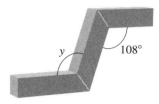

ILLUSTRATION 11

27. In Illustration 12, find angle *z* if *m* ∥ *n*.

28. Given $\overline{AB} \parallel \overline{CD}$ in Illustration 13, find the measure of **a.** angle 1, **b.** angle 2, **c.** angle 3.

ILLUSTRATION 12

ILLUSTRATION 13

Name each polygon:

29.

30.

31.

32.

33.

34.

35.

36.

12.2 Quadrilaterals

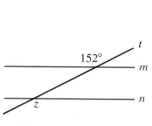

FIGURE 12.19
Parallelogram

A **parallelogram** is a quadrilateral with opposite sides parallel. In Figure 12.19, sides $\overline{AB}$ and $\overline{CD}$ are parallel, and sides $\overline{AD}$ and $\overline{BC}$ are parallel. Polygon *ABCD* is therefore a parallelogram.

Figure 12.20(a) shows the same parallelogram with a perpendicular line segment drawn from point *D* to side $\overline{AB}$. This line segment is an **altitude**.

Figure 12.20(b) shows the result of removing the triangle at the left side of the parallelogram and placing it at the right side. You now have a rectangle with sides of lengths *b* and *h*. Note that the area of this rectangle, *bh* square units, is the same as the area of the parallelogram. So the area of a parallelogram is given by the formula $A = bh$, where *b* is the length of the base and *h* is the length of the altitude drawn to that base. The perimeter is $2a + 2b$ or $2(a + b)$.

(a)

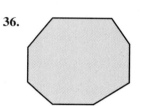

(b)

FIGURE 12.20

A **rectangle** is a parallelogram with four right angles. The area of the rectangle with sides of lengths b and h is given by the formula $A = bh$. (See Figure 12.21.)

Another way to find the area of a rectangle is to count the number of square units in it. In Figure 12.22, there are 15 squares in the rectangle, so the area is 15 square units.

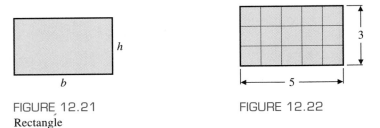

FIGURE 12.21
Rectangle

FIGURE 12.22

The formula for the area of each of the following quadrilaterals follows from the formula for the area of a rectangle.

A **square** (Figure 12.23) is a rectangle with the lengths of all four sides equal. Its area is given by the formula $A = b \cdot b = b^2$. The perimeter is $b + b + b + b$, or $4b$. Note that the length of the altitude is also b.

A **rhombus** (Figure 12.24) is a parallelogram with the lengths of all four sides equal. Its area is given by the formula $A = bh$. The perimeter is $b + b + b + b$, or $4b$.

A **trapezoid** (Figure 12.25) is a quadrilateral with only two sides parallel. Its area is given by the formula $A = \left(\dfrac{a + b}{2} \right)h$. The perimeter is $a + b + c + d$.

FIGURE 12.23
Square

FIGURE 12.24
Rhombus

FIGURE 12.25
Trapezoid

Summary of Formulas for Area and Perimeter of Quadrilaterals		
Quadrilateral	Area	Perimeter
Rectangle	$A = bh$	$P = 2(b + h)$
Square	$A = b^2$	$P = 4b$
Parallelogram	$A = bh$	$P = 2(a + b)$
Rhombus	$A = bh$	$P = 4b$
Trapezoid	$A = \left(\dfrac{a + b}{2} \right)h$	$P = a + b + c + d$

Note: Follow the rules for working with measurements in the rest of this chapter.

Example 1

15.5 m   19.8 m

27.2 m

FIGURE 12.26

Find the area and the perimeter of the parallelogram shown in Figure 12.26.

The formula for the area of a parallelogram is

$$A = bh$$

So $A = (27.2 \text{ m})(15.5 \text{ m})$

$$= 422 \text{ m}^2$$

The formula for the perimeter of a parallelogram is

$$P = 2(a + b)$$

So $P = 2(19.8 \text{ m} + 27.2 \text{ m})$

$$= 2(47.0 \text{ m})$$

$$= 94.0 \text{ m} \qquad \blacksquare$$

Example 2

98.7 ft

121.5 ft

FIGURE 12.27

A rectangular lot 121.5 ft by 98.7 ft must be fenced. (See Figure 12.27.) A fence is installed for $7.50 per running foot. Find the cost of fencing the lot.

The length of fencing needed equals the perimeter of the rectangle. The formula for the perimeter is

$$P = 2(b + h)$$

So $P = 2(121.5 \text{ ft} + 98.7 \text{ ft})$

$$= 2(220.2 \text{ ft})$$

$$= 440.4 \text{ ft}$$

$$\text{Cost} = \frac{\$7.50}{1 \text{ ft}} \times 440.4 \text{ ft} = \$3303 \qquad \blacksquare$$

Example 3

Find the cost of the fertilizer needed for the lawn in Example 2. One bag covers 2500 ft^2 and costs $18.95.

First, find the area of the rectangle. The formula for the area is

$$A = bh$$

So $A = (121.5 \text{ ft})(98.7 \text{ ft})$

$$= 12{,}\overline{0}00 \text{ ft}^2$$

The amount of fertilizer needed is found by dividing the total area by the area covered by one bag:

$$\frac{12{,}\overline{0}00 \text{ ft}^2}{2500 \text{ ft}^2} = 4.8 \text{ bags, or 5 bags.}$$

$$\text{Cost} = 5 \text{ bags} \times \frac{\$18.95}{1 \text{ bag}}$$

$$= \$94.75 \qquad \blacksquare$$

Exercises 12.2

Find the perimeter and the area of each quadrilateral:

1.

15.0 cm

15.0 cm

2.

8.00 cm

10.0 cm

3.

8.0 m

6.0 m

8.0 m

4.

10.0 dm

7.0 dm

5.0 dm

5.

10.0 m

8.0 m   11.0 m

6.0 m

16.0 m

6.

13.5 m

18.6 m

7.

23.9 in.

17.2 in.

20.8 in.

8.

13.7 m

11.9 m

13.7 m

9.

9.2 cm

9.2 cm

10.

13.5 m

8.01 m   6.91 m   7.21 m

21.3 m

In Exercises 11–12, use the formula A = bh:

11. $A = 24\overline{0}$ cm^2, $b = 10.0$ cm; find h.

12. $A = 792$ m^2, $h = 25.0$ m; find b.

13. The area of a parallelogram is 486 ft^2. The length of its base is 36.2 ft. Find its height.

14. The area of a rectangle is 280 cm^2. Its width is 14 cm. Find its length.

15. A piece of 16-gauge steel has been cut into the shape of a trapezoid with height 16.0 in. and bases 21.0 in. and 23.0 in. What is the area of the trapezoidal piece of steel?

16. Looking at the side of a welded metal storage bin, the shape is a trapezoid. The lengths of the bases are 52.3 cm and 68.3 cm, and the height is 41.4 cm. Find its area.

17. On a sectional chart used for aviation navigation, a military operating zone has the shape of a trapezoid. One base is 20.0 mi in length; the other base is 14.0 mi. The lengths of the other two sides are 12.0 mi and 13.42 mi.
a. What is the perimeter of the military operating zone?
b. What is its area if the distance between the parallel sides is 11.6 mi?

18. A pilot flies from an airport to a VOR (Very high frequency Omnidirectional Range) site 82.0 mi away, then 55.0 mi on to another airport. After this, the pilot flies 82.0 mi to an NDB (Non-Directional Beacon) station and then back to the original airport. If the quadrilateral shape flown is a parallelogram, what is the distance from the NDB station back to the airport and what is the total distance of this trip?

19. A rectangular metal duct has a width of 8.0 in. If the area of a cross section is 128 in^2, what is the height? What is the perimeter of a cross section?

20. A rectangular X-ray film measures 15 cm by 32 cm. What is its area?

21. Each hospital bed and its accessories use 96 ft^2 of floor space. How many beds can be placed in a ward 24 ft by 36 ft?

22. A respirator unit needs a rectangular floor space of 0.79 m by 1.2 m. How many units could be placed in a storeroom having $2\overline{0}$ m^2 of floor space?

23. A 108-ft^2 roll of fiberglass is 36 in. wide. What is its length in feet?

24. The cost of the fiberglass in Exercise 23 is \$9.16/yd^2. How much would this roll cost?

25. How many pieces of fiberglass, 36 in. wide and 72 in. long, can be cut from the roll in Exercise 23?

26. A mechanic plans to build a storage garage for 85 cars. Each car needs a space of 15.0 ft by 10.0 ft. **a.** Find the floor area of the garage. **b.** At a cost of $14/ft², find the cost of the garage.

27. The rear view mirror of a car measures 2.9 in. high and has length 9.75 in. What is the area of the rear view mirror if the shape is a parallelogram?

28. A machinist plans to build a screen around his shop area. The area is rectangular and measures 16.2 ft by 20.7 ft. **a.** How many linear feet of screen will be needed? **b.** If the screen is to be 8.0 ft high, how many square feet of screen will be needed?

29. A piece of sheet metal in the shape of a parallelogram has a rectangular hole in it, as shown in Illustration 1. Find **a.** the area of the piece that was punched out, and **b.** the area of the metal that is left.

ILLUSTRATION 1

30. A rectangular piece of sheet metal has an area of 10,680 in². Its length is 72.0 in. Find its width.

31. What is the acreage of a ranch that is a square, 25.0 mi on a side?

32. A rectangular field of corn is averaging 97 bu/acre. The field measures 1020 yd by 928 yd. How many bushels of corn will there be?

33. Find the amount of sheathing needed for the roof in Illustration 2. How many squares of shingles must be purchased? (1 square = 100 ft².)

ILLUSTRATION 2

34. A ceiling is 12 ft by 15 ft. How many 1-ft by 3-ft suspension panels are needed to cover the ceiling?

35. The Smith family plan to paint their home (shown in Illustration 3). The area of the openings not to be painted is 325 ft². The cost per square foot is $0.85. Find the cost of painting the house.

ILLUSTRATION 3

36. An 8-in.-thick wall uses 15 standard bricks (8 in. by $2\frac{1}{4}$ in. by $3\frac{3}{4}$ in.) for one square foot. (This includes mortar.) Find the number of bricks needed for a wall 8 in. thick, 18 ft long, and 8.5 ft high.

37. What is the display floor space of a parallelogram-shaped space that is 29.0 ft long and 8.7 ft deep?

38. Canvas that costs $\frac{3}{4}$¢/in² is used to make golf bags. Find the cost of 200 rectangular pieces of canvas, each 8 in. by 40 in.

39. By law, all businesses outside the Parkville city limits must fence their lots. How many feet of fence will be needed to fence the parallelogram-shaped lot shown in Illustration 4?

ILLUSTRATION 4

40. Find the area of the three trapezoid-shaped display floor spaces of the stores shown in Illustration 5.

ILLUSTRATION 5

41. In a plant, an inside walkway is being laid out to go from one corner of the building along the perimeter to the opposite corner, where the offices are located. If the building is 80.0 ft × 100.0 ft and the walkway is 4.00 ft wide, how many square feet of the building are unavailable for manufacturing?

42. A local manufacturer keeps cows on site to eat the grass in given areas. A new water trough is needed to replace the old one. Find the amount of material needed to build a new trough as shown in Illustration 6.

ILLUSTRATION 6

43. In the mid-20th century, engineers constructed a series of canals to move irrigation water from water sources to farming communities throughout the western United States. The volume of water a canal can move is partially dependent on its cross-sectional area. Suppose a canal is trapezoidal, 25.0 ft across the top and 15.0 ft across the bottom, with a planned depth of water of 10.0 ft as shown in Illustration 7. Find the cross-sectional area of the canal when it is full.

ILLUSTRATION 7

44. A game preserve manager fences off a pasture alongside a road with the dimensions shown in Illustration 8. Find the area of the pasture in acres.

ILLUSTRATION 8

45. A plot of land in the shape of a parallelogram contains 27,800 ft². The frontage along the road is 265 ft as shown in Illustration 9. How deep is the lot?

ILLUSTRATION 9

12.3 Triangles

Triangles are often classified in two ways:

1. by the number of equal sides
2. by the measures of the angles of the triangle

Triangles may be classified or named by the relative lengths of their sides. In each triangle in Figure 12.28, the lengths of the sides are represented by a, b, and c. An **equilateral triangle** has all three sides equal. All three angles are also equal. An **isosceles triangle** has two sides equal. The angles opposite these two sides are also equal. A **scalene triangle** has no sides equal. No angles are equal either.

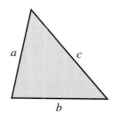

(a) Equilateral triangle   (b) Isosceles triangle   (c) Scalene triangle

FIGURE 12.28
Triangles named by sides

Triangles may also be classified or named in terms of the measures of their angles (see Figure 12.29). A **right triangle** has one right angle. An **acute triangle** has three acute angles. An **obtuse triangle** has one obtuse angle.

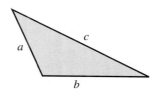

(a) Right triangle   (b) Acute triangle   (c) Obtuse triangle

FIGURE 12.29
Triangles named by angles

Pythagorean Theorem

In a right triangle, the side opposite the right angle is called the **hypotenuse**, which we label c. The other two sides, the sides opposite the acute angles, are called **legs**, which we label a and b. (See Figure 12.29a.) The **Pythagorean theorem** relates the lengths of the sides of any right triangle as follows:

Pythagorean Theorem

$$c^2 = a^2 + b^2 \qquad \text{or} \qquad c = \sqrt{a^2 + b^2}$$

The Pythagorean theorem states that *the square of the hypotenuse of a right triangle is equal to the sum of the squares of the lengths of the two legs*. Alternative forms of the Pythagorean theorem are

$$a^2 = c^2 - b^2 \quad \text{or} \quad a = \sqrt{c^2 - b^2}$$
$$\text{and} \quad b^2 = c^2 - a^2 \quad \text{or} \quad b = \sqrt{c^2 - a^2}$$

Before we use the Pythagorean theorem, you may wish to review square roots in Section 1.15.

Example 1

5.00 cm

c
12.0 cm

FIGURE 12.30

Find the length of the hypotenuse of the triangle in Figure 12.30.

Substitute 5.00 cm for a and 12.0 cm for b in the formula:

$$c = \sqrt{a^2 + b^2}$$
$$c = \sqrt{(5.00 \text{ cm})^2 + (12.0 \text{ cm})^2}$$
$$= \sqrt{25.0 \text{ cm}^2 + 144 \text{ cm}^2}$$
$$= \sqrt{169 \text{ cm}^2} = 13.0 \text{ cm}$$

 5 + 12

13

Note: You may need to use parentheses with some calculators.   ■

Example 2

13.7 m

c
28.1 m

FIGURE 12.31

Find the length of the hypotenuse of the triangle in Figure 12.31.

$$c = \sqrt{a^2 + b^2}$$
$$c = \sqrt{(13.7 \text{ m})^2 + (28.1 \text{ m})^2}$$
$$= 31.3 \text{ m}$$ ■

Example 3

105 mi
263 mi
b

FIGURE 12.32

Find the length of side b of the triangle in Figure 12.32.

$$b = \sqrt{c^2 - a^2}$$
$$b = \sqrt{(263 \text{ mi})^2 - (105 \text{ mi})^2}$$
$$= 241 \text{ mi}$$ ■

Example 4

Voltage applied
Voltage across coil
Voltage across resistance

FIGURE 12.33

The right triangle in Figure 12.33 gives the relationship in a circuit among the applied voltage, the voltage across a resistance, and the voltage across a coil. The voltage across the resistance is 79 V. The voltage across the coil is 82 V. Find the applied voltage.

Using the Pythagorean theorem, we have

$$\text{voltage applied} = \sqrt{(\text{voltage across coil})^2 + (\text{voltage across resistance})^2}$$
$$= \sqrt{(82 \text{ V})^2 + (79 \text{ V})^2}$$
$$= 110 \text{ V}$$ ■

Perimeter and Area

To find the perimeter of a triangle, find the sum of the lengths of the three sides. The formula is $P = a + b + c$, where P is the perimeter and a, b, and c are the lengths of the sides.

An **altitude** of a triangle is a line segment drawn perpendicular from one vertex to the opposite side. Sometimes this opposite side must be extended. See Figure 12.34.

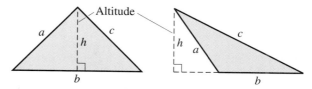

FIGURE 12.34

Look closely at a parallelogram (Figure 12.35a) to find the formula for the area of a triangle. Remember that the area of a parallelogram with sides of lengths a and b is given by $A = bh$. In this formula, b is the length of the base of the parallelogram, and h is the length of the altitude drawn to that base.

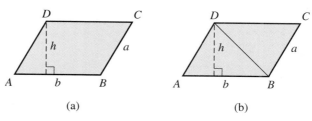

(a)   (b)

FIGURE 12.35

Next, draw a line segment from B to D in the parallelogram as in Figure 12.35(b). Two triangles are formed. We know from geometry that these two triangles have equal areas. Since the area of the parallelogram is bh square units, the area of one triangle is one-half the area of the parallelogram. So the formula for the area of a triangle is

---

Area of Triangle

$$A = \frac{1}{2}bh$$

where b is the length of the base of the triangle (the side to which the altitude is drawn) and h is the length of the altitude.

◆ Example 5

The length of the base of a triangle is 10.0 cm. The length of the altitude to that base is 6.00 cm. Find the area of the triangle. (See Figure 12.36.)

$$A = \frac{1}{2}bh$$

$$A = \frac{1}{2}(10.0 \text{ cm})(6.00 \text{ cm})$$

$$= 30.0 \text{ cm}^2$$

The area is 30.0 cm². ■

FIGURE 12.36

Example 6

In the corner of an office, a counter is built that is 16 ft long as shown in Figure 12.37. Find the area behind the counter if the two walls behind it are of equal length.

FIGURE 12.37

First, let $x =$ the length of each wall behind the counter. Using the Pythagorean theorem, we have

$$a^2 + b^2 = c^2$$
$$x^2 + x^2 = (16.0 \text{ ft})^2$$
$$2x^2 = 256 \text{ ft}^2$$
$$x^2 = 128 \text{ ft}^2 \qquad \text{Divide both sides by 2.}$$
$$x = 11.3 \text{ ft} \qquad \text{Take the square root of both sides.}$$

The area of the right triangle is then

$$A = \tfrac{1}{2}bh$$
$$= \tfrac{1}{2}(128 \text{ ft}^2) \qquad \text{Since } b = h = x \text{ in this right triangle, we can replace } bh \text{ by } x^2 = 128 \text{ ft}^2$$
$$\qquad \qquad \text{or replace each by 11.3 ft. The answers may differ due to rounding.} \quad ■$$
$$= 64.0 \text{ ft}^2$$

If only the lengths of the three sides are known, the area of a triangle is found by the following formula (called Heron's formula):

$$A = \sqrt{s(s - a)(s - b)(s - c)}$$

where a, b, and c are the lengths of the three sides and $s = \tfrac{1}{2}(a + b + c)$.

Example 7

Find the perimeter and the area (rounded to three significant digits) of the triangle in Figure 12.38.

FIGURE 12.38

$$P = a + b + c$$
$$P = 9 \text{ cm} + 15 \text{ cm} + 18 \text{ cm} = 42 \text{ cm}$$

To find the area, first find s.

$$s = \frac{1}{2}(a + b + c)$$

$$s = \frac{1}{2}(9 + 15 + 18) = \frac{1}{2}(42) = 21$$
$$A = \sqrt{s(s - a)(s - b)(s - c)}$$
$$A = \sqrt{21(21 - 9)(21 - 15)(21 - 18)}$$
$$= \sqrt{21(12)(6)(3)}$$
$$= \sqrt{4536}$$
$$= 67.3 \text{ cm}^2$$

67.34983296

■

Note: You may need to insert this right parenthesis to clarify the order of operations. The square root key may also include the left parenthesis; if not, you need to key it in.

FIGURE 12.39

The following relationship is often used in geometry and trigonometry.

The sum of the measures of the angles of any triangle is 180° (Figure 12.39).

Example 8

FIGURE 12.40

Two angles of a triangle have measures 80° and 40°. (See Figure 12.40.) Find the measure of the third angle of the triangle.

Since the sum of the measures of the angles of any triangle is 180°, we know that

$$40° + 80° + x = 180°$$
$$120° + x = 180°$$
$$x = 60°$$

So the measure of the missing angle is 60°. ■

Exercises 12.3

Use the rules for working with measurements. Find the length of the hypotenuse in each triangle:

1.

2.

3.

4.

5.

6.
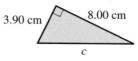

Find the length of the missing side in each triangle:

7.

8.

9.

10.

11.

12.
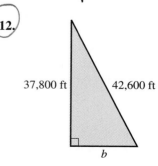

13. Find the length of the braces needed for the rectangular supports shown in Illustration 1.

45.0 in.

39.5 in.

ILLUSTRATION 1

14. Find the center-to-center distance between the two holes in Illustration 2.

13.6 cm

28.2 cm

ILLUSTRATION 2

15. Find the total length of the brace material needed in Illustration 3.

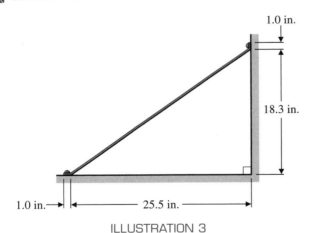

1.0 in.

18.3 in.

1.0 in.→   25.5 in.

ILLUSTRATION 3

16. Often, a machinist must cut a keyway in a shaft. The total depth of cut equals the keyway depth plus the height of a circular segment. The height of a circular segment is found by applying the Pythagorean theorem or by using the formula

$$h = r - \sqrt{r^2 - \left(\frac{l}{2}\right)^2}$$

where h is the height of the segment, r is the radius of the shaft, and l is the length of the chord (or the width

of the keyway). Find the total depth of cut shown in Illustration 4.

Height

Depth = 0.350 in.
Width = 1.750 in.
Dia. = 3.720 in.

ILLUSTRATION 4

17. A piece of 4.00-in.-diameter round stock is to be milled into a square piece of stock with the largest dimensions possible. (See Illustration 5.) What will be the length of the side of the square?

4.00 in.

ILLUSTRATION 5

18. Find the length of the rafter in Illustration 6.

Rafter

1.5 ft   1.5 ft   5.0 ft Rise

12.0 ft   12.0 ft

Span

ILLUSTRATION 6

19. Find the offset distance x (rounded to nearest tenth of an inch) of the 6-ft length of pipe shown in Illustration 7.

$22\frac{1}{16}$ in.   7 in.

18 in. x

$31\frac{15}{16}$ in.

ILLUSTRATION 7

20. A conduit is run in a building (see Illustration 8).
   a. Find the length of the conduit from A_1 to A_6.
   b. Find the straight-line distance from A_1 to A_6.

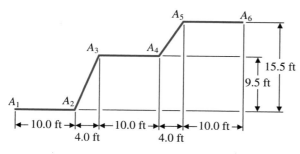

ILLUSTRATION 8

21. The voltage across a resistance is 85.2 V. The voltage across a coil is 78.4 V. Find the voltage applied in the circuit. (See Illustration 9.)

22. The voltage across a coil is 362 V. The voltage applied is 537 V. Find the voltage across the resistance. (See Illustration 9.)

ILLUSTRATION 9

23. The resistor current is 24 A. The total current is 32 A. Find the coil current. (See Illustration 10.)

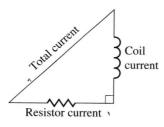

ILLUSTRATION 10

24. The resistor current is 50.2 A. The coil current is 65.3 A. Find the total current. (See Illustration 10.)

In Exercises 25–27, see Illustration 11:

25. Find the reactance of a circuit with impedance 165 Ω and resistance 105 Ω.

ILLUSTRATION 11

26. Find the impedance of a circuit with reactance 20.2 Ω and resistance 38.3 Ω.

27. Find the resistance of a circuit with impedance 4.5 Ω and reactance 3.7 Ω.

28. The base of a window is 7.2 m above the ground. The lower end of a ladder is 3.1 m from the side of the house. How long must a ladder be to reach the base of the window?

Find the area and perimeter of each isosceles triangle:

29.   **30.**

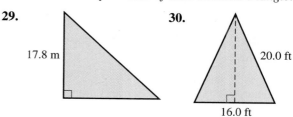

31. Find the area and perimeter of an equilateral triangle with one side 6.00 cm long.

32. Find the area and perimeter of an equilateral triangle with one side 18.0 m long.

Find the area and perimeter of each triangle:

33.

34.

35.

36.

37.

38.

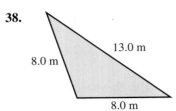

13.0 m
8.0 m
8.0 m

39. Two pieces of steel angle are welded to form right angles. The lengths of the two pieces are 6.0 ft and 9.0 ft, respectively. What is the distance between the two unwelded ends?

40. A triangular gusset (a triangular metal bracket to strengthen a joist) is 11.0 in. in height and 14.4 in. across the base. What is its area?

41. A helicopter is 62.0 mi due north of a VOR (Very high frequency Omnidirectional Range) station according to its DME (Distance Measuring Equipment). One hour later it is 41.0 mi due east of the VOR. How far has the helicopter flown?

42. An unusual architectural design requires triangular ducts that will be painted and exposed in the room. If the cross-sectional area is an equilateral triangle 3.6 ft^2, find the length of each side.

43. A steel plate is punched with a triangular hole as shown in Illustration 12. Find the area of the hole.

28.6 in.   11.0 in.
26.4 in.

ILLUSTRATION 12

44. A square hole is cut from the equilateral triangle in Illustration 13. Find the area remaining in the triangle.

10.0 cm
40.0 cm

ILLUSTRATION 13

Find the measure of the missing angle in each triangle (do not use a protractor):

45.

41°
58°   x

46.

x
62°

47.

x
8   8
x   x
8

48.

10   72°
x   y
10

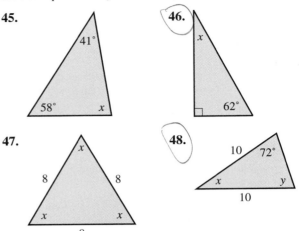

49. In a shaded corner outside of a manufacturing building, a decorative shrub garden as shown in Illustration 14 is to be planted. After the soil has been tilled, it will need to be fertilized. If one bag covers 75 ft^2, how many bags will be needed?

20.0 ft   36.0 ft

ILLUSTRATION 14

50. A helicopter is at a position from two VORs as in the diagram. Given the angles as shown in Illustration 15, find the third angle.

Helicopter

78.0°   62.0°
VOR   VOR

ILLUSTRATION 15

51. Slope may be defined as the steepness of an area and is sometimes given as a percent. For example, a hillside that falls 5 ft for each 100 ft of level length is said to have a 5% slope. The steeper an area of land, the more it is subject to erosion. On a hillside, a conservationist drove a stake into the ground at one point and a second stake 50.0 ft down the hillside, measured horizontally (level). The second stake is 4.50 ft lower than the first stake as shown in Illustration 16. **a.** What is the slope of the hill? **b.** If a sidewalk needs to be installed between the two stakes, what would be its length?

ILLUSTRATION 16

52. Starting at point A, a hiker walks due south for 7.00 mi to point B, then she walks due east for 5.00 mi to point C. How far must she walk to return directly from point C to point A?

53. When measuring land in odd-shaped lots, we often divide the lots into rectangles, triangles, and other geometric figures and calculate each area accordingly. For a lot with dimensions shown in Illustration 17, find its total area in acres.

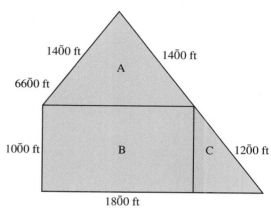

ILLUSTRATION 17

12.4 Similar Polygons

Polygons with the same shape are called **similar polygons**. Polygons are similar when the *corresponding angles are equal.* In Figure 12.41, polygon *ABCDE* is similar to polygon *A'B'C'D'E'* because the corresponding angles are equal. That is, $\angle A = \angle A'$, $\angle B = \angle B'$, $\angle C = \angle C'$, $\angle D = \angle D'$, and $\angle E = \angle E'$.

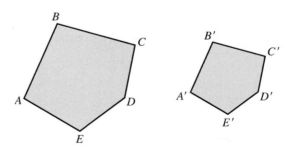

FIGURE 12.41
Similar polygons

When two polygons are similar, the lengths of the *corresponding sides are proportional.* That is,

$$\frac{AB}{A'B'} = \frac{BC}{B'C'} = \frac{CD}{C'D'} = \frac{DE}{D'E'} = \frac{EA}{E'A'}$$

◆   Example 1   The polygons in Figure 12.41 are similar, and $AB = 12$, $DE = 3$, $A'B' = 8$. Find $D'E'$.

Since the polygons are similar,

$$\frac{AB}{A'B'} = \frac{DE}{D'E'}$$

$$\frac{12}{8} = \frac{3}{D'E'}$$

$12(D'E') = (8)(3)$ The product of the means equals the product of the extremes.

$$D'E' = \frac{24}{12} = 2$$ ■

Two triangles are similar when two pairs of corresponding angles are equal, as in Figure 12.42. (If two pairs of corresponding angles are equal, then the third pair of corresponding angles must also be equal. Why?)

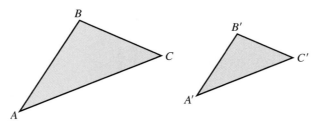

FIGURE 12.42
Similar triangles

Triangle ABC is similar to triangle $A'B'C'$ because $\angle A = \angle A'$, $\angle B = \angle B'$, and $\angle C = \angle C'$.

When two triangles are similar, the lengths of the corresponding sides are proportional. That is,

$$\frac{AB}{A'B'} = \frac{BC}{B'C'} = \frac{CA}{C'A'}$$

or

$$\frac{AB}{A'B'} = \frac{BC}{B'C'} \qquad \frac{BC}{B'C'} = \frac{CA}{C'A'} \qquad \text{and} \qquad \frac{AB}{A'B'} = \frac{CA}{C'A'}$$

◆   Example 2   Find DE and AE in Figure 12.43.

Triangles ADE and ABC are similar, because $\angle A$ is common to both and each triangle has a right angle. So the lengths of the corresponding sides are proportional.

$$\frac{AB}{AD} = \frac{BC}{DE}$$

$$\frac{24}{20} = \frac{18}{DE}$$

$$24(DE) = (20)(18)$$

$$DE = \frac{360}{24} = 15$$

FIGURE 12.43

Use the Pythagorean theorem to find AE.

$$AE = \sqrt{(AD)^2 + (DE)^2}$$
$$AE = \sqrt{(20)^2 + (15)^2}$$
$$= \sqrt{400 + 225} = \sqrt{625} = 25$$ ■

Exercises 12.4

Follow the rules for working with measurements beginning with Exercise 6.

1. In Illustration 1, suppose that $\overline{DE} \parallel \overline{BC}$. Find DE.

ILLUSTRATION 1

2. In Illustration 2, polygon *ABCD* is similar to polygon *FGHI*. Find **a.** $\angle H$ **b.** *FI* **c.** *IH* **d.** *BC*.

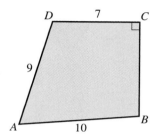

ILLUSTRATION 2

3. In Illustration 3, $\overline{AB} \parallel \overline{CD}$. Is triangle *ABO* similar to triangle *DCO*? Why or why not?

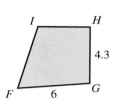

ILLUSTRATION 3

4. Find the lengths of $\overline{AO}$ and $\overline{CD}$ in Illustration 3.

5. In Illustration 4, quadrilaterals *ABCD* and *XYZW* are similar rectangles. $AB = 12$, $BC = 8$, and $XY = 8$. Find *YZ*.

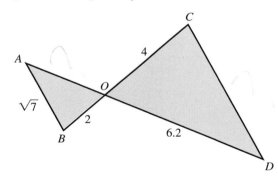

ILLUSTRATION 4

6. An inclined ramp is to be built so that it reaches a height of 6.00 ft over a 15.00-ft run. (See Illustration 5.) Braces are placed every 5.00 ft. Find the height of braces *x* and *y*.

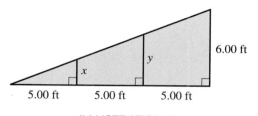

ILLUSTRATION 5

7. A tree casts a shadow $8\bar{0}$ ft long when a vertical rod 6.0 ft high casts a shadow 4.0 ft long. (See Illustration 6.) How tall is the tree?

ILLUSTRATION 6

8. A machinist must follow a part drawing with scale 1 to 16. Find the dimensions of the finished stock shown in Illustration 7. That is, find lengths *A*, *B*, *C*, and *D*.

ILLUSTRATION 7

9. Find **a.** the length of *DE* and **b.** the length of *BC* in Illustration 8.

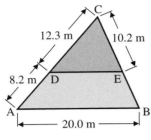

ILLUSTRATION 8

10. A collection of several canisters that fit inside each other is being manufactured. One of the larger sizes has a diameter of 6.00 in. and is 9.00 in. high. If one of the smaller sizes has a diameter of 4.00 in., what should its height be?

11. A right triangular support gusset is to be made similar to another right triangular gusset. The smaller gusset has sides with lengths 5.00 in. and 8.00 in. Find the length of the corresponding shorter side of the larger triangle if its longer side has length 17.0 in.

12. The perimeter of a regular pentagonal-shaped piece of flat steel is 25 in. If a welder cuts another piece of flat steel that is similar but with a 55-in. perimeter, what will be the length of each side?

13. A landing pad for a helicopter at a hospital has the shape of a regular hexagon with perimeter 3000 ft. Another hospital has a similar landing pad with perimeter 3000 ft. What is the length of each of its sides?

14. A rectangular runway is 6100 ft by 61 ft. A similar rectangular runway is 5200 ft. long. What is the width of this runway?

15. An older car has a fan belt assembly that is the shape of an isosceles triangle. The two equal sides are 10.0 in. each, and the third side is 6.0 in. An older truck has a similar fan belt arrangement, but the isosceles triangle has width 12.0 in. What is the length of the two equal sides of this triangle?

16. The side mirror of a small pickup truck is similar to that of a larger full-size pickup truck. If the smaller truck has a rectangular side mirror of width 5.0 in. and height 8.0 in., what is the height of the larger mirror if the width is 10.0 in.?

17. A small heater has a rectangular filter that is 16 in. by 20 in. Another larger heater requires a similar filter that has a 48-in. width. What is the length of this larger filter?

18. A polygon cross-sectional duct is to be exposed and painted. It is to be attached to a smaller duct of the same shape as shown in Illustration 9. If the dimensions of the ducts are *AB* = 12.0 in., *DE* = 20.0 in., and *A'B'* = 9.00 in., find *D'E'*.

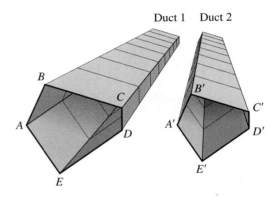

ILLUSTRATION 9

19. A 6.00-ft by 8.00-ft bookcase is to be built. It has horizontal shelves every foot. A support is to be notched in the shelves diagonally from one corner to the opposite corner. At what point should each of the shelves be notched? That is, find lengths *A, B, C, D,* and *E* in Illustration 10. How long is the crosspiece?

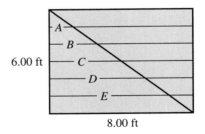

ILLUSTRATION 10

20. A vertical tower 132.0 ft high is anchored to the ground by guy wires as shown in Illustration 11. How long is each guy wire?

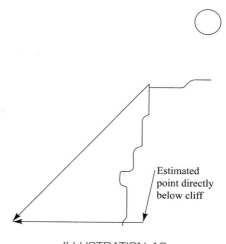

ILLUSTRATION 11

ILLUSTRATION 13

21. Illustration 12 shows the master mold for a symmetrical part for a model truck production. We need to reduce the dimensions to $\frac{1}{4}$ scale; that is, 1 in. $= \frac{1}{4}$ in. **a.** Find each given dimension to the new scale. **b.** At 2.00 oz/in^3, what is the weight of the finished model part?

23. A cat owner has a scratching post that is shaped as a right triangle as shown in Illustration 14. It is 24 in. tall with a base of 26 in. She wants to build a larger scratching post in the same shape that will be 36 in. tall. Lumber is sold in even-foot lengths (6 ft, 8 ft, 10 ft, etc.). Find the lengths of the three sides of the larger scratching post and determine what length of lumber she must purchase to build it.

ILLUSTRATION 12

 22. A team of rock climbers needs to estimate the height of a near-vertical cliff face. One holds a rod that is 36 in. tall vertically on the ground and measures its shadow as 25 in. long. Simultaneously, two other members of the group measure the length of the shadow cast by the cliff to be 117 ft from a point they estimate to be directly beneath the top of the cliff as shown in Illustration 13. Approximately how tall is the cliff?

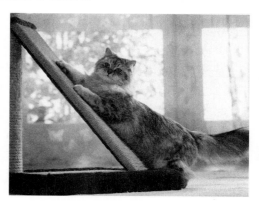

ILLUSTRATION 14

12.5 Circles

A **circle** is a plane curve consisting of all points at a given distance (called the **radius**, r) from a fixed point in the plane, called the **center**. (See Figure 12.44.) The **diameter**, d, of the circle is a line segment through the center of the circle with endpoints on the circle. Note that the length of the diameter equals the length of two radii—that is, $d = 2r$.

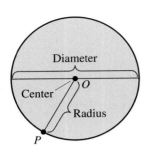

FIGURE 12.44
Circle

The **circumference** of a circle is the distance around the circle. The ratio of the circumference of a circle to the length of its diameter is a constant called π (pi). The number π cannot be written exactly as a decimal. Decimal approximations for π are 3.14 or 3.1416. When solving problems with π, use the π key on your calculator.

The following formulas are used to find the circumference and the area of a circle. C is the circumference and A is the area of a circle; d is the length of the diameter, and r is the length of the radius.

Circumference of circle:	Area of circle:
$C = 2\pi r$	$A = \pi r^2$
$C = \pi d$	$A = \dfrac{\pi d^2}{4}$

 Example 1

Find the area and the circumference of the circle shown in Figure 12.45.

The formula for the area of a circle given the radius is

$$A = \pi r^2$$
$$A = \pi (16.0 \text{ cm})^2$$
$$= 804 \text{ cm}^2$$

The formula for the circumference of a circle given the radius is

$$C = 2\pi r$$
$$C = 2\pi (16.0 \text{ cm})$$
$$= 101 \text{ cm}$$ ■

16.0 cm

FIGURE 12.45

 Example 2

The area of a circle is 576 m². Find the radius.

The formula for the area of a circle in terms of the radius is

$$A = \pi r^2$$
$$576 \text{ m}^2 = \pi r^2$$
$$\frac{576 \text{ m}^2}{\pi} = r^2 \qquad \text{Divide both sides by } \pi.$$
$$\sqrt{\frac{576 \text{ m}^2}{\pi}} = r \qquad \text{Take the square root of both sides.}$$
$$13.5 \text{ m} = r$$ ■

 Example 3

The circumference of a circle is 28.2 cm. Find the radius.

The formula for the circumference of a circle in terms of the radius is

$$C = 2\pi r$$
$$28.2 \text{ cm} = 2\pi r$$
$$\frac{28.2 \text{ cm}}{2\pi} = r \qquad \text{Divide both sides by } 2\pi.$$
$$4.49 \text{ cm} = r$$ ■

An angle whose vertex is at the center of a circle is called a **central angle**. Angle A in Figure 12.46 is a central angle.

A

FIGURE 12.46
Central angle

In general,

> The sum of the measures of all the central angles of any circle is 360°.

Common Terms and Relationships of a Circle

A **chord** is a line segment that has its endpoints on the circle.

A **secant** is any line that intersects a circle at two points.

A **tangent** is a line (or a line segment) that has only one point in common with a circle and lies totally outside the circle. In Figure 12.47, C is the center. $\overline{AB}$ is a chord. Line n is a secant. Line m is a tangent. $\overline{DE}$ is a diameter. $\overline{CF}$ is a radius.

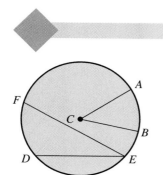

FIGURE 12.47

Arcs

An **inscribed angle** is an angle whose vertex is on the circle and whose sides are chords. The part of the circle between the two sides of an inscribed or central angle is called the **intercepted arc**. In Figure 12.48, C is the center and $\angle ACB$ is a central angle. $\angle DEF$ is an inscribed angle. $\widehat{AB}$ is the intercepted arc of $\angle ACB$. $\widehat{DF}$ is the intercepted arc of $\angle DEF$.

The following three relationships are often helpful to solve problems:

- The measure of a central angle in a circle is equal to the measure of its intercepted arc. (See Figure 12.49.)

- The measure of an inscribed angle in a circle is equal to one-half the measure of its intercepted arc. (See Figure 12.49.)

FIGURE 12.48
Arcs of a circle

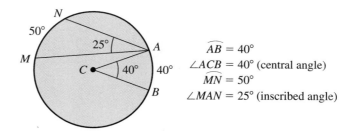

$\widehat{AB} = 40°$
$\angle ACB = 40°$ (central angle)
$\widehat{MN} = 50°$
$\angle MAN = 25°$ (inscribed angle)

FIGURE 12.49

- The measure of an angle formed by two intersecting chords in a circle is equal to one-half the sum of the measures of the intercepted arcs. (See Figure 12.50.)

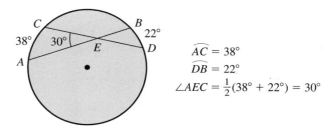

$\widehat{AC} = 38°$
$\widehat{DB} = 22°$
$\angle AEC = \frac{1}{2}(38° + 22°) = 30°$

FIGURE 12.50

Other Chords and Tangents

A diameter that is perpendicular to a chord bisects the chord. (See Figure 12.51.)

A line segment from the center of a circle to the point of tangency is perpendicular to the tangent. (See Figure 12.52.)

Two tangents drawn from a point outside a circle to the circle are equal. The line segment drawn from the center of the circle to this point outside the circle bisects the angle formed by the tangents. (See Figure 12.53.)

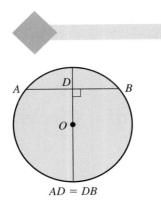

$AD = DB$

FIGURE 12.51

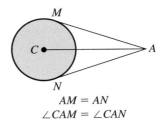

$CE \perp m$

FIGURE 12.52

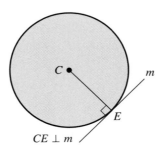

$AM = AN$

$\angle CAM = \angle CAN$

FIGURE 12.53

Exercises 12.5

Follow the rules for working with measurements.

*Find **a.** the circumference and **b.** the area of each circle:*

1.

5.00 in.

2.

20.0 m

3.

9.21 mm

4.

2.70 cm

5.

56.1 mi

6.

39.8 mm

Find the measure of each unknown angle:

7.

97°

92° x

8.

63.8°

149.1°

32.7°

x

9.

x 31.8°

111.1°

29.8°

143.9°

10. A round plate 12 in. in diameter is to have 10 holes drilled through it. To show this, draw half a circle and then "mirror" the other half as shown in Illustration 1. The $\frac{1}{2}$-in.-diameter holes are to be equally spaced on a 9-in. concentric circle (circles with the same center). What is angle A between the holes? Find angle B between the horizontal and each of the end holes.

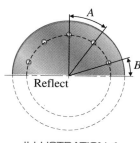

ILLUSTRATION 1

11. The area of a circle is 28.2 cm². Find its radius.

12. The area of a circle is 214 ft². Find its radius.

13. The circumference of a circle is 62.9 m. Find its radius.

14. The circumference of a circle is 17.2 in. Find its radius.

15. How many degrees are in a central angle whose arc is $\frac{1}{4}$ of a circle?

16. How many degrees are in a central angle whose arc is $\frac{2}{3}$ of a circle?

17. A welded circular metal tank has radius 24.0 in. A lid for this tank has the same radius. Find the area of the lid in square feet.

18. A circular hole is to be made in the side of a metal wall. If the area of the hole is to be 90.0 ft², what must the radius be?

19. The airspeed indicator of an airplane is circular with diameter 2.25 in. What are its area and circumference?

20. The side view of a tire resembles a doughnut. If the inner diameter is 15.0 in. and the outer diameter is 23.0 in., what is the area of the side of the tire?

21. If the rim diameter of a wheel of a vehicle is 16.0 in., what is its circumference?

22. Round metal duct has a cross-sectional area of 113 in². What is its diameter?

23. A wheel of radius 1.80 ft is used to measure a field. The wheel rotates 236 times while going the length of the field. How long is the field?

24. Find the length of the diameter of a circular silo with circumference 52.0 ft.

25. A rectangular piece of insulation is to be wrapped around a pipe 4.25 in. in diameter. (See Illustration 2.) How wide does the rectangular piece need to be?

ILLUSTRATION 2

26. How many 1.5-in.-diameter pipes will be needed to have approximately the same total cross-sectional area as one whose diameter is 5.0 in.?

27. A manifold is being designed to carry compressed gas from a tank to four processing stations where the gas is being used. (See Illustration 3.) The main line from the tank is 2.50 in. in diameter. The total cross-sectional area in the four outlet pipes must be the same as the cross-sectional area of the main line. For simplicity, we will not consider flow restriction due to friction, turbulence, or bends in the cylindrical lines. What diameter manifold discharge pipes are required?

ILLUSTRATION 3

28. A pipe has a 3.50-in. outside diameter and a 3.25-in. inside diameter. (See Illustration 4.) Find the area of its cross section.

ILLUSTRATION 4

29. In Illustration 5, find the area of the rectangular piece of metal after the two circles are removed.

14.00-cm diameter

20.00-cm diameter

30.00 cm

60.00 cm

ILLUSTRATION 5

30. Find the area and perimeter of the figure in Illustration 6.

25.0 in.

40.5 in.

ILLUSTRATION 6

31. Find the length of strapping needed for the pipe in Illustration 7.

1.50 in.

12.00-in. diameter pipe

1.50 in.

ILLUSTRATION 7

32. In a design for a workstation, in Illustration 8, each of the four circular sections is to be cut out and removed. **a.** Find the area of the workstation. **b.** How far would a worker have to reach to touch the center of the workstation?

8.00 ft

8.00 ft

18.0 in.

?

ILLUSTRATION 8

33. A boiler 5.00 ft in diameter is to be placed in a corner of a room shown in Illustration 9.
  a. How far from corner C are points A and B of the boiler?
  b. How long is a pipe from C to the center of the boiler M?

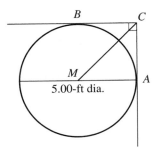

B   C

M

A

5.00-ft dia.

ILLUSTRATION 9

34. A pulley is connected to a spindle of a wheel by a belt. The distance from the spindle to the center of the pulley is 15.0 in. The diameter of the pulley is 15.0 in. What is the length of the belt? (See Illustration 10.)

15.0 in.   120.0°

15.0-in. dia.

ILLUSTRATION 10

35. Mary needs to punch 5 equally spaced holes in a circular metal plate (see Illustration 11). Find the measure of each central angle.

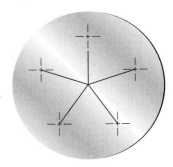

ILLUSTRATION 11

36. In Illustration 12, **a.** find the measure of $\angle 1$, where C is the center, **b.** find the measure of $\angle 2$, and **c.** find the measure of $\angle 3$, given that $\overline{AC} \parallel \overline{DB}$.

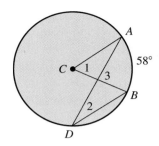

ILLUSTRATION 12

37. Find the measure of ∠1 in Illustration 13.

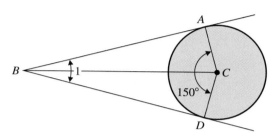

ILLUSTRATION 13

38. In Illustration 13, the length of *AC* is 5 and the distance between *B* and center *C* is 13. Find the length of $\overline{AB}$.

39. Illustration 14 shows a satellite at position *P* relative to a strange planet of radius 2000 miles. The angle between the tangent lines is 11.14°. The distance from the satellite to *Q* is 20,500 miles. Find the altitude $\overline{SP}$ of the satellite above the planet.

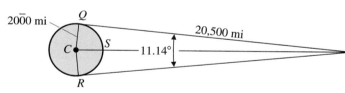

ILLUSTRATION 14

40. In Illustration 15, *CP* = 12.2 m and *PB* = 10.8 m. Find the radius of the circle, where *C* is the center.

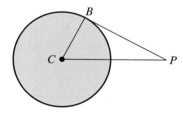

ILLUSTRATION 15

Exercises 41–44 refer to Illustration 16. $\overrightarrow{AB}$ *and* $\overrightarrow{AC}$ *are secants;* $\overline{CD}$ *and* $\overline{BF}$ *are chords.*

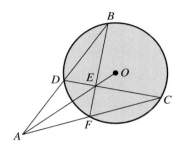

ILLUSTRATION 16

41. Suppose $\overarc{BD} = 85°$ and ∠*DEF* = 52°. Find the measure of $\overarc{CF}$.

42. Suppose $\overarc{BC} = 100°$ and $\overarc{DF} = 40°$. Find the measure of ∠*BAC*.

43. Suppose ∠*BEC* = 78° and $\overarc{BC} = 142°$. Find the measure of $\overarc{DF}$.

44. Suppose ∠*DCF* = 30°, ∠*EFC* = 52°, and $\overarc{CF} = 110°$. Find the measure of $\overarc{BD}$.

45. Inscribe an equilateral triangle in a circle. **a.** How many degrees are contained in each arc? **b.** How many degrees are contained in each inscribed angle? **c.** Draw a central angle to each arc. How many degrees are contained in each central angle?

46. Inscribe a square in a circle. **a.** How many degrees are contained in each arc? **b.** How many degrees are contained in each inscribed angle? **c.** Draw a central angle to each arc. How many degrees are contained in each central angle?

47. Inscribe a regular hexagon in a circle. **a.** How many degrees are contained in each arc? **b.** How many degrees are contained in each inscribed angle? **c.** Draw a central angle to each arc. How many degrees are contained in each central angle?

48. An arc of a circle is doubled. Is its central angle doubled? Is its chord doubled?

49. In designing a bracket for use in a satellite, weight is of major importance. (See Illustration 17.) Find **a.** the area of the part in in², **b.** the overall length of the part, **c.** the overall height of the part, and **d.** its total weight.

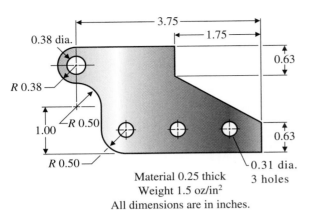

Material 0.25 thick
Weight 1.5 oz/in²
All dimensions are in inches.

ILLUSTRATION 17

50. A piece of aluminum flat bar stock for an equipment bracket on an aircraft is to be bent to the shape shown in Illustration 18. Disregarding material consumed in cutting and squaring ends, what is the total length of the material required? *Note:* In computing material length, the measurement to the mean thickness of material is used for greater accuracy.

All bent radii are given to the inside of the bend.

All dimensions are in inches.

ILLUSTRATION 18

51. Windmills are used to generate electricity. This windmill in Illustration 19 in Madison County, New York, has a blade that is $13\overline{0}$ ft long from the center of its mounting. **a.** When the blade makes one complete rotation, how far (in ft) does the tip of the blade travel? **b.** What is the surface area of the rotating blades measured in acres?

ILLUSTRATION 19

52. A homeowner uses a water sprinkler that rotates in a circle and sends out a jet of water 60.0 ft. If she applies an average of 1 in. of water to the area being irrigated, how many gallons of water will she apply? (1 ft³ = 7.48 gal)

12.6 Radian Measure

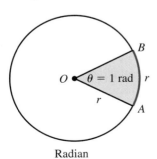

Radian

FIGURE 12.54

Radian measure, the metric unit of angle measure, is used in many applications, such as arc length and rotary motion. The **radian** (rad) unit is defined as the measure of an angle with its vertex at the center of a circle and with an intercepted arc on the circle equal in length to the radius. (In Figure 12.54, $\angle AOB$ forms the intercepted arc AB on the circle.)

In general, the radian is defined as the ratio of the length of arc that an angle intercepts on a circle to the length of its radius. In a complete circle or one complete revolution, the circumference $C = 2\pi r$. This means that for any circle the ratio of the circumference to the radius is constant (2π) because $\frac{C}{r} = 2\pi$. That is, the radian measure of one complete revolution is 2π rad. Technically an angle measured in radians is defined as the ratio of two lengths: the lengths of the arc and the radius of the circle. Because the length units in the ratio cancel, the radian is a dimensionless unit.

What is the relationship between radians and degrees?

One complete revolution = 360°

One complete revolution = 2π rad

Therefore,   360° = 2π rad

180° = π rad

This gives us the conversion factors as follows:

$$\frac{\pi \text{ rad}}{180°} = 1 \quad \text{and} \quad \frac{180°}{\pi \text{ rad}} = 1$$

For comparison purposes,

$$1 \text{ rad} = \frac{180°}{\pi} = 57.3°$$

$$1° = \frac{\pi \text{ rad}}{180°} = 0.01745 \text{ rad}$$

 Example 1   How many degrees are in an angle that measures $\frac{\pi}{2}$ rad?

Use the conversion factor $\dfrac{180°}{\pi \text{ rad}}$.

$$\frac{\pi}{2} \text{ rad} \times \frac{180°}{\pi \text{ rad}} = \frac{180°}{2} = 90°$$ ■

Example 2   How many radians are in an angle that measures 30°?

Use the conversion factor $\dfrac{\pi \text{ rad}}{180°}$.

$$30° \times \frac{\pi \text{ rad}}{\underset{6}{180°}} = \frac{\pi}{6} \text{ rad} \quad \text{or} \quad 0.524 \text{ rad}$$ ■

In general, round lengths and angles in radians (when not expressed in terms of π) to three significant digits and angles in degrees to the nearest tenth of a degree.

As a wheel rolls along a surface, the distance s that a point on the wheel travels equals the product of the radius r and angle θ, measured in radians, through which the wheel turns. (See Figure 12.55.)

FIGURE 12.55

$$s = r\theta \quad (\theta \text{ in rad})$$

Example 3

Find the distance a point on the surface of a pulley travels if its radius is 10.0 cm and the angle of the turn is $\frac{5}{4}$ rad.

$$s = r\theta$$

$$s = 10.0 \text{ cm} \times \frac{5}{4}$$

$$= 12.5 \text{ cm}$$

Example 4

Find the distance a point on the surface of a gear travels if its radius is 15 cm and the angle of the turn is 420°.

$$s = r\theta$$

Since the angle is given in degrees, you must change 420° to radians. Use the conversion factor $\frac{\pi \text{ rad}}{180°}$.

$$\overset{7}{\cancel{420°}} \times \frac{\pi \text{ rad}}{\underset{3}{\cancel{180°}}} = \frac{7\pi}{3} \text{ rad}$$

$$s = \overset{5}{\cancel{15}} \text{ cm} \times \frac{7\pi}{\cancel{3}} = 35\pi \text{ cm or } 11\overline{0} \text{ cm}$$

Example 5

A wheel with radius 5.40 cm travels a distance of 21.0 cm. Find angle θ **a.** in radians and **b.** in degrees that the wheel turns.

a.
$$s = r\theta$$

Solve for θ:

$$\frac{s}{r} = \theta$$

$$\frac{21.0 \text{ cm}}{5.40 \text{ cm}} = \theta$$

$$3.89 \text{ rad} = \theta$$

b. $3.89 \text{ } \cancel{\text{rad}} \times \dfrac{180°}{\pi \text{ } \cancel{\text{rad}}} = 223.0°$

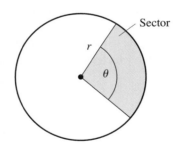

FIGURE 12.56
Sector of a circle

A **sector of a circle** is the region bounded by two radii of a circle and the arc intercepted by them. (See Figure 12.56.) The area of a given sector is proportional to the area of the circle itself; the area of the sector is a fraction of the area of the whole circle.

If the central angle of a sector is measured in degrees, the ratio of the measure of the central angle to 360° specifies the fraction of the area of the circle contained in the sector as follows.

> ## Area of a Sector of a Circle (with the central angle measured in degrees)
>
> $$A = \frac{\theta}{360°} \pi r^2$$
>
> where θ is the measure of the central angle in radians and r is the radius.

If the central angle is measured in radians, the ratio of the measure of the central angle to 2π specifies the fraction of the area of the circle contained in the sector as follows:

$$A = \frac{\theta}{2\pi} \cdot \pi r^2 = \frac{1}{2} r^2 \theta$$

That is,

Area of a Sector of a Circle (with the central angle measured in radians)

$$A = \frac{1}{2} r^2 \theta$$

where θ is the measure of the central angle in radians and r is the radius of the circle.

Example 6

Find the area of the sector of a circle of radius 15.0 cm with a central angle of 125.0°. (See Figure 12.57.)

$$A = \frac{\theta}{360°} \cdot r^2$$

$$A = \frac{125.0°}{360°} \cdot \pi (15.0 \text{ cm})^2$$

$$= 245 \text{ cm}^2 \qquad \blacksquare$$

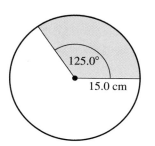

FIGURE 12.57

A segment of a circle is the region between a chord and an arc subtended by the chord. (See Figure 12.58.) Draw the radii to the ends of the chord as in Figure 12.59. The area of a segment equals the area of the sector minus the area of the isosceles triangle formed by the chord and two radii. Next, draw altitude h from the center, perpendicular to the chord. The area of the isosceles triangle is $A = \frac{1}{2} ch$, where c is the length of the chord. Using the Pythagorean theorem in $\triangle OMB$, we have

$$(OM)^2 + (MB)^2 = (OB)^2 \qquad \text{Pythagorean theorem}$$

$$h^2 + \left(\frac{c}{2}\right)^2 = r^2$$

$$h^2 = r^2 - \frac{c^2}{4} \qquad \text{Subtract } \frac{c^2}{4} \text{ from both sides.}$$

$$h^2 = \frac{4r^2 - c^2}{4} \qquad \text{Write with LCD} = 4.$$

$$h = \frac{\sqrt{4r^2 - c^2}}{2} \qquad \text{Take the square root of both sides.}$$

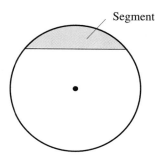

FIGURE 12.58
Segment of a circle

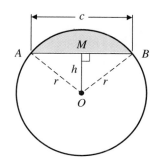

FIGURE 12.59

The area of the isosceles triangle is then

$$A = \frac{1}{2} ch$$

$$= \frac{1}{2}c\left(\frac{\sqrt{4r^2 - c^2}}{2}\right) \qquad \text{Substitute for } h \text{ from above.}$$

$$= \frac{c\sqrt{4r^2 - c^2}}{4} \qquad \text{Simplify.}$$

The area of the segment is the area of the sector minus the area of the isosceles triangle:

$$A = \frac{1}{2} r^2\theta - \frac{c\sqrt{4r^2 - c^2}}{4}$$

where r is the radius, θ is the measure of the central angle, and c is the length of the chord.

If only the length of the chord and the radius of the circle are known, trigonometry is required. This application is treated in Chapter 13.

Example 7

The chord in Figure 12.60 has a length of 26.6 cm. The radius of the circle is 15.0 cm. The measure of the central angle is 125.0°, as in Example 6. Find the area of the segment.

The area of the isosceles triangle is

$$A = \frac{c\sqrt{4r^2 - c^2}}{4}$$

$$A = \frac{(26.6 \text{ cm}) \sqrt{4(15.0 \text{ cm})^2 - (26.6 \text{ cm})^2}}{4}$$

$$= 92.3 \text{ cm}^2$$

Using the result from Example 6, the area of the segment is then

$$245 \text{ cm}^2 - 92.3 \text{ cm}^2 = 153 \text{ cm}^2 \qquad ■$$

FIGURE 12.60

Exercises 12.6

In general, round lengths and angles in radians (when not expressed in terms of π) to three significant digits and angles in degrees to the nearest tenth of a degree.

1. π rad _____°

2. 1.7 rad = _____°

3. 21.0° = _____ rad

4. 45.0° = _____ rad

5. Change $\frac{\pi}{3}$ rad to degrees.

6. Change 150.0° to radians.

7. Change 135.0° to radians.

8. Change $\frac{\pi}{12}$ rad to degrees.

9. How many radians are contained in a central angle that is $\frac{2}{3}$ of a circle?

10. What percent of 2π rad is $\frac{\pi}{2}$ rad?

11. Find the number of radians in a central angle whose arc is $\frac{2}{5}$ of a circle.

12. What percent of 2 rad is $\frac{\pi}{12}$ rad?

Complete the table using the formula s = rθ (θ in radians):

	Radius, r	Angle, θ	Distance, s
13.	25.0 cm	$\dfrac{2\pi}{5}$ rad	
14.	30.0 cm	$\dfrac{4\pi}{3}$ rad	
15.	6.00 cm	45.0°	
16.	172 mm	$\dfrac{\pi}{4}$ rad	
17.	18.0 cm	330.0°	
18.	3.00 m	250.0°	
19.	40.0 cm	rad	112 cm
20.	0.0081 mm	rad	0.011 mm
21.	0.500 m	°	0.860 m
22.	0.0270 m	°	0.0283 m
23.		$\dfrac{2\pi}{3}$ rad	18.5 cm
24.		315.0°	106 m

25. A pulley is turning at an angular velocity of 10.0 rad per second. How many revolutions is the pulley making each second? (*Hint:* One revolution equals 2π rad.)

26. The radius of a wheel is 20.0 in. It turns through an angle of 2.75 rad. What is the distance a point travels on the surface of the wheel?

27. The radius of a gear is 22.0 cm. It turns through an angle of 240.0°. What is the distance a point travels on the surface of the gear?

28. A wheel of diameter 6.00 m travels a distance of 31.6 m. Find the angle θ (in radians) that the wheel turns.

29. A wheel of diameter 15.2 cm turns through an angle of 3.40 rad. Find the distance a point travels on the surface of the wheel.

30. In Illustration 1, find
  a. the length of arc s
  b. the area of the sector
  c. the area of the segment

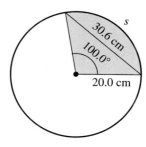

ILLUSTRATION 1

31. In Illustration 2, find
  a. the length of arc s
  b. the area of the sector
  c. the area of the segment

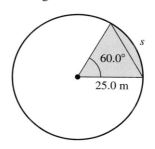

ILLUSTRATION 2

32. Given two concentric circles (circles with the same center) with central angle 45.0°, $r_1 = 4.00$ m, and $r_2 = 8.00$ m, find the shaded area in Illustration 3.

ILLUSTRATION 3

12.7 Prisms

Up to now, you have studied the geometry of two dimensions. **Solid geometry** is the geometry of three dimensions: length, width, and depth.

    A **prism** is a solid whose sides are parallelograms and whose bases are one pair of parallel polygons that have the same size and shape. (See Figure 12.61.) (Recall that a closed

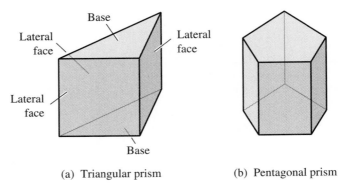

(a) Triangular prism   (b) Pentagonal prism

FIGURE 12.61
Basic parts of a prism. The name of the polygon used as the base names the type of prism.

figure made up of straight lines in two dimensions is called a *polygon*.) The two parallel polygons (which may be any type of polygon) are called the *bases* of the prism. The remaining polygons will be parallelograms and are called *lateral faces*. A *right prism* has lateral faces that are rectangles and are therefore perpendicular to the bases.

The name of the polygon used as the base names the type of prism. For example, a prism with bases that are triangles is called a *triangular* prism (see Figure 12.61a). A prism with bases that are pentagons is called a *pentagonal* prism (see Figure 12.61b), and so on.

> The **lateral surface area** of a prism is the sum of the areas of the lateral faces of the prism.
>
> The **total surface area** of a prism is the sum of the areas of the lateral faces and the areas of the bases of the prism.

Example 1

FIGURE 12.62

Find the lateral surface area of the triangular right prism in Figure 12.62.

To find the lateral surface area, find the area of each lateral face. Then find the sum of these areas.

The area of the rectangular face located on the right front side (see Figure 12.63a) is

$$A = lw$$
$$A = (12.0 \text{ cm})(5.0 \text{ cm})$$
$$= 6\overline{0} \text{ cm}^2$$

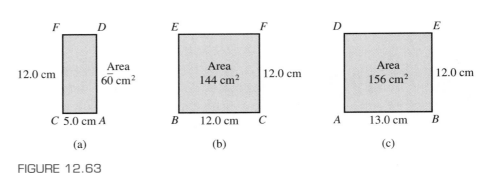

(a)   (b)   (c)

FIGURE 12.63

The area of the square face on the left front side (see Figure 12.63b) is

$$A = s^2$$
$$A = (12.0 \text{ cm})^2$$
$$= 144 \text{ cm}^2$$

To find the area of the third face (the back side of the prism), first find length AB. Since AB is also the hypotenuse of the right triangle ABC, use the Pythagorean theorem as follows:

$$c = \sqrt{a^2 + b^2}$$
$$c = \sqrt{(12.0 \text{ cm})^2 + (5.0 \text{ cm})^2}$$
$$= \sqrt{144 \text{ cm}^2 + 25 \text{ cm}^2}$$
$$= \sqrt{169 \text{ cm}^2}$$
$$= 13.0 \text{ cm}$$

The area of the third face (see Figure 12.63c) is

$$A = lw$$
$$A = (13.0 \text{ cm})(12.0 \text{ cm})$$
$$= 156 \text{ cm}^2$$

Therefore, the lateral surface area is

$$60 \overline{}\text{cm}^2 + 144 \text{ cm}^2 + 156 \text{ cm}^2 = 36\overline{0} \text{ cm}^2 \qquad ■$$

Example 2

Find the total surface area of the prism in Example 1.

To find the total surface area, first find the area of the bases. Then add this result to the lateral surface area from Example 1. The bases have the same size and shape, so just find the area of one base and then double it.

5.0 cm

12.0 cm

FIGURE 12.64

The area of one base as shown in Figure 12.64 is

$$B = A = \frac{1}{2}bh$$
$$B = \frac{1}{2}(12.0 \text{ cm})(5.0 \text{ cm})$$
$$= 3\overline{0} \text{ cm}^2$$

Double this to find the area of both bases.

$$2(3\overline{0} \text{ cm}^2) = 6\overline{0} \text{ cm}^2$$

Add this area to the lateral surface area to find the total area.

$$36\overline{0} \text{ cm}^2 + 6\overline{0} \text{ cm}^2 = 42\overline{0} \text{ cm}^2$$

So the total surface area is $42\overline{0} \text{ cm}^2$.   ■

The volume of a prism is found by the following formula:

Volume of Prism

$$V = Bh$$

where B is the area of one of the bases and h is the altitude (perpendicular distance between the parallel bases).

FIGURE 12.65

See Figure 12.65.

Example 3   Find the volume of the prism in Example 1.

To find the volume of the prism, use the formula $V = Bh$. B is the area of the base which we found to be $3\overline{0}$ cm^2. The altitude of a lateral face is 12.0 cm.

$$V = Bh$$
$$V = (3\overline{0} \text{ cm}^2)(12.0 \text{ cm})$$
$$= 360 \text{ cm}^3$$ ■

Example 4   Find the volume of the prism in Figure 12.66.

Use the formula $V = Bh$. The base is a parallelogram with sides of length 10.0 cm and 4.0 cm. The altitude of the base is 3.0 cm. First, find B, the area of the base.

$$B = bh$$
$$B = (10.0 \text{ cm})(3.0 \text{ cm})$$
$$= 3\overline{0} \text{ cm}^2$$

The altitude of a lateral face of the prism is 7.0 cm. Therefore,

$$V = Bh$$
$$V = (3\overline{0} \text{ cm}^2)(7.0 \text{ cm})$$
$$= 210 \text{ cm}^3$$ ■

7.0 cm

10.0 cm

3.0 cm

4.0 cm

FIGURE 12.66

Example 5   A rectangular piece of steel is 24.1 in. by 13.2 in. by 8.20 in. (Figure 12.67). Steel weighs 0.28 lb/in^3. Find its weight, in pounds.

Find the volume using the formula for the volume of a prism, $V = Bh$. First, find B, the area of the base of the prism.

$$B = lw$$
$$B = (13.2 \text{ in.})(24.1 \text{ in.})$$

The volume is then

$$V = Bh$$
$$V = [(13.2 \text{ in.})(24.1 \text{ in.})](8.20 \text{ in.})$$
$$= 2610 \text{ in}^3$$

Since steel weighs 0.28 lb/in^3, the total weight is

$$2610 \text{ in}^3 \times \frac{0.28 \text{ lb}}{1 \text{ in}^3} = 730 \text{ lb}$$ ■

8.20 in.

13.2 in.

24.1 in.

FIGURE 12.67

Example 6   A parking lot 256 ft by 124 ft will be repaved with asphalt $2\frac{1}{2}$ in. thick. **a.** How many cubic yards will be needed? **b.** If one cubic foot of asphalt weighs 137 lb, what is the total weight that will need to be trucked to the site?

a. The volume of asphalt is found by the formula

$$V = lwh$$
$$V = (256 \text{ ft})(124 \text{ ft})\left(2\frac{1}{2} \text{ in.} \times \frac{1 \text{ ft}}{12 \text{ in.}}\right) \quad \text{Note the conversion factor to obtain ft}^3.$$
$$= 6610 \text{ ft}^3 \times \frac{1 \text{ yd}^3}{27 \text{ ft}^3} \quad \text{Then convert this result to yd}^3.$$
$$= 245 \text{ yd}^3$$

b. $= 245 \ \cancel{yd^3} \times \dfrac{137 \ lb}{\cancel{ft^3}} \times \dfrac{27 \ \cancel{ft^3}}{1 \ \cancel{yd^3}}$ Note the use of conversion factors.

$= 906{,}000 \ lb$ ■

Exercises 12.7

Follow the rules for working with measurements:

1. a. Find the lateral surface area of the prism shown in Illustration 1.
   b. Find the total surface area of the prism.
   c. Find the volume of the prism.

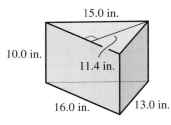

ILLUSTRATION 1

2. a. Find the lateral surface area of the rectangular prism shown in Illustration 2.
   b. Find the total surface area of the prism.
   c. Find the volume of the prism.
   d. What is the name given to this geometric solid?

ILLUSTRATION 2

3. A gusset is the shape of a right triangular prism. If the right triangular base has dimensions 3.0 in. by 4.0 in. by 5.0 in. and the height is 6.0 in., what is the total surface area of the prism? Find its volume.

4. In a drawing, a ceiling-to-floor bay window area is being added to a room. The shape of the added floor space area is an isosceles trapezoid with bases 12.0 ft and 6.00 ft and with slant sides 3 ft 6 in. **a.** Find the area of the floor space added to the room. **b.** If the room has 9 ft 6 in. ceilings, find the additional volume of the room. **c.** How many square yards of vinyl floor material would be purchased if it is available only in 6-ft widths and the owner wants no seams within the added bay window area?

5. The baggage compartment of a helicopter is a rectangular prism. The dimensions of the baggage compartment are 3.0 ft by 4.0 ft by 5.0 ft. What is the volume of the compartment?

6. A piece of 16.0-in. by 20.0-in. metal duct is a rectangular prism. If it is 4.00 ft long, what is the lateral surface area? What is its volume?

7. a. What is the area of the four sides to be painted in Illustration 3? (Assume no windows.)
   b. What is the area of roof to be covered with shingles?
   c. What is the volume of concrete needed to pour a floor 16 cm deep?
   d. What is the total surface area of the figure? (Include painted surface, roof, and floor.)

ILLUSTRATION 3

8. Find the volume of the wagon box in Illustration 4.

ILLUSTRATION 4

9. Find the volume of the gravity bin in Illustration 5.

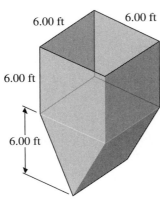

6.00 ft   6.00 ft

6.00 ft

6.00 ft

ILLUSTRATION 5

10. Steel weighs 0.28 lb/in^3. What is the weight of a rectangular piece of steel 0.3125 in. by 12.0 in. by 20.0 in.?

11. A steel rod of cross-sectional area 5.0 in^2 weighs 42.0 lb. Find its length. (Steel weighs 0.28 lb/in^3.)

12. The rectangular lead sleeve shown in Illustration 6 has a cored hole 2.0 in. by 3.0 in. How many cubic inches of lead are in this sleeve?

2.0 in.

6.0 in.

12.0 in.

4.0 in.   3.0 in.

ILLUSTRATION 6

13. A triangular display pedestal is to be made of corrugated paper. Using Illustration 7, what size sheet needs to be designated on the drawing?

4.00 ft   4.00 ft

3.00 ft

4.00 ft

1-inch glue strip   1-inch glue strip

?

?

ILLUSTRATION 7

14. A concrete cube 3 ft on a side is being designed. To reduce the weight of the final product, it will be "voided" using a cardboard box. If 4 in. of concrete must be maintained and there is no concrete on the bottom, find the dimensions needed for a drawing of the cardboard box as shown in Illustration 8.

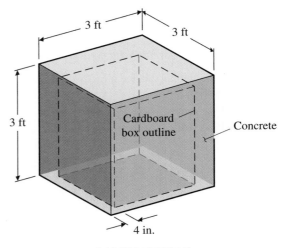

3 ft   3 ft

3 ft

Cardboard box outline   Concrete

4 in.

ILLUSTRATION 8

15. From what size sheet of cardboard can the cardboard box in Exercise 14 be cut? Assume a one-inch glue strip as shown in Illustration 9.

Top

1-inch glue strip (typical)

?

Front   Side   Back   Side

Bottom

?

ILLUSTRATION 9

16. A Victorian building has one room in the shape of a rhombus with a ceiling height of 10.0 ft. Find the volume of air in the room shown in Illustration 10.

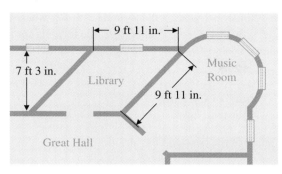

ILLUSTRATION 10

17. A swimming pool company is preparing drawings of a
 newly designed uniform 4.00-ft-deep lap pool as shown
in Illustration 11. One overflow "scupper" valve is required for every 12,000 gal of water. The pool is 40.0 ft
wide and 80.0 ft long. Water weighs 62.4 lb/ft^3 and
8.34 lb/gal. How many scuppers are required?

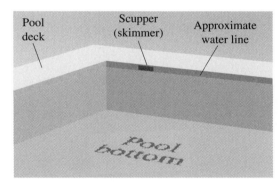

ILLUSTRATION 11

18. A box manufacturer must design a rectangular box to
hold four cylindrical containers, each 12.0 in. in diameter and 15.0 in. tall. Each box needs a double-thickness
bottom and vertical dividers between the containers.
All materials are $\frac{1}{8}$ in. thick. Find the total amount of
material required to make this box.

19. A rectangular swimming pool is 20.0 ft wide and
40.0 ft long. The bottom slopes evenly from the shallow
end, which is 3.00 ft deep, to the deep end, which is
6.00 ft deep. What is the volume of water (in ft^3) that is
required to fill the pool?

20. An aquaculture facility is installed in a school. To take
best advantage of the space, the teacher decides to construct the fish tank from concrete blocks in a corner and
line the sides and bottom with heavy vinyl. The tank is
in the shape of a right triangle with interior measurements of 9.00 ft, 12.0 ft, and 15.0 ft. The walls are built
3.00 ft high. The rule of thumb in projects of this sort is
to order 10% more than the actual required amount in
order to account for waste and overlapping. At a cost of
$0.26/ft^2, how much will the vinyl cost to line the inte-
rior walls and bottom?

12.8 Cylinders

A **circular cylinder** is a geometric solid with a curved lateral surface and circles as parallel bases. The *axis* of a cylinder is the line segment between the centers of the bases. The
altitude, h, is the shortest (perpendicular) distance between the bases. If the axis is perpendicular to the bases, the cylinder is called a *right circular cylinder* and the axis is the same
length as the altitude. See Figure 12.68. You may also think of a right circular cylinder as
the solid formed by rotating a rectangle about one of its sides.

(a) Cylinder   (b) Right circular cylinder

FIGURE 12.68

The volume of a right circular cylinder is found by the formula $V = Bh$, where B is the area of the base. The base is a circle with area $B = \pi r^2$. Therefore, the formula for the volume of a cylinder is written as follows.

Volume of Cylinder

$$V = \pi r^2 h$$

where r is the radius of the base and h is the altitude.

◆ **Example 1**

Find the volume of the right circular cylinder in Figure 12.69.

The diameter is 24.0 m, so the radius is 12.0 m.

$$V = \pi r^2 h$$
$$V = \pi(12.0 \text{ m})^2(40.0 \text{ m})$$
$$= 18,100 \text{ m}^3$$

■

24.0 m

40.0 m

FIGURE 12.69

◆ **Example 2**

Find the diameter of a cylindrical tank 23.8 ft high with a capacity of 136,000 gallons ($1 \text{ ft}^3 = 7.48$ gal).

First, find the volume of the cylinder in ft^3.

$$136,000 \text{ gal} \times \frac{1 \text{ ft}^3}{7.48 \text{ gal}} = 18,200 \text{ ft}^3$$

Since V and h are known, find r using the formula

$$V = \pi r^2 h$$

$$r^2 = \frac{V}{\pi h} \qquad \text{Divide both sides by } \pi h.$$

$$r = \sqrt{\frac{V}{\pi h}} \qquad \text{Take the square root of both sides.}$$

$$r = \sqrt{\frac{18,200 \text{ ft}^3}{\pi(23.8 \text{ ft})}}$$

$$= 15.6 \text{ ft}$$

Diameter is $2r$. So the diameter is $2(15.6 \text{ ft}) = 31.2$ ft. ■

The *lateral surface area of a right circular cylinder* can be visualized as a can without ends. Cut through the side of the can and then flatten it out, as shown in Figure 12.70.

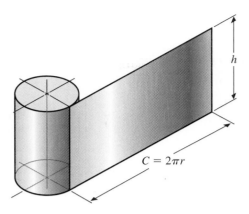

FIGURE 12.70
Lateral surface area of a cylinder

The lateral surface area of a right cylinder is a rectangle with base $2\pi r$ and altitude h. The formula for the lateral surface area is

Lateral Surface Area of Cylinder

$$A = 2\pi rh$$

The *total surface area of a cylinder* is the area of the bases plus the lateral surface area.

Example 3

Find the total surface area of the right circular cylinder in Figure 12.71.

The area of one base is

$$A = \pi r^2 = \pi(1.78\text{ m})^2$$
$$= 9.95\text{ m}^2$$

The area of both bases, then, is

$$2(9.95\text{ m}^2) = 19.9\text{ m}^2$$

The lateral surface area $= 2\pi rh$
$$= 2\pi(1.78\text{ m})(2.32\text{ m})$$
$$= 25.9\text{ m}^2$$

The total surface area $= 19.9\text{ m}^2 + 25.9\text{ m}^2$
$$= 45.8\text{ m}^2$$ ■

1.78 m
2.32 m

FIGURE 12.71

Example 4

A manufacturer needs to plan for storage of 5-gallon oil drums. Find the height of a 5.00-gallon oil drum with a diameter of 12.00 in. Note: The volume of 1 gallon is 231 in³.

The volume of the oil drum is found by using the formula

$$V = \pi r^2 h \qquad \text{Next, solve for } h.$$

$$h = \frac{V}{\pi r^2} \qquad \text{Divide both sides by } \pi r^2.$$

$$h = \frac{5\text{ gal} \times \dfrac{231\text{ in}^3}{1\text{ gal}}}{\pi(6.00\text{ in.})^2} \qquad \text{Substitute the data and simplify.}$$

$$= 10.2\text{ in.}$$ ■

Exercises 12.8

Follow the rules for working with measurements.

Find the volume of each cylinder:

1.

12.0 mm

30.0 mm

2.

13.2 m

17.9 m

3. How many litres does a cylindrical tank of height 39.2 m with radius 8.20 m hold? (1 m³ = 1000 L)

4. A steel cylindrical tank must hold 7110 gal of dyed water for a cloth process. Due to space constraints, the cylindrical tank is made 11.0 ft in diameter. How tall must the tank be? (Water weighs 8.34 lb/gal and 62.4 lb/ft³.)

5. A technician draws plans for a 40$\overline{0}$,000-gallon cylindrical tank with radius 20.0 ft. What should the height be? (1 ft³ = 7.48 gal)

6. An oil filter for a small car is cylindrical with radius 1.80 in. and height 3.60 in. What is the volume of the oil filter?

7. An air filter for an old car is cylindrical. If the inner radius is 4.00 in. and the outer radius is 5.00 in., what is the volume of the 2.00-in.-tall air filter?

8. An engine has 8 cylinders. Each cylinder has a bore of 4.70 in. in diameter and a stroke of 5.25 in. Find its total piston displacement.

9. A cylindrical tank is 25 ft 9 in. long and 7 ft 6 in. in diameter. How many cubic feet does it hold?

10. A 3.0-in.-diameter cylindrical rod is 16 in. long. Find its volume.

11. A cylindrical piece of steel is 10.0 in. long. Its volume is 25.3 in³. Find its diameter.

12. If a metal cylindrical storage tank has a volume of 30$\overline{0}$0 ft³ and a radius of 8.00 ft, what is its height? What is its total surface area?

13. Copper tubing $\frac{1}{2}$ in. in I.D. (inside diameter) is 12.0 ft long. What is the volume of the refrigerant contained in the tubing?

14. A rectangular steel plate 3.76 in. by 9.32 in. by 1.00 in. thick with a 2.00-in.-diameter hole in the center is being drawn. **a.** Find the volume of steel left after the hole is cut. **b.** If this steel weighs 30.0 oz/in³, what weight should be given on the drawing?

15. Find **a.** the lateral surface area and **b.** the total surface area of the right circular cylinder shown in Illustration 1.

39.7 mm

8.21 mm

ILLUSTRATION 1

16. Find **a.** the lateral surface area and **b.** the total surface area of the cylinder shown in Illustration 2.

39.2 cm

51.7 cm

ILLUSTRATION 2

17. Find the total amount (area) of paper used for labels for 1000 cans like the one shown in Illustration 3.

2.18 cm

7.38 cm

ILLUSTRATION 3

18. How many square feet of sheet metal are needed for the sides of the cylindrical tank shown in Illustration 4? (Allow 2.0 in. for seam overlap.)

2.0 ft

10.5 ft

ILLUSTRATION 4

19. A cylindrical piece of stock is turned on a lathe from 3.10 in. down to 2.24 in. in diameter. The cut is 5.00 in. long. What is the volume of the metal removed?

20. What is the volume of lead in the "pig" shown in Illustration 5? What is the volume of the mold?

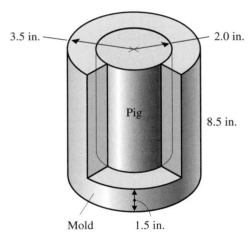

3.5 in.   2.0 in.

Pig

8.5 in.

Mold   1.5 in.

ILLUSTRATION 5

21. A cylinder bore is increased in diameter from 2.78 in. to 2.86 in. The cylinder is 5.50 in. high. How much has the surface area of the walls been increased?

22. Each cylinder bore of a 6-cylinder engine has a diameter of 2.50 in. and a height of 4.90 in. What is the lateral surface area of the six cylinder bores?

23. The sides of a cylindrical silo 15 ft in diameter and 26 ft high are to be painted. Each gallon of paint will cover $2\overline{0}0$ ft². How many gallons of paint will be needed?

24. How many square feet of sheet metal are needed to form the trough shown in Illustration 6?

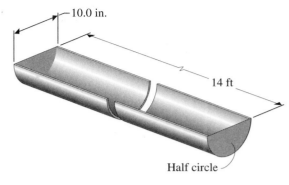

10.0 in.

14 ft

Half circle

ILLUSTRATION 6

25. Find the number of kilograms of metal needed for 2,700,000 cans with ends of the type shown in Illustration 7. The metal has a density of 0.000147 g/cm².

2.38 cm

7.22 cm

ILLUSTRATION 7

26. A cylindrical cooling tank has an outside diameter of 5.00 ft. The walls on all sides are 5.00 in. thick and the tank is 12.0 ft tall. How many gallons of water will this tank hold?

27. A concrete forming "paper" tube is used to void a column as shown in Illustration 8. The walls must be 3.00 in. thick and the outside column must be 37.0 in. in radius. The column is 20.0 ft tall. **a.** Find the diameter of the paper tube. **b.** If concrete weighs 148 lb/ft³, what would you put on the drawing for the final column weight? **c.** At 186 ft²/gal per coat of paint, how much paint would you put in the bill of materials as required for one coat per each column exterior surface. **d.** If the column is not voided and poured solid, how much would it weigh?

37.0 in.

20.0 ft

3.00 in.

Paper
tube

ILLUSTRATION 8

28. A machine shop needs a new parts washer. An ad in a shop journal shows the tank in Illustration 9, but someone forgot to state how many gallons the tank can hold. Find its capacity if it is filled to within 9.00 in. of the top. (1 gal = 231 in^3)

2.00 ft

2.00 ft

6.50 ft

ILLUSTRATION 9

29. Two identical steel cylindrical tanks must hold a total of 22,020 gal of seal oil from a leather treatment plant. The tanks will be made in the shop and installed upstream of the sand filter. They can be only 11.0 ft in diameter. Ignoring the wall thickness, find the length of each tank. (1 gal = 231 in^3)

30. The Trans-Alaska oil pipeline, also known as the Alyeska pipeline, is approximately 800 mi long. Assuming that it has a uniform 48-in. inside diameter, how many ft^3 of oil would be required to fill the pipeline?

31. A silo is a cylinder-shaped structure that is used to store feeds and grain. If the inside dimensions of a silo are 40.0 ft in height and 20.0 ft in diameter, how many bushels of grain can be stored? (1 bu = 1.2445 ft^3)

32. A circular sediment tank at a wastewater treatment plant has a diameter of 110 ft. If the liquid inside the tank is 8.50 ft deep, how many ft^3 of waste will the tank hold?

12.9 Pyramids and Cones

A **pyramid** is a geometric solid whose base is a polygon and whose lateral faces are triangles with a common vertex. The common vertex is called the *apex* of the pyramid. If a pyramid has a base that is a triangle, then it is called a *triangular pyramid*. If a pyramid has a base that is a square, then it is called a *square pyramid*. In general, a pyramid is named by the shape of its base. Two types of pyramids are shown in Figure 12.72.

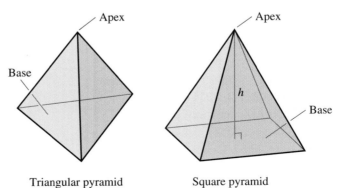

Apex

Apex

Base

h

Base

Triangular pyramid

Square pyramid

FIGURE 12.72
A pyramid is named by the shape of its base.

The volume of a pyramid is found using the following formula:

Volume of Pyramid

$$V = \frac{1}{3}Bh$$

where B is the area of the base and h is the height of the pyramid.

The height of a pyramid is the shortest (perpendicular) distance between the apex and the base of the pyramid.

◆ Example 1

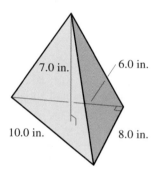

7.0 in.　　　6.0 in.

10.0 in.　　　8.0 in.

FIGURE 12.73

Find the volume of the pyramid in Figure 12.73.

The base is a right triangle with legs 6.0 in. and 8.0 in. Therefore,

$$B = \frac{1}{2}(6.0 \text{ in.})(8.0 \text{ in.}) = 24 \text{ in}^2$$

The height is 7.0 in. Therefore,

$$V = \frac{1}{3}Bh = \frac{1}{3}(24 \text{ in}^2)(7.0 \text{ in.}) = 56 \text{ in}^3$$　■

A **cone** is a geometric solid whose base is a circle. You may also think of a right circular cone as the solid formed by rotating a right triangle about one of its legs and which has a curved lateral surface that comes to a point called the *vertex* (Figure 12.74). The *axis* of a cone is a line segment from the vertex to the center of the base.

The height, h, of a cone is the shortest (perpendicular) distance between the vertex and the base. A *right circular cone* is a cone in which the height is the distance from the vertex to the center of the base. The *slant height* of a right circular cone is the length of a line segment that joins the vertex to any point on the circle that forms the base of the cone.

The *volume of a circular cone* is given by the formula $V = \frac{1}{3}Bh$. Since the base, B, is always a circle, its area is πr^2, where r is the radius of the base. Thus, the formula for the volume of a right circular cone is written as follows.

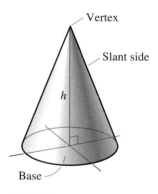

Vertex

Slant side

h

Base

FIGURE 12.74
Cone

Volume of Cone

$$V = \frac{1}{3}\pi r^2 h$$

The *lateral surface area of a right circular cone* is found using the following formula:

Lateral Surface Area

$$A = \pi rs$$

where r is the radius of the base and s is the slant height of the cone.

The *total surface area of a right circular cone* is the sum of the lateral surface area and the area of the base.

Example 2

Find the volume of the right circular cone in Figure 12.75.

$$V = \frac{1}{3}\pi r^2 h$$

$$V = \frac{1}{3}\pi (6.1 \text{ m})^2 (17.2 \text{ m})$$

$$= 670 \text{ m}^3 \qquad ■$$

 Example 3

Find the lateral surface area of the right circular cone in Figure 12.75.

The formula for the lateral surface area is $A = \pi r s$. The slant height is not given. However, a right triangle is formed by the axis, the radius, and the slant height. Therefore, to find the slant height, s, use the formula

$$c = \sqrt{a^2 + b^2}$$

$$\text{Then} \quad s = \sqrt{(6.1 \text{ m})^2 + (17.2 \text{ m})^2}$$

$$= 18.2 \text{ m}$$

The lateral surface area can then be found as follows.

$$A = \pi r s$$

$$= \pi (6.1 \text{ m})(18.2 \text{ m})$$

$$= 350 \text{ m}^2 \qquad ■$$

17.2 m

6.1 m

FIGURE 12.75

The **frustum of a pyramid** is the section of a pyramid between the base and a plane parallel to the base, as shown in Figure 12.76.

Frustum

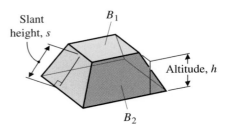

Slant height, s

B_1

Altitude, h

B_2

FIGURE 12.76
Frustum of pyramid

The altitude of the frustum is the perpendicular distance between the two bases. The *volume of the frustum of a pyramid* is

$$V = \frac{1}{3}h\left(B_1 + B_2 + \sqrt{B_1 B_2}\right)$$

where h is the altitude and B_1 and B_2 are the areas of the bases.

The *lateral surface area of the frustum of a pyramid* is

$$A = \frac{1}{2}s(P_1 + P_2)$$

where s is the slant height and P_1 and P_2 are the perimeters of the bases.

Example 4   Find the lateral surface area and the volume of the frustum in Figure 12.77.

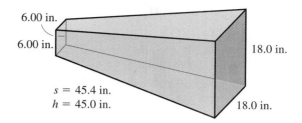

6.00 in.

6.00 in.

18.0 in.

$s = 45.4$ in.
$h = 45.0$ in.

18.0 in.

FIGURE 12.77
Air duct connection

$P_1 = 4(6.00 \text{ in.}) = 24.0 \text{ in.}$

$P_2 = 4(18.0 \text{ in.}) = 72.0 \text{ in.}$

$A = \dfrac{1}{2}s(P_1 + P_2)$

$A = \dfrac{1}{2}(45.4 \text{ in.})(24.0 \text{ in.} + 72.0 \text{ in.})$

$= 2180 \text{ in}^2$

$B_1 = (6.00 \text{ in.})^2 = 36.0 \text{ in}^2$

$B_2 = (18.0 \text{ in.})^2 = 324 \text{ in}^2$

$V = \dfrac{1}{3}h\left(B_1 + B_2 + \sqrt{B_1B_2}\right)$

$V = \dfrac{1}{3}(45.0 \text{ in.})\left(36.0 \text{ in}^2 + 324 \text{ in}^2 + \sqrt{(36.0 \text{ in}^2)(324 \text{ in}^2)}\right)$

$= 7020 \text{ in}^3$ ■

The **frustum of a cone** is the section of the cone between the base and a plane parallel to the base, as shown in Figure 12.78.

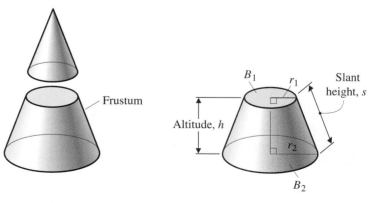

Frustum

B_1 r_1 Slant
height, s

Altitude, h

r_2

B_2

FIGURE 12.78
Frustum of cone

The altitude of the frustum is the perpendicular distance between the two bases. The *volume of the frustum of a cone* is

$$V = \frac{1}{3}h\left(B_1 + B_2 + \sqrt{B_1 B_2}\right)$$

where h is the altitude and B_1 and B_2 are the areas of the bases.

The *lateral surface area of a frustum of a right circular cone* (area of the curved surface) is

$$A = \pi s(r_1 + r_2)$$

where s is the slant height and r_1 and r_2 are the radii of the bases.

Exercises 12.9

Follow the rules for working with measurements.

Find the volume of each figure in Exercises 1–10:

1.

6.70 in.

8.10 in.

8.10 in.

2. 22.6 cm   42.5 cm

42.0 cm

3.

16.2 m

10.8 m

18.8 m

4.   30.2 ft

95.0 ft   208 ft

5.   36.0 mm

101 mm

115 mm

6.   29.7 ft

91.3 ft   109.4 ft

91.3 ft

109.4 ft

83.2 ft

7.

19.2 mm

29.2 mm

15.4 mm

24.6 mm

Hint: To find B, use

$$B = A = \sqrt{s(s - a)(s - b)(s - c)}$$

8.   7.9 mm

4.3 mm

4.3 mm

6.8 mm   2.7 mm

5.9 mm

Hint: To find the area of the base of the pyramid, find the sum of the areas of the rectangle and the triangle.

9.

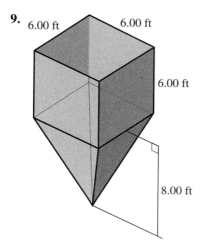

6.00 ft   6.00 ft

6.00 ft

8.00 ft

10.

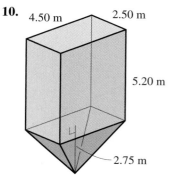

4.50 m   2.50 m

5.20 m

2.75 m

Find **a.** *the volume and* **b.** *the lateral surface area of each right circular cone in Exercises 11–12:*

11.

15.0 cm

16.0 cm

12.

42.3 cm

20.2 cm

13. A loading chute in a flour mill goes directly into a feeding bin. The feeding bin is in the shape of an inverted right circular cone, as shown in Illustration 1. How many bushels of wheat can be placed in the feeding bin? (0.804 bu $= 1$ ft^3)

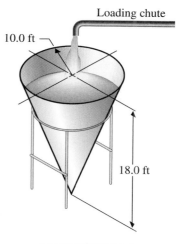

Loading chute

10.0 ft

18.0 ft

ILLUSTRATION 1

14. The circular tank in Illustration 2 is made of $\frac{1}{2}$-in. steel weighing 19.8 lb/ft^2. **a.** What is the total weight of the top? **b.** What is the total weight of the top, sides, and bottom of the tank?

13.0 ft

4.80 ft

18.1 ft

24.2 ft

ILLUSTRATION 2

15. The nose of an airplane has a right circular cone in the center of the propeller. If the cone has a slant height of 13.0 in. with a base radius of 5.50 in., what is the lateral surface area?

16. A welder decides to make a pyramid out of flat steel. The height of the pyramid is to be 3.0 ft and the square base has edges of length 2.0 ft. Find the volume of the pyramid.

17. Gravel is piled in the shape of a cone. The circumference of the base is 224 ft. The slant height is 45 ft. Find the volume of gravel. If gravel weighs 3200 lb/yd^3, how many 22-ton truckloads are needed to transport the gravel?

18. Find the weight of the display model shown in Illustration 3. The model is made of pine. Pine weighs 31.2 lb/ft^3.

2.50 ft

2.50 ft

ILLUSTRATION 3

19. Find the volume of the frustum of the pyramid shown in Illustration 4.

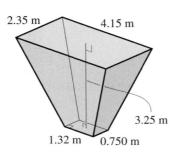

2.35 m   4.15 m

3.25 m

1.32 m   0.750 m

ILLUSTRATION 4

20. Find the volume and lateral surface area of the frustum of the cone shown in Illustration 5.

25.0 ft

45.0 ft

15.0 ft

ILLUSTRATION 5

Find the volume and the lateral surface area of each storage bin:

21.

15.0 ft

22.9 ft   25.0 ft

40.0 ft

25.0 ft

22.

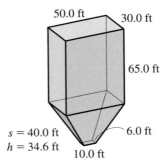

50.0 ft   30.0 ft

65.0 ft

$s = 40.0$ ft
$h = 34.6$ ft   6.0 ft

10.0 ft

23. A hopper must be designed to contain plastic resin pellets for an injection mold machine. (See Illustration 6.) The cylindrical portion of the tank is 18.0 in. in diameter. The spout is a conical frustum. For the hopper to hold 5.00 ft^3 of resin pellets, find the length l of the cylindrical part.

18.0-in.
dia.

l

15.0 in.

3.00-in. dia.

ILLUSTRATION 6

24. A piece of 1-in. (diameter) round stock is tapered so that its tip is a cone. (See Illustration 7.) If the taper begins 3.00 in. from the end of the stock, find the volume of stock that was removed in order to produce the tapered end.

1.00 in.

3.00 in.

ILLUSTRATION 7

25. A welder is assigned to fabricate luggage storage compartments to fit in the luggage storage area of an aircraft. What is the cubic foot displacement of the compartment shown in Illustration 8?

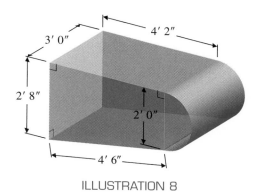

ILLUSTRATION 8

26. Two different sizes of square metal duct are joined by a piece that is a frustum of a pyramid. Find the lateral surface area if the small end is square with one side of length 10.0 in. and the larger square end is of length 15.0 in. on one side with slant height 8.00 in.

27. A lamp manufacturer decides to offset various rising costs by cutting the size of a lampshade, which in turn decreases the amount of material required. The original shade has a top diameter of 13.5 in., a bottom diameter of 15.0 in., and a slant height of 15.0 in. If 1.00 in. is cut from each of these three dimensions, will at least 10% of the material cost be saved?

28. Find the volume of the trash can that is being manufactured as shown in Illustration 9. In advertising it, how many quarts should it be advertised as holding? (1 gal = 231 in^3)

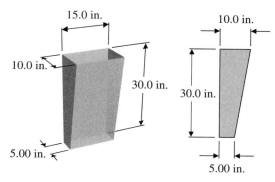

ILLUSTRATION 9

29. As part of a larger piece of work, an artist wishes to cast a solid plaster obelisk in the shape of a tall, slender square pyramid. She builds a form with a square base 26.0 in. on a side with a height of 128 in. into which she pours the plaster. How many gallons of plaster must she mix to fill the form? (1 gal = 231 in^3)

30. The artist in Exercise 29 wishes to pour a second obelisk in the shape of a right circular cone with the same dimensions (diameter 26.0 in. and height 128 in.). How many gallons of plaster must she mix to fill this form? (1 gal = 231 in^3)

12.10 Spheres

A **sphere** (Figure 12.79) is a geometric solid formed by a closed curved surface, with all points on the surface the same distance from a given point (the center). The given distance from any point on the surface to the center is called the *radius*. You may also think of a sphere as the solid formed by rotating a circle about its diameter.

The volume of a sphere is found by using the following formula.

Sphere

FIGURE 12.79

Volume of Sphere

$$V = \frac{4}{3}\pi r^3$$

where r is the radius of the sphere.

The surface area of a sphere is found by using the following formula.

Surface Area of Sphere

$$A = 4\pi r^2$$

Example 1   Find the surface area of a sphere of radius 2.80 cm. (See Figure 12.80.)

$$A = 4\pi r^2$$
$$A = 4\pi(2.80 \text{ cm})^2$$
$$= 4\pi(2.80)^2 \text{ cm}^2$$
$$= 98.5 \text{ cm}^2$$

■

$r = 2.80$ cm

FIGURE 12.80

 Example 2   Find the volume of the sphere in Example 1. The formula for the volume of a sphere is

$$V = \frac{4}{3}\pi r^3$$
$$V = \frac{4}{3}\pi(2.80 \text{ cm})^3$$
$$= \frac{4}{3}\pi(2.80)^3 \text{ cm}^3$$
$$= 92.0 \text{ cm}^3$$

■

Exercises 12.10

Follow the rules for working with measurements.

Find **a.** *the surface area and* **b.** *the volume of each sphere:*

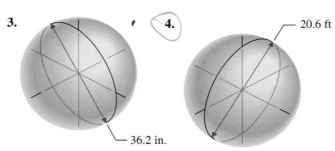

1.

2.   — 18.7 cm

— 8.00 m

3.   ' 4.   — 20.6 ft

— 36.2 in.

5. A balloon 30.1 m in radius is to be filled with helium. How many m^3 of helium are needed to fill it?

6. An experimental balloon is to have a diameter of 5.72 m. How much material is needed for this balloon?

7. A welder has a large pan with the shape of a hemisphere that is used for scrap metal pieces. If the radius of this pan is 9.00 in., what is the volume?

8. An experimental aircraft has a Plexiglas covering over the cockpit that is hemispherical. If the radius of the hemisphere is 2.00 ft, what is the surface area?

9. A cooling water tower is in need of a larger ball float to shut off the inlet water when the tank is full. The present float is 6.00-in. in diameter and exerts 16.0 lb of force. Calculations show that a ball having twice the volume would provide the necessary force. What is the diameter of the new larger float?

10. A small ball bearing has diameter 0.625 in. What is the volume of the bearing?

11. How many gallons of water can be stored in the spherical portion of the water tank shown in Illustration 1? (7.48 gal $= 1$ ft^3)

ILLUSTRATION 1

 12. A city drains 150,000 gal of water from a full spherical tank with a radius of 26 ft. How many gallons of water are left in the tank? (1 ft^3 = 7.48 gal)

13. A spherical tank for liquefied petroleum is 16.0 in. in diameter. **a.** What is the ratio of surface area to the volume of the tank? **b.** Find the same ratio for a tank 24.0 in. in diameter. **c.** Find the same ratio for any tank of radius r.

14. Find the volume of the cylindrical silo with a hemispherical top shown in Illustration 2.

125 ft

35.0 ft

ILLUSTRATION 2

15. You are designing a dome house in the shape of a hemisphere with a 40.0-ft outside diameter. **a.** How many square feet of stucco should be listed on the bill of materials as needed to cover this home if 15% of it is windows and doors? **b.** Find the volume, which is needed to calculate the heat pump size. **c.** If the walls are 12 in. thick, what is the weight of this house used for the footing design? Concrete weighs 148 lb/ft^3. Exclude the weight of the windows and doors. **d.** Two coats of sealer are required to cover and seal the exposed external concrete. If 1 gal covers 110 ft^2, how many gallons are required for the bill of materials?

16. A canister of coolant used to charge an air conditioner is cylindrical with a hemispherical top. What is the volume of the canister if the canister is 1.50 ft tall with radius 4.00 in.?

17. A meteorologist needs to launch a weather balloon that is roughly spherical. How many ft^3 of helium will be required to inflate the balloon to a diameter of 9.0 ft?

18. A park ranger orders a hot air balloon to use for tourists in the park. She orders a balloon that is roughly spherical and has a diameter of 40.0 ft. **a.** Assuming it is spherical, what will be the volume of the balloon in ft^3? **b.** Assuming the manufacturer of the balloon allows 10% more than the required amount of fabric for seams and waste, how much fabric will the manufacturer need to make the park ranger's balloon?

Chapter 12 Group Activities

1. In Section 12.3, we dealt with triangles, and in Section 12.4, we talked about similar polygons. Together with the use of ratios, we have some interesting tools at our disposal. In groups, figure out how tall something around your campus is such as a flagpole, a building, or tower. Try to determine its height. Wait until a shadow is cast from a building and measure how far from the base it is. Then align the shadow of a person or a yardstick so that the tip of the person's shadow and the tip of the building's shadow are touching. Illustration 1 will help to illustrate this.

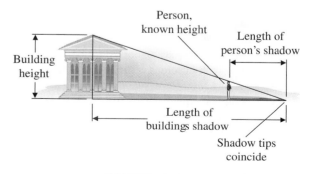

ILLUSTRATION 1

Here is a model problem to help with this activity. We want to find the height of a tree. At a certain time, the tree's shadow measures 40 ft. A person who is 5 ft tall casts a shadow that is 18 ft long at the same time, as shown in Illustration 2.

Solution:

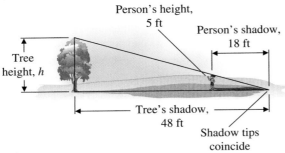

Person's height, 5 ft
Person's shadow, 18 ft
Tree height, *h*
Tree's shadow, 48 ft
Shadow tips coincide

ILLUSTRATION 2

Therefore, the ratio from similar triangles is $\frac{h}{48} = \frac{5}{18}$. Now, $h = \frac{5 \cdot 48}{18} = \frac{40}{3} \approx 13.3$ ft, so the tree is 13.3 ft tall.

2. Design a two-story house with a circular room on one corner and a trapezoidal bay window in the living room. Find the total gross area inside the house, the inside area of each room, and the total net inside or floor area.

3. A box manufacturer is designing a box to hold from six to twelve 1-gal containers that are 4 in. in diameter. How much material is required to make a double-bottomed box with lid to hold 6, 8, 9, 10, or 12 containers? Which has the best ratio of materials to number of containers?

4. How many 8-in. circular disks can be cut from a $\frac{1}{8}$-in.-thick 4-ft by 8-ft sheet of stainless steel if $\frac{1}{8}$ in. is wasted for each cut? How much material is wasted? What is the percent of scrap? Is there more than one way to design this? Does it increase or decrease the percentage of scrap?

5. Design an in-ground swimming pool with both a shallow and deep end. Make a scale drawing. **a.** Calculate its perimeter. **b.** Determine how many 8-in.-square tiles are needed to surround the pool on the top outside edge and the inside of the pool. **c.** Determine the volume of the pool and how many gallons of water are needed to fill it within 6 in. of the top.

6. Design STOP signs for around campus. The signs are to be scaled $\frac{3}{4}$ the size of the real signs on $8\frac{1}{2}$-by-11-in. paper. Choose a scale for the drawing so that the drawing fills most of the paper. The drawing may be done by using a regular scale drawing or by using CAD. There are several components: **a.** Before making any drawings or models, measure a real STOP sign. Each team member should do this independently and then come together, reaching a consensus on the size. **b.** Choose a scale to meet the requirements stated in the problem. **c.** Calculate the length of the sides using the scale and proportions. **d.** Determine the size of the interior angles using complementary, supplementary, and other angle information.

Chapter 12 Summary

Glossary of Basic Terms

Acute angle.   An angle with a measure less than 90°. (p. 400)

Acute triangle.   A triangle with three acute angles. (p. 411)

Adjacent angles.   Two angles with a common vertex and a common side lying between them. (p. 400)

Alternate angles.   Angles on opposite sides of the transversal with different vertices. (p. 402)

Altitude.   A line segment drawn perpendicular to the base of a triangle or a quadrilateral. (pp. 405, 413)

Angle.   Formed by two lines that have a common point. (p. 398)

Center of a circle.   The fixed point that is the same given distance, *r*, from all points on a circle. (p. 423)

Central angle.   An angle whose vertex is at the center of a circle. (p. 424)

Chord.   A line segment that has its endpoints on the circle. (p. 425)

Circle.   A plane curve consisting of all points at a given distance, called the *radius,* from a fixed point in the plane, called the *center.* (p. 423)

Circular cylinder.   A geometric solid with a curved lateral surface and circles as parallel bases. Its *axis* is the line segment between the centers of the bases. The altitude, *h*, is the shortest (perpendicular) distance between the bases. If the axis is perpendicular to the bases, the cylinder is called a *right circular cylinder* and the axis is the same length as the altitude. (p. 441)

Circumference.   The distance around a circle. (p. 424)

Complementary angles.   Two angles for which the sum of their measures is 90°. (p. 401)

Cone.   A geometric solid whose base is a circle and a curved lateral surface that comes to a point called the *vertex*. The *axis* of a cone is a line segment from the vertex to the center of the base. A *right circular cone* is a cone in which the height is the distance from the vertex to the center of the base. The *slant height* of a right circular cone is the length of a line segment that joins the vertex to any point on the circle that forms the base of the cone. (p. 447)

Corresponding angles.   Exterior-interior angles on the same side of the transversal. (p. 402)

Diameter.   A line segment through the center of a circle with endpoints on the circle. (p. 423)

Equilateral triangle.   A triangle with all three sides equal. All three angles are also equal. (p. 410)

Exterior angles.   Angles formed outside the lines by the transversal. (p. 401)

Frustum.   The section between the base and a plane parallel to the base of a pyramid or a cone. (pp. 448, 449)

Heptagon.   A polygon with seven sides. (p. 403)

Hexagon.   A polygon with six sides. (p. 403)

Hypotenuse.   The side opposite the right angle in a right triangle. (p. 411)

Inscribed angle.   An angle whose vertex is on the circle and whose sides are chords. (p. 425)

Intercepted arc.   The part of the circle between two sides of an inscribed or central angle. (p. 425)

Interior angles.   Angles formed inside the lines by the transversal. (p. 401)

Intersect.   Two lines intersect if they have only one point in common. (p. 400)

Isosceles triangle.   A triangle with two sides equal. The angles opposite the equal sides are also equal. (p. 410)

Lateral surface area.   The sum of the areas of the lateral faces of a prism or a pyramid or the area of the curved side of a cylinder or a cone. (p. 436)

Legs.   The sides of a right triangle opposite the acute angles. (p. 411)

Nonagon.   A polygon with nine sides. (p. 403)

Obtuse angle.   An angle with a measure greater than 90° but less than 180°. (p. 400)

Obtuse triangle.   A triangle with one obtuse angle. (p. 411)

Octagon.   A polygon with eight sides. (p. 403)

Parallel.   Two lines are parallel (∥) if they do not intersect even when extended. (p. 400)

Parallelogram.   A quadrilateral with opposite sides equal. (p. 405)

Pentagon.   A polygon with five sides. (p. 403)

Perpendicular.   Two lines are perpendicular ($\perp$) if they intersect and form equal adjacent angles. (p. 401)

Plane geometry.   The study of the properties, measurement, and relationships of points, angles, lines, and curves in two dimensions: length and width. (p. 398)

Polygon.   A closed figure whose sides are straight line segments. (p. 403)

Prism.   A solid whose sides are parallelograms and whose bases are one pair of parallel polygons that have the same size and shape. The two parallel polygons (which may be any type of polygon) are called the *bases* of the prism. The remaining polygons will be parallelograms and are called *lateral faces*. A *right prism* has lateral faces that are rectangles and therefore are perpendicular to the bases. (p. 435)

Pyramid.   A geometric solid whose base is a polygon and whose lateral surfaces are triangles with a common vertex, called the *apex*. (p. 446)

Pythagorean theorem.   The square of the hypotenuse of a right triangle is equal to the sum of the squares of the lengths of the two legs, or $c^2 = a^2 + b^2$. (p. 411)

Quadrilateral.   A polygon with four sides. (p. 403)

Radian.   The measure of an angle with its vertex at the center of a circle and with an intercepted arc on the circle equal in length to the radius. (p. 430)

Radius.   The distance, r, between the center of a circle and all points on the circle. (p. 423)

Rectangle.   A parallelogram with four right angles. (p. 406)

Regular polygon.   A polygon with all of its sides and interior angles equal. (p. 403)

Rhombus.   A parallelogram with the lengths of all four sides equal. (p. 406)

Right angle.   An angle with a measure of 90°. (p. 400)

Right triangle.   A triangle with one right angle. (p. 411)

Scalene triangle.   A triangle with no sides equal. No angles are equal either. (p. 410)

Secant.   Any line that intersects a circle in two points. (p. 425)

Sector of a circle.   The region of a circle bounded by two radii and the arc intercepted by them. (p. 432)

Sides of an angle.   The parts of the lines that form an angle. (p. 398)

Similar polygons.   Polygons with the same shape. (p. 419)

Solid geometry.   The geometry of three dimensions: length, width, and depth. (p. 435)

Sphere.   A geometric solid formed by a closed curved surface, with all points on the surface the same distance from a given point (the center). The given distance from any point on the surface to the center is called the *radius*. (p. 453)

Square.   A rectangle with the lengths of all four sides equal. (p. 406)

Supplementary angles.   Two angles for which the sum of their measures is 180°. (p. 401)

Tangent.   A line or a line segment that has only one point in common with a circle and lies totally outside the circle. (p. 425)

Total surface area.   The area of the bases plus the lateral surface area, or all surface areas of the figure. (p. 436)

Transversal.   A line that intersects two or more lines in different points in the same plane. (p. 401)

Trapezoid.   A quadrilateral with only two sides parallel. (p. 406)

Triangle.   A polygon with three sides. (p. 403)

Vertex.   The common point of two lines that form an angle. (p. 398)

Vertical angles.   The angles opposite each other when two lines intersect. (p. 401)

12.1 Angles and Polygons

1. **Review this section on the use of a protractor.**

2. **If two parallel lines are cut by a transversal, then**
    a. the *corresponding* angles are *equal.*
    b. the *alternate-interior* angles are *equal.*
    c. the *alternate-exterior* angles are *equal.*
    d. the *interior angles* on the same side of the transversal are *supplementary.* (p. 402)

12.2 Quadrilaterals

1. **Summary of formulas for area and perimeter of quadrilaterals:** (p. 406)

Quadrilateral	Area	Perimeter
Rectangle	$A = bh$	$P = 2(b + h)$
Square	$A = b^2$	$P = 4b$
Parallelogram	$A = bh$	$P = 2(a + b)$
Rhombus	$A = bh$	$P = 4b$
Trapezoid	$A = \left(\dfrac{a + b}{2}\right)h$	$P = a + b + c + d$

12.3 Triangles

1. **Pythagorean theorem:** the square of the hypotenuse of a right triangle is equal to the sum of the squares of the lengths of the two legs. That is, $c^2 = a^2 + b^2$ or $c = \sqrt{a^2 + b^2}$. (p. 411)

2. **Formulas for area and perimeter of a triangle:** (pp. 413–414)

$$A = \tfrac{1}{2}bh \quad P = a + b + c$$
$$A = \sqrt{s(s - a)(s - b)(s - c)},$$
where $s = \tfrac{1}{2}(a + b + c)$

3. **The sum of the measures of the angles of any triangle is 180°. (p. 415)**

12.4 Similar Polygons

1. **Polygons are similar when the corresponding angles are equal.**

2. **When two polygons are similar, the lengths of their corresponding sides are proportional. (p. 419)**

12.5 Circles

1. **Formulas for circumference and area of a circle: (p. 424)**

Circumference	Area
$C = 2\pi r$	$A = \pi r^2$
$C = \pi d$	$A = \dfrac{\pi d^2}{4}$

2. **The sum of the measures of all the central angles of any circle is 360°. (p. 425)**

3. **Arcs:**
    a. The measure of a central angle in a circle is equal to the measure of its intercepted arc.
    b. The measure of an inscribed angle in a circle is equal to one-half the measure of its intercepted arc.
    c. The measure of an angle formed by two intersecting chords in a circle is equal to one-half the sum of the measures of the intercepted arcs. (p. 425)

4. **Chords and tangents:**
    a. A diameter that is perpendicular to a chord bisects the chord.
    b. A line segment from the center of a circle to the point of tangency is perpendicular to the tangent.
    c. Two tangents drawn from a point outside a circle to the circle are equal. The line segment drawn from the center of the circle to this point outside the circle bisects the angle formed by the tangents. (p. 426)

12.6 Radian Measure

1. **Area of a sector of a circle with the central angle measured in degrees:**

$$A = \frac{\theta}{360°} \pi r^2 \text{ (p. 432)}$$

2. **Area of a sector of a circle with the central angle measured in radians:**

$$A = \frac{1}{2} r^2 \theta \text{ (p. 433)}$$

12.7 Prisms

1. **Formulas for prisms:**
   a. The *lateral surface area of a prism* is the sum of the areas of the lateral faces of the prism.
   b. The *total surface area of a prism* is the sum of the areas of the lateral faces and the areas of the bases of the prism.
   c. The *volume of a prism* is $V = Bh$, where B is the area of one of the bases and h is the altitude of the prism. (pp. 436–437)

12.8 Cylinders

1. **Formulas for cylinders:**
   a. The *volume of a right circular cylinder* is $V = \pi r^2 h$, where r is the radius of the base and h is the altitude.
   b. The *lateral surface area* of a right circular cylinder is $A = 2\pi rh$, where r is the radius of the base and h is the altitude.
   c. The *total surface area* of a right circular cylinder is the area of the bases plus the lateral surface area. (pp. 442, 443)

12.9 Pyramids and Cones

1. **Volume of a pyramid:** $V = \frac{1}{3} Bh$, where B is the area of the base and h is the height of the pyramid. The height of a pyramid is the shortest (perpendicular) distance between the apex and the base of the pyramid. (p. 447)

2. **Formulas for right circular cones:**
   a. The *volume of a right circular cone* is $V = \frac{1}{3}\pi r^2 h$, where r is the radius of the base and h is the altitude.
   b. The *lateral surface area of a right circular cone* is $A = \pi rs$, where r is the radius of the base and s is the slant height.
   c. The *total surface area of a right circular cone* is the sum of the lateral surface area and the area of the base. (p. 447)

3. **Formulas for the frustum of a pyramid:**
   a. Volume: $V = \frac{1}{3}h\left(B_1 + B_2 + \sqrt{B_1 B_2}\right)$,

   where h is the altitude and B_1 and B_2 are the areas of the bases.

   b. Lateral surface area: $A = \frac{1}{2}s(P_1 + P_2)$,

   where s is the slant height and P_1 and P_2 are the perimeters of the bases. (p. 448)

4. **Formulas for the frustum of a cone:**
   a. Volume: $V = \frac{1}{3}h\left(B_1 + B_2 + \sqrt{B_1 B_2}\right)$,

   where h is the altitude and B_1 and B_2 are the areas of the bases.

   b. Lateral surface area: $A = \pi s(r_1 + r_2)$, where s is the slant height and r_1 and r_2 are the radii of the bases. (p. 450)

12.10 Spheres

1. **Formulas for spheres:**
   a. Volume: $V = \frac{4}{3}\pi r^3$, where r is the radius of the sphere.

   b. Surface area: $A = 4\pi r^2$, where r is the radius of the sphere. (p. 453)

Chapter 12 Review

Classify each angle as right, acute, or obtuse:

1.

2.

For Exercises 3–5, see Illustration 1:

ILLUSTRATION 1

3. In Illustration 1, *l* is parallel to *m* and ∠5 = 121°. Find the measure of each angle.

4. In Illustration 1, ∠4 and ∠5 are called ___?___ angles.

5. Suppose ∠1 = 4*x* + 5 and ∠2 = 2*x* + 55. Find the value of *x*.

6. Name the polygon that has **a.** 4 sides, **b.** 5 sides, **c.** 6 sides, **d.** 3 sides, and **e.** 8 sides.

Find the perimeter and the area of each quadrilateral:

7.

6.00 cm   5.00 cm

12.00 cm

8.

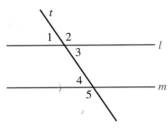

79.2 m

101.0 m

9.

10.21 cm

7.91 cm   9.59 cm

15.63 cm

10. The area of a rectangle is 79.6 m². The length is 10.3 m. What is the width?

11. The area of a parallelogram is 2.53 cm². Find its height if the base is 1.76 cm.

Find the area and the perimeter of each triangle:

12.

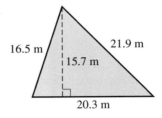

16.5 m   21.9 m   15.7 m

20.3 m

13.

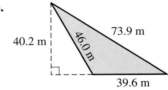

40.2 m   46.0 m   73.9 m

39.6 m

14.

2.42 cm   4.60 cm

3.91 cm

Find the length of the hypotenuse of each triangle:

15.

29.1 m

30.2 m

16.

1.72 cm

3.44 cm

17. Find the measure of the missing angle in Illustration 2.

?

58°   36°

ILLUSTRATION 2

18. In Illustration 3, suppose $\overline{DE} \parallel \overline{BC}$. Find length BC.

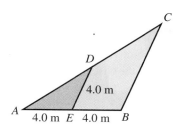

ILLUSTRATION 3

19. Find the area and circumference of the circle in Illustration 4.

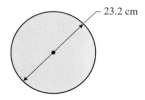

ILLUSTRATION 4

20. The area of a circle is 462 cm². Find its radius.

21. How many degrees are in a central angle whose arc is $\frac{3}{5}$ of a circle?

22. Change 24.0° to radians.

23. Change $\frac{\pi}{18}$ rad to degrees.

24. The radius of a wheel is 75.3 cm. The wheel turns 0.561 rad. Find the distance the wheel travels.

25. A wheel of diameter 25.8 cm travels a distance of 20.0 cm. Find the angle θ (in radians) that the wheel turns.

26. A wheel of radius 16.2 cm turns an angle of 1028°. Find the distance a point travels on the surface of the wheel.

For Exercises 27–28, see Illustration 5:

27. Find **a.** the lateral surface area and **b.** the total surface area of the prism.

28. Find the volume of the prism.

ILLUSTRATION 5   ILLUSTRATION 6

29. Find the volume of the right circular cylinder shown in Illustration 6.

30. A metallurgist needs to cast a molten alloy in the shape of a right circular cylinder. The dimensions of the mold are as shown in Illustration 7. Find the amount (volume) of molten alloy needed.

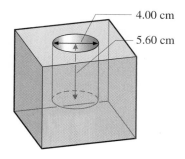

ILLUSTRATION 7

31. Find **a.** the lateral surface area and **b.** the total surface area of the cylinder that was cast in Exercise 30.

32. Find the volume of the pyramid shown in Illustration 8.

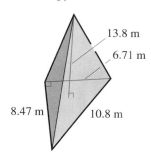

ILLUSTRATION 8

33. Find **a.** the volume and **b.** the lateral surface area of the right circular cone shown in Illustration 9.

ILLUSTRATION 9

34. Find **a.** the volume and **b.** the surface area of the sphere shown in Illustration 10.

ILLUSTRATION 10

35. Find the volume and lateral surface area of the frustum of the cone shown in Illustration 11.

18.0 in.

45.0 in.

25.0 in.

ILLUSTRATION 11

36. An open (no top) rectangular box with length 6.00 ft, width 4.00 ft, and height 2.25 ft is to be painted on all interior and exterior sides, including the bottom, with a waterproof material. Each gallon covers 11.0 ft² and costs $40. **a.** Find how many gallon containers must be purchased and the cost of painting this box. **b.** If the box is $\frac{4}{5}$ full of water, find the weight of the water. (Water weighs 62.4 lb/ft³.)

Chapter 12 Test

1. Find the area of a rectangle 18.0 ft long and 6.00 ft wide.

2. Find the perimeter of a square lot 160 m on a side.

Given the trapezoid in Illustration 1, find

3. its area

4. its perimeter

8.00 cm

8.60 cm

5.00 cm

15.0 cm

ILLUSTRATION 1

5. Find the length of side *a* in Illustration 2.

B

a

30.0 km

C   24.0 km   A

ILLUSTRATION 2

Given a circle of radius 20.0 cm, find

6. its area

7. its circumference

8. Change 240° to radians.

9. Change $\frac{7\pi}{4}$ rad to degrees.

10. Find the volume of a rectangular box 12.0 ft × 8.00 ft × 9.00 ft.

11. Find the total surface area of the box in Exercise 10.

12. Find the volume of a cylindrical tank 20.0 m in diameter and 30.0 m high.

13. Find **a.** the volume and **b.** the lateral surface area of the bin in Illustration 3.

8.00 ft

12.0 ft

6.00 ft

ILLUSTRATION 3

14. Find the volume of the frustum of the cone in Illustration 4.

ILLUSTRATION 4

15. A welder is assigned to fabricate the sheet metal water trough shown in Illustration 5. **a.** Find its cubic foot capacity. **b.** What is the length of side y?

ILLUSTRATION 5

Cumulative Review

Chapters 1-12

1. Add $-8 + (+7) + (-3)$.

2. Given $P = 2(l + w)$, where $l = 4\frac{1}{8}$ in. and $w = 2\frac{3}{4}$ in., find P.

3. The mass of a full-size automobile is **a.** 100 kg, **b.** 1500 kg, **c.** 10 kg, or **d.** 15,000 kg.

4. Find **a.** precision and **b.** greatest possible error: 20,400 L

5. Simplify: $\dfrac{14x^3 - 56x^2 - 28x}{7x}$

6. Find the product: $(-2x^2 + 7x - 3)(4x + 5)$

7. Solve: $4 - 2x = 18$

8. $E = mv^2$; find m if $E = 952$ and $v = 7.00$.

9. A 200-bu wagon holds 3.4 tons of grain. Express the weight of the grain in pounds per bushel.

10. A 160-lb object, 28.0 in. from the fulcrum of a lever, balances a second object 80.0 in. from the other end of the lever. What is the weight of the second object?

11. Solve for y: $5x - 8y = 10$

12. Find the slope of the line containing the points $(2, -1)$ and $(5, -8)$.

13. Solve: $5x - y = 12$
$$y = 2x$$

14. The sum of the resistance of two resistors is 1300 Ω. The larger has three times the resistance of the smaller. Find the resistance of each.

Find each product mentally:

15. $(3x - 5)(2x + 7)$ **16.** $(4x - 3)^2$

Factor each expression completely:

17. $5x^3 - 15x$ **18.** $x^2 - 3x - 28$

19. Solve: $10x^2 - 5x = 105$

20. Solve: $2x^2 - x = 3$

Solve each equation using the quadratic formula (when necessary, round the results to three significant digits):

21. $5x^2 + 13x - 6 = 0$ **22.** $4x^2 - 10x - 29 = 0$

23. Draw the graph of $y = 2x^2 - 3x - 2$ and label the vertex.

24. In Illustration 1, $m \parallel n$ and $\angle 4 = 82°$. Find the measure of the other angles.

ILLUSTRATION 1

25. In Illustration 1, if $\angle 4 = 2x - 3$ and $\angle 5 = x + 6$, find the value of x.

Find the area and perimeter of each triangle:

26.

22.2 cm   24.3 cm   17.8 cm   29.8 cm

27.

1.82 m   2.04 m

28. Find the area and the circumference of the circle in Illustration 2.

15.6 cm

ILLUSTRATION 2

29. The area of a circle is 168 cm². Find its radius.

30. Find the volume and total surface area of the cylinder shown in Illustration 3.

8.92 cm   15.8 cm

ILLUSTRATION 3

13 Right Triangle Trigonometry

Mathematics at Work

Automotive collision repair technicians repair, repaint, and refinish automotive vehicle bodies; straighten vehicle frames; and replace damaged glass and other automotive parts that cannot be economically repaired. Using modern techniques including diagnostics, electronic equipment, computer support equipment, and other specialized equipment, the technician's primary task is to restore damaged vehicles to their original condition. Training and education for this work are available at many community colleges and trade schools. Various automobile manufacturers and their participating dealers also sponsor programs at postsecondary schools across the United States. Good reading, mathematics, computer, and communications skills are needed.

Voluntary certification is available through the National Institute for Automotive Service Excellence (ASE) and is the recognized standard of achievement for automotive collision repair technicians. For more information, go to the website listed below.

Automotive Collision Repair Technician
Automotive collision repair technician working in a spray paint booth.

Objectives

- Using the basic terms of a right triangle, write the trigonometric ratios for the sine, cosine, and tangent of an angle.

- Find the value of a trigonometric ratio using a scientific calculator.

- Use a trigonometric ratio to find angles.

- Use a trigonometric ratio to find sides.

- Solve a right triangle.

- Solve application problems involving trigonometric ratios and right triangles.

13.1 Trigonometric Ratios

Many applications in science and technology require the use of triangles and trigonometry. Early applications of trigonometry, beginning in the second century B.C., were in astronomy, surveying, and navigation. Applications that you may study include electronics, the motion of projectiles, light refraction in optics, and sound.

    In this chapter, we consider only right triangles. A right triangle has one right angle, two acute angles, a hypotenuse, and two legs. The right angle, as shown in Figure 13.1, is usually labeled with the capital letter C. The vertices of the two acute angles are usually labeled with the capital letters A and B. The hypotenuse is the side opposite the right angle, the longest side of a right triangle, and is usually labeled with the lowercase letter c. The legs are the sides opposite the acute angles. The leg (side) opposite angle A is labeled a, and the leg opposite angle B is labeled b. Note that each side of the triangle is labeled with the lowercase of the letter of the angle opposite that side.

    The two legs are also named as the side *opposite* angle A and the side *adjacent to* (or next to) angle A or as the side opposite angle B and the side adjacent to angle B. See Figure 13.2.

FIGURE 13.1
Right triangle

 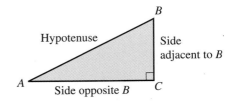

FIGURE 13.2
Key parts of a right triangle

Pythagorean Theorem

In any right triangle,
$$c^2 = a^2 + b^2$$

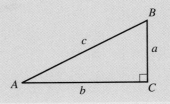

That is, the square of the length of the hypotenuse is equal to the sum of the squares of the lengths of the legs. The following equivalent formulas are often more useful:

$$c = \sqrt{a^2 + b^2} \qquad \text{used to find the length of the hypotenuse}$$
$$a = \sqrt{c^2 - b^2} \qquad \text{used to find the length of leg } a$$
$$b = \sqrt{c^2 - a^2} \qquad \text{used to find the length of leg } b$$

Recall that the Pythagorean theorem was developed in detail in Section 12.3.

 Example 1

Find the length of side b in Figure 13.3.

Using the formula to find the length of leg b, we have

$$b = \sqrt{c^2 - a^2}$$
$$b = \sqrt{(60.0 \text{ cm})^2 - (25.0 \text{ cm})^2}$$
$$= 54.5 \text{ cm} \qquad ■$$

FIGURE 13.3

 Example 2

Find the length of side c in Figure 13.4.

Using the formula to find the hypotenuse c, we have

$$c = \sqrt{a^2 + b^2}$$
$$c = \sqrt{(29.7 \text{ m})^2 + (34.2 \text{ m})^2}$$
$$= 45.3 \text{ m} \qquad ■$$

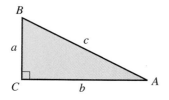

FIGURE 13.4

A **ratio** is the comparison of two quantities by division. The ratios of the sides of a right triangle can be used to find an unknown part—or parts—of that right triangle. Such a ratio is called a **trigonometric ratio** and expresses the relationship between an acute angle and the lengths of two of the sides of a right triangle.

The **sine of angle A**, abbreviated "sin A," equals the ratio of the length of the side opposite angle A, which is a, to the length of the hypotenuse, c.

The **cosine of angle A**, abbreviated "cos A," equals the ratio of the length of the side adjacent to angle A, which is b, to the length of the hypotenuse, c.

The **tangent of angle A**, abbreviated "tan A," equals the ratio of the length of the side opposite angle A, which is a, to the length of the side adjacent to angle A, which is b.

That is, in a right triangle (Figure 13.5), we have the following ratios.

FIGURE 13.5

Trigonometric Ratios

$$\sin A = \frac{\text{length of side opposite angle } A}{\text{length of hypotenuse}} = \frac{a}{c}$$

$$\cos A = \frac{\text{length of side adjacent to angle } A}{\text{length of hypotenuse}} = \frac{b}{c}$$

$$\tan A = \frac{\text{length of side opposite angle } A}{\text{length of side adjacent to angle } A} = \frac{a}{b}$$

Similarly, the ratios can be defined for angle B.

$$\sin B = \frac{\text{length of side opposite angle } B}{\text{length of hypotenuse}} = \frac{b}{c}$$

$$\cos B = \frac{\text{length of side adjacent to angle } B}{\text{length of hypotenuse}} = \frac{a}{c}$$

$$\tan B = \frac{\text{length of side opposite angle } B}{\text{length of side adjacent to angle } B} = \frac{b}{a}$$

 Example 3

Find the three trigonometric ratios for angle A in the triangle in Figure 13.6.

$$\sin A = \frac{\text{length of side opposite angle } A}{\text{length of hypotenuse}} = \frac{a}{c} = \frac{144 \text{ m}}{156 \text{ m}} = 0.9231$$

$$\cos A = \frac{\text{length of side adjacent to angle } A}{\text{length of hypotenuse}} = \frac{b}{c} = \frac{60.0 \text{ m}}{156 \text{ m}} = 0.3846$$

$$\tan A = \frac{\text{length of side opposite angle } A}{\text{length of side adjacent to angle } A} = \frac{a}{b} = \frac{144 \text{ m}}{60.0 \text{ m}} = 2.400 \qquad ■$$

FIGURE 13.6

The values of the trigonometric ratios of various angles can be found with calculators. You will need a calculator that has sin, cos, and tan keys.

Very Important Note: When working with the trigonometric functions on your calculator, make certain that it is set in the degree mode. If your calculator has a DRG key, it is used to change angle measurement modes from degrees to radians to grads. In the degree mode, the circle or one complete revolution is divided into 360°. In the radian mode, the circle or one complete revolution is divided into 2π rad. In the grad mode, the circle or one complete revolution is divided into 400^g. We will be working exclusively in the degree mode.

 Example 4

Find sin 37.5° rounded to four significant digits.

 SIN   37.5 $=$

0.608761429

Thus, sin 37.5° = 0.6088 rounded to four significant digits.   ■

 Example 5

Find cos 18.63° rounded to four significant digits.

 COS   18.63 $=$

0.947601273

Thus, cos 18.63° = 0.9476 rounded to four significant digits.   ■

 Example 6

Find tan 81.7° rounded to four significant digits.

 TAN   81.7 $=$

6.854750833

Thus, tan 81.7° = 6.855 rounded to four significant digits.   ■

A calculator may also be used to find the *angle* when the value of the trigonometric ratio is known. The procedure is shown in the examples below.

 Example 7 Find angle A to the nearest tenth of a degree when sin A = 0.6372.

 .6372 ⬭

39.583346

Note: Make certain that your calculator is in the degree mode.

Thus, angle A = 39.6° rounded to the nearest tenth of a degree. ■

Example 8 Find angle B to the nearest tenth of a degree when tan B = 0.3106.

 .3106

17.25479431

Thus, angle B = 17.3° rounded to the nearest tenth of a degree. ■

Example 9 Find angle A to the nearest hundredth of a degree when cos A = 0.4165.

COS⁻¹ .4165 ⬭

65.3861858

Thus, angle A = 65.39° rounded to the nearest hundredth of a degree. ■

Exercises 13.1 _____

Refer to right triangle ABC in Illustration 1 for Exercises 1–10:

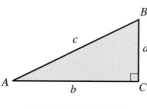

ILLUSTRATION 1

1. The side opposite angle A is ____?____.
2. The side opposite angle B is ____?____.
3. The hypotenuse is ____?____.
4. The side adjacent to angle A is ____?____.
5. The side adjacent to angle B is ____?____.
6. The angle opposite side a is ____?____.
7. The angle opposite side b is ____?____.
8. The angle opposite side c is ____?____.
9. The angle adjacent to side a is ____?____.
10. The angle adjacent to side b is ____?____.

Use right triangle ABC in Illustration 1 and the Pythagorean theorem to find each unknown side, rounded to three significant digits:

11. c = 75.0 m, a = 45.0 m
12. a = 25.0 cm, b = 60.0 cm
13. a = 29.0 mi, b = 47.0 mi
14. a = 12.0 km, c = 61.0 km
15. c = 18.9 cm, a = 6.71 cm
16. a = 20.2 mi, b = 19.3 mi
17. a = 171 ft, b = 203 ft
18. c = 35.3 m, b = 25.0 m
19. a = 202 m, c = 404 m
20. a = 1.91 km, b = 3.32 km

21. $b = 1520$ km, $c = 2160$ km

22. $a = 203{,}000$ ft, $c = 521{,}000$ ft

23. $a = 45{,}800$ m, $b = 38{,}600$ m

24. $c = 3960$ m, $b = 3540$ m

Use the triangle in Illustration 2 for Exercises 25–30:

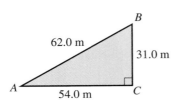

ILLUSTRATION 2

25. Find $\sin A$.   **26.** Find $\cos A$.   **27.** Find $\tan A$.

28. Find $\sin B$.   **29.** Find $\cos B$.   **30.** Find $\tan B$.

Find the value of each trigonometric ratio rounded to four significant digits:

31. $\sin 49.6°$ **32.** $\cos 55.2°$ **33.** $\tan 65.3°$

34. $\sin 69.7°$ **35.** $\cos 29.7°$ **36.** $\tan 14.6°$

37. $\sin 31.64°$ **38.** $\tan 13.25°$ **39.** $\cos 75.31°$

40. $\cos 84.83°$ **41.** $\tan 3.05°$ **42.** $\sin 6.74°$

43. $\sin 37.62°$ **44.** $\cos 18.94°$ **45.** $\tan 21.45°$

46. $\sin 11.31°$ **47.** $\cos 47.16°$ **48.** $\tan 81.85°$

Find each angle rounded to the nearest tenth of a degree:

49. $\sin A = 0.7941$ **50.** $\tan A = 0.2962$

51. $\cos B = 0.4602$ **52.** $\cos A = 0.1876$

53. $\tan B = 1.386$ **54.** $\sin B = 0.3040$

55. $\sin B = 0.1592$ **56.** $\tan B = 2.316$

57. $\cos A = 0.8592$ **58.** $\cos B = 0.3666$

59. $\tan A = 0.8644$ **60.** $\sin A = 0.5831$

Find each angle rounded to the nearest hundredth of a degree:

61. $\tan A = 0.1941$ **62.** $\sin B = 0.9324$

63. $\cos A = 0.3572$ **64.** $\cos B = 0.2597$

65. $\sin A = 0.1506$ **66.** $\tan B = 2.500$

67. $\tan B = 3.806$ **68.** $\sin A = 0.4232$

69. $\cos B = 0.7311$ **70.** $\cos A = 0.6427$

71. $\sin B = 0.3441$ **72.** $\tan A = 0.5536$

73. In Exercises 25–30, there are two pairs of ratios that are equal. Name them.

13.2 Using Trigonometric Ratios to Find Angles

A trigonometric ratio may be used to find an angle of a right triangle, given the lengths of any two sides.

  Example 1

In Figure 13.7, find angle A using a calculator, as follows.

We know the sides opposite and adjacent to angle A. So we use the tangent ratio:

$$\tan A = \frac{\text{length of side opposite angle } A}{\text{length of side adjacent to angle } A}$$

$$\tan A = \frac{28.5 \text{ m}}{21.3 \text{ m}} = 1.338$$

Next, find angle A to the nearest tenth of a degree when $\tan A = 1.338$. The complete set of operations on a calculator follows.

28.5 21.3

> 53.22672246

Thus, angle $A = 53.2°$ rounded to the nearest tenth of a degree.   ■

FIGURE 13.7

When calculations involve a trigonometric ratio, we shall use the following rule for significant digits.

Angles expressed to the nearest	The length of each side of the triangle contains
1°	Two significant digits
0.1°	Three significant digits
0.01°	Four significant digits

An example of each case is shown in Figure 13.8.

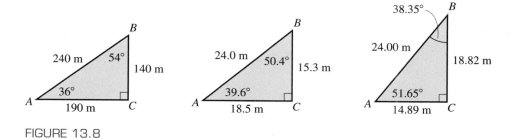

FIGURE 13.8

Example 2

Find angle *B* in the triangle in Figure 13.9.

We know the hypotenuse and the side adjacent to angle *B*. So let's use the cosine ratio.

$$\cos B = \frac{\text{length of side adjacent to angle } B}{\text{length of hypotenuse}}$$

$$\cos B = \frac{35.20 \text{ cm}}{45.85 \text{ cm}}$$

Find angle *B* using a calculator as follows.

35.2 ÷ 45.85 = COS⁻¹ ANS =

39.85033989

Thus, angle *B* = 39.85° rounded to the nearest hundredth of a degree. ■

FIGURE 13.9

The question is often raised, "Which of the three trig ratios do I use?" First, notice that each trigonometric ratio consists of two sides and one angle, or three quantities in all. To find the solution to such an equation, two of the quantities must be known. We will answer the question in two parts.

Which Trig Ratio to Use

1. If you are finding an angle, two sides must be known. Label these two known sides as *side opposite* the angle you are finding, *side adjacent* to the angle you are finding, or *hypotenuse*. Then choose the trig ratio that has these two sides.

continued

> **2.** If you are finding a side, one side and one angle must be known. Label the known side and the unknown side as *side opposite* the known angle, *side adjacent* to the known angle, or *hypotenuse.* Then choose the trig ratio that has these two sides.

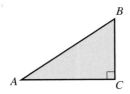

FIGURE 13.10

A useful and time-saving fact about right triangles (Figure 13.10) is that the sum of the acute angles of any right triangle is 90°.

$$A + B = 90°$$

Why? We know that the sum of the interior angles of any triangle is 180°. A right triangle, by definition, contains a right angle, whose measure is 90°. That leaves 90° to be divided between the two acute angles.

Note, then, that if one acute angle is given or known, the other acute angle may be found by subtracting the known angle from 90°. That is,

$$A = 90° - B$$
$$B = 90° - A$$

◆ **Example 3**

Find angle A and angle B in the triangle in Figure 13.11.

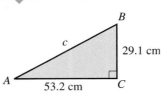

FIGURE 13.11

$$\tan A = \frac{\text{length of side opposite angle } A}{\text{length of side adjacent to angle } A}$$

$$\tan A = \frac{29.1 \text{ cm}}{53.2 \text{ cm}} = 0.5470$$

$$A = 28.7°$$

Angle $B = 90° - 28.7° = 61.3°$.   ■

Exercises 13.2 _____

Using Illustration 1, find the measure of each acute angle for each right triangle:

1. $a = 36.0$ m, $b = 50.9$ m
2. $a = 72.0$ cm, $c = 144$ cm
3. $b = 39.7$ cm, $c = 43.6$ cm
4. $a = 171$ km, $b = 695$ km
5. $b = 13.6$ m, $c = 18.7$ m
6. $b = 409$ km, $c = 612$ km
7. $a = 29.7$ m, $b = 29.7$ m, $c = 42.0$ m
8. $a = 36.2$ mm, $b = 62.7$ mm, $c = 72.4$ mm
9. $a = 2902$ km, $b = 1412$ km

10. $b = 21.34$ m, $c = 47.65$ m
11. $a = 0.6341$ cm, $c = 0.7982$ cm
12. $b = 4.372$ m, $c = 5.806$ m
13. $b = 1455$ ft, $c = 1895$ ft
14. $a = 25.45$ in., $c = 41.25$ in.
15. $a = 243.2$ km, $b = 271.5$ km
16. $a = 351.6$ m, $b = 493.0$ m
17. $a = 16.7$ m, $c = 81.4$ m
18. $a = 847$ m, $b = 105$ m
19. $b = 1185$ ft, $c = 1384$ ft
20. $a = 48.7$ cm, $c = 59.5$ cm

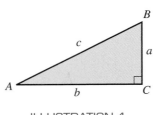

ILLUSTRATION 1

21. $a = 845$ km, $b = 2960$ km
22. $b = 2450$ km, $c = 3570$ km
23. $a = 8897$ m, $c = 9845$ m
24. $a = 58.44$ mi, $b = 98.86$ mi

13.3 Using Trigonometric Ratios to Find Sides

We also use a trigonometric ratio to find a side of a right triangle, given one side and the measure of one of the acute angles.

 Example 1

FIGURE 13.12

Find side *a* in the triangle in Figure 13.12.

With respect to the known angle *B*, we know the hypotenuse and are finding the adjacent side. So we use the cosine ratio.

$$\cos B = \frac{\text{length of side adjacent to angle } B}{\text{length of hypotenuse}}$$

$$\cos 24.5° = \frac{a}{258 \text{ ft}}$$

$$a = (\cos 24.5°)(258 \text{ ft}) \qquad \text{Multiply both sides by 258 ft.}$$

Side *a* can be found by using a calculator as follows.

  *

Thus, side *a* = 235 ft rounded to three significant digits. ■

 Example 2

Find sides *b* and *c* in the triangle in Figure 13.13.

If we find side *b* first, we are looking for the side adjacent to angle *A*, the known angle. We are given the side opposite angle *A*. Thus, we should use the tangent ratio.

$$\tan A = \frac{\text{length of side opposite angle } A}{\text{length of side adjacent to angle } A}$$

$$\tan 52.3° = \frac{29.7 \text{ m}}{b}$$

$$b(\tan 52.3°) = 29.7 \text{ m} \qquad \text{Multiply both sides by } b.$$

$$b = \frac{29.7 \text{ m}}{\tan 52.3°} \qquad \text{Divide both sides by } \tan 52.3°.$$

FIGURE 13.13

Thus, side *b* = 23.0 m rounded to three significant digits.

_____

Note: You might need to insert a right parenthesis to clarify the order of operations. The trigonometry keys may also include the left parenthesis.

To find side c, we are looking for the hypotenuse, and we have the opposite side given. Thus, we should use the sine ratio.

$$\sin A = \frac{\text{length of side opposite angle } A}{\text{length of hypotenuse}}$$

$$\sin 52.3° = \frac{29.7 \text{ m}}{c}$$

$c(\sin 52.3°) = 29.7 \text{ m}$ Multiply both sides by c.

$$c = \frac{29.7 \text{ m}}{\sin 52.3°}$$ Divide both sides by $\sin 52.3°$.

$$= 37.5 \text{ m}$$

The Pythagorean theorem may be used to check your work.   ■

Exercises 13.3

Find the unknown sides of each right triangle (see Illustration 1):

1. $a = 36.7$ m, $A = 42.1°$
2. $b = 73.6$ cm, $B = 19.0°$
3. $a = 236$ km, $B = 49.7°$
4. $b = 28.9$ ft, $A = 65.2°$
5. $c = 49.1$ cm, $A = 36.7°$
6. $c = 236$ m, $A = 12.9°$
7. $b = 23.7$ cm, $A = 23.7°$
8. $a = 19.2$ km, $B = 63.2°$
9. $b = 29,200$ km, $A = 12.9°$
10. $c = 36.7$ mi, $B = 68.3°$

11. $a = 19.72$ m, $A = 19.75°$
12. $b = 125.3$ cm, $B = 23.34°$
13. $c = 255.6$ mi, $A = 39.25°$
14. $c = 7.363$ km, $B = 14.80°$
15. $b = 12,350$ m, $B = 69.72°$
16. $a = 3678$ m, $B = 10.04°$
17. $a = 1980$ m, $A = 18.4°$
18. $a = 9820$ ft, $B = 35.7°$
19. $b = 841.6$ km, $A = 18.91°$
20. $c = 289.5$ cm, $A = 24.63°$

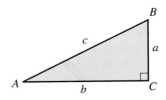

ILLUSTRATION 1

21. $c = 185.6$ m, $B = 61.45°$
22. $b = 21.63$ km, $B = 82.06°$
23. $c = 256$ cm, $A = 25.6°$
24. $a = 18.3$ mi, $A = 71.2°$

 13.4 Solving Right Triangles

The phrase **solving a right triangle** refers to finding the measures of the various parts of a triangle that are not given. We proceed as we did in the last two sections.

 Example 1

Solve the right triangle in Figure 13.14.

We are given the measure of one acute angle and the length of one leg.

$A = 90° − B$
$A = 90° − 36.7° = 53.3°$

We then can use either the sine or the cosine ratio to find side c.

$$\sin B = \frac{\text{length of side opposite angle } B}{\text{length of hypotenuse}}$$

$$\sin 36.7° = \frac{19.2 \text{ m}}{c}$$

$c(\sin 36.7°) = 19.2 \text{ m}$ Multiply both sides by c.

$$c = \frac{19.2 \text{ m}}{\sin 36.7°}$$ Divide both sides by $\sin 36.7°$.

$$= 32.1 \text{ m}$$

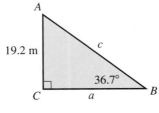

FIGURE 13.14

Now we may use either a trigonometric ratio or the Pythagorean theorem to find side a.

Solution by a Trigonometric Ratio:

$$\tan B = \frac{\text{length of side opposite angle } B}{\text{length of side adjacent to angle } B}$$

$$\tan 36.7° = \frac{19.2 \text{ m}}{a}$$

$a(\tan 36.7°) = 19.2 \text{ m}$ Multiply both sides by a.

$$a = \frac{19.2 \text{ m}}{\tan 36.7°} \quad \text{Divide both sides by } \tan 36.7°.$$

$$= 25.8 \text{ m}$$

Solution by the Pythagorean Theorem:

$$a = \sqrt{c^2 - b^2}$$
$$a = \sqrt{(32.1 \text{ m})^2 - (19.2 \text{ m})^2}$$
$$= 25.7 \text{ m}$$

Can you explain the difference in these two results?   ■

Example 2

Solve the right triangle in Figure 13.15.

We are given the measure of one acute angle and the length of the hypotenuse.

$$A = 90° - B$$
$$A = 90° - 45.7° = 44.3°$$

FIGURE 13.15

To find side b, we must use the sine or the cosine ratio, since the hypotenuse is given.

$$\sin B = \frac{\text{length of side opposite angle } B}{\text{length of hypotenuse}}$$

$$\sin 45.7° = \frac{b}{397 \text{ km}}$$

$(\sin 45.7°)(397 \text{ km}) = b$ Multiply both sides by 397 km.

$$284 \text{ km} = b$$

Again, we can use either a trigonometric ratio or the Pythagorean theorem to find side a.

Solution by a Trigonometric Ratio:

$$\cos B = \frac{\text{length of side adjacent to angle } B}{\text{length of hypotenuse}}$$

$$\cos 45.7° = \frac{a}{397 \text{ km}}$$

$(\cos 45.7°)(397 \text{ km}) = a$ Multiply both sides by 397 km.

$$277 \text{ km} = a$$

Solution by the Pythagorean Theorem:

$$a = \sqrt{c^2 - b^2}$$
$$a = \sqrt{(397 \text{ km})^2 - (284 \text{ km})^2}$$
$$= 277 \text{ km}$$

Example 3

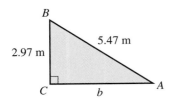

FIGURE 13.16

Solve the right triangle in Figure 13.16.

We are given two sides of the right triangle.

To find angle A or angle B, we could use either the sine or cosine, since the hypotenuse is given.

$$\sin A = \frac{\text{length of side opposite angle } A}{\text{length of hypotenuse}}$$

$$\sin A = \frac{2.97 \text{ m}}{5.47 \text{ m}} = 0.5430$$

$$A = 32.9°$$

Then

$$B = 90° - A$$
$$B = 90° - 32.9° = 57.1°$$

The unknown side b can be found by using the Pythagorean theorem.

$$b = \sqrt{c^2 - a^2}$$
$$b = \sqrt{(5.47 \text{ m})^2 - (2.97 \text{ m})^2}$$
$$= 4.59 \text{ m}$$

■

Exercises 13.4

Using Illustration 1, solve each right triangle:

1. $A = 50.6°$, $c = 49.0$ m
2. $a = 30.0$ cm, $b = 40.0$ cm
3. $B = 41.2°$, $a = 267$ ft
4. $A = 39.7°$, $b = 49.6$ km
5. $b = 72.0$ mi, $c = 78.0$ mi
6. $B = 22.4°$, $c = 46.0$ mi
7. $A = 52.1°$, $a = 72.0$ mm
8. $B = 42.3°$, $b = 637$ m
9. $A = 68.8°$, $c = 39.4$ m
10. $a = 13.6$ cm, $b = 13.6$ cm

11. $a = 12.00$ m, $b = 24.55$ m
12. $B = 38.52°$, $a = 4315$ m
13. $A = 29.19°$, $c = 2975$ ft
14. $B = 29.86°$, $a = 72.62$ m
15. $a = 46.72$ m, $b = 19.26$ m
16. $a = 2436$ ft, $c = 4195$ ft
17. $A = 41.1°$, $c = 485$ m
18. $a = 1250$ km, $b = 1650$ km
19. $B = 9.45°$, $a = 1585$ ft
20. $A = 14.60°$, $b = 135.7$ cm

ILLUSTRATION 1

21. $b = 269.5$ m, $c = 380.5$ m
22. $B = 75.65°$, $c = 92.75$ km
23. $B = 81.5°$, $b = 9370$ ft
24. $a = 14.6$ mi, $c = 31.2$ mi

13.5 Applications Involving Trigonometric Ratios

Trigonometric ratios can be used to solve many applications similar to those problems we solved in the preceding sections. However, instead of having to find all the parts of a right triangle, we usually need to find only one.

Example 1   The roof in Figure 13.17 has a rise of 7.50 ft and a run of 18.0 ft. Find angle A.

FIGURE 13.17

We know the length of the side opposite angle A and the length of the side adjacent to angle A. So we use the tangent ratio.

$$\tan A = \frac{\text{length of side opposite angle } A}{\text{length of side adjacent to angle } A}$$

$$\tan A = \frac{7.50 \text{ ft}}{18.0 \text{ ft}} = 0.4167$$

$$A = 22.6°$$ ■

The **angle of depression** is the angle between the horizontal and the line of sight to an object that is *below* the horizontal. The **angle of elevation** is the angle between the horizontal and the line of sight to an object that is *above* the horizontal.

In Figure 13.18, angle A is the angle of depression for an observer in the helicopter sighting down to the building on the ground, and angle B is the angle of elevation for an observer in the building sighting up to the helicopter.

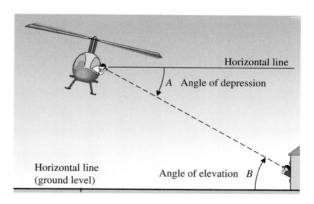

FIGURE 13.18

Example 2   A ship's navigator measures the angle of elevation to the beacon of a lighthouse to be 10.1°. He knows that this particular beacon is 225 m above sea level. How far is the ship from the lighthouse?

First, you should sketch the problem, as in Figure 13.19. Since this problem involves finding the length of the side adjacent to an angle when the opposite side is known, we use the tangent ratio.

FIGURE 13.19

$$\tan A = \frac{\text{length of side opposite angle } A}{\text{length of side adjacent to angle } A}$$

$$\tan 10.1° = \frac{225 \text{ m}}{b}$$

$b(\tan 10.1°) = 225 \text{ m}$ Multiply both sides by b.

$$b = \frac{225 \text{ m}}{\tan 10.1°}$$ Divide both sides by $\tan 10.1°$.

$$= 1260 \text{ m}$$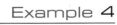

Example 3

In ac (alternating current) circuits, the relationship between impedance Z (in ohms), the resistance R (in ohms), and the phase angle θ is shown by the right triangle in Figure 13.20(a). If the resistance is 250 Ω and the phase angle is 41°, find the impedance.

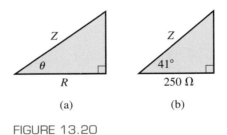

(a)   (b)

FIGURE 13.20

Here, we know the adjacent side and the angle and wish to find the hypotenuse (see Figure 13.20b). So we use the cosine ratio.

$$\cos \theta = \frac{\text{length of side adjacent to angle } \theta}{\text{length of hypotenuse}}$$

$$\cos 41° = \frac{250 \text{ Ω}}{Z}$$

$Z(\cos 41°) = 250 \text{ Ω}$ Multiply both sides by Z.

$$Z = \frac{250 \text{ Ω}}{\cos 41°}$$ Divide both sides by $\cos 41°$.

$$= 330 \text{ Ω}$$ ■

Example 4

You are machining the part shown in Figure 13.21. Before you begin, you must find angle 1 and length AB.

FIGURE 13.21

First, complete triangle *ABE* by drawing the dashed lines as shown.

Length *BE* = length *CD* = 50.0 cm − 37.5 cm = 12.5 cm

From right triangle *ABE*, we have

$$\tan \angle 2 = \frac{\text{length of side opposite } \angle 2}{\text{length of side adjacent to } \angle 2} = \frac{BE}{AE}$$

$$\tan \angle 2 = \frac{12.5 \text{ cm}}{10.0 \text{ cm}} = 1.25$$

$$\angle 2 = 51.3°$$

From right triangle *ABE*, we have

$$\sin \angle 2 = \frac{\text{length of side opposite } \angle 2}{\text{length of hypotenuse}}$$

$$\sin 51.3° = \frac{12.5 \text{ cm}}{AB}$$

$AB(\sin 51.3°) = 12.5$ cm   Multiply both sides by *AB*.

$$AB = \frac{12.5 \text{ cm}}{\sin 51.3°}$$ Divide both sides by sin 51.3°.

$$AB = 16.0 \text{ cm}$$

Since ∠1 and ∠2 are supplementary, ∠1 = 180° − ∠2 = 180° − 51.3° = 128.7°. ■

Exercises 13.5

1. A conveyor is used to lift paper to a shredder. The most efficient operating angle of elevation for the conveyor is 35.8°. The paper is to be elevated 11.0 m. What length of conveyor is needed?

2. Maria is to weld a support for a 23-m conveyor so that it will operate at a 20° angle. What is the length of the support? See Illustration 1.

ILLUSTRATION 1

3. A bullet is found embedded in the wall of a room 2.3 m above the floor. The bullet entered the wall going upward at an angle of 12°. How far from the wall was the bullet fired if the gun was held 1.2 m above the floor?

4. The recommended safety angle of a ladder against a building is 78°. A $1\overline{0}$-m ladder will be used. How high up on the side of the building will the ladder safely reach? (See Illustration 2.)

ILLUSTRATION 2

5. A piece of conduit 38.0 ft long cuts across the corner of a room, as shown in Illustration 3. Find length x and angle A.

ILLUSTRATION 3

6. Find the width of the river in Illustration 4.

ILLUSTRATION 4

7. A roadbed rises 220 ft for each 2300 ft of horizontal. (See Illustration 5.) Find the angle of inclination of the roadbed. (This is usually referred to as *% of grade*.)

ILLUSTRATION 5

8. A smokestack is $18\overline{0}$ ft high. A guy wire must be fastened to the stack 20.0 ft from the top. The guy wire makes an angle of 40.0° with the ground. Find the length of the guy wire.

9. A railroad track has an angle of elevation of 1.0°. What is the difference in height (in feet) of two points on the track 1.00 mi apart?

10. A machinist needs to drill four holes 1.00 in. apart in a straight line in a metal plate, as shown in Illustration 6. If the first hole is placed at the origin and the line forms an angle of 32.0° with the vertical axis, find the coordinates of the other three holes (A, B, and C).

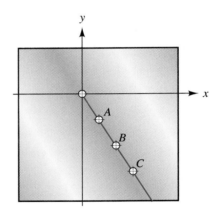

ILLUSTRATION 6

11. Enrico has to draft a triangular roof to a house. (See Illustration 7.) The roof is 30.0 ft wide. If the rafters are 17.0 ft long, at what angle will the rafters be laid at the eaves? Assume no overhang.

ILLUSTRATION 7

12. A small plane takes off from an airport and begins to climb at a 10.0° angle of elevation at $500\overline{0}$ ft/min. After 3.00 min, how far along the ground will the plane have flown?

13. A gauge is used to check the diameter of a crankshaft journal. It is constructed to make measurements on the basis of a right triangle with a 60.0° angle. Distance AB in Illustration 8 is 11.4 cm. Find radius BC of the journal.

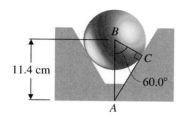

11.4 cm

ILLUSTRATION 8

14. Round metal duct runs alongside some stairs from the floor to the ceiling. If the ceiling is 9.00 ft high and the angle of elevation between the floor and duct is 37.0°, how long is the duct?

15. The cables attached to a TV relay tower are $11\overline{0}$ m long. They meet level ground at an angle of 60.0°, as in Illustration 9. Find the height of the tower.

ILLUSTRATION 9

16. A lunar module is resting on the moon's surface directly below a spaceship in its orbit, 12.0 km above the moon. (See Illustration 10.) Two lunar explorers find that the angle from their position to that of the spaceship is 82.9°. What distance are they from the lunar module?

ILLUSTRATION 10

17. In ac (alternating current) circuits, the relationship between the impedance Z (in ohms), the resistance R (in ohms), the phase angle θ, and the reactance X (in ohms) is shown by the right triangle in Illustration 11.
  a. Find the impedance if the resistance is 350 Ω and the phase angle is 35°.
  b. Suppose the resistance is 550 Ω and the impedance is $70\overline{0}$ Ω. What is the phase angle?
  c. Suppose the reactance is 182 Ω and the resistance is 240 Ω. Find the impedance and the phase angle.

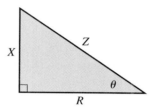

ILLUSTRATION 11

18. A right circular conical tank with its point down (Illustration 12) has a height of 4.00 m and a radius of 1.20 m. The tank is filled to a height of 3.70 m with liquid. How many litres of liquid are in the tank? (1000 litres = 1 m³)

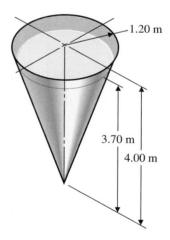

ILLUSTRATION 12

19. Use the right triangle in Illustration 13:
  a. Find the voltage applied if the voltage across the coil is 35.6 V and the voltage across the resistance is 40.2 V.
  b. Find the voltage across the resistance if the voltage applied is 378 V and the voltage across the coil is 268 V.
  c. Find the voltage across the coil if the voltage applied is 448 V and the voltage across the resistance is 381 V.

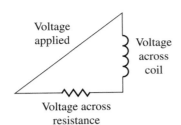

ILLUSTRATION 13

20. Machinists often use a coordinate system to drill holes by placing the origin at the most convenient location. A bolt circle is the circle formed by completing an arc through the centers of the bolt holes in a piece of metal. Find the coordinates of the centers of eight equally spaced $\frac{1}{4}$-in. holes on a bolt circle of radius 6.00 in., as shown in Illustration 14.

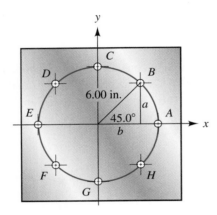

ILLUSTRATION 14

21. Twelve equally spaced holes must be drilled on a 14.500-in.-diameter bolt circle. (See Illustration 15.) What is the straight-line center-to-center distance between two adjacent holes?

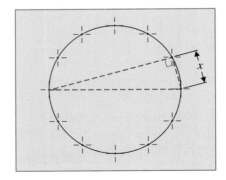

ILLUSTRATION 15

22. Dimension x in the dovetail shown in Illustration 16 is a check dimension. Precision steel balls of diameter 0.1875 in. are used in this procedure. What should this check dimension be?

ILLUSTRATION 16

23. Find angle θ of the taper in Illustration 17.

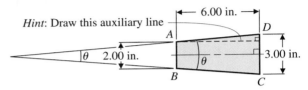

ILLUSTRATION 17

24. You need to use a metal screw with a head angle of angle A, which is not less than 65° and no larger than 70°. The team leader wants you to find angle A from the sketch shown in Illustration 18 and determine if the head angle will be satisfactory. Find the head angle A.

ILLUSTRATION 18

25. Find **a.** distance x and **b.** distance BD in Illustration 19. Length $BC = 5.50$ in.

ILLUSTRATION 19

26. Find length AB along the roofline of the building in Illustration 20. The slope of the roof is 45.0°.

ILLUSTRATION 20

27. Find length x and angle A in Illustration 21.

ILLUSTRATION 21

28. From the base of a building, measure out a horizontal distance of 215 ft along the ground. The angle of elevation to the top of the building is 63.0°. Find the height of the building.

29. A mechanical draftsperson needs to find the distance across the corners of a hex-bolt. See Illustration 22. If the distance across the flats is 2.25 cm, find the distance across the corners.

ILLUSTRATION 22

30. A hydraulic control valve has two parallel angular passages that must connect to two threaded ports, as shown in Illustration 23 with all lengths in inches. What are the missing dimensions necessary for the location of the two ports?

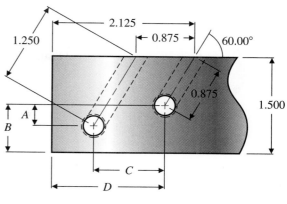

ILLUSTRATION 23

31. A benchmark has been covered up with dirt and needs to be found. The CAD drawing in Illustration 24 shows that it is located 33.0 ft from a property stake at an angle of 112.0°. What distance b from the stake parallel to the curb line and distance a in from the curb line is this benchmark?

ILLUSTRATION 24

32. A mating part is being designed with two pins attached to a flat block to match the item in Illustration 25. The pins must fit into the holes shown. What is the distance from point C (center of small hole) to point D (center of larger hole)? Also find angle x.

ILLUSTRATION 25

2.50 in.

1.00 in.

7.50 in.

2.00 in.

33. Solar heating and electric panels should face the sun so that you gain as much sunlight as possible. Fixed panels are mounted at the best winter angle to the horizontal according to the following formula: $\theta = 0.90$ (your latitude in degrees) $+ 29°$. A homeowner in Ithaca, New York (latitude 42.44°N) wishes to mount a solar heating panel that is 4.0 ft wide and 8.0 ft long facing south with its 4.0-ft base mounted on a flat roof at the appropriate winter angle. **a.** Find the height above the roof of the top of the solar panel. **b.** Find the length of the base of the support if the top of the solar panel is supported by vertical posts.

34. A lean-to is a simple shelter with three walls, a sloping roof, and an open front facing away from the prevailing winds. The back wall is short compared to the front opening. If the lean-to at a campsite has a front opening that is 6.0 ft tall, a back wall that is 2.0 ft tall, and a floor that is 8.0 ft deep, what angle does the roofline make with the ground?

Chapter 13 Group Activities

1. Estimate the height of a building or some other tall object by first marking a point A some distance from the base of the object. Then measure the distance from the base of the object to point A. Then measure the angle (as best you can) formed at point A by a horizontal line to the base of the object and a line to the top of the object. Then calculate its height by using right triangle trigonometry. Have different groups repeat the exercise by choosing other points at different distances from the base of the object. Compare the results and explain any differences.

2. In Section 13.1, we saw sin A, cos A, tan A, and the Pythagorean theorem $c^2 = a^2 + b^2$. Sometimes it is useful to change equations and formulas involving sine, cosine, and tangent to get other equations and formulas involving sine, cosine, and tangent. Below are some key trigonometric identities involving sine, cosine, and tangent.

$$\tan x = \frac{\sin x}{\cos x} \quad \sin^2 x + \cos^2 x = 1$$

If we start with an expression that includes sine, cosine, and tangent and change it to another expression, we have what is called an identity.

Verify the following identity by starting with the left-hand side and substituting other basic identities from the chapter until you obtain the expression on the right-hand side.

Example

$$(\cos x + \sin x)^2 - 2 = -1(\sin x - \cos x)^2$$

Verification

$$(\cos x + \sin x)^2 - 2$$
$$= \cos^2 x + 2 \cos x \sin x + \sin^2 x - 2$$
$$= \sin^2 x + \cos^2 x + 2 \cos x \sin x - 2$$
$$= -1 + 2 \cos x \sin x$$
$$= -1(-2 \cos x \sin x + 1)$$
$$= -1(\sin^2 x - 2 \cos x \sin x + \cos^2 x)$$
$$= -1(\sin x - \cos x)^2$$

This was a difficult example. The following will not be as challenging for your group.

Verify the following identity by starting with the left-hand side and substituting basic identities from the chapter until you obtain the right-hand side.

a. $\sin x + \tan x = \tan x(\cos x + 1)$

b. $\tan x + \cos x = \dfrac{\cos^2 x + \sin x}{\cos x}$

c. $\tan x(\cos x + 1) = \tan x + \sin x$

d. $\tan x - \sin x = \tan x(1 - \cos x)$

See whether you can develop some other identities.

Chapter 13 Summary

Glossary of Basic Terms

Angle of depression.   The angle between the horizontal and the line of sight to an object that is *below* the horizontal. (p. 477)

Angle of elevation.   The angle between the horizontal and the line of sight to an object that is *above* the horizontal. (p. 477)

Cosine of angle *A* (cos *A*).   Equals the ratio of the length of the side adjacent to angle *A* to the length of the hypotenuse. (p. 467)

Ratio.   The comparison of two quantities by division. (p. 467)

Sine of angle *A* (sin *A*).   Equals the ratio of the length of the side opposite angle *A* to the length of the hypotenuse. (p. 467)

Solving a right triangle.   Find the measures of the various parts of a triangle that are not given. (p. 474)

Tangent of angle *A* (tan *A*).   Equals the ratio of the length of the side opposite angle *A* to the length of the side adjacent to angle *A*. (p. 467)

Trigonometric ratio.   Expresses the relationship between an acute angle and the lengths of two of the sides of a right triangle. (p. 467)

13.1 Trigonometric Ratios

1. **Pythagorean theorem:** In any right triangle, $c^2 = a^2 + b^2$. (p. 466)

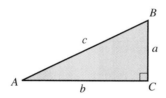

2. **Trigonometric ratios:**

$$\sin A = \frac{\text{length of side opposite angle } A}{\text{length of hypotenuse}}$$

$$\cos A = \frac{\text{length of side adjacent to angle } A}{\text{length of hypotenuse}}$$

$$\tan A = \frac{\text{length of side opposite angle } A}{\text{length of side adjacent to angle } A}$$

(p. 467)

13.2 Using Trigonometric Ratios to Find Angles

1. **Trigonometric ratios and significant digits:** When calculations involve a trigonometric ratio, use the following for significant digits. (p. 471)

Angles expressed to the nearest	The length of each side of the triangle contains
1°	Two significant digits
0.1°	Three significant digits
0.01°	Four Significant Digits

2. **Which trig ratio to use:**
   a. If you are finding an angle, two sides must be known. Label these two known sides as *side opposite* the angle you are finding, *side adjacent* to the angle you are finding, or *hypotenuse*. Then choose the trigonometric ratio that has these two sides.
   b. If you are finding a side, one side and one angle must be known. Label the known side and the unknown side as *side opposite* the known angle, *side adjacent* to the known angle, or *hypotenuse*. Then choose the trigonometric ratio that has these two sides. (pp. 471–472)

3. **Right triangle acute angle relationships:** In any right triangle, the sum of the acute angles is 90°; that is, $A + B = 90°$. (p. 472)

Chapter 13 Review

For Exercises 1–7, see Illustration 1:

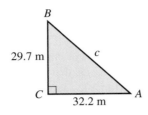

ILLUSTRATION 1

1. What is the length of the side opposite angle A in the right triangle?
2. What is the angle adjacent to the side whose length is 29.7 m?
3. Side c is known as the ___?___ .
4. What is the length of side c?
5. $\frac{\text{length of side opposite angle } A}{\text{length of hypotenuse}}$ is what ratio?
6. $\cos A = \frac{?}{\text{length of hypotenuse}}$ 7. $\tan B = \frac{?}{?}$

Find the value of each trigonometric ratio rounded to four significant digits:

8. $\cos 36.2°$ 9. $\tan 48.7°$ 10. $\sin 23.72°$

Find each angle rounded to the nearest tenth of a degree:

11. $\sin A = 0.7136$ 12. $\tan B = 0.1835$
13. $\cos A = 0.4104$ 14. Find angle A in Illustration 2.
15. Find angle B in Illustration 2.

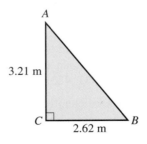

ILLUSTRATION 2

16. Find side b in Illustration 3.
17. Find side c in Illustration 3.

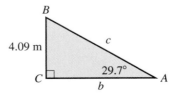

ILLUSTRATION 3

Solve each right triangle:

18.

19.

20.

21. A satellite is directly overhead one observer station when it is at an angle of 68.0° from another observer station. (See Illustration 4.) The distance between the two stations is 2000 m. What is the height of the satellite?

ILLUSTRATION 4

22. A ranger at the top of a fire tower observes the angle of depression to a fire on level ground to be 3.0°. If the tower is 275 ft tall, what is the ground distance from the base of the tower to the fire?

23. Find the angle of slope of the symmetrical roof in Illustration 5.

12.0 ft

96.0 ft

ILLUSTRATION 5

Chapter 13 Test

Find the value of each trigonometric ratio rounded to four significant digits:

1. sin 35.5°   **2.** cos 16.9°   **3.** tan 57.1°

Find each angle rounded to the nearest tenth of a degree:

4. cos A = 0.5577   **5.** tan B = 0.8888

6. sin A = 0.4166

7. Find angle B in Illustration 1.

8. Find side a in Illustration 1.

9. Find side c in Illustration 1.

B

a c

34.2°

C 15.8 cm A

ILLUSTRATION 1

10. Find angle A in Illustration 2.

11. Find angle B in Illustration 2.

12. Find side b in Illustration 2.

B

245 ft   305 ft

C b A

ILLUSTRATION 2

13. A tower 50.0 ft high has a guy wire that is attached to its top and anchored in the ground 15.0 ft from its base. Find the length of the guy wire.

14. Find length x in the retaining wall in Illustration 3.

15. Find angle A in the retaining wall in Illustration 3.

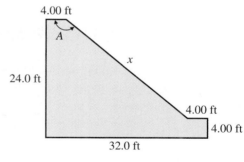

4.00 ft

A

x

24.0 ft

4.00 ft

4.00 ft

32.0 ft

ILLUSTRATION 3

14

Trigonometry with Any Angle

Mathematics at Work

Land surveyors establish official land, airspace, and water boundaries. They write descriptions of land for deeds, leases, and other legal documents; define airspace for airports; and measure construction and mineral sites. Land surveyors also manage and plan the work of survey parties that measure distances, directions, and angles between points and elevations of points, lines, and contours on, above, and below the earth's surface. Other surveyors provide data relevant to the shape, contour, location, elevation, or dimension of land or land features. Surveying technicians assist land surveyors by operating surveying instruments, such as the theodolite (used to measure horizontal and vertical angles) and electronic distance-measuring equipment, and collecting information in the field and by performing computations and computer-aided drafting in offices. New technology, such as the satellite Global Positioning System (GPS) that locates points on the earth to a high degree of precision, is continually changing the nature of the work of surveyors and surveying technicians.

Surveying Technician

Surveying technician working on a city street project.

Land surveyors and surveying technicians often spend a lot of time outdoors and work longer hours during the summer, when weather and light conditions are related to the demand for specific surveying services. The work is often strenuous, requires long periods of walking carrying heavy equipment, and often requires traveling long distances from home.

In the past, many people with little formal training in surveying started as members of survey crews and worked their way up to become licensed surveyors. However, advancing technology and more stringent licensing standards are increasing formal education requirements. Specific requirements vary by state; many states have a continuing education requirement. Generally, the quickest path to licensure is a combination of four years of college, up to four years of experience under the supervision of an experienced surveyor, and passing the licensing examinations.

Surveying technicians often complete surveying technology programs in a community or technical college with emphasis placed on knowledge and hands-on skills needed for computer-aided drafting, construction layout, engineering surveys, and land surveying. Often, these associate degree programs may be transferred to universities that offer a four-year degree program that also prepares the student to take the licensing examination. For more information, go to the website listed below.

www.cengage.com/mathematics/ewen

Objectives

■ Draw sine and cosine graphs by plotting points.

■ Find the amplitude, the period, and the phase shift of sine and cosine graphs.

■ Solve oblique triangles using the law of sines.

■ Solve oblique triangles that have two possible solutions using the law of sines.

■ Solve oblique triangles using the law of cosines.

■ Solve application problems involving oblique triangles and the law of sines and the law of cosines.

14.1 Sine and Cosine Graphs

Up to this point, we have considered only the trigonometric ratios of angles between 0° and 90°, because we were working only with right triangles. For many applications, we need to consider values greater than 90°. In this section, we use a calculator to find the values of the sine and cosine ratios of angles greater than 90°. Then we use these values to construct various sine and cosine graphs.

The procedure for finding the value of the sine or cosine of an angle greater than 90° is the same as for angles between 0° and 90°, as shown in Section 13.1.

Example 1   Find sin 255° rounded to four significant digits.

> SIN 255 =

> -0.965925826

Thus, sin 255° = −0.9659 rounded to four significant digits.   ■

Let's graph $y = \sin x$ for values of x between 0° and 360°. First, find a large number of values of x and y that satisfy the equation and plot them in the xy plane. For convenience, we will choose values of x in multiples of 30° and round the values of y to two significant digits.

x	0°	30°	60°	90°	120°	150°	180°	210°	240°	270°	300°	330°	360°
y	0	0.50	0.87	1.0	0.87	0.50	0	−0.50	−0.87	−1.0	−0.87	−0.50	0

Now choose a convenient scale for the x axis so that one unit equals 30°, and mark the x axis between 0° and 360°. Then choose a convenient scale for the y axis so that one unit equals 0.1, and mark the y axis between +1.0 and −1.0. Plot the points corresponding to the ordered pairs (x, y) from the table above. Then connect the points with a smooth, continuous curve. The graph is shown in Figure 14.1.

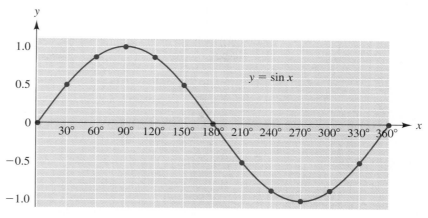

FIGURE 14.1

The graph of $y = \cos x$ for values of x between $0°$ and $360°$ can be found in a similar manner and is shown below.

x	$0°$	$30°$	$60°$	$90°$	$120°$	$150°$	$180°$	$210°$	$240°$	$270°$	$300°$	$330°$	$360°$
y	1.0	0.87	0.50	0	-0.50	-0.87	-1	-0.87	-0.50	0	0.50	0.87	1.0

Plot the points corresponding to these ordered pairs and connect them with a smooth curve, as shown in Figure 14.2.

FIGURE 14.2

 Example 2 Graph $y = 4 \sin x$ for values of x between $0°$ and $360°$.

Prepare a table listing values of x in multiples of $30°$. To find each value of y, find the sine of the angle, multiply this value by 4, and round to two significant digits.

x	$0°$	$30°$	$60°$	$90°$	$120°$	$150°$	$180°$	$210°$	$240°$	$270°$	$300°$	$330°$	$360°$
y	0	2.0	3.5	4	3.5	2.0	0	-2.0	-3.5	-4	-3.5	-2.0	0

Here, let's choose the scale for the y axis so that one unit equals 0.5 and mark the y axis between +4.0 and −4.0. Plot the points corresponding to these ordered pairs in the table and then connect them with a smooth curve, as shown in Figure 14.3.

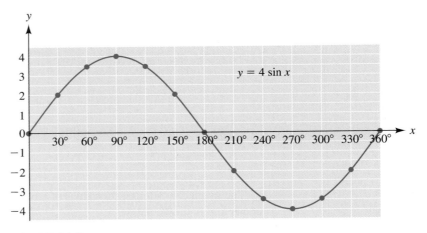

FIGURE 14.3   ■

Note that the graphs for $y = \sin x$ and $y = 4 \sin x$ are similar. That is, each starts at $(0°, 0)$, reaches its maximum at $x = 90°$, crosses the x axis at $(180°, 0)$, reaches its minimum at $x = 270°$, and meets the x axis at $(360°, 0)$. In general, the graphs of equations in the form

$$y = A \sin x \quad \text{and} \quad y = A \cos x, \quad \text{where} \quad A > 0$$

reach a maximum value of A and a minimum value of $-A$. The value of A is usually called the **amplitude**.

One of the most common applications of waves is in alternating current. In a generator, a coil of wire is rotated in a magnetic field, which produces an alternating electric current. See Figure 14.4.

In a simple generator, the current i changes as the coil rotates according to the equation

$$i = I \sin x$$

where I is the maximum current and x is the angle through which the coil rotates.

Similarly, the voltage v also changes as the coil rotates according to the equation

$$v = V \sin x$$

where V is the maximum voltage and x is the angle through which the coil rotates.

FIGURE 14.4
A coil of wire rotating in a magnetic field produces an alternating electric current.

◆ Example 3

The maximum voltage V in a simple generator is 25 V. The changing voltage v as the coil rotates is given by

$$v = 25 \sin x$$

Graph the equation in multiples of 30° for one complete revolution of the coil.

First, prepare a table. To find each value of y, find the sine of the angle, multiply this value by 25, and round to two significant digits.

x	0°	30°	60°	90°	120°	150°	180°	210°	240°	270°	300°	330°	360°
y	0	13	22	25	22	13	0	−13	−22	−25	−22	−13	0

Let's choose the scale for the y axis so that one unit equals 5 V. Mark the y axis between $+25$ V and -25 V. Plot the points corresponding to the ordered pairs in the table. Then connect these with a smooth curve, as shown in Figure 14.5.

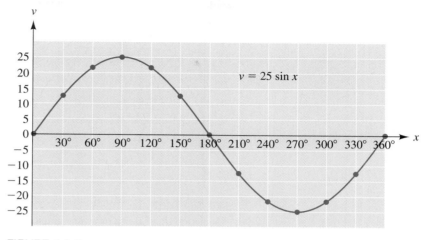

FIGURE 14.5   ■

If you were to continue finding ordered pairs in the previous tables by choosing values of x greater than $360°$ and less than $0°$, you would find that the y values repeat themselves and that the graphs form *waves*, as shown in Figure 14.6.

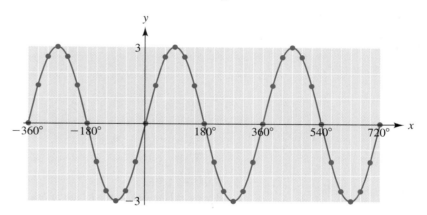

(a) $y = 3 \sin x$ for x between $-360°$ and $720°$

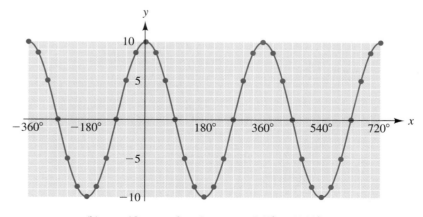

(b) $y = 10 \cos x$ for x between $-360°$ and $720°$

FIGURE 14.6

In general, the **period** of a sine or cosine graph is the *x* distance between any point on the graph and the corresponding point in the next cycle where the graph starts repeating itself. See Figure 14.7.

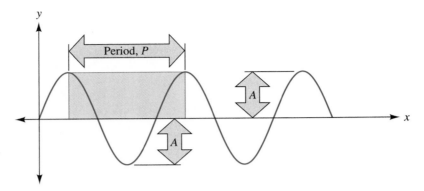

FIGURE 14.7
The period, *P*, of a sine or cosine graph is the *x* distance between any two successive corresponding points on the graph.

If the horizontal, or *x* axis, variable is *distance,* the length of one complete wave of the sine and cosine graphs along the *x* axis is called the **wavelength** and is given by the symbol λ, the Greek letter lambda. See Figure 14.8.

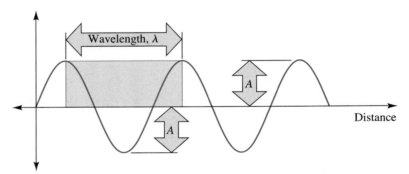

FIGURE 14.8
The wavelength, λ, is the length of one complete wave when the horizontal axis is distance.

If the horizontal, or *x* axis, variable is *time,* the time required for one complete wave of the sine and cosine graphs to pass a given point is called the **period**, *T*. See Figure 14.9.

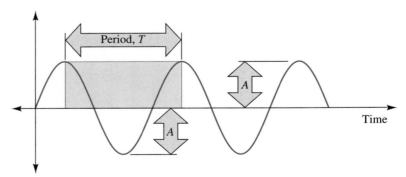

FIGURE 14.9
The period, *T,* is the length of time required for one complete wave to pass a given point when the horizontal axis is time.

The **frequency** f is the number of waves that pass a given point on the time axis each second. That is,

$$f = \frac{1}{T}$$

The unit of frequency is the hertz (Hz), where

1 Hz = 1 wave/s   or   1 cycle/s

Common multiples of the hertz are the kilohertz (kHz, 10^3 Hz) and the megahertz (MHz, 10^6 Hz).

Example 4

Find the period of a wave whose frequency is 250 Hz.

Given $f = \dfrac{1}{T}$

$\quad fT = 1$ Multiply both sides by T.

$\quad T = \dfrac{1}{f}$ Divide both sides by f.

$\quad T = \dfrac{1}{250 \text{ Hz}}$ Substitute $f = 250$ Hz.

$\quad = \dfrac{1}{250} \cdot \dfrac{1}{\text{Hz}}$

$\quad = \dfrac{1}{250} \cdot \dfrac{1}{\dfrac{\text{wave}}{\text{s}}}$ $\dfrac{1}{\dfrac{\text{wave}}{\text{s}}} = 1 \div \dfrac{\text{wave}}{\text{s}} = 1 \cdot \dfrac{\text{s}}{\text{wave}} = \dfrac{\text{s}}{\text{wave}}$

$\quad = \dfrac{1}{250} \cdot \dfrac{\text{s}}{\text{wave}}$

$\quad = 4.0 \times 10^{-3}$ s   That is, one wave passes a given point each 4.0×10^{-3} s or 4.0 ms.

Frequency and wavelength are related to wave velocity by the formula

$$v = \lambda f$$

where v is the wave velocity, λ is the wavelength, and f is the frequency.

Example 5

The FM band of a radio station is 90.9 MHz (megahertz). The speed of a radio wave is the same as the speed of light, which is 3.00×10^8 m/s. Find its wavelength.

$v = \lambda f$

$\lambda = \dfrac{v}{f}$ Divide both sides by f.

$\lambda = \dfrac{3.00 \times 10^8 \text{ m/s}}{90.9 \text{ MHz}}$ M = mega = 10^6

$\quad = \dfrac{3.00 \times 10^8 \text{ m/s}}{90.9 \times 10^6 \text{ Hz}}$ 1 Hz = 1 cycle/s = 1 wave/s

$\quad = \dfrac{3.00 \times 10^8 \text{ m/s}}{90.9 \times 10^6 \text{ waves/s}}$ $\dfrac{\text{m}}{\text{s}} \div \dfrac{\text{waves}}{\text{s}} = \dfrac{\text{m}}{\text{s}} \cdot \dfrac{\text{s}}{\text{waves}} = \dfrac{\text{m}}{\text{wave}}$

$\quad = 3.30$ m   That is, the length of each wave is 3.30 m.

Exercises 14.1

Find each value rounded to four significant digits:

1. sin 137°   **2.** sin 318°   **3.** cos 246°

4. cos 295°   **5.** sin 205.8°   **6.** sin 106.3°

7. cos 166.5°   **8.** cos 348.2°   **9.** tan 217.6°

10. tan 125.5°   **11.** tan 156.3°   **12.** tan 418.5°

Graph each equation for values of x between 0° and 720° in multiples of 30°:

13. $y = 6 \sin x$ **14.** $y = 2 \sin x$

15. $y = 5 \cos x$ **16.** $y = 4 \cos x$

Graph each equation for values of x between 0° and 360° in multiples of 15°:

17. $y = \sin 2x$ **18.** $y = \cos 2x$

Graph each equation for values of x between 0° and 360° in multiples of 10°:

19. $y = 4 \cos 3x$ **20.** $y = 2 \sin 3x$

The maximum voltage in a simple generator is V. The changing voltage v as the coil rotates is given by

$$v = V \sin x$$

Graph this equation in multiples of 30° for two complete revolutions of the coil for each value of V:

21. $V = 36$ V   **22.** $V = 48$ V

The maximum current in a simple generator is I. The changing current i as the coil rotates is given by

$$i = I \sin x$$

Graph this equation in multiples of 30° for two complete revolutions of the coil for each value of I:

23. $I = 5.0$ A   **24.** $I = 7.5$ A

25. From the graph in Exercise 21, estimate the value of v at $x = 45°$ and $x = 295°$.

26. From the graph in Exercise 22, estimate the value of v at $x = 135°$ and $x = 225°$.

27. From the graph in Exercise 23, estimate the value of i at $x = 135°$ and $x = 225°$.

28. From the graph in Exercise 24, estimate the value of i at $x = 45°$ and $x = 190°$.

29. Find the period of a wave whose frequency is 5.0 kHz.

30. Find the period of a wave whose frequency is 1.1 MHz.

31. Find the frequency of a wave whose period is 0.56 s.

32. Find the frequency of a wave whose period is 25 μs.

33. A radar unit operates at a wavelength of 3.4 cm. Radar waves travel at the speed of light, which is 3.0×10^8 m/s. What is the frequency of the radar waves?

34. A local AM radio station broadcasts at 1400 kHz. What is the wavelength of its radio waves? (They travel at the speed of light, which is 3.0×10^8 m/s.)

35. Find the speed of a wave having frequency 4.50/s and wavelength 0.500 m.

36. Find the wavelength of water waves with frequency 0.55/s and speed 1.4 m/s.

37. Radio antennas are made so that the electromagnetic radio waves "fit" on the wires. Think of a radio antenna as the x axis of a sine or cosine graph, depending on the kind of antenna. The ribbon-like FM antenna that comes with the purchase of most stereo receivers is actually a "half-wave folded dipole antenna" so that its length is roughly one-half of the wavelength of a frequency in the middle of the FM band. Because all electromagnetic waves move at the speed of light (3.00×10^8 m/s), how long is the typical FM ribbon antenna for a frequency of 98.0 MHz? *Hint:* The length of a half-wave antenna is one-half of a wavelength.

14.2 Period and Phase Shift

As we saw in Figure 14.7, the period is the length of one complete cycle of the sine or cosine graph. In general, the period for $y = A \sin Bx$ and for $y = A \cos Bx$ may be found by the formula

$$P = \frac{360°}{B}$$

Example 1   Find the period and amplitude of $y = 2 \cos 3x$ and draw its graph.

$$P = \frac{360°}{3} = 120°$$

Draw the cosine graph with amplitude 2 and period 120°, as in Figure 14.10.

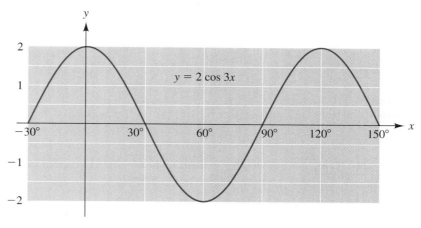

FIGURE 14.10

Example 2   Find the period and amplitude of $y = 5 \sin 4x$ and draw one period of its graph.

$$P = \frac{360°}{4} = 90°$$

Draw the sine graph with amplitude 5 and period 90°, as in Figure 14.11.

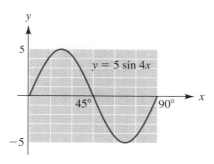

FIGURE 14.11

If the graph of a sine curve does not pass through the origin (0°, 0) or if the graph of a cosine curve does not pass through the point (0°, A), where A is the amplitude, the curve is **out of phase**. If the curve is out of phase, the **phase shift** is the horizontal distance between two successive corresponding points of the curve $y = A \sin Bx$ (or $y = A \cos Bx$) and the out-of-phase curve. (See Figure 14.12.)

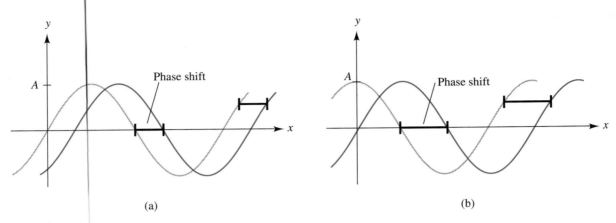

FIGURE 14.12
The phase shift is the horizontal distance between two successive points of either graph.

 Example 3   Graph $y = 2 \sin x$ and $y = 2 \sin (x - 45°)$ on the same set of coordinate axes.
For $y = 2 \sin (x - 45°)$,

x	−45°	0°	45°	90°	135°	180°	225°	270°	315°	360°	405°	450°
y	−2	−1.4	0	1.4	2	1.4	0	−1.4	−2	−1.4	0	1.4

Graph both equations, as in Figure 14.13.

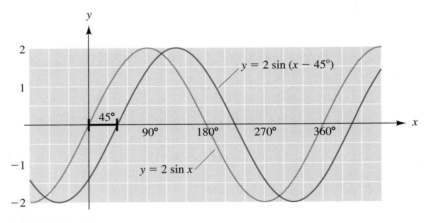

FIGURE 14.13   ■

Graphing equations in the form $y = A \sin (Bx + C)$ or $y = A \cos (Bx + C)$ involves a phase shift.

Phase Shift

The effect of C in each equation is to shift the curve $y = A \sin Bx$ or $y = A \cos Bx$

1. to the *left* $\frac{C}{B}$ units if $\frac{C}{B}$ is positive.
2. to the *right* $\frac{C}{B}$ units if $\frac{C}{B}$ is negative.

Example 4

Graph $y = 4 \sin (3x + 60°)$.

The amplitude is 4. The period is $P = \frac{360°}{3} = 120°$. The phase shift is $\frac{C}{B} = \frac{60°}{3} = 20°$, or 20° to the *left*.

Graph $y = 4 \sin 3x$ and shift the curve 20° to the left, as shown in Figure 14.14.

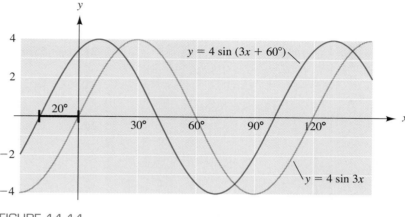

FIGURE 14.14

Example 5

Graph $y = 6 \cos (2x - 90°)$.

The amplitude is 6. The period is $\frac{360°}{2} = 180°$. The phase shift is $\frac{C}{B} = \frac{-90°}{2} = -45°$, or 45° to the *right*.

Graph $y = 6 \cos 2x$ and shift the curve 45° to the right, as shown in Figure 14.15.

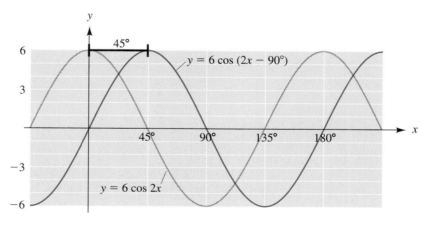

FIGURE 14.15

Note that the graph of $y = 6 \cos (2x - 90°)$ is the same as the graph of $y = 6 \sin 2x$. Each sine or cosine graph may be expressed in terms of the other trigonometric function with the appropriate phase shift.

Exercises 14.2

Find the period and amplitude, and graph at least two periods of each equation:

1. $y = 3 \sin 3x$ **2.** $y = 7 \cos 4x$

3. $y = 8 \cos 6x$ **4.** $y = 9 \sin 5x$

5. $y = 10 \sin 9x$ **6.** $y = 15 \cos 10x$

7. $y = 6 \cos \dfrac{1}{2}x$ **8.** $y = 4 \sin \dfrac{1}{3}x$

9. $y = 3.5 \sin \dfrac{2}{3}x$ **10.** $y = 1.8 \cos \dfrac{3}{4}x$

11. $y = 4 \sin \dfrac{5}{2}x$ **12.** $y = 6 \cos \dfrac{4}{3}x$

Find the period, amplitude, and phase shift, and graph at least two periods of each equation:

13. $y = \sin (x + 30°)$ **14.** $y = \cos (x + 45°)$

15. $y = 2 \cos (x - 60°)$ **16.** $y = 3 \sin (x - 120°)$

17. $y = 4 \sin (3x + 180°)$ **18.** $y = 5 \cos (2x + 60°)$

19. $y = 10 \sin (4x - 120°)$ **20.** $y = 12 \cos (4x + 180°)$

21. $y = 5 \sin \left(\dfrac{1}{2}x + 90° \right)$ **22.** $y = 6 \cos \left(\dfrac{3}{4}x - 240° \right)$

23. $y = 10 \cos \left(\dfrac{1}{4}x + 180° \right)$

24. $y = 15 \sin \left(\dfrac{2}{3}x - 120° \right)$

14.3 Solving Oblique Triangles: Law of Sines

An **oblique triangle** is a triangle with no right angle. We use the common notation of labeling vertices of a triangle by the capital letters A, B, and C, and using the small letters a, b, and c as the sides opposite angles A, B, and C, respectively.

Recall that an **acute angle** is an angle with a measure less than 90°. An **obtuse angle** is an angle with a measure greater than 90° but less than 180°.

The trigonometric ratios used in Chapter 13 apply *only* to right triangles. So we must use other ways to solve oblique triangles. One law that we use to solve oblique triangles is the law of sines.

Law of Sines

$$\frac{a}{\sin A} = \frac{b}{\sin B} = \frac{c}{\sin C}$$

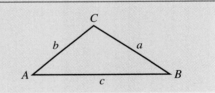

That is, *for any triangle, the ratio of the length of any side to the sine of the opposite angle equals the ratio of the length of any other side to the sine of its opposite angle.*

When using this law, you must form a proportion by choosing two of the three ratios in which three of the four terms are known. In order to use the law of sines, you must know

a. two angles and a side opposite one of them (actually, knowing two angles and any side is enough, because in knowing two angles, the third is easily found) or

b. two sides and an angle opposite one of them.

Example 1

If $C = 28.0°$, $c = 46.8$ cm, and $B = 101.5°$, solve the triangle.*

First, find side b, in Figure 14.16.

$$\frac{c}{\sin C} = \frac{b}{\sin B}$$

$$\frac{46.8 \text{ cm}}{\sin 28.0°} = \frac{b}{\sin 101.5°}$$

$b(\sin 28.0°) = (\sin 101.5°)(46.8 \text{ cm})$ Multiply both sides by the LCD.

$b = \dfrac{(\sin 101.5°)(46.8 \text{ cm})}{\sin 28.0°}$ Divide both sides by $\sin 28.0°$.

$= 97.7$ cm

FIGURE 14.16

You may use a calculator to do this calculation as follows:

SIN 101.5) ✕ 46.8 ÷ SIN 28 =

97.68531219

$$A = 180° − B − C = 180° − 101.5° − 28.0° = 50.5°$$

To find side a,

$$\frac{c}{\sin C} = \frac{a}{\sin A}$$

$$\frac{46.8 \text{ cm}}{\sin 28.0°} = \frac{a}{\sin 50.5°}$$

$a(\sin 28.0°) = (\sin 50.5°)(46.8 \text{ cm})$ Multiply both sides by the LCD.

$a = \dfrac{(\sin 50.5°)(46.8 \text{ cm})}{\sin 28.0°}$ Divide both sides by $\sin 28.0°$.

$= 76.9$ cm

The solution is $a = 76.9$ cm, $b = 97.7$ cm, and $A = 50.5°$. ■

A wide variety of applications may be solved using the law of sines.

 Example 2

Find the lengths of rafters AC and BC for the roofline shown in Figure 14.17.

First, find angle C.

$$C = 180° − A − B = 180° − 35.0° − 65.0° = 80.0°$$

To find side AC,

$$\frac{AC}{\sin B} = \frac{AB}{\sin C}$$

$$\frac{AC}{\sin 65.0°} = \frac{24.0 \text{ ft}}{\sin 80.0°}$$

$AC(\sin 80.0°) = (\sin 65.0°)(24.0 \text{ ft})$ Multiply both sides by the LCD.

$AC = \dfrac{(\sin 65.0°)(24.0 \text{ ft})}{\sin 80.0°}$ Divide both sides by $\sin 80.0°$.

$= 22.1$ ft

FIGURE 14.17

---

*As in previous sections, follow the rules for working with measurements; round sides to three significant digits and angles to the nearest tenth of a degree.

To find side BC,

$$\frac{BC}{\sin A} = \frac{AB}{\sin C}$$

$$\frac{BC}{\sin 35.0°} = \frac{24.0 \text{ ft}}{\sin 80.0°}$$

$BC(\sin 80.0°) = (\sin 35.0°)(24.0 \text{ ft})$ Multiply both sides by the LCD.

$$BC = \frac{(\sin 35.0°)(24.0 \text{ ft})}{\sin 80.0°}$$ Divide both sides by $\sin 80.0°$.

$$= 14.0 \text{ ft}$$ ■

Exercises 14.3

Solve each triangle using the labels as shown in Illustration 1 (round lengths of sides to three significant digits and angles to the nearest tenth of a degree):

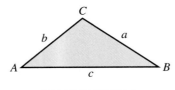

ILLUSTRATION 1

1. $A = 68.0°$, $a = 24.5$ m, $b = 17.5$ m
2. $C = 76.3°$, $b = 142$ cm, $c = 155$ cm
3. $A = 61.5°$, $B = 75.6°$, $b = 255$ ft
4. $B = 41.8°$, $C = 59.3°$, $c = 24.7$ km
5. $A = 14.6°$, $B = 35.1°$, $c = 43.7$ cm
6. $B = 24.7°$, $C = 136.1°$, $a = 342$ m
7. $A = 54.0°$, $C = 43.1°$, $a = 26.5$ m
8. $B = 64.3°$, $b = 135$ m, $c = 118$ m
9. $A = 20.1°$, $a = 47.5$ mi, $c = 35.6$ mi
10. $B = 75.2°$, $A = 65.1°$, $b = 305$ ft
11. $C = 48.7°$, $B = 56.4°$, $b = 5960$ m
12. $A = 118.0°$, $a = 5750$ m, $b = 4750$ m
13. $B = 105.5°$, $c = 11.3$ km, $b = 31.4$ km
14. $A = 58.2°$, $a = 39.7$ mi, $c = 27.5$ mi
15. $A = 16.5°$, $a = 206$ ft, $b = 189$ ft
16. $A = 35.0°$, $B = 49.3°$, $a = 48.7$ m
17. Find the distance AC across the river shown in Illustration 2.

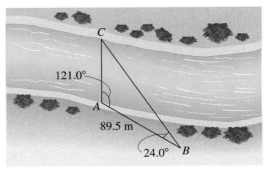

ILLUSTRATION 2

18. Find the lengths of rafters AC and BC of the roof shown in Illustration 3.

ILLUSTRATION 3

19. Find the distance AB between the ships shown in Illustration 4.

ILLUSTRATION 4

20. Find the height of the cliff shown in Illustration 5.

29.0° 44.5°

75.0 ft

ILLUSTRATION 5

21. A contractor needs to grade the slope of a subdivision lot to place a house on level ground. (See Illustration 6.) The present slope of the lot is 12.5°. The contractor needs a level lot that is 105 ft deep. To control erosion, the back of the lot must be cut to a slope of 24.0°. How far from the street, measured along the present slope, will the excavation extend?

24.0° 12.5°

105 ft

ILLUSTRATION 6

22. A weather balloon is sighted from points *A* and *B*, which are 4.00 km apart on level ground. The angle of elevation of the balloon from point *A* is 29.0°. Its angle of elevation from point *B* is 48.0°.
a. Find the height (in m) of the balloon if it is between points *A* and *B*.
b. Find its height (in m) if point *B* is between point *A* and the weather balloon.

14.4 Law of Sines: The Ambiguous Case

The solution of a triangle in which two sides and an angle opposite one of the sides are given needs special care. In this case, there may be one, two, or no triangles formed from the given information. Let's look at the possibilities.

 Example 1
Construct a triangle given that $A = 32°$, $a = 18$ cm, and $b = 24$ cm.

As you can see from Figure 14.18, there are two triangles that satisfy these conditions: triangles *ABC* and *AB'C*. In one case, angle *B* is acute. In the other, angle *B'* is obtuse.

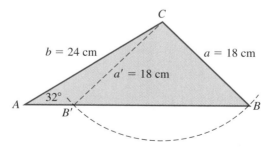

b = 24 cm   *a* = 18 cm

a' = 18 cm

32°

A   *B'*   *B*

FIGURE 14.18
Two possible triangles can be drawn; one with an obtuse angle and two acute angles and one with all acute angles.

Example 2   Construct a triangle given that $A = 40°$, $a = 12$ cm, and $b = 24$ cm.

As you can see from Figure 14.19, there is no triangle that satisfies these conditions. Side a is just not long enough to reach AB.

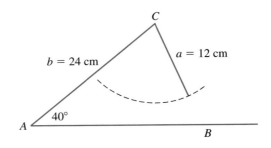

FIGURE 14.19
No complete triangle can be drawn.   ■

Example 3   Construct a triangle given that $A = 50°$, $a = 12$ cm, and $b = 8$ cm.

As you can see from Figure 14.20, there is only one triangle that satisfies these conditions. Side a is too long for two solutions.

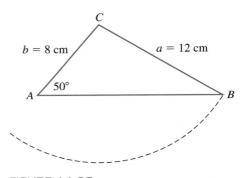

FIGURE 14.20
Only one possible triangle can be drawn.   ■

Let's summarize the possible cases when two sides and an angle opposite one of the sides are given. Assume that angle A and adjacent side b are given. From these two parts, the altitude ($h = b \sin A$) is determined and fixed.

If angle A is *acute*, we have four possible cases as shown in Figure 14.21.

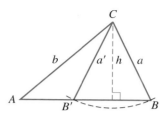

(a) When $h < b < a$, we have only one possible triangle. That is, when the side opposite the given acute angle is greater than the known adjacent side, there is only one possible triangle.

(b) When $h < a < b$, we have two possible triangles. That is, when the side opposite the given acute angle is less than the known adjacent side but greater than the altitude, there are two possible triangles.

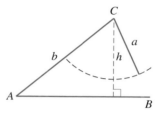

(c) When $a = h$, we have one possible triangle. That is, when the side opposite the given acute angle equals the altitude, there is only one possible triangle—a right triangle.

(d) When $a < h$, there is no possible triangle. That is, when the side opposite the given acute angle is less than the altitude, there is no possible triangle.

FIGURE 14.21
The four possible cases when angle A is acute.

If angle A is *obtuse*, we have two possible cases, as shown in Figure 14.22.

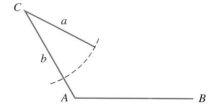

(a) When $a > b$, we have one possible triangle. That is, when the side opposite the given obtuse angle is greater than the known adjacent side, there is only one possible triangle.

(b) When $a \leq b$, there is no possible triangle. That is, when the side opposite the given obtuse angle is less than or equal to the known adjacent side, there is no possible triangle.

FIGURE 14.22
The two possible cases when angle A is obtuse.

Note: If the given parts are not angle A, side opposite a, and side adjacent b as in our preceding discussion, then you must substitute the given angle and sides accordingly. This is why it is so important to understand the general word description corresponding to each case.

The following table summarizes the possibilities for the ambiguous case of the law of sines.

If the given angle is acute: **We have:**

a. altitude < side adjacent < side opposite   one triangle

b. altitude < side opposite < side adjacent   two triangles

c. side opposite = altitude   one right triangle

d. side opposite < altitude   no triangle

If the given angle is obtuse: **We have:**

a. side opposite > side adjacent   one triangle

b. side opposite ≤ side adjacent   no triangle

Example 4   If $A = 25.0°$, $a = 50.0$ m, and $b = 80.0$ m, solve the triangle.

First, find h.

$$h = b \sin A = (80.0 \text{ m})(\sin 25.0°) = 33.8 \text{ m}$$

Since $h < a < b$, we have two solutions. First, let's find *acute* angle B in Figure 14.23.

FIGURE 14.23

$$\frac{a}{\sin A} = \frac{b}{\sin B}$$

$$\frac{50.0 \text{ m}}{\sin 25.0°} = \frac{80.0 \text{ m}}{\sin B}$$

$(50.0 \text{ m})(\sin B) = (\sin 25.0°)(80.0 \text{ m})$ Multiply both sides by the LCD.

$$\sin B = \frac{(\sin 25.0°)(80.0 \text{ m})}{50.0 \text{ m}}$$ Divide both sides by 50.0 m.

$$\sin B = 0.6762$$

$$B = 42.5°$$

You may use a calculator to do this calculation as follows.

42.54656809

$$C = 180° - A - B = 180° - 25.0° - 42.5° = 112.5°$$

To find side c,

$$\frac{c}{\sin C} = \frac{a}{\sin A}$$

$$\frac{c}{\sin 112.5°} = \frac{50.0 \text{ m}}{\sin 25.0°}$$

$c(\sin 25.0°) = (\sin 112.5°)(50.0 \text{ m})$ Multiply both sides by the LCD.

$$c = \frac{(\sin 112.5°)(50.0 \text{ m})}{\sin 25.0°}$$ Divide both sides by sin 25.0°.

$$= 109 \text{ m}$$

Next, let's find *obtuse* angle B in Figure 14.24.

$$\frac{a}{\sin A} = \frac{b}{\sin B}$$

$$\frac{50.0 \text{ m}}{\sin 25.0°} = \frac{80.0 \text{ m}}{\sin B}$$

$$(50.0 \text{ m})(\sin B) = (\sin 25.0°)(80.0 \text{ m})$$

$$\sin B = \frac{(\sin 25.0°)(80.0 \text{ m})}{50.0 \text{ m}}$$

$$\sin B = 0.6762$$

$$B = 180° - 42.5° = 137.5°$$

FIGURE 14.24

Note: If B is acute, B is $\boxed{\text{SIN}^{-1}}$ of 0.6762.

If B is obtuse, B is $180° - \boxed{\text{SIN}^{-1}}$ of 0.6762.

$$C = 180° - A - B = 180° - 25.0° - 137.5° = 17.5°$$

To find side c,

$$\frac{c}{\sin C} = \frac{a}{\sin A}$$

$$\frac{c}{\sin 17.5°} = \frac{50.0 \text{ m}}{\sin 25.0°}$$

$$c(\sin 25.0°) = (\sin 17.5°)(50.0 \text{ m})$$

$$c = \frac{(\sin 17.5°)(50.0 \text{ m})}{\sin 25.0°}$$

$$= 35.6 \text{ m}$$

The two solutions are $c = 109$ m, $B = 42.5°$, $C = 112.5°$ and $c = 35.6$ m, $B = 137.5°$, $C = 17.5°$. ■

Example 5 If $A = 59.0°$, $a = 205$ m, and $b = 465$ m, solve the triangle.

First, find h.

$$h = b \sin A = (465 \text{ m})(\sin 59.0°) = 399 \text{ m}$$

Since $a < h$, there is no possible triangle.

What would happen if you tried to apply the law of sines anyway?

$$\frac{a}{\sin A} = \frac{b}{\sin B}$$

$$\frac{205 \text{ m}}{\sin 59.0°} = \frac{465 \text{ m}}{\sin B}$$

$$(205 \text{ m})(\sin B) = (\sin 59.0°)(465 \text{ m})$$

$$\sin B = \frac{(\sin 59.0°)(465 \text{ m})}{205 \text{ m}} = 1.944$$

Note: $\sin B = 1.944$ is impossible, because $-1 \leq \sin B \leq 1$. Recall that the graph of $y = \sin x$ has an amplitude of 1, which means that the values of $\sin x$ vary between 1 and -1. Your calculator will also indicate an error when you try to find angle B. ■

As a final check to make certain that your solution is correct, verify that the following geometric triangle property is satisfied:

> In any triangle, the largest side is opposite the largest angle and the smallest side is opposite the smallest angle.

Exercises 14.4

*For each general triangle, **a.** determine the number of solutions and **b.** solve the triangle, if possible, using the labels as shown in Illustration 1 (round lengths to three significant digits and angles to the nearest tenth of a degree):*

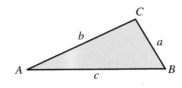

ILLUSTRATION 1

1. $A = 38.0°$, $a = 42.3$ m, $b = 32.5$ m
2. $C = 47.6°$, $a = 85.2$ cm, $c = 96.1$ cm
3. $A = 25.6°$, $b = 306$ m, $a = 275$ m
4. $B = 41.2°$, $c = 1860$ ft, $b = 1540$ ft
5. $A = 71.6°$, $b = 48.5$ m, $a = 15.7$ m
6. $B = 40.3°$, $b = 161$ cm, $c = 288$ cm
7. $C = 71.2°$, $a = 245$ cm, $c = 238$ cm
8. $A = 36.1°$, $b = 14.5$ m, $a = 12.5$ m
9. $B = 105.0°$, $b = 33.0$ mi, $a = 24.0$ mi
10. $A = 98.3°$, $a = 1420$ ft, $b = 1170$ ft
11. $A = 31.5°$, $a = 376$ m, $c = 406$ m
12. $B = 50.0°$, $b = 4130$ ft, $c = 4560$ ft
13. $C = 60.0°$, $c = 151$ m, $b = 181$ m
14. $A = 30.0°$, $a = 4850$ mi, $c = 3650$ mi
15. $B = 8.0°$, $b = 451$ m, $c = 855$ m
16. $C = 8.7°$, $c = 89.3$ mi, $b = 61.9$ mi

17. The owner of a triangular lot wishes to fence it along the lot lines. Lot markers at A and B have been located, but the lot marker at C cannot be found. The owner's attorney gives the following information by phone: $AB = 355$ ft, $BC = 295$ ft, and $A = 36.0°$. What is the length of AC?

18. The average distance from the sun to the earth is 1.5×10^8 km, and that from the sun to Venus is 1.1×10^8 km. Find the distance between the earth and Venus when the angle between the earth and the sun and the earth and Venus is $24.7°$. (Assume that the earth and Venus have circular orbits around the sun.)

19. A manufacturer has moved into a new building and wants to hang a sign outside the building based on the drawing in Illustration 2. How long is the lower support, and what angle does the upper support make with the building?

ILLUSTRATION 2

14.5 Solving Oblique Triangles: Law of Cosines

A second law used to solve oblique triangles is the law of cosines.

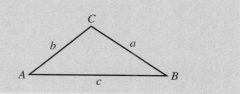

Law of Cosines

$$a^2 = b^2 + c^2 - 2bc \cos A$$
$$b^2 = a^2 + c^2 - 2ac \cos B$$
$$c^2 = a^2 + b^2 - 2ab \cos C$$

That is, *for any triangle, the square of any side equals the sum of the squares of the other two sides minus twice the product of these two sides and the cosine of their included angle.*

To use the law of cosines, you must know

a. two sides and the included angle or

b. all three sides.

Example 1   If $A = 115.2°$, $b = 18.5$ m, and $c = 21.7$ m, solve the triangle. (See Figure 14.25.)

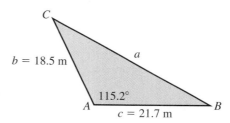

FIGURE 14.25

To find side a,

$$a^2 = b^2 + c^2 - 2bc \cos A$$
$$a^2 = (18.5 \text{ m})^2 + (21.7 \text{ m})^2 - 2(18.5 \text{ m})(21.7 \text{ m})(\cos 115.2°)$$
$$a = 34.0 \text{ m}$$

You may use a calculator to do this calculation as follows:

18.5 $\boxed{x^2}$ $\boxed{+}$ 21.7 $\boxed{x^2}$ $\boxed{-}$ 2 $\boxed{\times}$ 18.5 $\boxed{\times}$ 21.7 $\boxed{\times}$ $\boxed{\cos}$

115.2 $\boxed{=}$ $\boxed{\sqrt{}}$ $\boxed{\text{ANS}}$ $\boxed{=}$

$$\boxed{33.98526435}$$

To find angle B, use the law of sines, because it requires less computation.

$$\frac{a}{\sin A} = \frac{b}{\sin B}$$

$$\frac{34.0 \text{ m}}{\sin 115.2°} = \frac{18.5 \text{ m}}{\sin B}$$

$(34.0 \text{ m})(\sin B) = (\sin 115.2°)(18.5 \text{ m})$ Multiply both sides by the LCD.

$\sin B = \dfrac{(\sin 115.2°)(18.5 \text{ m})}{34.0 \text{ m}}$ Divide both sides by 34.0 m.

$\sin B = 0.4923$

$B = 29.5°$

$C = 180° - A - B = 180° - 115.2° - 29.5° = 35.3°$

The solution is $a = 34.0$ m, $B = 29.5°$, and $C = 35.3°$.   ■

 Example 2   If $a = 125$ cm, $b = 285$ cm, and $c = 382$ cm, solve the triangle. (See Figure 14.26.)

FIGURE 14.26

When three sides are given, you are advised to find the angle opposite the largest side first. Why?

To find angle C,

$$c^2 = a^2 + b^2 - 2ab \cos C$$

$(382 \text{ cm})^2 = (125 \text{ cm})^2 + (285 \text{ cm})^2 - 2(125 \text{ cm})(285 \text{ cm}) \cos C$

$(382 \text{ cm})^2 - (125 \text{ cm})^2 - (285 \text{ cm})^2 = -2(125 \text{ cm})(285 \text{ cm}) \cos C$

$\dfrac{(382 \text{ cm})^2 - (125 \text{ cm})^2 - (285 \text{ cm})^2}{-2(125 \text{ cm})(285 \text{ cm})} = \cos C$

$-0.6888 = \cos C$

$133.5° = C$

You may use a calculator to do this calculation as follows:

| 133.5318657 |

To find angle A, let's use the law of sines.

$$\frac{c}{\sin C} = \frac{a}{\sin A}$$

$$\frac{382 \text{ cm}}{\sin 133.5°} = \frac{125 \text{ cm}}{\sin A}$$

$$(382 \text{ cm})(\sin A) = (\sin 133.5°)(125 \text{ cm})$$

$$\sin A = \frac{(\sin 133.5°)(125 \text{ cm})}{382 \text{ cm}}$$

$$\sin A = 0.2374$$

$$A = 13.7°$$

$$B = 180° - A - C = 180° - 13.7° - 133.5° = 32.8°$$

The solution is $A = 13.7°$, $B = 32.8°$, and $C = 133.5°$. ■

Example 3

Find the lengths of guy wires AC and BC for a tower located on a hillside, as shown in Figure 14.27(a). The height of the tower is 50.0 m; $\angle ADC = 120.0°$; $AD = 20.0$ m; $BD = 15.0$ m.

First, let's use triangle ACD in Figure 14.27(b) to find length AC. Using the law of cosines,

$$(AC)^2 = (AD)^2 + (DC)^2 - 2(AD)(DC) \cos ADC$$
$$(AC)^2 = (20.0 \text{ m})^2 + (50.0 \text{ m})^2 - 2(20.0 \text{ m})(50.0 \text{ m}) \cos 120.0°$$
$$AC = 62.4 \text{ m}$$

(a)

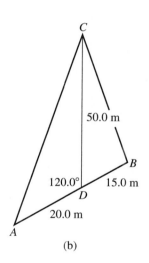
(b)

FIGURE 14.27

Next, use triangle CDB and the law of cosines to find length BC. Note that $\angle CDB = 180° - 120.0° = 60.0°$.

$$(BC)^2 = (BD)^2 + (DC)^2 - 2(BD)(DC) \cos CDB$$
$$(BC)^2 = (15.0 \text{ m})^2 + (50.0 \text{ m})^2 - 2(15.0 \text{ m})(50.0 \text{ m}) \cos 60.0°$$
$$BC = 44.4 \text{ m}$$ ■

Exercises 14.5 _____

Solve each triangle using the labels as shown in Illustration 1 (round lengths of sides to three significant digits and angles to the nearest tenth of a degree):

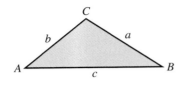

ILLUSTRATION 1

1. $A = 55.0°$, $b = 20.2$ m, $c = 25.0$ m
2. $B = 14.5°$, $a = 37.6$ cm, $c = 48.2$ cm
3. $C = 115.0°$, $a = 247$ ft, $b = 316$ ft
4. $A = 130.0°$, $b = 15.2$ km, $c = 9.50$ km
5. $a = 38{,}500$ mi, $b = 67{,}500$ mi, $c = 47{,}200$ mi
6. $a = 146$ cm, $b = 271$ cm, $c = 205$ cm
7. $B = 19.3°$, $a = 4820$ ft, $c = 1930$ ft
8. $C = 108.5°$, $a = 415$ m, $b = 325$ m
9. $a = 19.5$ m, $b = 36.5$ m, $c = 25.6$ m
10. $a = 207$ mi, $b = 106$ mi, $c = 142$ mi
11. Find the distance a across the pond in Illustration 2.

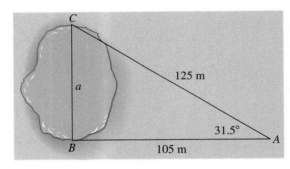

ILLUSTRATION 2

12. Find the length of rafter AC in Illustration 3.

ILLUSTRATION 3

13. Find angles A and C in the roof in Illustration 4.

ILLUSTRATION 4

14. **a.** Find angles A and ACB in the roof in Illustration 5.
    $AC = BC$
    b. Find length CD.

ILLUSTRATION 5

15. A piece of sheet metal is to be cut in the shape of a triangle with sides of 24.0 in., 12.0 in., and 21.0 in. Find the measures of the angles.

16. Three pieces of steel angle are welded to form a triangle. If two pieces are welded at a 42.0° angle and the lengths of these two pieces are 36.0 in. and 20.0 in., what is the length of the third piece?

17. A plane flies 70.0 mi due north from its base airport. Then it makes a 70.0° turn northeast and flies another 90.0 mi. How far will it be to go straight back to the base airport?

18. The taxiways for a small airport make a triangle with the runway. The runway is 61$\overline{0}$0 ft long, and one of the taxiways is 33$\overline{0}$0 ft. How long is the other taxiway if the angle between the 61$\overline{0}$0-ft runway and 33$\overline{0}$0-ft taxiway is 62.0°?

19. A car has been sideswiped, causing much damage. The distance from the driver's side front wheel to the driver's side rear wheel is 120.0 in., and the length between the rear wheels is 54.0 in. If the angle between these measurements is 110.0°, what is the distance from the driver's side front tire to the passenger's side rear tire?

20. A car seat is reclined at 140.0°. If the length of the cushion is 20.0 in. and the length of the back of the seat is 34.0 in. including the headrest, what is the distance from the front tip of the cushion to the tip of the headrest?

21. A room is shaped like a kite. The lengths of two adjacent walls are 20.0 ft and 28.0 ft. If the angle between these walls is 130.0°, how long must duct be that stretches from corner to opposite corner of this room?

22. The refrigerant line from the outside condensing unit to the air handler must be bent in the shop and taken to the job site. The sketch in Illustration 6 was made and taken

to the shop. Notice the air handler is on the base line of the triangle. Find the three angles.

ILLUSTRATION 6

23. In the framework shown in Illustration 7, we know that $AE = CD$, $AB = BC$, $BD = BE$, and $\overline{AC} \parallel \overline{ED}$. Find **a.** $\angle BEA$, **b.** $\angle A$, **c.** length BE, and **d.** length DE.

ILLUSTRATION 7

24. In the framework shown in Illustration 8, we know that $AB = DE$, $BC = CD$, $AH = FE$, $HG = GF$. Find **a.** length HB, **b.** $\angle AHB$, **c.** length GC, and **d.** length AG.

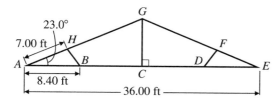

ILLUSTRATION 8

25. A triangular lot has sides 1580 ft, 2860 ft, and 1820 ft long. Find its largest angle.

26. A ship starts at point A and travels 125 mi northeast. It then travels 150 mi due east and arrives at point B. If the ship had sailed directly from A to B, what distance would it have traveled?

27. See Illustration 9. Deloney and Jackson Streets meet at a $45°$ angle. A lot extends $5\overline{0}$ yards along Jackson and $4\overline{0}$ yards along Deloney. Find the length of the back border.

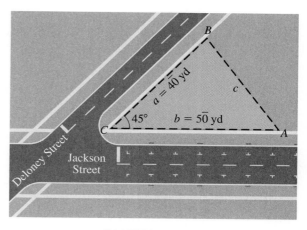

ILLUSTRATION 9

28. See Illustration 10. From its home port (H), a ship sails 52 miles at a bearing of $147°$ to point A and then sails 78 miles at a bearing of $213°$ to point B. How far is point B from home port, and at what bearing will it return?

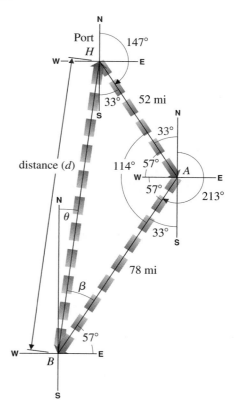

ILLUSTRATION 10

29. An 8.00-ft antenna must be mounted on the roof of a warehouse. Find the length of a guy wire to be attached 1.00 ft from the top of the antenna to the edge of the roof as shown in Illustration 11.

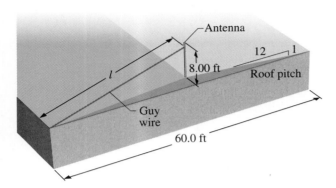

ILLUSTRATION 11

30. Snakes, an important part of the food chain, cannot chew their prey, so they generally swallow their prey whole. A special hinged jaw permits some snakes to open their mouth at an angle of as much as 150° so that they can swallow animals larger than the size of their own heads. If a particular snake's jaws are 3.0 in. long and it opens its mouth at 150°, how wide is the jaw opening? That is, what is the largest prey it can swallow?

31. A kite is flying from the top of a hill as shown in Illustration 12. How far is the kite from the person standing at point A?

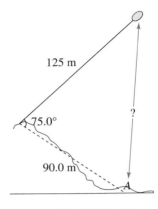

ILLUSTRATION 12

32. A game preserve manager is fencing the triangular plot of land in Illustration 13. Find the most acute angle.

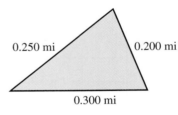

ILLUSTRATION 13

◆ Chapter 14 Group Activities

1. In Sections 14.1 and 14.2, you saw how to graph $y = \sin x$ and $y = \cos x$. Graph the following and make observations:

  a. $y = \sin x$ **b.** $y = \sin(x - 60°)$

  c. $y = \sin(x - 300°)$

Graph the following. How is the graph changing?

  d. $y = \sin(x + 60°)$ **e.** $y = \sin(x + 300°)$

Graph the following. How is the graph changing here?

  f. $y = (\sin x) + 1$ **g.** $y = (\sin x) + 3$

Graph the following. How is the graph changing?

  h. $y = (\sin x) - 1$ **i.** $y = (\sin x) - 3$

2. Graph the equations above, but replace sine with cosine. Do you recognize any patterns in the graphs above? *Hint:* See Chapter 11 Accent on Teamwork.

3. Use the law of cosines to find the distance between two points A and B that cannot be measured directly because of an obstacle, such as a building or across a pond or a trench as follows: Mark a point C that is on level ground to both points A and B. Measure the distances between points A and C and points B and C. Then measure the angle (as best you can) formed at point C between lines AC and BC. Then calculate the distance between A and B using the law of cosines. Have different groups repeat the exercise by choosing other points, compare the results, and explain any differences.

Chapter 14 Summary

Glossary of Basic Terms

Acute angle. An angle with a measure less than 90°. (p. 500)

Amplitude. The maximum *y* value for the sine and cosine graphs. (p. 492)

Frequency. The number of waves that pass a given point on the time axis each second. (p. 495)

Oblique triangle. A triangle with no right angle. (p. 500)

Obtuse angle. An angle with a measure greater than 90° but less than 180°. (p. 500)

Out of phase. When the graph of the sine curve does not pass through the origin (0°, 0) or when the graph of the cosine curve does not pass through the point (0°, *A*), where *A* is the amplitude. (p. 497)

Period. The *x* distance between any point on a sine or cosine graph and the corresponding point in the next cycle where the graph starts repeating itself. Also, the time required for one complete wave to pass a given point on the horizontal axis when the horizontal, or *x* axis, variable is *time*. (p. 494)

Phase shift. The horizontal distance between two successive corresponding points of the curve $y = A \sin Bx$ (or $y = A \cos Bx$) and the out-of-phase curve. (p. 497)

Wavelength. The length of one complete wave of the sine and cosine graphs along the horizontal axis when the horizontal, or *x* axis, variable is *distance*. (p. 494)

14.1 Sine and Cosine Graphs

1. **Frequency and period:** $f = 1/T$. (p. 495)
2. **Frequency and wavelength:** $v = \lambda f$. (p. 495)

14.2 Period and Phase Shift

1. **Period:** the period for $y = A \sin Bx$ and for $y = \cos Bx$ is $P = 360°/B$. (p. 496)
2. **Phase shift:** the phase shift for $y = A \sin (Bx + C)$ and for $y = \cos (Bx + C)$ is
   a. to the *left C/B* units if *C/B* is positive.
   b. to the *right C/B* units if *C/B* is negative. (p. 498)

14.3 Solving Oblique Triangles: Law of Sines

1. **Law of Sines:** For any triangle, the ratio of the length of any side to the sine of the opposite angle equals the ratio of the length of any other side to the sine of its opposite angle. (p. 500) Or

$$\frac{a}{\sin A} = \frac{b}{\sin B} = \frac{c}{\sin C}$$

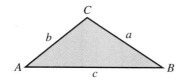

14.4 Law of Sines: The Ambiguous Case

1. **The following table summarizes the possibilities for the ambiguous case of the law of sines:** (p. 506)

If the given angle is acute:	We have:
a. altitude < side adjacent < side opposite	one triangle
b. altitude < side opposite < side adjacent	two triangles
c. side opposite = altitude	one right triangle
d. side opposite < altitude	no triangle

If the given angle is obtuse:	We have:
a. side opposite > side adjacent	one triangle
b. side opposite ≤ side adjacent	no triangle

2. **One check when solving an oblique triangle:** the largest side is opposite the largest angle and the smallest side is opposite the smallest angle. (p. 508)

14.5 Solving Oblique Triangles: Law of Cosines

1. **Law of cosines:** For any triangle, the square of any side equals the sum of the squares of the other two sides minus twice the product of these two sides and the cosine of their included angle. (p. 509) Or

$$a^2 = b^2 + c^2 - 2bc \cos A$$
$$b^2 = a^2 + c^2 - 2ac \cos B$$
$$c^2 = a^2 + b^2 - 2ab \cos C$$

Chapter 14 Review

Find each value rounded to four significant digits:

1. tan 143°   **2.** sin 209.8°   **3.** cos 317.4°

Graph each equation for values of x between 0° and 360° in multiples of 15°:

4. $y = 6 \cos x$ **5.** $y = 3 \sin 2x$

Find the period and amplitude, and graph at least two periods of each equation:

6. $y = 5 \sin 3x$ **7.** $y = 3 \cos 4x$

Find the period, amplitude, and phase shift, and graph at least two periods of each equation:

8. $y = 4 \cos (x + 60°)$ **9.** $y = 6 \sin (2x - 180°)$

Solve each triangle using the labels as shown in Illustration 1 (round lengths of sides to three significant digits and angles to the nearest tenth of a degree):

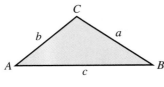

ILLUSTRATION 1

10. $B = 52.7°, b = 206$ m, $a = 175$ m
11. $A = 61.2°, C = 75.6°, c = 88.0$ cm
12. $B = 17.5°, a = 345$ m, $c = 405$ m
13. $a = 48.6$ cm, $b = 31.2$ cm, $c = 51.5$ cm

14. $A = 29.5°, b = 20.5$ m, $a = 18.5$ m
15. $B = 18.5°, a = 1680$ m, $b = 1520$ m
16. $a = 575$ ft, $b = 1080$ ft, $c = 1250$ ft
17. $C = 73.5°, c = 58.2$ ft, $b = 81.2$ ft
18. Find **a.** angle B and **b.** length x in Illustration 2.

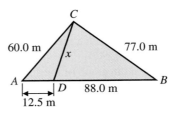

ILLUSTRATION 2

19. The centers of five holes are equally spaced around a circle of diameter 16.00 in. Find the distance between the centers of two successive holes.

20. In the roof truss in Illustration 3, $AB = DE$, $BC = CD$, and $AE = 36.0$ m. Find the lengths **a.** AF, **b.** BF, **c.** CF, and **d.** BC.

ILLUSTRATION 3

Chapter 14 Test

Find each value rounded to four significant digits:

1. cos 182.9°   **2.** tan 261°

3. Find the period, amplitude, and phase shift, and draw at least two periods of $y = 2 \sin (3x + 45°)$.

4. Find angle B in Illustration 1.

5. Find angle C in Illustration 1.

6. Find side c in Illustration 1.

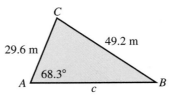

ILLUSTRATION 1

7. Find angle C in Illustration 2.

8. Find angle B in Illustration 2.

9. Find angle A in Illustration 2.

ILLUSTRATION 2

10. Find the length of the rafter shown in Illustration 3.

11. Find angle A in Illustration 3.

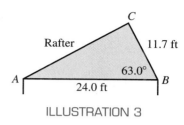

ILLUSTRATION 3

◆ Cumulative Review

Chapters 1–14

1. Given the formula $R = \dfrac{Vt}{I}$, where $V = 32$, $t = 5$, and $I = 20$, find R.

2. Simplify: $-\dfrac{5}{6} - \left(-\dfrac{3}{5}\right)$

3. Write 41,800 in scientific notation.

4. 90 kg = _____ lb

5. Read the scale in Illustration 1.

ILLUSTRATION 1

6. For the measurement 0.0018 mm, find **a.** the precision, **b.** the greatest possible error, **c.** the relative error, and **d.** the percent of error rounded to the nearest hundredth of a percent.

7. Simplify: $(2a^2 - 5a + 3) + (4a^2 + 3a - 1)$

8. Solve: $6 + 3(x - 2) = 24$

9. Solve: $\dfrac{x}{2} - \dfrac{2}{7} = \dfrac{1}{3}$

10. The area of a rectangle with constant width varies directly as its length. The area is 30.8 m² when the length is 12.8 m. Find the area when the length is 42.5 m.

11. Solve for y: $-5x - 3y = -8$

12. Solve: $7x - y = 4$
            $14x - 2y = 8$

13. Factor: $x^2 - 2x - 168$ 14. Factor: $3x^2 - 6x - 189$

15. Solve: $3x^2 - 13x = 10$ 16. Solve: $2x^2 - x - 8 = 0$

17. Find **a.** the area and **b.** the perimeter of the triangle in Illustration 2.

ILLUSTRATION 2

18. The area of a rectangle is 307 ft². The length is 22.4 ft. Find the width.

19. Find the volume of the frustum of the rectangular pyramid in Illustration 3.

ILLUSTRATION 3

20. Find tan 67.2° rounded to four significant digits.

21. If cos A = 0.6218, find angle A in degrees.

From the triangle in Illustration 4, find:

22. ∠B   **23.** side a   **24.** side c

ILLUSTRATION 4

25. A roof has a rise of 8.00 ft, and a span of 24.0 ft. Find its pitch and the distance rounded to the nearest inch, from the eave to its peak.

26. Find cos 191.13° rounded to four significant digits.

27. Draw the graph of $y = \dfrac{3}{2} \cos 2x$ for values of x between 0° and 360° in multiples of 15°.

28. Find the period and amplitude and draw at least two periods of the graph of the equation $y = 2 \sin \dfrac{1}{2}x$.

29. Given the triangle in Illustration 5, find angle C.

ILLUSTRATION 5   ILLUSTRATION 6

30. Find angle A in Illustration 6.

15 Basic Statistics

Mathematics at Work

Firefighters help protect the public by fighting fires and by rapidly responding to a wide variety of emergency situations. They are often the first emergency personnel at the scene of an accident or a medical emergency. Fighting fires is dangerous and very complex work that requires a high degree of organization, training, and teamwork, with specific duties assigned to each person.

Firefighters receive training in emergency medical procedures because most calls to which they respond involve medical emergencies. Many fire departments require them to be certified as emergency medical technicians.

Firefighters work in a wide variety of settings (urban, suburban, and rural areas; industrial plants and sites; airports; national forests and parks) as well as in specialty areas such as fire prevention, fire investigation, chemical spills, and hazardous materials.

Applicants for municipal fire fighting jobs generally must pass a written exam; tests for strength, physical stamina, coordination, and agility; and a medical exam that includes drug screening. The completion of community college fire science courses may improve an applicant's chances for appointment. In recent years, increasing numbers of applicants have had some postsecondary education. For more information, go to the website listed below.

© Blue Shadows/Alamy

Firefighter
Firefighters attack a house fire with a hose line.

www.cengage.com/mathematics/ewen

Objectives

■ Use bar graphs, circle graphs, line graphs, and frequency distributions to present data.

■ Find and use the mean, the median, and the mode of a data set.

■ Find and use percentiles to describe ranked data.

■ Find the range and the sample standard deviation of a set of data.

■ Find the sample standard deviation of a set of grouped data.

■ Use statistical process control tools and techniques to determine whether a process is in control.

■ Use histograms, run charts, and scattergrams to view data.

■ Use a normal distribution to find the number or percent of data within a given interval.

■ Find the sample space for a given event and the probability that a given event will happen.

■ Determine when two events are independent and find the probability of two independent events.

15.1   Bar Graphs

Statistics is the branch of mathematics that deals with the collection, analysis, interpretation, and presentation of masses of numerical data. In this chapter, we will first study the different ways in which data can be presented using graphs. Then, in the later sections, we will study some of the different ways to describe small sets of data. We will study only the most basic parts to help you read and better understand newspapers, magazines, and some of the technical reports in your field of interest. The chapter includes an examination of statistical process control, a technique that is widely used in manufacturing.

A graph is a picture that shows the relationship between several types of collected information. A graph is very useful when there are large quantities of information to analyze. There are many ways of graphing. A **bar graph** is a graph with parallel bars whose lengths are proportional to the frequency of the given quantities in a data set. Look closely at the bar graph in Figure 15.1.

Example 1

What are the monthly earnings of a data-processing technician with two years of experience?

Find the data-processing technician in the "Classification of workers" column. Read the right end of the bar on the "Monthly salary" scale: $3340.   ■

Classification of workers

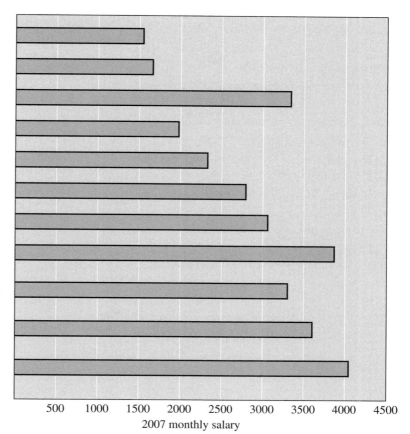

FIGURE 15.1
Bar graph

Exercises 15.1

Find the monthly earnings of the following workers from the bar graph in Figure 15.1:

1. Auto mechanic
2. Office worker and typist
3. Clerk in retail store
4. Skilled technician, community college graduate
5. Unskilled worker
6. Semiskilled worker
7. Architectural draftsperson
8. Semiprofessional technician, community college graduate
9. Semiprofessional technician after 5 years experience
10. Senior engineering technician after 10 years experience

Find the following information from Illustration 1:

11. How many barrels per day were used by France?
12. How many barrels per day were used by Japan?

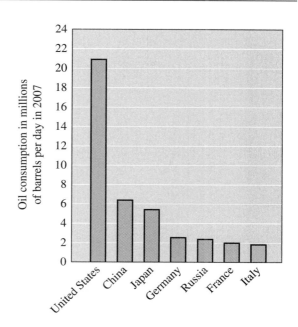

ILLUSTRATION 1

13. What country used the most barrels per day?

14. What country used the fewest barrels per day?

15. How many barrels per day were used by Russia?

16. How many barrels per day were used by Italy?

17. How many barrels per day were used by the United States?

18. How many barrels per day were used by China?

19. How many barrels per day were used by Germany?

20. What was the total number of barrels per day used by all the countries listed?

21. A survey of 100 families was taken to find the number of times the families had gone out to eat in the past month. The data are given in Illustration 2. Draw a bar graph for this survey.

22. Illustration 3 shows the average test scores on chapter tests given in a mathematics class. Draw a bar graph for these scores.

23. Illustration 4 gives the number of workers employed in the given industries in 2006. Draw a bar graph for these data.

24. Illustration 5 lists the 2007 male life expectancy for the given countries. Draw a bar graph for these data.

25. Illustration 6 gives the 2007 female life expectancy for the given countries. Draw a bar graph for these data.

Times out in past month	Number of families
0	2
1	15
2	51
3	17
4	10
5 or more	5

ILLUSTRATION 2

Chapter	1	2	3	4	5	6	7	8	9	10	11	12	13	14
Score	78	81	75	77	84	81	79	70	72	73	75	69	81	72

ILLUSTRATION 3

Industry	Mining	Manufac-turing	Educational services	Construction	Transpor-tation	Health	Retail trade	Financial
Number (in 1000s)	556.4	14,197	2198	5535	4465	14,920	15,319	8363

ILLUSTRATION 4

Country	Egypt	Nigeria	Israel	Japan	France	Sweden	Canada	Mexico	U.S.	Russia
Years	69.0	46.8	76.5	78.7	77.4	78.4	77.0	73.1	75.2	59.1

ILLUSTRATION 5

Country	Egypt	Nigeria	Israel	Japan	France	Sweden	Canada	Mexico	U.S.	Russia
Years	74.2	48.3	83.3	85.6	84.0	83.0	83.9	78.8	81.0	73.0

ILLUSTRATION 6

15.2 Circle Graphs

Another type of graph used quite often to give results of surveys is the circle graph (see Figure 15.2). A **circle graph** is a circle whose radii divide the circle into sectors that are proportional in angle and area in relative size to the quantities represented. The circle graph is used to show the relationship between the parts and the whole.

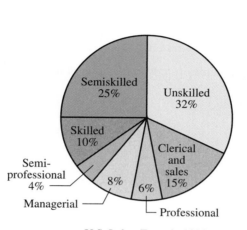

U.S. Labor Force in 1930   U.S. Labor Force in 2000

FIGURE 15.2
Circle graphs

To make a circle graph with data given in percents, first draw the circle. Since there are 360 degrees in a circle, multiply the percent of an item by 360 to find what part of the circle is used by that item.

Example 1

Draw a circle graph with the following data.

In 1930, 58% of the people working had a grade school education or less, 32% had a high school education, and 10% had a college education.

$$58\% \text{ of } 360° = 0.58 \times 360° = 208.8°, \text{ or about } 209°$$
$$32\% \text{ of } 360° = 0.32 \times 360° = 115.2°, \text{ or about } 115°$$
$$10\% \text{ of } 360° = 0.10 \times 360° = 36°$$

With a protractor draw central angles of 209°, 115°, and 36°. Then label the sections (see Figure 15.3). ■

Sometimes data are not written in percent form. When a circle graph is to be drawn from data that are not in percent form, the data must be converted to percents. Once the data are in this form, the steps in drawing the graph are the same as those already given.

Education of Labor Force in 1930

FIGURE 15.3

Example 2   Draw a circle graph with the following data.

Suggested semester credit-hour requirements for a community college curriculum in engineering technology are as follows.

Course	Semester hours
Mathematics (technical)	10
Applied science	10
Technical courses in major	34
General education courses	12
	66

Write each area of study as a percent of the whole program.

Mathematics:
$$\frac{10}{66} = \frac{r}{100}$$

$66r = 1000$ The product of the means equals the product of the extremes.

$r = 15.2\%$ Divide both sides by 66.

$15.2\% \times 360° = 0.152 \times 360°$

$= 55°$ (rounded to nearest whole degree)

Science: same as mathematics, 55°

Technical courses:
$$\frac{34}{66} = \frac{r}{100}$$

$66r = 3400$

$r = 51.5\%$

$51.5\% \times 360° = 0.515 \times 360° = 185°$

General education:
$$\frac{12}{66} = \frac{r}{100}$$

$66r = 1200$

$r = 18.2\%$

$18.2\% \times 360° = 0.182 \times 360° = 66°$

Then draw the central angles and label the sections (see Figure 15.4).

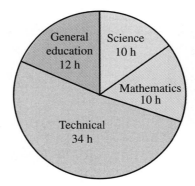

Engineering Technology Semester
Credit-Hour Requirements

FIGURE 15.4

Exercises 15.2

1. Find 26% of 360°.
2. Find 52% of 360°.
3. Find 15.2% of 360°.
4. Find 37.1% of 360°.
5. Find 75% of 360°.
6. Find 47.7% of 360°.

7. Of 744 students, 452 are taking mathematics. What angle of a circle would show the percent of students taking mathematics?

8. Of 2017 students, 189 are taking technical physics. What angle of a circle would show the percent of students taking technical physics?

9. Of 5020 TV sets, 208 are found to be defective. What angle of a circle would show the percent of defective TV sets?

10. Candidate A was one of four candidates in an election. Of 29,106 votes cast, 4060 were for Candidate A. What angle of a circle would show the percent of votes *not* cast for Candidate A?

11. A department spends $16,192 of its $182,100 budget for supplies. What angle of a circle would show the percent of money the department spends on things other than supplies?

12. In one month, the sales of calculators were as follows:

Brand A: 29   Brand D: 75

Brand B: 52   Brand E: 43

Brand C: 15

What central angle of a circle graph would show Brand B's sales as a percent of the total sales for the month?

Draw a circle graph for Exercises 13–20:

13. United States work-related deaths by cause:

Transportation	43%
Assaults	14%
Contact with object or equipment	18%
Falls	13%
Exposure to harmful substances	9%
Fire	3%

14. United States estimated population by age in 2007:

9 and under	13.5%
10–19	13.9%
20–34	20.3%
35–54	28.9%
55–74	17.3%
75 and older	6.1%

15. The suggested semester credit-hour requirements for a community college curriculum in industrial technology is shown in Illustration 1.

Course	Semester hours
Mathematics	6
Applied science	8
Technical specialties	34
General education courses	12
	60

ILLUSTRATION 1

16. A company interviewed its 473 employees to find the toughest day to work of a five-day work week as shown in Illustration 2.

Day	Number
Monday	251
Tuesday	33
Wednesday	57
Thursday	43
Friday	89

ILLUSTRATION 2

17. Total forest by continent in year 2000 in thousands of hectares:

Africa	649,866
Asia	547,793
Europe	1,039,251
North and Central America	549,304
Oceania	197,623
South America	885,618

18. The highest level of education for persons age 25 and older in the United States in 2004 can be found in the data shown in Illustration 3.

8th grade or less	4.3%	Some high school	8.6%
High school graduate	32.0%	Some college	17%
Associate degree	8.4%	Bachelor's degree	18.1%
Master's degree	8.7%	Doctoral degree	1.3%
Professional degree	1.6%		

ILLUSTRATION 3

19. In 2004 the population data of the following regions of the world in millions were as follows: North America, 494; South America, 351; Europe, 726; Asia, 3759; Africa, 827; Oceania, 34.

20. The 2020 projected population data for the regions of the world are as follows in millions: North America, 435.5; South America, 433.1; Africa, 1155.6; Europe, 704.5; Asia, 5875.8; Oceania, 38.3.

15.3 Line Graphs

A **line graph** is a graph formed by segments of straight lines that join the plotted points that represent given data. The line graph is used to show changing conditions, often over a certain time interval.

    Example 1   An industrial technician must keep a chemical at a temperature below 60°F. He must also keep an hourly record of its temperature and record each day's temperatures on a line graph. The following table shows the data he collected.

Time	8:00	9:00	10:00	11:00	12:00	1:00	2:00	3:00	4:00	5:00
Temp. (°F)	60°	58°	54°	51°	52°	57°	54°	52°	54°	58°

When drawing line graphs, (a) use graph paper, because it is already subdivided both vertically and horizontally; (b) choose horizontal and vertical scales so that the line uses up most of the space allowed for the graph; (c) name and label each scale so that all marks on the scale are the same distance apart and show equal intervals; (d) plot the points from the given data; (e) connect each pair of points in order by a straight line. When you have taken all these steps, you will have a line graph (see Figure 15.5).

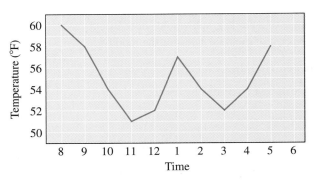

FIGURE 15.5
Line graph

Exercises 15.3

1. The data in Illustration 1 are from the records of the industrial technician in Example 1. These data were recorded on the following day. Draw a line graph for them.

Time	8:00	9:00	10:00	11:00	12:00	1:00	2:00	3:00	4:00	5:00
Temp. (°F)	59°	57°	55°	54°	53°	57°	55°	53°	56°	59°

ILLUSTRATION 1

2. An inspector recorded the number of faulty calculators and the hour in which they passed by his station, as shown in Illustration 2. Draw a line graph for these data.

3. Illustration 3 lists the major league's winning batting average for the years 1997 through 2004. Draw a line graph for these data.

Time	7–8	8–9	9–10	10–11	11–12	1–2	2–3	3–4	4–5	5–6
Number of faulty calculators	1	2	2	3	6	2	4	4	7	10

ILLUSTRATION 2

Year	1997	1998	1999	2000	2001	2002	2003	2004
Batting average	0.372	0.363	0.379	0.372	0.350	0.370	0.359	0.372

ILLUSTRATION 3

A technician is often asked to read graphs drawn by a machine. The machine records measurements by plotting them on a graph. Any field in which quality control or continuous information is needed might use this way of recording measurements. Illustration 4 shows a microbarograph used by the weather service to record atmospheric pressure in inches. For example, the reading on Monday at 8:00 P.M. was 29.34 in.

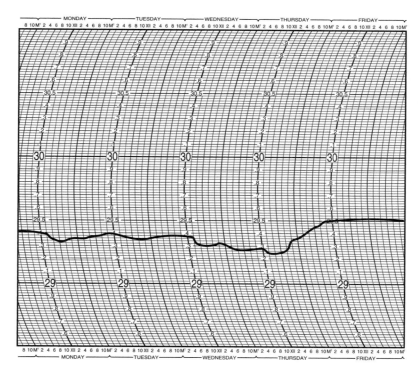

ILLUSTRATION 4
Microbarograph (atmospheric pressure in inches)

Use the microbarograph in Illustration 4 to answer Exercises 4–8.

4. What was the atmospheric pressure recorded for Tuesday at 2:00 P.M.?

5. What was the highest atmospheric pressure recorded? When was it recorded?

6. What was the lowest atmospheric pressure recorded?

7. What was the atmospheric pressure recorded for Thursday at 10:00 P.M.?

8. What was the atmospheric pressure recorded for Monday at noon?

A hygrothermograph is used by the weather services to record temperature and relative humidity (see Illustration 5). The lower part of the graph is used to measure relative humidity from 0% to 100%. The upper part of the graph is used to measure temperature from 10°F to 110°F. For example, at 8:00 P.M., the temperature was 86°F, and the relative humidity was 82%.

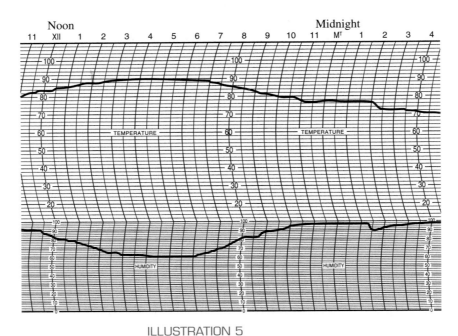

ILLUSTRATION 5

Hygrothermography (temperature in °F and relative humidity in %)

Use the hygrothermograph in Illustration 5 to answer Exercises 9–14.

9. What was the relative humidity at 12:00 midnight?

10. What was the temperature at 3:30 A.M.?

11. What was the highest temperature recorded?

12. What was the lowest temperature recorded?

13. What was the relative humidity at 2:00 A.M.?

14. What was the lowest relative humidity recorded?

15. According to National Oceanic and Atmospheric Administration (NOAA) records, the concentration of CO_2 gas concentrations in the atmosphere measured at the Mauna Loa Observatory in Hawaii has increased steadily for the past 50 years as shown in Illustration 6. **a.** What was the change in average annual concentrations in CO_2 (measured in parts per million by volume, ppmv) from 1960 to 2000? **b.** In what year did the average annual concentrations exceed 360 ppmv?

ILLUSTRATION 6

15.4 Other Graphs

A graph can be a curved line, as shown in Figure 15.6. This graph shows typical power gain for class B push-pull amplifiers with 9-volt power supply.

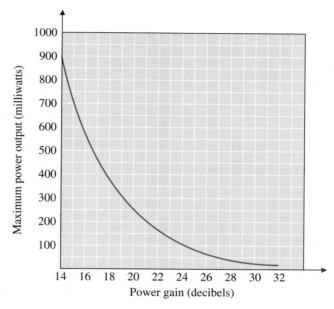

FIGURE 15.6
Curved line graph

Example 1

What is the power output in Figure 15.6 when the gain is 22 decibels (dB)?

Find 22 on the horizontal axis and read up until you meet the graph. Read left to the vertical axis and read 160 milliwatts (mW). ■

One way to avoid having to use a curved line for a graph is to use *semilogarithmic* graph paper. It has a logarithmic (nonuniform) scale for one axis and a uniform scale for the other axis. Figure 15.7 shows the data from Figure 15.6 plotted on semilogarithmic graph paper.

FIGURE 15.7
Semilogarithmic graph

Example 2   Find the power gain in Figure 15.7 when the power output is 60 mW.

Find 60 on the vertical axis. Read across until you meet the graph; read down to the horizontal axis and read 27 dB.   ■

We can see that each type of graph has advantages. The least and greatest changes are easier to read from the curved-line graph, but it is easier to read specific values from the straight-line graph.

Exercises 15.4

Use Figure 15.6 to find the answers for Exercises 1–5:

1. What is the highest power output? What is the power gain at the highest power output?

2. What is the power gain when the power output is 600 mW?

3. What is the power output when the power gain is 25 dB?

4. Between what two decibel readings is the greatest change in power output found?

5. Between what two decibel readings is the least change in power output found?

Use Figure 15.7 to find the answers for Exercises 6–10:

6. What is the highest power output? What is the power gain at the highest power output?

7. What is the power gain when the power output is 600 mW?

8. What is the power output when the power gain is 25 dB?

9. Between what two decibel readings is the greatest change in power output found?

10. Between what two decibel readings is the least change in power output found?

15.5   Mean Measurement

We have already seen in other chapters that with each technical measurement, a certain amount of error is made. One way in which a technician can offset this error is to use what is called the mean of the measurements or the mean measurement. The **mean measurement** (or **mean**) is the average of a set of measurements. To find the mean measurement, the technician takes several measurements. The mean measurement is then found by dividing the sum of these measurements by the number of measurements taken.

$$\text{mean measurement} = \frac{\text{sum of the measurements}}{\text{number of measurements}}$$

Example 1   A machinist measured the thickness of a metal disk with a micrometer at four different places. She found the following values:

2.147 in., 2.143 in., 2.151 in., 2.148 in.

Find the mean measurement.

Step 1 Add the measurements:

2.147 in.

2.143 in.

2.151 in.

<u>2.148 in.</u>

8.589 in.

Step 2 Divide the sum of the measurements by the number of measurements.

$$\text{mean measurement} = \frac{\text{sum of measurements}}{\text{number of measurements}}$$

$$= \frac{8.589 \text{ in.}}{4}$$

$$= 2.14725 \text{ in.}$$

So the mean measurement is 2.147 in.

Note that the mean measurement is written so that it has the same precision as each of the measurements.   ■

Exercises 15.5

Find the mean measurement for each set of measurements:

1. 6091; 505$\overline{0}$; 7102; 4111; 606$\overline{0}$; 591$\overline{0}$; 7112; 5855; 628$\overline{0}$; 10,171; 902$\overline{0}$; 10,172

2. 2.7; 8.1; 9.3; 7.2; 10.6; 11.4; 12.9; 13.5; 16.1; 10.9; 12.7; 15.9; 20.7; 21.9; 30.6; 42.9

3. 205$\overline{0}$; 1951; 2132; 2232; 2147; 1867; 1996; 1785

4. 0.018; 0.115; 0.052; 0.198; 0.222; 0.189; 0.228; 0.346; 0.196; 0.258; 0.337; 0.532

5. 1.005; 1.102; 1.112; 1.058; 1.068; 1.115; 1.213

6. 248; 625; 324; 125; 762; 951; 843; 62; 853; 192; 346; 367; 484; 281; 628; 733; 801; 97; 218

7. 21; 53; 78; 42; 63; 28; 57; 83; 91; 32; 18

8. 0.82; 0.31; 1.63; 0.79; 1.08; 0.78; 1.14; 1.93; 0.068

9. 1.69; 2.38; 4.17; 7.13; 3.68; 2.83; 4.17; 8.29; 4.73; 3.68; 6.18; 1.86; 6.32; 4.17; 2.83; 1.08; 9.62; 7.71

10. 3182; 444$\overline{0}$; 2967; 7632; 1188; 6653; 2161; 8197; 5108; 9668; 5108; 6203; 1988; 4033; 1204; 3206; 4699; 3307; 7226

11. 47.61 cm; 48.23 cm; 47.92 cm; 47.81 cm

12. 9234 m; 9228 m; 9237 m; 9235 m; 9231 m

13. 0.2617 in.; 0.2614 in.; 0.2624 in.; 0.2620 in.; 0.2619 in.; 0.2617 in.

14. 6.643 mm; 6.644 mm; 6.647 mm; 6.645 mm; 6.650 mm

15. The mileage on six vehicles leased for one year was recorded as follows: 25,740 mi, 32,160 mi, 41,005 mi, 21,612 mi, 35,424 mi, 25,810 mi. What is the mean measurement?

16. A trucking company had hauls of 2018 km, 2101 km, 2005 km, 2025 km, 2035 km. What is the mean measurement of the hauls?

17. Over an eight-day period of time the high temperature of each day was recorded in degrees Fahrenheit as follows: 69, 81, 74, 83, 67, 71, 75, 63. What is the mean measurement?

18. A pharmacist weighed ten different capsules of the same compound and recorded these measurements: 3414 mg, 3433 mg, 3431 mg, 3419 mg, 3441 mg, 3417 mg, 3427 mg, 3434 mg, 3435 mg, 3432 mg. What is the mean measurement?

19. A technician measured the power usage of six appliances and recorded the following results: 108 kW, 209 kW, 176 kW, 162 kW, 188 kW, 121 kW. What is the mean measurement?

20. A trucking company had seven items with the following weights: 728 lb, 475 lb, 803 lb, 915 lb, 1002 lb, 256 lb, 781 lb. What is the mean measurement?

21. According to the U.S. Energy Information Administration, the total worldwide coal production (in millions of tons, where 1 U.S. short ton = 2000 lb) by year for the decade ending in 2004 was:

1995	5,096	2000	4,935
1996	5,106	2001	5,233
1997	5,132	2002	5,265
1998	5,046	2003	5,648
1999	4,941	2004	6,079

What was the mean annual worldwide coal production?

 22. As trees grow in diameter, the wood produced in spring and early summer is less dense and lighter in color than the wood produced in late summer and fall. The result is rings of light and dark wood known as annual tree rings. The thickness of the annual ring is determined largely by the weather that the tree experienced during that particular year. Hot, dry years tend to produce thin rings, whereas cool, wet years tend to produce thick rings. In a cross section of the 100+-year-old oak tree in Illustration 1, the annual rings for the 10 years from 1920 to 1929 were measured as follows:

Annual tree ring thickness (in mm)

1920	6	1925	7
1921	8	1926	6
1922	8	1927	6
1923	8	1928	9
1924	8	1929	11

What was the mean growth of the annual rings for the 1920s?

ILLUSTRATION 1
Photo courtesy of Bobbi Townsend, Cornell University

 15.6 **Other Average Measurements and Percentiles**

There are other procedures to determine an average measurement besides finding the mean measurement. The **median measurement** is the measurement that falls in the middle of a group of measurements arranged in order of size. That is, one-half of the measurements are larger than or equal to the median, and one-half of the measurements are less than or equal to the median.

Example 1   Find the median of the following set of measurements.

2.151 mm, 2.148 mm, 2.146 mm, 2.143 mm, 2.149 mm

Step 1   Arrange the measurements in order of size.

2.151 mm

2.149 mm

2.148 mm

2.146 mm

2.143 mm

Step 2 Find the middle measurement.

Since there are five measurements, the third measurement, 2.148 mm, is the median. ■

In Example 1, there was an odd number of measurements. When there is an even number of measurements, there is no one middle measurement. In this case, the median is found by taking the mean of the two middle measurements.

Example 2   Find the median measurement of the following set of measurements.

$$54°, 57°, 59°, 55°, 53°, 57°, 50°, 56°$$

Step 1   Arrange the measurements in order of size.

59°

57°

57°

56°

55°

54°

53°

50°

Step 2   Since there are eight measurements, find the mean of the two middle measurements.

$$\frac{55° + 56°}{2} = \frac{111°}{2} = 55.5°$$

So the median measurement is 55.5°. ■

Another kind of average often used is the mode. The **mode** is the measurement that appears most often in a set of measurements. In Example 2, 57° is the mode, because two of the measurements have this value. However, the mode can present problems. There can be more than one mode, and the mode may or may not be near the middle.

Example 3   Find the mode of the following set of measurements.

3.8 cm, 3.2 cm, 3.7 cm, 3.5 cm, 3.8 cm, 3.9 cm, 3.5 cm, 3.1 cm

The measurements 3.5 cm and 3.8 cm are both modes because each appears most often—twice. ■

Related to averages is the idea of measuring the position of a piece of data relative to the rest of the data. **Percentiles** are numbers that divide a given data set into 100 equal parts. The **nth percentile** is the number P_n such that n percent of the data (ranked from smallest to largest) is at or below P_n. For example, if you score in the 64th percentile on some standardized test, this means that you scored higher than 64% of those who took the test and you scored lower than 36% of those who took the test.

Example 4   The following list gives 50 pieces of ranked data (ranked from smallest to largest).

Ranked data				
16	49	82	121	147
19	50	88	125	148
23	51	89	126	150
27	52	99	129	155
31	57	101	130	156
32	64	103	131	161
32	71	104	138	163
39	72	107	142	169
43	78	118	143	172
47	79	120	145	179

   a. Find the 98th percentile.
   b. Find the 75th percentile.
   c. Find the 26th percentile.

Solution

   a. The 98th percentile is 172 (the 49th piece of data: $0.98 \times 50 = 49$). Ninety-eight percent of the data is at or smaller in value than 172.

   b. The 75th percentile is 142 (the 38th piece of data: $0.75 \times 50 = 37.5$ or 38). Seventy-five percent of the data is at or smaller in value than 142.

   c. The 26th percentile is 51 (the 13th piece of data: $0.26 \times 50 = 13$). Twenty-six percent of the data is at or smaller in value than 51.   ■

Exercises 15.6 _____

1–20. *Find the median measurement for each set of measurements in Exercises 1–20 of Exercises 15.5.*

Find the following percentiles for the data listed in Example 4:

21. 94th percentile

22. 80th percentile

23. 55th percentile

24. 12th percentile

25. 5th percentile

26. 50th percentile

Find the mode for each set of measurements in Exercises 27–36.

27. 2.81 mm, 2.90 mm, 2.78 mm, 2.85 mm, 2.82 mm, 2.85 mm, 2.81 mm, 2.85 mm

28. 105, 110, 211, 313, 415, 475

29. 7291, 7288, 7285, 7287, 7285, 7291

30. 41,265; 42,051; 43,006; 42,051; 41,258; 42,051

31. 28, 30, 41, 30, 19, 25, 28, 19, 28, 42, 36

32. 5129 mi, 7025 mi, 6291 mi, 7025 mi, 8207 mi, 5129 mi, 6292 mi

33. 6.2 in., 3.8 in., 5.2 in., 7.1 in., 6.2 in., 3.9 in.

34. 25,607 km, 26,108 km, 27,203 km, 26,512 km, 27,203 km, 26,607 km

35. 0.018 in., 0.024 in., 0.022 in., 0.019 in., 0.024 in., 0.023 in.

36. 625 lb, 571 lb, 652 lb, 537 lb, 553 lb, 652 lb, 718 lb, 652 lb

37. The cross section of the oak tree mentioned in the previous section showed the following annual ring thicknesses (in mm) from 1940 to 1949:

1940	11	1945	10
1941	9	1946	9
1942	8	1947	8
1943	11	1948	9
1944	7	1949	7

What was **a.** the median annual ring thickness for the 1940s, **b.** the mode, and **c.** the midrange?

38. As shown in Section 15.5, total worldwide coal production (in millions of tons) by year for the decade ending in 2004 was:

1995	5,096	2000	4,935
1996	5,106	2001	5,233
1997	5,132	2002	5,265
1998	5,046	2003	5,648
1999	4,941	2004	6,079

What was the median annual worldwide coal production?

15.7 Range and Standard Deviation

The mean measurement gives the technician the average value of a group of measurements, but the mean does not give any information about how the actual data vary in values. Some type of measurement that gives the amount of variation is often helpful in analyzing a set of data.

One way of describing the variation in the data is to find the range. The **range** is the difference between the highest value and the lowest value in a set of data.

Example 1

Find the range of the following measurements:

54°, 57°, 59°, 55°, 53°, 57°, 50°, 56°

The range is the difference between the highest value, 59°, and the lowest value, 50°. The range is $59° - 50° = 9°$. ■

The range gives us an idea of how much the data are spread out, but another measure, the standard deviation, is often more helpful. The **standard deviation** tells how the data typically vary from the mean. Suppose we are given data sets $A = 4, 5, 5, 6$ and $B = 2, 3, 7, 8$. Both sets have a mean of 5, but the data in set A are "nearer" to the mean than are the data in set B. A mathematical way of describing this is to use standard deviation. There are two types of standard deviation: population standard deviation, denoted by σ, and sample standard deviation, denoted by s. We shall use sample standard deviation in this text; that is,

$$s = \sqrt{\frac{\text{sum of (measurement} - \text{mean)}^2}{\text{number of measurements} - 1}}$$

$$\text{In set } A, \ s = \sqrt{\frac{(4-5)^2 + (5-5)^2 + (5-5)^2 + (6-5)^2}{3}} = \sqrt{\frac{2}{3}} = 0.82$$

$$\text{In set } B, \ s = \sqrt{\frac{(2-5)^2 + (3-5)^2 + (7-5)^2 + (8-5)^2}{3}} = \sqrt{\frac{26}{3}} = 2.9$$

In set A, the data deviate from the mean by 0.82; in set B, the deviation is 2.9.

Example 2   Find the sample standard deviation for the data given in Example 1.

Step 1   Find the mean.

$$\text{mean} = \frac{\text{sum of measurements}}{\text{number of measurements}}$$

$$\text{mean} = \frac{54° + 57° + 59° + 55° + 53° + 57° + 50° + 56°}{8}$$

$$= \frac{441°}{8} = 55.1°$$

Step 2   Find the difference between each piece of data and the mean.

$$54 - 55.1 = -1.1$$
$$57 - 55.1 = 1.9$$
$$59 - 55.1 = 3.9$$
$$55 - 55.1 = -0.1$$
$$53 - 55.1 = -2.1$$
$$57 - 55.1 = 1.9$$
$$50 - 55.1 = -5.1$$
$$56 - 55.1 = 0.9$$

Step 3   Square each difference and find the sum of the squared amounts.

$$(-1.1)^2 = 1.21$$
$$(1.9)^2 = 3.61$$
$$(3.9)^2 = 15.21$$
$$(-0.1)^2 = 0.01$$
$$(-2.1)^2 = 4.41$$
$$(1.9)^2 = 3.61$$
$$(-5.1)^2 = 26.01$$
$$(0.9)^2 = \underline{0.81}$$
$$54.88$$

Step 4 $$s = \sqrt{\frac{54.88}{7}} = 2.80°$$ ■

Example 3   Find the sample standard deviation for the following data:

2015 mi, 1926 mi, 3251 mi, 4007 mi, 1821 mi, 5238 mi, 9111 mi, 7212 mi, 5778 mi, 6661 mi

Step 1   Find the mean.

$$\text{mean} = \frac{2015 \text{ mi} + 1926 \text{ mi} + 3251 \text{ mi} + 4007 \text{ mi} + 1821 \text{ mi} + 5238 \text{ mi} + 9111 \text{ mi} + 7212 \text{ mi} + 5778 \text{ mi} + 6661 \text{ mi}}{10}$$

$$= \frac{47,020}{10} = 4702$$

Step 2 Find the difference between each piece of data and the mean.

$$2015 - 4702 = -2687$$
$$1926 - 4702 = -2776$$
$$3251 - 4702 = -1451$$
$$4007 - 4702 = -695$$
$$1821 - 4702 = -2881$$
$$5238 - 4702 = 536$$
$$9111 - 4702 = 4409$$
$$7212 - 4702 = 2510$$
$$5778 - 4702 = 1076$$
$$6661 - 4702 = 1959$$

Step 3 Square each difference and find the sum of the squared amounts.

$$(-2687)^2 = 7,219,969$$
$$(-2776)^2 = 7,706,176$$
$$(-1451)^2 = 2,105,401$$
$$(-695)^2 = 483,025$$
$$(-2881)^2 = 8,300,161$$
$$(536)^2 = 287,296$$
$$(4409)^2 = 19,439,281$$
$$(2510)^2 = 6,300,100$$
$$(1076)^2 = 1,157,776$$
$$(1959)^2 = \underline{3,837,681}$$
$$56,836,866$$

Step 4

$$s = \sqrt{\frac{\text{sum of (measurement} - \text{mean)}^2}{\text{number of measurements} - 1}}$$
$$= \sqrt{\frac{56,836,866}{10 - 1}} = 2513$$ ■

Exercises 15.7

1–20. Find the range for each set of measurements in Exercises 1–20 in Exercises 15.5.

21–40. Find the sample standard deviation for each set of measurements in Exercises 1–20 in Exercises 15.5.

15.8 Grouped Data

Finding the mean of a large number of measurements can take much time and can be subject to mistakes. Grouping the measurements (the data) can make the work in finding the mean much easier.

Grouped data are data arranged in groups that are determined by setting up intervals. An **interval** contains all data between two given numbers a and b. We will show such an interval here by the form $a-b$. For example, 2–8 means all numbers between 2 and 8.

The number a is called the *lower limit* and b is called the *upper limit* of the interval. The number midway between a and b, $\frac{a+b}{2}$, is called the *midpoint* of the interval. In the above example, the lower limit is 2, the upper limit is 8, and the midpoint is $\frac{2+8}{2} = 5$.

While there are no given rules for choosing these intervals, the following general rules are helpful.

General Rules for Choosing Intervals for Grouped Data

1. The number of intervals chosen should be between 6 and 20.
2. The length of all intervals should be the same and should always be an odd number.
3. The midpoint of each interval should have the same number of digits as each of the measurements that fall within that interval. The lower limit and the upper limit of each interval will have one more digit than the measurements within the interval. In this way, no actual measurement will have exactly the same value as any of these limits. It will therefore be clear to which interval each measurement belongs.
4. The lower limit of the first interval should be lower than the lowest measurement value, and the upper limit of the last interval should be higher than the highest measurement value.

Once the intervals have been chosen, form a frequency distribution. A **frequency distribution** is a list of each interval, its midpoint, and the number of measurements (frequency) that lie in that interval.

◆   Example 1

Make a frequency distribution for the recorded high temperatures for the days from November 1 to January 31 as given in Table 15.1.

First, choose the number and size of the group intervals to be used. We must have enough group intervals to cover the range of the data (the difference between the highest and the lowest values). Here, the range is $55° - 2° = 53°$. Since 53 is close to 54, let us choose the odd number 9 as the interval length. This means that we will need $54 \div 9 = 6$ group intervals. This satisfies our general rule for the number of intervals.

Our first interval is 1.5–10.5 with a midpoint of 6. Here, 1.5 is the lower limit and 10.5 is the upper limit of the interval. We then make the frequency distribution as shown in Table 15.2.

Table 15.1

November	High temperature (°F)	December	High temperature (°F)	January	High temperature (°F)
1	42	1	20	1	29
2	45	2	27	2	29
3	36	3	32	3	30
4	41	4	45	4	26
5	29	5	26	5	2
6	40	6	24	6	45
7	29	7	28	7	41
8	18	8	45	8	12
9	45	9	13	9	31
10	49	10	32	10	26
11	30	11	41	11	25
12	38	12	49	12	15
13	20	13	32	13	52
14	41	14	23	14	42
15	26	15	46	15	22
16	15	16	31	16	30
17	46	17	12	17	19
18	50	18	31	18	19
19	31	19	40	19	19
20	36	20	9	20	55
21	31	21	42	21	23
22	38	22	40	22	17
23	22	23	15	23	26
24	29	24	24	24	12
25	39	25	28	25	16
26	52	26	27	26	21
27	29	27	29	27	39
28	25	28	8	28	20
29	30	29	36	29	23
30	36	30	45	30	22
		31	12	31	9

Table 15.2

Temperature (°F)	Midpoint x	Tally	Frequency f
1.5–10.5	6	////	4
10.5–19.5	15	IHI IHI ////	14
19.5–28.5	24	IHI IHI IHI IHI ///	23
28.5–37.5	33	IHI IHI IHI IHI ///	23
37.5–46.5	42	IHI IHI IHI IHI //	22
46.5–55.5	51	IHI /	6
			92

■

To find the mean from the frequency distribution, (a) multiply the frequency of each interval by the midpoint of that interval, xf; (b) add the products of xf; and (c) divide by the number of data, sum of f.

$$\text{mean} = \frac{\text{sum of } xf}{\text{sum of } f}$$

◆ Example 2   Find the mean of the data given in Example 1.

A frequency distribution table (Table 15.3) gives the information for finding the mean.

Table 15.3

Temperature (°F)	Midpoint x	Frequency f	Product xf
1.5–10.5	6	4	24
10.5–19.5	15	14	210
19.5–28.5	24	23	552
28.5–37.5	33	23	759
37.5–46.5	42	22	924
46.5–55.5	51	6	306
		92	2775

The mean temperature is found as follows.

$$\text{mean} = \frac{\text{sum of } xf}{\text{sum of } f} = \frac{2775}{92} = 30.2°\text{F}$$

■

Note: If the mean of the data in Example 2 were found by summing the actual temperatures and dividing by the number of temperatures, the mean would be 29.86°F, or 29.9°F. There may be a small difference between the two calculated means. This is because we are using the midpoints of the intervals rather than the actual data. However, since the mean is easier to find by this method, the small difference in values is acceptable.

◆ Example 3

Find the mean of the data given in Example 1, this time using an interval length of 5.

The range of the data is 53, which is close to 55, a number that is divisible by 5. Since $55 \div 5 = 11$, we will use 11 intervals, each of length 5. Now make a frequency distribution with 1.5–6.5 as the first interval, using 4 as the first midpoint. The frequency distribution then becomes as shown in Table 15.4.

Table 15.4

Temperature (°F)	Midpoint X	Frequency f	Product xf
1.5–6.5	4	1	4
6.5–11.5	9	3	27
11.5–16.5	14	9	126
16.5–21.5	19	9	171
21.5–26.5	24	15	360
26.5–31.5	29	20	580
31.5–36.5	34	7	238
36.5–41.5	39	11	429
41.5–46.5	44	11	484
46.5–51.5	49	3	147
51.5–56.5	54	3	162
		92	2728

Find the mean temperature:

$$\text{mean} = \frac{\text{sum of } xf}{\text{sum of } f} = \frac{2728}{92} = 29.7°F$$

Exercises 15.8

1. From the following grouped data, find the mean.

Interval	Midpoint X	Frequency f	Product xf
41.5–48.5		12	
48.5–55.5		15	
55.5–62.5		20	
62.5–69.5		25	
69.5–76.5		4	
76.5–83.5		2	

2. Make a frequency distribution of the following scores from a mathematics test and use it to find the mean score.

85, 73, 74, 69, 87, 81, 68, 76, 78, 75, 88, 85, 67, 83, 82, 95, 63, 84, 94, 66, 84, 78, 96, 67, 63, 59, 100, 90, 100, 94, 79, 79, 74

3. A laboratory technician records the life span (in months) of rats treated at birth with a fertility hormone. From the following frequency distribution, find the mean life span.

Life span (months)	Midpoint x	Frequency f	Product xf
−0.5–2.5		12	
2.5–5.5		18	
5.5–8.5		22	
8.5–11.5		30	
11.5–14.5		18	

Cost ($)	Midpoint x	Frequency f	Product xf
−0.5–24.5		2	
24.5–49.5		17	
49.5–74.5		25	
74.5–99.5		51	
99.5–124.5		38	
124.5–149.5		32	

4. The life expectancy of a fluorescent light bulb is given by the number of hours that it will burn. From the following frequency distribution, find the mean life of this type of bulb.

Life of bulb (hours)	Midpoint x	Frequency f	Product xf
−0.5–499.5		2	
499.5–999.5		12	
999.5–1499.5		14	
1499.5–1999.5		17	
1999.5–2499.5		28	
2499.5–2999.5		33	
2999.5–3499.5		14	
3499.5–3999.5		5	

7. The number of passengers and their luggage weight in pounds on Flight 2102 have been tabulated in the following frequency distribution. Find the mean luggage weight.

Weight (lb)	Midpoint x	Frequency f	Product xf
0.5–9.5		1	
9.5–18.5		3	
18.5–27.5		22	
27.5–36.5		37	
36.5–45.5		56	
45.5–54.5		19	
54.5–63.5		17	
63.5–72.5		10	
72.5–81.5		5	
81.5–90.5		2	

5. The shipment times in hours for a load of goods from a factory to market are tabulated in the following frequency distribution. Find the mean shipment time.

Shipment time (hours)	Midpoint x	Frequency f	Product xf
22.5–27.5		2	
27.5–32.5		41	
32.5–37.5		79	
37.5–42.5		28	
42.5–47.5		15	
47.5–52.5		6	

8. The income of the residents in a neighborhood was tabulated. The results are shown in the following frequency distribution. Find the mean income.

Income ($)	Midpoint x	Frequency f	Product xf
2,500–12,500		1	
12,500–22,500		2	
22,500–32,500		15	
32,500–42,500		25	
42,500–52,500		8	

6. The cost of goods stolen from a department store during the month of December has been tabulated by dollar amounts in the following frequency distribution. Find the mean cost of the thefts.

9. The number of defective parts per shipment has been tabulated in the following frequency distribution. Find the mean of defective parts per shipment.

Number of defective parts per shipment	Frequency f
0.5–3.5	1
3.5–6.5	7
6.5–9.5	20
9.5–12.5	9
12.5–15.5	32
15.5–18.5	3

10. The following dollar amounts are traffic fines collected in one day in a city. Make a frequency distribution and use it to find the mean amount of the fines.

$30, $28, $15, $14, $32, $67, $45, $30, $17, $25, $30, $19, $27, $32, $51, $45, $36, $42, $72, $50, $18, $41, $23, $32, $35, $46, $50, $61, $82, $78, $39, $42, $27, $20

11. The length of hospital stays for patients at a local hospital has been tabulated, and the results are shown in the following frequency distribution. Find the mean length for a hospital stay.

Length of stay (days)	Frequency
0.5–1.5	50
1.5–2.5	32
2.5–3.5	18
3.5–4.5	10
4.5–5.5	8
5.5–6.5	5
6.5–7.5	26
7.5–8.5	17
8.5–9.5	22

12. The frequency of repair for the trucks owned by a trucking firm over a five-year period has been tabulated. The results are shown in the following frequency distribu-

tion. Find the mean number of repairs over the five-year period.

Times repaired	Frequency
1.5–2.5	22
2.5–3.5	53
3.5–4.5	71
4.5–5.5	108
5.5–6.5	102
6.5–7.5	120
7.5–8.5	146
8.5–9.5	135
9.5–10.5	98
10.5–11.5	84
11.5–12.5	42
12.5–13.5	12
13.5–14.5	8

13. The scores that golfers shot on 18 holes at a local course were tabulated. The results are shown in the following frequency distribution. Find the mean score.

Score	Frequency
68.5–73.5	5
73.5–78.5	7
78.5–83.5	10
83.5–88.5	12
88.5–93.5	20
93.5–98.5	22
98.5–103.5	25
103.5–108.5	32
108.5–113.5	17
113.5–118.5	12
118.5–123.5	9

14. The corn yield in bushels per acre for a certain hybrid planted by farmers during the year was tabulated in the following frequency distribution. Find the mean yield.

Yield (bu/acre)	Frequency
45.5–54.5	2
54.5–63.5	1
63.5–72.5	3
72.5–81.5	6
81.5–90.5	27
90.5–99.5	43
99.5–108.5	201
108.5–117.5	197
117.5–126.5	483
126.5–135.5	332
135.5–144.5	962
144.5–153.5	481
153.5–162.5	512
162.5–171.5	193
171.5–180.5	185
180.5–189.5	92
189.5–198.5	87
198.5–207.5	53
207.5–216.5	38

15. The following are the squad sizes of the football teams in a regional area. Make a frequency distribution and use it to find the mean.

> 108, 115, 97, 68, 72, 63, 19, 24, 202, 38, 43, 52, 83, 74, 39, 40, 51, 22, 37, 43, 48, 19, 23, 56, 72, 63, 23, 31, 43

16. The number of miles traveled by an experimental tire before it became unfit for use is recorded below. Make a frequency distribution and use it to find the mean.

> 8,457; 22,180; 15,036; 32,168; 9,168; 25,068; 32,192; 38,163; 18,132; 34,186; 36,192; 37,072; 14,183; 42,183; 19,182; 33,337; 38,162; 28,048; 20,208; 34,408; 35,108; 40,002; 29,208; 32,225; 33,207

15.9 Standard Deviation for Grouped Data

For grouped data, the sample standard deviation is found similarly to how the mean is found. A frequency table is used. Columns are inserted to show the difference, D, between the midpoints and the mean ($D = x -$ mean); the square of D, D^2; and the frequency times D^2, D^2f. The following formula gives the sample standard deviation for grouped data.

> **Sample Standard Deviation for Grouped Data**
>
> $$s = \sqrt{\frac{\text{sum of } D^2f}{n - 1}}$$

where n is the number of pieces of data.

Example 1   Given the grouped data in Table 15.5, find **a.** the mean and **b.** the sample standard deviation.

Table 15.5	
Interval (cm)	Frequency f
3.5–12.5	3
12.5–21.5	3
21.5–30.5	7
30.5–39.5	6
39.5–48.5	4
48.5–57.5	1

First, add the x, xf, D, D^2, and D^2f columns in Table 15.5 as shown in Table 15.6 and complete the corresponding entries in the frequency distribution as shown in Table 15.6.

Table 15.6

Interval (cm)	Midpoint x	Frequency f	Product xf	x − mean D	D²	D²f
3.5–12.5	8	3	24	−21	441	1323
12.5–21.5	17	3	51	−12	144	432
21.5–30.5	26	7	182	−3	9	63
30.5–39.5	35	6	210	6	36	216
39.5–48.5	44	4	176	15	225	900
48.5–57.5	53	1	53	24	576	576
		n = 24	696			3510

a. The mean $= \dfrac{\text{sum of } xf}{n} = \dfrac{696}{24} = 29.0$ cm

b. $s = \sqrt{\dfrac{\text{sum of } D^2f}{n-1}} = \sqrt{\dfrac{3510}{24-1}} = 12.4$ cm

So the mean measurement is 29.0 cm, and the data typically tend to vary from the mean by 12.4 cm. ■

Many scientific calculators have statistical functions. These can be used to find the mean (usually denoted by $\bar{x}$) and the sample standard deviation (denoted by s). If you have a calculator with these functions, you should read its manual.

 Example 2   Given the grouped data in Table 15.7, find **a.** the mean and **b.** the sample standard deviation.

Table 15.7

Mass (nearest kg)	Frequency f
1.5–4.5	10
4.5–7.5	15
7.5–10.5	8
10.5–13.5	24
13.5–16.5	125
16.5–19.5	62
19.5–22.5	89
22.5–25.5	51
25.5–28.5	28
28.5–31.5	17

First, add the x, xf, D, D^2, and D^2f columns in Table 15.7 as shown in Table 15.8 and complete the corresponding entries in the frequency distribution as shown in Table 15.8.

Table 15.8

Mass (nearest kg)	Midpoint x	Frequency f	Product xf	x − mean D	D^2	D^2f
1.5–4.5	3	10	30	−15	225	2250
4.5–7.5	6	15	90	−12	144	2160
7.5–10.5	9	8	72	−9	81	648
10.5–13.5	12	24	288	−6	36	864
13.5–16.5	15	125	1875	−3	9	1125
16.5–19.5	18	62	1116	0	0	0
19.5–22.5	21	89	1869	3	9	801
22.5–25.5	24	51	1224	6	36	1836
25.5–28.5	27	28	756	9	81	2268
28.5–31.5	30	17	510	12	144	2448
		429	7830			14,400

a. The mean $x = \dfrac{\text{sum of } xf}{n} = \dfrac{7830}{429} = 18.3$ or 18 kg

b. $s = \sqrt{\dfrac{\text{sum of } D^2f}{n-1}} = \sqrt{\dfrac{14,400}{429-1}} = 5.80$ or 6 kg

So, the mean is 18 kg, and the data typically tend to vary from the mean by 6 kg. ■

Exercises 15.9

1–16. Find the sample standard deviation for each set of measurements in Exercises 1–16 of Exercises 15.8.

17. Measuring the annual tree rings for the 100+-year-old oak tree in Exercise 22 in Section 15.5 produces the following thicknesses (in mm) for 1930 to 1939:

1930	9	1935	8
1931	9	1936	10
1932	8	1937	10
1933	7	1938	11
1934	7	1939	9

a. What is the range of the annual ring thicknesses for the 1930s?

b. What is the standard deviation?

15.10 Statistical Process Control

One of the many uses of statistics is in random sampling of processed goods to improve quality control. **Statistical process control** is a primary analysis tool for quality improvement that helps companies collect, organize, interpret, and track a wide variety of information during production of materials, delivery of services, and monitoring their normal work processes, business-related processes, and customer satisfaction. By watching the production process, technicians can make changes early rather than waiting until a large number of defective goods has been produced.

Control charts are used to help find the information to improve quality control. Different kinds of control charts give different information; three types of control charts are listed below.

Median Chart: easy to use; shows the variation of the process. It is usually used to compare the output of several processes or various stages of the same process.

Individual Reading Chart: used for expensive measurements or when the output at any point in time remains relatively constant. Such a chart does not isolate individual steps of the process, so it can be hard to find out why there is a variation.

Mean Control Chart: shows the sample means plotted over time, to show whether the process is changing and whether it is in control. We will study this type of chart in this section. The chart has a center line at the *target value* of the process mean or at the process mean as determined by the data. Dashed lines represent *control lines,* which are located at the mean plus or minus three times the standard deviation divided by the square root of the number of samples. There are two cases when the process is out of control: (i) when any point falls outside the central limits and (ii) when any run of nine or more consecutive points falls on the same side of the center line. Figure 15.8 shows examples of two processes that are out of control (parts a and b) and one that is in control (part c).

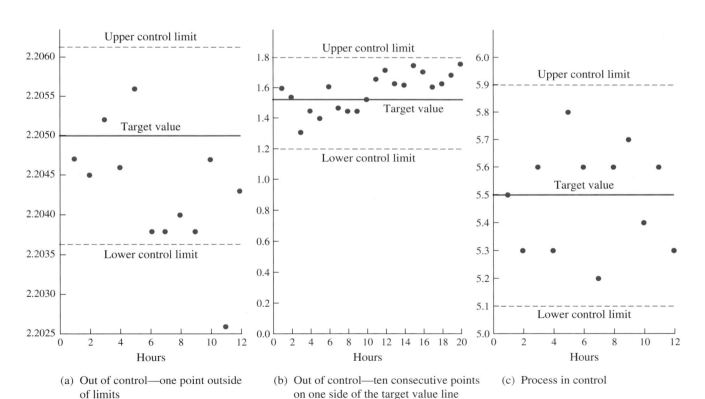

(a) Out of control—one point outside of limits

(b) Out of control—ten consecutive points on one side of the target value line

(c) Process in control

FIGURE 15.8
Three mean control charts

Example 1   A manufacturer of golf balls checks a sample of 100 balls every hour. The compression of the ball has a target value of 90. The standard deviation is ±4.5. Construct a mean control chart using this and the information in the following table.

Hour	1:00	2:00	3:00	4:00	5:00	6:00	7:00	8:00	9:00	10:00	11:00	12:00
Mean	91.1	90.8	90.5	90.2	90.1	89.8	89.5	89.2	88.9	88.6	89.5	89.5

The lower and upper control limits are found by the mean ±3 standard deviations divided by the square root of the number of samples as follows:

$$90 \pm \frac{3(4.5)}{\sqrt{100}} = 90 \pm \frac{1.35}{10} = 90 \pm 1.35 = 91.35 \text{ or } 88.65$$

The mean control chart is shown in Figure 15.9. The process is out of control. One point is beyond the control limits; this happened at the tenth hour. At that time, some source caused the process to go out of control. It is then up to technicians to locate the trouble and fix it.

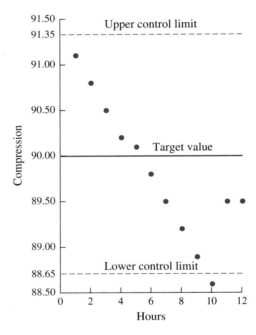

FIGURE 15.9
Mean control chart

Exercises 15.10

1. A certain manufacturing process has a target value of 1.20 cm and a standard deviation of ± 0.15 cm. A sample of nine measurements is made each hour.
   a. Draw a mean control chart using data from Illustration 1.
   b. Is the process out of control? If it is, at what time does the mean control chart signal lack of control?

Hour	1	2	3	4	5	6
Mean	1.30	1.15	1.10	1.25	1.08	1.11
Hour	7	8	9	10	11	12
Mean	1.18	1.15	1.07	1.12	1.08	1.16
Hour	13	14	15	16	17	18
Mean	1.12	1.21	1.25	1.50	1.30	1.26
Hour	19	20	21	22	23	24
Mean	1.29	1.19	1.26	1.31	1.17	1.15

ILLUSTRATION 1

2. The depth of a silicon wafer is targeted at 1.015 mm. If properly functioning, the process produces items with mean 1.015 mm and has a standard deviation of ± 0.004 mm. A sample of 16 items is measured once each hour. The sample means for the past 12 h are given in Illustration 2. From the data, make a mean control chart and determine whether the process is in control.

Hour	1	2	3	4	5	6
Mean	1.016	1.013	1.015	1.017	1.013	1.014
Hour	7	8	9	10	11	12
Mean	1.017	1.016	1.014	1.013	1.016	1.017

ILLUSTRATION 2

3. A sporting goods manufacturer makes baseballs. The target mass of a baseball is 145.5 grams, with a standard deviation of ± 3.5 grams. A technician selects 100 balls at random per hour and records the mean mass of the samples. Illustration 3 lists the mass in grams for a 36-hour period.
   a. Make a mean control chart for the data shown.
   b. Is the process out of control? If it is, at what time does the mean control chart signal lack of control?

Hour	1	2	3	4	5	6
Mass	146.2	145.3	145.2	144.8	146.3	144.6
Hour	7	8	9	10	11	12
Mass	145.0	146.1	144.8	145.1	145.4	143.7
Hour	13	14	15	16	17	18
Mass	145.0	146.3	145.2	145.9	146.0	145.7
Hour	19	20	21	22	23	24
Mass	146.3	144.8	144.9	144.9	145.6	143.8
Hour	25	26	27	28	29	30
Mass	145.7	145.8	144.8	144.9	144.6	145.3
Hour	31	32	33	34	35	36
Mass	146.3	144.0	146.2	145.5	145.4	144.6

ILLUSTRATION 3

4. The illumination of a light bulb is targeted at 1170 lumens. The standard deviation is ± 16.6. A technician randomly selects 15 bulbs per hour and records the mean illumination each hour. Use the data in the following table to make a mean control chart and determine if the process is in control.

Hour	1	2	3	4	5	6
Illumination	1156	1141	1145	1180	1183	1180
Hour	7	8	9	10	11	12
Illumination	1177	1191	1193	1171	1188	1172
Hour	13	14	15	16	17	18
Illumination	1179	1161	1159	1173	1190	1143
Hour	19	20	21	22	23	24
Illumination	1187	1191	1181	1144	1158	1181
Hour	25	26	27	28	29	30
Illumination	1192	1191	1165	1168	1195	1181

ILLUSTRATION 4

5. The target load weight for a ladder is 250 lb. A technician selects 5 ladders at random per day to test load limit and records the mean load for each day. Use the following data to make a mean control chart and determine if the process is in control.

Day	1	2	3	4	5	6
Pounds	238	275	260	258	260	238
Day	7	8	9	10	11	12
Pounds	261	250	270	265	241	248
Day	13	14	15	16	17	18
Pounds	273	265	260	253	240	245

ILLUSTRATION 5

6. A capsule is targeted to contain 50 mg of garlic. A technician selects 25 capsules at random out of a batch of 2000 and records the mean amount of garlic in the samples. Use the following data to make a mean control chart and determine if the process is in control.

Batch	1	2	3	4	5	6
Garlic	53	51	45	47	52	57
Batch	7	8	9	10	11	12
Garlic	42	45	48	53	32	52
Batch	13	14	15	16	17	18
Garlic	57	55	48	51	50	44
Batch	19	20	21	22	23	24
Garlic	47	57	50	47	52	55

ILLUSTRATION 6

15.11   Other Graphs for Statistical Data

In Sections 15.1 through 15.4, we saw how we can view statistical data using various types of graphs. In this section, you will be introduced to three more types.

A **histogram** is a bar graph that reflects the frequency of the number displayed in a frequency distribution.

 Example 1   Use the grouped data in Table 15.6 on page 545 to draw a histogram.

For grouped data, use the intervals for the base and use the frequency for the height. The intervals are 3.5–12.5, 12.5–21.5, 21.5–30.5, 30.5–39.5, 39.5–48.5, and 48.5–57.5. The corresponding frequencies (heights) are 3, 3, 7, 6, 4, and 1. See Figure 15.10.

FIGURE 15.10
Histogram

Example 2

A **run chart** is a line graph in which data are collected over a period of time.

In Table 15.1 on page 539, choose the December readings and create a run chart. Pair each day in December with the high temperature reading. Plot the points in order from left to right. See Figure 15.11.

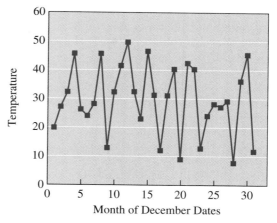

FIGURE 15.11
Run chart

A **scattergram** is a graph of two variables as distinct points that is useful in trying to determine whether a relationship between the two variables can be inferred.

Example 3

Use the information from Table 15.9 showing hours of study for a final exam and the grade received to construct a scattergram. See Figure 15.12.

Table 15.9

Study time	Grade received
3	71
4	90
6	93
5	98
3	70
4	82
3	87
5	93
3	82
4	95
3	82
4	95
2	71
5	89
2	74
3	86
4	83
5	95
2	65
3	82

FIGURE 15.12
Scattergram

The scattergram shows a positive linear correlation, since the grade received increases as the number of study hours increases.   ■

A graph with the points descending, as in Figure 15.13a, shows a negative linear correlation. The points in the graph in Figure 15.13b show no linear correlation.

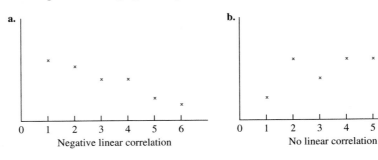

FIGURE 15.13

Exercises 15.11

In Exercises 1–4, draw a histogram for the data displayed in each frequency distribution.

1.

Interval	Frequency f
41.5–48.5	12
48.5–55.5	15
55.5–62.5	20
62.5–69.5	25
69.5–76.5	4
76.5–83.5	2

ILLUSTRATION 1

2. The number of passengers and their luggage weight in pounds on Flight 2102 have been tabulated in the frequency distribution in Illustration 2.

Weight (lb)	Frequency f
0.5–9.5	1
9.5–18.5	3
18.5–27.5	22
27.5–36.5	37
36.5–45.5	56
45.5–54.5	19
54.5–63.5	17
63.5–72.5	10
72.5–81.5	5
81.5–90.5	2

ILLUSTRATION 2

3. The life expectancy of a fluorescent light bulb is given by the number of hours that it will burn, tabulated in the frequency distribution in Illustration 3.

Life of bulb (hours)	Frequency f
−0.5–499.5	2
499.5–999.5	12
999.5–1499.5	14
1499.5–1999.5	17
1999.5–2499.5	28
2499.5–2999.5	33
2999.5–3499.5	14
3499.5–3999.5	5

ILLUSTRATION 3

4. The shipment times in hours for a load of goods from a factory to market are tabulated in the frequency distribution in Illustration 4.

Shipment time (hours)	Frequency f
22.5–27.5	2
27.5–32.5	41
32.5–37.5	79
37.5–42.5	28
42.5–47.5	15
47.5–52.5	6

ILLUSTRATION 4

Use the data displayed in Table 15.1 on page 539 to draw a run chart for the daily high temperature.

5. using the November data.

6. using the January data.

Precision Manufacturing produces 20,000 machine nuts daily. There is an allowance of 1% error. Quality control checks 250 nuts each day for 40 days. The number of machine nuts that are not acceptable each day is given in Illustration 5.

7. Draw a run chart for number of defective parts by week using Illustration 5.

8. Draw a run chart for number of defective parts by day of the week using Illustration 5.

Week	Mon	Tue	Wed	Thu	Fri	Weekly totals
1	18	65	42	22	67	214
2	8	70	35	42	51	206
3	8	40	50	30	20	148
4	8	42	44	68	5	167
5	81	0	53	13	67	214
6	57	70	22	39	40	228
7	45	72	22	62	8	209
8	20	35	5	41	6	107
Daily totals	245	394	273	317	264	1493

ILLUSTRATION 5

In Exercises 9–12,

a. Draw a scattergram for the data given.

b. Does the scattergram have a positive, a negative, or no linear correlation?

9. Mindy's basketball coach kept records on each team member in minutes played and points scored. Mindy's statistics follow: Game 1, 20 minutes, 5 points; Game 2, 12 minutes, 8 points; Game 3, 24 minutes, 13 points; Game 4, 16 minutes, 6 points; Game 5, 8 minutes, 4 points; Game 6, 22 minutes, 14 points; Game 7, 28 minutes, 16 points; Game 8, 30 minutes, 20 points.

10. A poultry-eviscerating line processed the following number of boxes of turkey for each hour of the day.

Hour	7–8	8–9	9–10	10–11	11–12	1–2	2–3	3–4
Boxes of turkey	8	9	9	8	6	8	7	5

11. x is paired with y by the following table:

x	4	8	9	11	13	15
y	2	14	6	18	2	20

12. x is paired with y by the following table:

x	−6	−5	−4	0	2	3	5
y	−3	−3	−2	−2	−1	−1	0

15.12 Normal Distribution

The **normal distribution** of large data sets tends to group data around the mean and/or the median in a way that the result resembles a bell-shaped curve, as shown in Figure 15.14. The shape of the normal distribution curve will depend on the size of the standard deviation. The empirical rule states that approximately 68% of the data will be within one standard deviation of the mean, 95% of the data will fall within two standard deviations, and 99.7% of the data will fall between three standard deviations (see Figure 15.15).

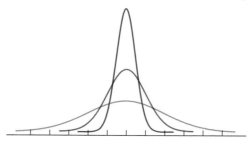

FIGURE 15.14
Different standard deviations

FIGURE 15.15
s represents one standard deviation.

◆ Example 1

Given the data in Table 15.10, determine the mean, median, mode, and standard deviation. Draw the normal distribution curve.

◆ Table 15.10

Data	Frequency	Data	Frequency	Data	Frequency
70	1	74	175	78	30
71	10	75	250	79	10
72	20	76	160	80	2
73	117	77	93		Total 868

The mean is 75, the median is 75, and the mode is 75, and the standard deviation is 1.5. In this example:

> One standard deviation below the mean is $75 - 1.5 = 73.5$, and one standard deviation above the mean is $75 + 1.5 = 76.5$. This includes $175 + 250 + 160 = 585$ of the 868 scores, or 67%.
> Two standard deviations below the mean is $75 - 2(1.5) = 72$, and two standard deviations above the mean is $75 + 2(1.5) = 78$. This includes $20 + 117 + 175 + 250 + 160 + 93 + 30 = 845$ of the 868 scores, or 97%.
> Three standard deviations below the mean is $75 - 3(1.5) = 70.5$, and three standard deviations above the mean is $75 + 3(1.5) = 79.5$. This includes $10 + 20 + 117 + 175 + 250 + 160 + 93 + 30 + 10 = 865$ of the 868 scores, or 99.7%. See Figure 15.16.

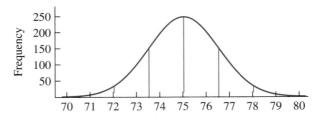

FIGURE 15.16
Normal distribution   ■

◆ Example 2

A clothing manufacturer is going to produce 100,000 women's blouses in a normal distribution of sizes with a mean of 10 and a standard deviation of 2.5. Given that the sizes are 2, 4, 6, 8, 10, 12, 14, 16, and 18, how many of each size should the manufacturer make?

68% of the blouses should be within one standard deviation of the mean 10. One standard deviation below this mean is $10 - 2.5 = 7.5$, and one standard deviation above this mean is $10 + 2.5 = 12.5$, which includes sizes 8, 10, and 12. Thus, he would manufacture 68,000 of the blouses in sizes 8, 10, and 12.

95% of the blouses should be within two standard deviations of the mean 10. Two standard deviations below this mean is $10 - 2(2.5) = 5$, and two standard deviations above it is $10 + 2(2.5) = 15$. This includes all of the blouses in sizes 8, 10, and 12 plus all in sizes 6 and 14. Thus, the manufacturer would produce 95,000 blouses in sizes 6 through 14, including the 68,000 in sizes 8, 10, and 12. $95,000 - 68,000 = 27,000$ blouses in sizes 6 and 14.

99.7% of the blouses should be within three standard deviations of the mean 10. Three standard deviations below this mean is $10 - 3(2.5) = 2.5$, and three standard deviations above it is $10 + 3(2.5) = 17.5$. This includes all blouses in sizes 6, 8, 10, and 12, plus those in sizes 4 and 16. The manufacturer would make 99,700 blouses, which include the 95,000 in the sizes 6, 8, 10, 12, and 14, plus those in sizes 4 and 16. $99,700 - 95,000 = 4,700$ blouses in sizes 4 and 16.

The remaining 300 blouses would be in sizes 2 and 18.   ■

Exercises 15.12 _____

1. Given a normal distribution with a mean of 85 and a sample standard deviation of 15, how much of the data should be in the interval between 55 and 115?

2. Given the 20 numbers 32, 34, 35, 41, 42, 43, 44, 45, 48, 49, 51, 53, 55, 55, 57, 58, 59, 63, 65, and 71, find **a.** the mean, **b.** the median, **c.** the mode, and **d.** the sample standard deviation. **e.** Construct a normal distribution.

*In Exercises 3–6, use the frequency distribution to find **a.** the mean and **b.** the sample standard deviation. **c.** Does the data form a normal distribution? Remember: if the data fall within 2% of the empirical rule of 68%, 95%, and 99.7% for one, two, or three standard deviations, respectively, they form a normal distribution.*

3.

Interval	Midpoint x	Frequency f
41.5–48.5		10
48.5–55.5		50
55.5–62.5		200
62.5–69.5		188
69.5–76.5		40
76.5–83.5		20

4.

Weight (lb)	Midpoint x	Frequency f
0.5–9.5		10
9.5–18.5		40
18.5–27.5		120
27.5–36.5		210
36.5–45.5		360
45.5–54.5		340
54.5–63.5		220
63.5–72.5		100
72.5–81.5		30
81.5–90.5		10

5.

Life of bulb (hours)	Frequency f
0–999.5	105
999.5–1999.5	480
1999.5–2999.5	2050
2999.5–3999.5	4100
3999.5–4999.5	2450
4999.5–5999.5	420
5999.5–6999.5	155

6.

Shipment time (hours)	Frequency f
22.5–27.5	20
27.5–32.5	410
32.5–37.5	790
37.5–42.5	700
42.5–47.5	500
47.5–52.5	40

7. In Example 2, find the number of blouses to be made of each size if the mean were 12 and the sample standard deviation were 2.8.

8. A trouser manufacturer has an order for 80,000 men's trousers in a normal distribution of waist sizes 28, 30, 32, 34, 36, 38, 40, 42, and 44 with a mean of 36 and a sample standard deviation of 2.1. Find the number of trousers to be made of each waist size.

 15.13 Probability

Probability is another useful mathematical tool that a technician can use to make decisions. A **sample space** is the set of all possible outcomes of an event.

 Example 1

Find the sample space of rolling one die.

The die has 6 sides with dots on each of the six sides and this gives the sample space

$$[1, 2, 3, 4, 5, 6]$$ ■

 Example 2

Find the sample space of rolling two dice.

This gives a sample space of the following 36 possible outcomes:

$$[(1, 1), (1, 2), (1, 3), (1, 4), (1, 5), (1, 6), (2, 1), (2, 2), (2, 3), (2, 4), (2, 5), (2, 6),$$
$$(3, 1), (3, 2), (3, 3), (3, 4), (3, 5), (3, 6), (4, 1), (4, 2), (4, 3), (4, 4), (4, 5), (4, 6),$$
$$(5, 1), (5, 2), (5, 3), (5, 4), (5, 5), (5, 6), (6, 1), (6, 2), (6, 3), (6, 4), (6, 5), (6, 6)]$$ ■

 Example 3

Find the sample space when two balls are drawn out of a bag containing 4 balls: 3 red and 1 white.

Two balls drawn at a time will have a sample space of

$$[(red, red), (red, red), (red, red), (red, white), (white, red), (white, red)]$$ ■

The **probability** p is the likelihood that an event will happen and is given by the ratio of number of actual events happening n to the total number of possible events in the sample space s.

$$p = \frac{n}{s}$$

 Example 4

Find the probability that a 3 will be rolled in Example 1.

Here, $n = 1$ and $s = 6$. So

$$p = \frac{n}{s}$$

$$p = \frac{1}{6}$$ ■

Example 5   Find the probability that a total of 3 will be rolled in Example 2.

Here, $n = 2$ [the pairs $(2, 1)$ and $(1, 2)$] and $s = 36$. So

$$p = \frac{n}{s}$$

$$p = \frac{2}{36} = \frac{1}{18}$$ ■

 Example 6   Find the probability in Example 3 that two red balls will be drawn.

Here, $n = 3$ [three pairs of (red, red)] and $s = 6$. Then

$$p = \frac{n}{s}$$

$$p = \frac{3}{6} = \frac{1}{2}$$ ■

All probabilities are between 0 and 1 inclusive. That is, $0 \leq p \leq 1$. The probability of an event that must happen is 1. The probability of an event that is impossible is 0. The sum of all the probabilities in a sample space is 1.

 Example 7   Find the probability of rolling a 7 when rolling one die.

A regular die has a sample space of $[1, 2, 3, 4, 5, 6]$. The probability that a 7 will happen is 0.   ■

 Example 8   Find the probability of a head when a two-headed coin is flipped.

The sample space is $[1]$. The probability that a head will occur when flipped is 1.   ■

Exercises 15.13

In Exercises 1–6, find each sample space:

1. The hearts from a standard deck of 52 cards.

2. The cards from a standard deck of cards that are less than 3.

3. The red face cards from a standard deck of cards.

4. The cards taken two at a time from a standard deck of cards that are less than 3.

5. Marbles taken two at a time from a bag with 2 red marbles and 1 white marble.

6. Pieces of paper taken two at a time with the numbers 1–7 written on them.

7. From the sample space in Exercise 1, what is the probability that an ace will be drawn?

8. From the sample space in Exercise 2, what is the probability that one ace will be drawn?

9. From the sample space in Exercise 3, what is the probability that the queen of hearts will be drawn?

10. From the sample space in Exercise 1, what is the probability that the 7 of spades will be drawn?

11. From the sample space in Exercise 2, what is the probability that the card drawn will be a jack?

12. From the sample space in Exercise 3, what is the probability that the card drawn will be greater than 10?

13. From the sample space in Exercise 4, what is the probability that the sum of two cards drawn will be 4?

14. From the sample space in Exercise 5, what is the probability that two red marbles will be drawn?

15. From the sample space in Exercise 6, what is the probability that the sum of the two numbers drawn will be 7?

16. A bag contains 6 red and 4 white marbles. One marble is drawn. **a.** What is the probability that a white marble will be drawn? **b.** What is the probability that a red marble will be drawn? **c.** What is the sum of the answers in part **a** and part **b**?

17. During a manufacturing process, 52 defective items are found out of 10,000 produced. What is the probability that when an item is selected, it will be defective?

18. In a classroom, there are 18 female students and 7 male students. If a teacher picks a student at random, what is the probability that the student will be a female?

15.14 Independent Events

Events can happen in many different ways. One of these is when the events are independent. Two events are **independent** if the probability of one event does not change the probability of the second event.

Find whether drawing 1 red marble then replacing it and drawing a second red marble from a bag containing 1 red marble and 1 white marble are independent events.

The same number of marbles exists the second time as the first, so the probability is the same and the events are independent.   ■

Find whether rolling a die and getting a 3 and rolling a second die and getting a 5 are independent events.

In each case, the probability $p = \dfrac{1}{6}$ so the events are independent.   ■

Find whether the events of drawing 2 marbles from a bag containing 3 red marbles and 1 white marble taken in order are independent events.

When a marble, either red or white, is removed from the bag, the number of marbles left is different from the first time. The probabilities are different and the events are not independent.   ■

The probability of two independent events occurring in a given order can be found by finding the product of the probabilities of each separate event. That is,

$$p(A \text{ and } B) = p(A) \cdot p(B)$$

Find the probability in Example 1 that both marbles drawn will be red.

$$p(\text{red and red}) = p(\text{red}) \cdot p(\text{red}) = \frac{1}{2} \cdot \frac{1}{2} = \frac{1}{4}$$ ■

Find the probability in Example 2 that a 3 and 5 will be rolled.

$$p(3 \text{ and } 5) = p(3) \cdot p(5) = \frac{1}{6} \cdot \frac{1}{6} = \frac{1}{36}$$ ■

Exercises 15.14

1. A bag contains 1 red marble, 1 blue marble, 1 green marble, and 1 white marble. What is the probability of drawing 1 red marble, replacing it, and then drawing a red marble again?

2. A card is chosen from a deck of 52 cards. It is put back in the deck and a second card is chosen. What is the probability of drawing an ace and a 10?

3. A bag contains 8 green marbles, 3 white marbles, and 5 red marbles. A marble is removed and then placed back in the bag, and a second marble is picked. What is the probability of drawing a green marble and a red marble?

4. A large box of vegetables contains 4 tomatoes, 3 heads of lettuce, and 7 onions. A vegetable is chosen at random and then replaced in the box, and a second vegetable is picked. What is the probability of choosing a tomato and an onion?

5. Six out of ten motorcyclists wear safety helmets in states that do not require them. If two motorcyclists are chosen, what is the probability that both wear a safety helmet?

6. A card is chosen from a deck of cards and placed back in the deck, and a second card is chosen from the deck. What is the probability of drawing the ace of spades and the queen of hearts?

7. A coin is tossed, and a die is thrown. What is the probability of a head and a 5?

8. A card is drawn and replaced four times from a deck of 52 cards. What is the probability of drawing 4 clubs?

9. A card is drawn and replaced four times from a deck of 52 cards. What is the probability of drawing 4 aces?

10. A pair of dice is rolled. What is the probability of rolling 12?

11. Three dice are rolled. What is the probability of rolling three 3's?

12. A card is chosen from a deck of cards and then placed back in the deck, and a second card is chosen. What is the probability of drawing 2 kings?

13. A spinner has numbers 1–7 marked equally on the face. If the spinner is spun 2 times, what is the probability of having a 6 and a 4?

14. A spinner has numbers 1–7 marked equally on the face. If the spinner is spun 3 times, what is the probability of an even and an odd and a 4?

15. A bag of marbles contains 5 yellow marbles, 4 white marbles, 3 blue marbles, and 7 red marbles. A marble is drawn and replaced. What is the probability of having a red marble, a white marble, and a blue marble?

16. A card is drawn from a deck of 52 cards and then replaced. What is the probability of having a heart, the 10 of spades, and a jack?

 Chapter 15 Group Activities

1. Go outside and count the number of cars in a small parking lot. Record the colors of the cars (red, blue, green, etc.) and the makes of the cars (Honda, Toyota, Ford, Chevrolet, etc.). Now in your groups, make a circle graph and a bar graph as in Sections 15.1 and 15.2 to illustrate for the car color and for the make. What is the median car color?

2. Record the height of everyone in your class. Form groups and find the mean height of the entire class. (Mean is discussed in Section 15.5.) Next, find the mean height of only those in your group. Call this height the sample mean. Do these two means differ?

Explain why or why not. Next, gather and record the sample means from all the other groups and find the average of these sample means. What are your findings? Is the average of the sample means the same as the mean of the entire class? This same procedure can be used with sample standard deviation in all calculations.

3. In pairs, flip different coins with one person flipping the coin at least 100 times onto a flat surface and the second person recording "heads" or "tails." On the basis of the data collected, what was the probability of each? Were there any significant differences between different coins? If so, explain.

Next, repeat the same process with thumbtacks, except record point "up" or "down." Before you begin, each person should record his or her own best guess of the probability of "ups" and "downs." On the basis of the data collected, what was the probability of each? Were there any significant differences between different thumbtacks? If so, explain.

As a group, discuss how to display the data in the most logical way so that other groups will easily understand what you did, what you found, and your conclusions. Then, write and submit your written group report.

How would you present the report orally to the class or another group? If time permits, prepare and present your report orally to the class.

Chapter 15 Summary

Glossary of Basic Terms

Bar graph. A graph with parallel bars whose lengths are proportional to the frequency of the given quantities in a data set. (p. 520)

Circle graph. A circle whose radii divide the circle into sectors that are proportional in angle and area relative in size to the quantities represented. (p. 523)

Control charts. Charts used to help find information to improve quality control. (p. 547)

Frequency distribution. A list of each interval, its midpoint, and the number of measurements (frequency) that lie in that interval. (p. 538)

Grouped data. Data arranged in groups that are determined by setting up intervals. (p. 538)

Histogram. A bar graph that reflects the frequency of the number displayed in a frequency distribution. (p. 550)

Independent events. Two events for which the probability of one event does not change the probability of the second event. (p. 558)

Interval. All data between two given numbers such as a and b. The smaller number a is called the *lower limit* of the interval. The larger number b is called the *upper limit* of the interval. The *midpoint* of the interval is the number midway between a and b; that is, $\frac{a + b}{2}$. (p. 538)

Line graph. A graph formed by segments of straight lines that join the plotted points that represent given data. (p. 526)

Mean measurement (or mean). The average of a set of measurements. (p. 530)

Median measurement. The measurement that falls in the middle of a group of measurements arranged in order of size. (p. 532)

Mode. The measurement that appears most often in a set of measurements. (p. 533)

Normal distribution. The distribution of large data sets tends to group around the mean and/or median in a way that resembles a bell-shaped curve. (p. 553)

nth percentile. The number P_n such that n percent of the data (ranked from smallest to largest) is smaller than P_n. (p. 533)

Percentiles. Numbers that divide a given data set into 100 equal parts. (p. 533)

Probability. The likelihood that an event will happen; the ratio of number of actual events happening to the total number of possible events in the sample space. (p. 556)

Range. The difference between the highest value and the lowest value in a set of data. (p. 535)

Run chart. A line graph in which data are collected over a period of time. (p. 551)

Sample space. The set of all possible outcomes of an event. (p. 556)

Scattergram. A graph of two variables as distinct points that is useful in trying to determine whether a relationship between the two variables can be inferred. (p. 551)

Standard deviation. Describes how much the data typically vary from the mean. (p. 535)

Statistical process control. A primary analysis tool for quality improvement that helps companies to collect, organize, interpret, and track a wide variety of information during production of materials, delivery of services, and monitoring their normal work processes, business-related processes, and customer satisfaction. (p. 546)

Statistics. The branch of mathematics that deals with the collection, analysis, interpretation, and presentation of masses of numerical data. (p. 520)

15.5 Mean Measurement

1. **Mean measurement:**

$$\text{mean measurement} = \frac{\text{sum of measurements}}{\text{number of measurements}}.$$

(p. 530)

15.7 Range and Standard Deviation:

1. **Sample standard deviation:**

$$s = \sqrt{\frac{\text{sum of (measurement} - \text{mean})^2}{\text{number of measurements} - 1}}.$$

(p. 535)

15.8 Grouped Data

1. **General rules for choosing intervals for grouped data:**
   a. The number of intervals chosen should be between 6 and 20.
   b. The length of all intervals should be the same and should always be an odd number.
   c. The midpoint of each interval should have the same number of digits as each of the measurements that fall within that interval. The lower limit and the upper limit of each interval will have one more digit than the measurements within the interval. In this way, no actual measurement will have exactly the same value as any of these limits. It will therefore be clear to which interval each measurement belongs.
   d. The lower limit of the first interval should be lower than the lowest measurement value, and the upper limit of the last interval should be higher than the highest measurement value. (p. 536)

2. **To find the mean from a frequency distribution:**
   a. Multiply the frequency of each interval by the midpoint of that interval, xf.
   b. Add the products of xf.
   c. Divide by the number of data, sum of f. That is,

$$\text{mean} = \frac{\text{sum of } xf}{\text{sum of } f}. \text{ (p. 540)}$$

15.9 Standard Deviation for Grouped Data

1. **Sample standard deviation for grouped data:**

$$s = \sqrt{\frac{\text{sum of } D^2 f}{n - 1}}. \qquad \text{(p. 544)}$$

15.10 Statistical Process Control

1. *Control charts* are used to help find the information to improve quality control. Review the following three types of control charts discussed in this section: median charts, individual reading chart, and mean control chart. (p. 547)

15.12 Normal Distribution

1. The *normal distribution* of large data sets tends to group data around the mean and/or median in a way that resembles a bell-shaped curve. The empirical rule states that approximately 68% of the data will be within one standard deviation of the mean, 95% of the data will be within two standard deviations of the mean, and 99.7% of the data will be within three standard deviations of the mean. (p. 553)

15.13 Probability

1. The *probability p* is the likelihood that an event will happen and is given by the ratio of number of actual events happening n to the total number of possible events in the sample space s, or

$$p = \frac{n}{s}. \text{ (p. 556)}$$

15.14 Independent Events

1. The *probability of two independent events* is the product of the probabilities of each separate event; that is, $p(A \text{ and } B) = p(A) \cdot p(B)$. (p. 558)

Chapter 15 Review

1. Find 35% of 360°.

2. Find 56.1% of 360°.

3. Draw a circle graph using the following data. In 2000, 32,870,000 students attended primary schools; 15,647,000 attended grades 9 to 12; and 12,807,000 attended college.

4. Draw a line graph using the data in Exercise 3.

5. In Illustration 5 on page 528, what was the temperature at 10:00 P.M.?

For Exercises 6–8, use the following data.

A technician, using a very precise tool, measured a piece of metal to be used in a satellite. He recorded the following measurements: 7.0036 mm; 7.0035 mm; 7.0038 mm; 7.0035 mm; 7.0036 mm.

6. What was the mean measurement?

7. What was the median?

8. What was the sample standard deviation?

9. Given the frequency distribution in Illustration 1, find **a.** the mean and **b.** the sample standard deviation.

Interval	Frequency f
10.5–21.5	4
21.5–32.5	17
32.5–43.5	10
43.5–54.5	28
54.5–65.5	13
65.5–76.5	12
76.5–87.5	9

ILLUSTRATION 1

For Exercises 10–12, use the data below:

A student's test and quiz scores for a quarter were recorded as follows: 72, 83, 79, 85, 91, 93, 80, 95, 82.

10. What is the mean of the scores recorded?

11. What is the median?

12. What is the sample standard deviation?

For Exercises 13 and 14, use the frequency distribution shown in Illustration 2.

13. Find the mean of the data.

14. Find the sample standard deviation.

Interval	Midpoint	Frequency
6.5–9.5	8	3
9.5–12.5	11	10
12.5–15.5	14	4
15.5–18.5	17	9
18.5–21.5	20	15
21.5–24.5	23	28
24.5–27.5	26	3
27.5–30.5	29	2

ILLUSTRATION 2

15. A spinner with numbers 1–5 equally spaced on the face is spun.
    a. What is the sample space?
    b. What is the probability that an odd number will be spun?

16. A bag contains 4 white marbles, 3 red marbles, 1 green marble, and 10 black marbles. A marble is drawn and replaced in the bag each time. What is the probability of drawing a red marble and a red marble and a black marble?

Chapter 15 Test

1. See Illustration 1. What country has the longest life expectancy?

2. See Illustration 1. What country has a life expectancy of 48 years?

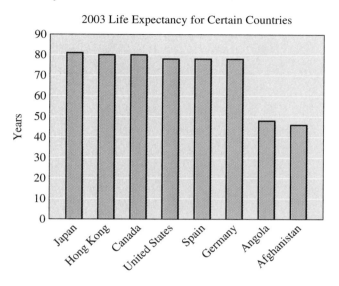

2003 Life Expectancy for Certain Countries

ILLUSTRATION 1

3. Find 38% of 360°.

4. Draw a circle graph using the following data on cargo traffic between cities in the United States in billions of ton-miles: rail, 975; road, 866; air, 8.7; inland water, 435; pipeline, 587.

5. Draw a line graph using the following decade data on population (in thousands) of the United States: 1940 – 132,594; 1950 – 152,271; 1960 – 180,671; 1970 – 204,879; 1980 – 227,757; 1990 – 249,246; 2000 – 281,422.

6. Using the data in Exercise 5, find the population of the United States in 1975.

7. See Illustration 2. What is the power output when the power gain is 28 dB?

8. See Illustration 2. What is the power gain when the power output is 250 mW?

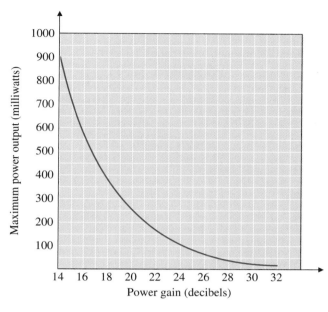

ILLUSTRATION 2

9. Draw a histogram using the data in the frequency distribution in Illustration 3.

Interval	Frequency
6.5–9.5	3
9.5–12.5	6
12.5–15.5	12
15.5–18.5	20
18.5–21.5	14
21.5–24.5	8
24.5–27.5	2

ILLUSTRATION 3

10. x is paired with y by the following table. **a.** Draw a scattergram using this data. **b.** Is a positive, a negative, or no linear correlation shown?

x	0	1	2	3	4	5
y	4	2	6	3	8	4

11. What is the sample space of cards lower than 4 in a deck of 52 playing cards?

For Exercises 12–14, use the data in Illustration 4.

12. Write a frequency distribution.

13. Find the mean using grouped data.

14. Find the sample standard deviation.

> 6.0, 3.1, 0.6, 1.8, 2.1, 1.5, 4.1, 3.7, 3.3, 3.5, 2.5, 5.2, 2.5,
> 1.1, 3.2, 3.7, 2.7, 1.7, 4.4, 4.6, 4.0, 3.9, 2.9, 2.0, 1.9, 5.9,
> 2.4, 3.5, 0.9, 2.4, 0.6, 3.4, 0.5, 3.0, 3.0, 3.9, 3.3, 1.1, 3.2,
> 3.3, 2.5, 3.0, 3.7, 3.5, 4.2, 3.5, 1.6, 5.6, 5.2, 3.0, 3.5, 2.0,
> 2.6, 3.4, 3.3, 3.0, 3.0, 1.4

ILLUSTRATION 4

15. A card is chosen at random from a deck of 52 playing cards. What is the probability that it is a spade?

16. A coin is tossed, and a die is rolled. What is the probability of a head and an even number?

16 Binary and Hexadecimal Numbers

Mathematics at Work

Telecommunications technicians install, troubleshoot, and maintain the connections used on copper and fiber optic communications cables as well as cell phone and wireless communications. This exciting career offers excellent opportunities. The telecommunications technician needs a solid background in electronics, telecommunications, and networking to employ hands-on troubleshooting and problem-solving skills in an office or field working environment. An understanding of business telephone systems, transmission line and multiplexing systems, structured cabling, and fiber optics is needed. Communications and computer information systems skills are also critical. For more information, go to the website listed below.

© Bill Lyons/Alamy

Telecommunications Technician
Technician servicing switching equipment.

www.cengage.com/mathematics/ewen

565

Objectives

■ Change a binary number to decimal form.

■ Add, subtract, and multiply binary numbers.

■ Change a decimal number to binary form.

■ Change a hexadecimal number to decimal form.

■ Change a decimal number to hexadecimal form.

■ Add and subtract hexadecimal numbers.

■ Change a binary number to hexadecimal form.

■ Change a hexadecimal number to binary form.

16.1 Introduction to Binary Numbers

The decimal system of numbers has ten symbols, or digits: 0, 1, 2, 3, 4, 5, 6, 7, 8, and 9. This system of numeration is based on the historical and natural way man has used his fingers to count objects and maintain a record of his possessions by groups of tens.

Computers use a **binary number system**, which has only two symbols or numerals: 0 and 1. These can represent the two positions in a transistor, "off" and "on." Off is assigned 0, and on is assigned 1.

Table 16.1 should help you to understand the relationship between these two systems by comparing place values in the decimal system with place values in the binary system.

Table 16.1

Decimal System						
Millions	Hundred thousands	Ten thousands	Thousands	Hundreds	Tens	Ones
10^6	10^5	10^4	10^3	10^2	10^1	10^0
$10 \times 10 \times 10 \times 10 \times 10 \times 10$	$10 \times 10 \times 10 \times 10 \times 10$	$10 \times 10 \times 10 \times 10$	$10 \times 10 \times 10$	10×10	10	1
1,000,000	100,000	10,000	1,000	100	10	1

Binary System						
Sixty-fours	Thirty-twos	Sixteens	Eights	Fours	Twos	Ones
2^6	2^5	2^4	2^3	2^2	2^1	2^0
$2 \times 2 \times 2 \times 2 \times 2 \times 2$	$2 \times 2 \times 2 \times 2 \times 2$	$2 \times 2 \times 2 \times 2$	$2 \times 2 \times 2$	2×2	2	1
64	32	16	8	4	2	1
1000000_2	100000_2	10000_2	1000_2	100_2	10_2	1_2

Note: In base ten, there are powers of ten and ten numerals.
In base two, there are powers of two and two numerals.

The binary equivalents of the decimal numbers 0–17 are given in Table 16.2.

Table 16.2

Decimal form	Binary form	Decimal form	Binary form
0	0	9	1001
1	1	10	1010
2	10	11	1011
3	11	12	1100
4	100	13	1101
5	101	14	1110
6	110	15	1111
7	111	16	10000
8	1000	17	10001

Some notation is needed to distinguish between decimal and binary numbers. The notation generally used is a subscript after the number to indicate the base of the system in which the number is written. For instance, 1101_{10} is a decimal number, and 1101_2 is a binary number.

Example 1

1101_{10} is a decimal number and means:

1 thousand + 1 hundred + 0 tens + 1 one
$$= 1 \times 10^3 + 1 \times 10^2 + 0 \times 10^1 + 1 \times 10^0$$
$$= 1000 + 100 + 0 + 1$$
$$= 1101_{10}$$

■

Example 2

1101_2 is a binary number and means:

1 eight + 1 four + 0 twos + 1 one
$$= 1 \times 2^3 + 1 \times 2^2 + 0 \times 2^1 + 1 \times 2^0$$
$$= 8 + 4 + 0 + 1$$
$$= 13_{10}$$

■

Example 3

What is 1100111_2 in base 10?
$$1100111_2 = 1 \times 2^6 + 1 \times 2^5 + 0 \times 2^4 + 0 \times 2^3 + 1 \times 2^2 + 1 \times 2^1 + 1 \times 2^0$$
$$= 64 + 32 + 0 + 0 + 4 + 2 + 1$$
$$= 103_{10}$$

■

Exercises 16.1

Change each binary number to decimal form:

1. 11	**2.** 101	**3.** 110	**10.** 11111
4. 1100	**5.** 1001	**6.** 11101	**11.** 1001110
7. 110011	**8.** 10001	**9.** 101111	**12.** 10010010

10. 11111	**11.** 1001110	**12.** 10010010
13. 111011	**14.** 1000001	**15.** 10011100
16. 1110001	**17.** 10001100	**18.** 1100111
19. 111111	**20.** 11100011	

16.2 Addition of Binary Numbers

Addition of binary numbers is relatively easy because only two numerals are used. The addition facts for binary addition are as follows:

$$
\begin{array}{cccc}
0 & 0 & 1 & 1 \\
\underline{+0} & \underline{+1} & \underline{+0} & \underline{+1} \\
0 & 1 & 1 & 10
\end{array}
$$

In binary numbers, $1 + 1 = 10$ is read, "One plus one equals *one-zero*."

 Example 1

Add:
$$
\begin{array}{r}
1\,1 \\
101 \\
\underline{11} \\
1000
\end{array}
$$

For convenience, write the binary number to be carried at the top of the next column to the left.

Check the result by decimal addition:

$$
\begin{array}{r}
101 = 5 \\
\underline{11 = 3} \\
1000 = 8
\end{array}
$$

■

 Example 2

Add. Check by decimal addition:
$$
\begin{array}{r}
1\,1 \\
10010 = 18 \\
\underline{111 = \ 7} \\
11001 = 25
\end{array}
$$

■

 Example 3

Add. Check by decimal addition:
$$
\begin{array}{r}
111001 \\
\underline{11011} \\
1010100
\end{array}
$$

$$
\begin{array}{r}
11\ \ 11 \\
111001 = 57 \\
\underline{11011 = 27} \\
1010100 = 84
\end{array}
$$

Note: In the second column from the left, $1 + 1 + 1 = 10 + 1 = 11$.

■

 Example 4

Add. Check by decimal addition:

$$
\begin{array}{r}
1111 \\
11101 = 29 \\
\underline{1111 = 15} \\
101100 = 44
\end{array}
$$

■

Example 5 Add. Check by decimal addition:

$$
\begin{array}{r}
101 \\
1010 = 10 \\
11 = 3 \\
101 = 5 \\
\underline{1011 = 11} \\
11101 = 29
\end{array}
$$

Note: In the second column from the right, $1 + 1 + 1 + 1 = 10 + 1 + 1 = 11 + 1 = 100$. Here, you must carry 10. Write 10 at the top of the next columns so that the 0 is above the next column to the left and the 1 is above the second column to the left. ■

Exercises 16.2

Add the following binary numbers and check your result by decimal addition:

1. 110
  10

2. 101
  101

3. 111
  100

4. 110
  11

5. 101
  111

6. 1011
  101

7. 1001
  111

8. 11010
  111

9. 10101
  1100

10. 11100
  111

11. 11011
  1001

12. 111010
  1101

13. 101110
  11001

14. 101001
  11111

15. 101010
  11011

16. 1110111
  111001

17. 1011001
  11100

18. 1000111
  101011

19. 11001110
  1011001

20. 11101011
  1100111

21. 10001
  10101

22. 10001
  11001

23. 110110
  11011

24. 111111
  11111

25. 100101
  11011

26. 1011
  101
  1001

27. 11101
  1001
  11101

28. 11010
  10101
  11100
  1101

29. 10001
  1011
  11010
  1001

30. 1111
  111
  11
  1

 16.3 # Subtraction of Binary Numbers

Two methods for subtraction can be used. The first method is to use the following subtraction facts: $0 - 0 = 0$, $1 - 1 = 0$, $1 - 0 = 1$ and $0 - 1$ is found by "borrowing" and having $10 - 1 = 1$.

 Example 1 Subtract:
$$
\begin{array}{r}
0\,1 \\
1\cancel{1}011 \\
-110 \\
\hline
10101
\end{array}
$$
■

Example 2

Subtract:
$$\begin{array}{r} 1 \\ 01\ \cancel{1}1 \\ \cancel{1}0\cancel{1}00 \\ -\,1001 \\ \hline 1011 \end{array}$$

■

The second method for subtraction of binary numbers is using the 1's complement. $a - b$ can be written as $a + (-b)$. In binary form, the negative is called the complement, and we will use the 1's complement. To find the 1's complement of a binary number, reverse each digit of the number being subtracted. For example, the 1's complement of 11100101 is 00011010. Subtraction can then be done as addition.

The following summarizes how to subtract binary numbers using the 1's complement. Given $a - b$, find the 1's complement of b by reversing all the digits of b to form $-b$. Find $a + (-b)$. If the sum has more than n digits, add the extra digit to the result. If the sum does not have more than n digits, leave the result. However, if the result starts with a 1, then the result is a negative number and the complement is used.

Example 3

Find $100011 - 1111$.

$$\begin{array}{cccc} 100011 - 1111 = & 100011 = & 100011 = & 100011 \\ & -1111 & -001111 & +110000 \\ \hline & & & 1010011 \end{array}$$

Since there are more than six digits in the result, you must add 1. The final answer is $10011 + 1$ or 10100. ■

Example 4

Find $10 - 1000$.

$$\begin{array}{ccc} 10 = & 0010 = & 0010 \\ -1000 & -1000 & +0111 \\ \hline & & 1001 \end{array}$$

The result does not have more than four digits, but it does start with 1, which means that the answer is negative and can be found by the complement of the result. The answer is -0110 or -110. ■

Exercises 16.3

Subtract the following binary numbers and check in the binary system:

1. 110
       10

2. 111
       101

3. 1011
       101

4. 110
       11

5. 1001
       111

6. 11001
       10101

7. 100
       11

8. 1000
       101

9. 100101
       11011

10. 11100
       111

11. 110001
       10101

12. 10100
       1010

13. 11000
       101

14. 1101101
       1010101

15. 10001
       1111

16. 10100
       1001

17. 10010
       1001

18. 100010
       1101

19. 10100100
       1011101

20. 10100000
       1101001

Use the 1's complement method to subtract the following binary numbers.

21. 11
       01

22. 101
       100

23. 01
       11

24. 100
    101

25. 1011
    1001

26. 1110
    1010

33. 1011101
    10001011

34. 1001011
    1110111

27. 1000
    1011

28. 10000
    1111

29. 101101
    1001010

35. 111110001100
    111000001010

36. 100010011001
    100001011110

30. 10110011
    1001010

31. 11110001
    110111

32. 10011
    1000100

16.4 Multiplication of Binary Numbers

The multiplication facts for binary numbers are

$$0 \times 0 = 0$$
$$1 \times 0 = 0$$
$$0 \times 1 = 0$$
$$1 \times 1 = 1$$

Since these facts are the same as for decimal numbers, binary multiplication is very simple to perform.

 Example 1

Multiply: 110
              × 11
              110
            110
            10010

Note: The positioning of the binary numbers is the same as in multiplication of decimal numbers.

 Example 2

Multiply: 1011 → 11
          101 → 5
          1011   55
          0000
          1011
          110111 → 55

Exercises 16.4

Multiply the following binary numbers:

1. 101
    11

2. 101
    10

3. 110
    10

4. 111
    11

5. 101
    101

6. 100
    110

7. 110
    110

8. 110
    101

9. 1011
    11

10. 1101
    11

11. 1100
    110

12. 1001
    101

13. 1110
    110

14. 10110
    1101

15. 111001
    1011

16. 11111
    1111

17. 1110
    1001

18. 10001
    1100

19. 110101
    101

20. 1011010
    101110

16.5 Conversion from Decimal to Binary System

We stated earlier that the computer performs computations in the binary system. Does this mean that you must change all decimal numbers to binary numbers before you can use the computer? The answer is no. Most computers have converters that automatically change numbers in the decimal system to the binary system.

How does a computer perform a simple computation? Consider multiplying 5×7. Figure 16.1 traces the flow of the computation through the computer.

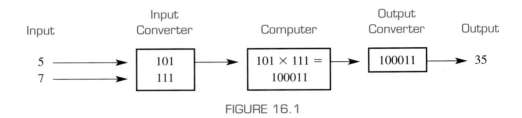

FIGURE 16.1

Notice that the input converter changes a number from decimal form to binary form; the computer performs the computations; and the output converter changes a number from binary form back to decimal form.

The work done by the input converter is a technique that the technician must understand. The process of conversion from decimal to binary form is not difficult. Read the following procedure carefully.

Changing a Number from Decimal Form to Binary Form

1. Write the given number in decimal form.
2. Divide it by 2.
3. Write the quotient below the decimal and write the remainder at the right, even if it is 0.
4. Divide the first quotient by 2 and write the remainder at the right; repeat the process until the final quotient is 0.
5. Obtain the binary form of the decimal number by using the remainders from each step in order from bottom to top.

Example 1   Write 13 in binary form.

Thus, $13_{10} = 1101_2$.

■

Example 2   Write 84 in binary form.

$$
\begin{array}{c|c}
2\underline{|84} & \text{remainders} \\
2\underline{|42} & 0 \\
2\underline{|21} & 0 \\
2\underline{|10} & 1 \\
2\underline{|\ 5} & 0 \\
2\underline{|\ 2} & 1 \\
2\underline{|\ 1} & 0 \\
0 & 1 \quad \text{Read digits up!}
\end{array}
$$

Thus, $84_{10} = 1010100_2$.   ■

Exercises 16.5 _____

Write each decimal number in binary form:

				9. 24	**10.** 56	**11.** 32	**12.** 80
1. 14	**2.** 39	**3.** 63	**4.** 17	**13.** 37	**14.** 85	**15.** 100	**16.** 162
5. 72	**6.** 40	**7.** 20	**8.** 47	**17.** 111	**18.** 128	**19.** 113	**20.** 170

 ## 16.6 Conversion from Binary to Decimal System

Remember that in decimal numbers the powers of ten and of $\frac{1}{10}$ are determined by the places of the digits in the number.

$$
10^6 \quad 10^5 \quad 10^4 \quad 10^3 \quad 10^2 \quad 10^1 \quad 10^0 \quad . \quad \frac{1}{10^1} \quad \frac{1}{10^2} \quad \frac{1}{10^3} \quad \frac{1}{10^4} \quad \frac{1}{10^5} \quad \frac{1}{10^6}
$$

$$
\underset{\substack{\uparrow \\ \text{decimal} \\ \text{point}}}{}
$$

Similarly, in the binary system, the powers of 2 and of $\frac{1}{2}$ are determined by the places of the digits in the number.

$$
2^6 \quad 2^5 \quad 2^4 \quad 2^3 \quad 2^2 \quad 2^1 \quad 2^0 \quad . \quad \frac{1}{2^1} \quad \frac{1}{2^2} \quad \frac{1}{2^3} \quad \frac{1}{2^4} \quad \frac{1}{2^5} \quad \frac{1}{2^6}
$$

$$
\underset{\substack{\uparrow \\ \text{binary} \\ \text{point}}}{}
$$

Note: $2^0 = 1$; remember, also, that the names of the places to the left of the binary point are names of powers of 2 and to the right of the binary point are names of powers of $\frac{1}{2}$.

Read binary numbers by naming the digits in the order they occur from left to right and read the binary point as "point." For example,

101.01 is read, "one-zero-one-point-zero-one."
110.101 is read, "one-one-zero-point-one-zero-one."

> ### Changing a Number from Binary Form to Decimal Form
>
> Find the place value of each digit in the given binary number and multiply it by its corresponding power of 2 or $\frac{1}{2}$.

Example 1

Write 1010.11_2 in decimal form.

$$1010.11 = 1 \times 2^3 + 0 \times 2^2 + 1 \times 2^1 + 0 \times 2^0 + 1 \times \frac{1}{2} + 1 \times \frac{1}{4}$$

$$= 8 + 0 + 2 + 0 + \frac{1}{2} + \frac{1}{4}$$

$$= 10\frac{3}{4} = 10.75 \qquad ■$$

Example 2

Write 11011.001_2 in decimal form.

$$11011.001 = 1 \times 2^4 + 1 \times 2^3 + 0 \times 2^2 + 1 \times 2^1 + 1 \times 2^0 + 0 \times \frac{1}{2} + 0 \times \frac{1}{4} + 1 \times \frac{1}{8}$$

$$= 16 + 8 + 0 + 2 + 1 + 0 + 0 + \frac{1}{8}$$

$$= 27\frac{1}{8} = 27.125 \qquad ■$$

Exercises 16.6

Change each binary number to decimal form.

1. 10.1	**2.** 1.1	**3.** 10.01
4. 11.11	**5.** 101.11	**6.** 110.11
7. 100.001	**8.** 110.101	**9.** 100.101

10. 1101.1	**11.** 1100.11	**12.** 1001.101
13. 111.111	**14.** 1100.1011	**15.** 11010.1001
16. 101.1	**17.** 100.1101	**18.** 1.11111
19. 11.1101	**20.** 11010.1101	

16.7 Hexadecimal System

The use of binary numbers when working with large decimal numbers is very cumbersome. The use of hexadecimal numbers simplifies this problem. Hexadecimal numbers are those numbers with base 16 as compared with decimal (base 10) and binary (base 2). The primary use of a hexadecimal number system is as a more compact, friendly representation of binary coded values often used in digital electronics and computer programming.

55 in decimal form is $5 \, (10^1) + 5 \, (10^0) = 55$
55 in binary form is $110111_2 = 1 \, (2^5) + \cdots + 1 \, (2^0)$
55 in hexadecimal form is $37_{16} = 3 \, (16^1) + 7 \, (16^0)$

The binary system uses 2 numerals 0 and 1. The decimal system uses 10 numerals 0, 1, 2, 3, 4, 5, 6, 7, 8, 9. The **hexadecimal number system** uses 16 numerals. We will use the numerals 0, 1, 2, 3, 4, 5, 6, 7, 8, 9, and the following:

$$A = 10 \qquad B = 11 \qquad C = 12 \qquad D = 13 \qquad E = 14 \qquad F = 15$$

The hexadecimal number 2A is written in decimal form as

$$2(16^1) + A(16^0) = 2(16^1) + 10(16^0)$$
$$= \quad 32 \quad + \quad 10$$
$$= 42$$

Changing a Number from Hexadecimal Form to Decimal Form

Find the place value of each digit of the given binary number and multiply it by its corresponding power of 16.

Example 1

Write the hexadecimal number DB2 in decimal form.

$$DB2 = D(16^2) + B(16^1) + 2(16^0)$$
$$= 13(16^2) + 11(16^1) + 2(16^0)$$
$$= 13(256) + 11(16) + 2(1)$$
$$= 3328 \quad + 176 \quad + 2$$
$$= 3506$$

■

Example 2

Write the hexadecimal number 5DF2 in decimal form.

$$5DF2 = 5(16^3) + D(16^2) + F(16^1) + 2(16^0)$$
$$= 5(16^3) + 13(16^2) + 15(16^1) + 2(16^0)$$
$$= 5(4096) + 13(256) + 15(16) + 2(1)$$
$$= 20{,}480 + 3{,}328 + 240 + 2$$
$$= 24{,}050$$

■

To write a decimal number in hexadecimal form, you can use the same basic method as writing a decimal number in binary form. In the binary form, you used repeated divisions by 2 and used the remainders. In the hexadecimal form, you use repeated divisions by 16 and use the remainders keeping in mind that you now get remainders 0 through 15. You use the hexadecimal letter representations for 10 through 15.

Changing a Number from Decimal Form to Hexadecimal Form

1. Write the given number in decimal form.
2. Divide it by 16.
3. Write the quotient below the decimal and write the remainder at the right, even if it is 0.
4. Divide the first quotient by 16 and write the remainder at the right; repeat the process until the final quotient is 0.
5. Obtain the hexadecimal form of the decimal number by using the remainders from each step in order from bottom to top.

◆   Example 3   Change 258 to hexadecimal form.

$$\begin{array}{r|r} 16 & 258 \\ \hline 16 & 16 \\ \hline 16 & 1 \\ \hline & 0 \end{array} \quad \begin{array}{c} \text{remainders} \\ 2 \\ 0 \\ 1 \end{array}$$

Read digits up.

Thus, $258_{10} = 102_{16}$.   ■

◆   Example 4   Write 3527 in hexadecimal form.

$$\begin{array}{r|r} 16 & 3527 \\ \hline 16 & 220 \\ \hline 16 & 13 \\ \hline & 0 \end{array} \quad \begin{array}{cc} \text{remainders} \\ 7 & \\ 12 & C \\ 13 & D \end{array}$$

Read digits up.

Thus, $3527_{10} = DC7_{16}$.   ■

◆   Example 5   Write 42508 in hexadecimal form.

$$\begin{array}{r|r} 16 & 42508 \\ \hline 16 & 2656 \\ \hline 16 & 166 \\ \hline 16 & 10 \\ \hline & 0 \end{array} \quad \begin{array}{cc} \text{remainders} \\ 12 & C \\ 0 & \\ 6 & \\ 10 & A \end{array}$$

Read digits up.

Thus, $42508_{10} = A60C_{16}$.   ■

Exercises 16.7

Change each hexadecimal number to decimal form:

1. 25	**2.** 37	**3.** 125	**4.** 208
5. 1E	**6.** A3	**7.** C5	**8.** 7B
9. 407	**10.** 579	**11.** A22	**12.** C51
13. BC2	**14.** AE8	**15.** B2B	**16.** DD4
17. CDE	**18.** ACA	**19.** 2A5B	**20.** 4D7A

Change each decimal number to hexadecimal form:

21. 235	**22.** 579	**23.** 58	**24.** 97
25. 3352	**26.** 7369	**27.** 89,504	**28.** 92,713
29. 33,558	**30.** 52,185		

16.8   Addition and Subtraction of Hexadecimal Numbers

Use Table 16.3 to add and subtract hexadecimal numbers.

Addition of Hexadecimal Numbers

To add two hexadecimal numbers, find the first number in the left vertical column, find the second number in the top horizontal row, and find their sum at the intersection in Table 16.3. For example, to add 3 and D, locate 3 in the left column and D in the top row and find 10 at the intersection. Continue this process until all pairs of digits have been added.

Table 16.3 Hexadecimal Table

+	0	1	2	3	4	5	6	7	8	9	A	B	C	D	E	F
0	0	1	2	3	4	5	6	7	8	9	A	B	C	D	E	F
1	1	2	3	4	5	6	7	8	9	A	B	C	D	E	F	10
2	2	3	4	5	6	7	8	9	A	B	C	D	E	F	10	11
3	3	4	5	6	7	8	9	A	B	C	D	E	F	10	11	12
4	4	5	6	7	8	9	A	B	C	D	E	F	10	11	12	13
5	5	6	7	8	9	A	B	C	D	E	F	10	11	12	13	14
6	6	7	8	9	A	B	C	D	E	F	10	11	12	13	14	15
7	7	8	9	A	B	C	D	E	F	10	11	12	13	14	15	16
8	8	9	A	B	C	D	E	F	10	11	12	13	14	15	16	17
9	9	A	B	C	D	E	F	10	11	12	13	14	15	16	17	18
A	A	B	C	D	E	F	10	11	12	13	14	15	16	17	18	19
B	B	C	D	E	F	10	11	12	13	14	15	16	17	18	19	1A
C	C	D	E	F	10	11	12	13	14	15	16	17	18	19	1A	1B
D	D	E	F	10	11	12	13	14	15	16	17	18	19	1A	1B	1C
E	E	F	10	11	12	13	14	15	16	17	18	19	1A	1B	1C	1D
F	F	10	11	12	13	14	15	16	17	18	19	1A	1B	1C	1D	1E

Example 1   Add the hexadecimal numbers B52 and 3A4 and check using decimals.

$$
\begin{array}{ll}
\text{B52} & 2 + 4 = 6 \\
\underline{\text{3A4}} & 5 + A = F \\
\text{EF6} & B + 3 = E
\end{array}
$$

Check using decimals:

$$
\begin{array}{ll}
\text{B52} & 11\,(16^2) + 5\,(16^1) + 2\,(16^0) = 2898 \\
\underline{\text{3A4}} & 3\,(16^2) + 10\,(16^1) + 4\,(16^0) = \underline{932} \\
\text{EF6} & 14\,(16^2) + 15\,(16^1) + 6\,(16^0) = 3830
\end{array}
$$

■

Example 2   Add the hexadecimal numbers and check using decimals.

$$
\begin{array}{ll}
\text{F2E7} & 7 + 2 = 9 \\
\underline{\text{E3B2}} & E + B = 19 \text{ (carry 1)} \\
\text{1D699} & 2 + 3 + 1 = 6 \\
& F + E = 1D
\end{array}
$$

Check using decimals.

$$
\begin{array}{lll}
\text{F2E7} & 15(16^3) + 2(16^2) + 14(16^1) + 7(16^0) & = 62{,}183 \\
\underline{\text{E3B2}} & 14(16^3) + 3(16^2) + 11(16^1) + 2(16^0) & = \underline{58{,}290} \\
\text{1D699} & 1\,(16^4) + 13(16^3) + 6(16^2) + 9(16^1) + 9\,(16^0) & = 120{,}473
\end{array}
$$

■

Subtraction of Hexadecimal Numbers

As with binary numbers, we again have two methods available for subtracting two hexadecimal numbers. The first method can be used only when the number you are subtracting is less than or equal to the number you are subtracting it from. Use the hexadecimal table

(Table 16.3). Find the number being subtracted in the first column and the number it is being subtracted from in the table to the right; the answer will be above the second number in the top row. Note that you may need to borrow; use the table when borrowing (subtracting) 1 from A, B, C, D, E, and F.

Example 3

Subtract: 58A3 − 2E57

$$
\begin{array}{l}
\;4\;1\;9 \\
5\cancel{8}\cancel{A}3 \\
\underline{-\,2E57} \\
2A4C
\end{array}
$$

3 − 7 means need to "borrow 1" from A; A − 1 = 9; 13 − 7 = C from Table 16.3.
9 − 5 = 4
8 − E means need to "borrow 1" from 5; 5 − 1 = 4; 18 − E = A from Table 16.3.
4 − 2 = 2   ■

The second method has no conditions. It uses the idea that a subtraction can be written as an addition. The complement of a hexadecimal number is the negative of the number. The 15's complement will be used. The following are *complementary pairs* of digits: 0 and F, 1 and E, 2 and D, 3 and C, 4 and B, 5 and A, 6 and 9, 7 and 8. Change each digit in the number being subtracted to its complement and add the resulting hexadecimal numbers. Using this method, you may find the result gives you more digits than when you started. When this happens, add the extra digit 1 to the result to get the answer.

Example 4

Subtract: 58A3 − 2E57

$$
\begin{array}{r}
\;1 \\
58A3 = 58A3 \\
\underline{-\,2E57 \quad +D1A8} \\
12A4B \\
\underline{1} \\
2A4C
\end{array}
$$

3 + 8   = B
A + A   = 14, carry 1
8 + 1 + 1 = A
5 + D   = 12
There is an extra digit; therefore, add 1 as shown.   ■

Example 5

Subtract: 1E05 − 314C

$$
\begin{array}{r}
1E05 = 1E05 \\
\underline{-\,314C \quad +CEB3} \\
ECB8
\end{array}
$$

Since the number being subtracted is larger, the answer is negative, so we use the complement of each digit, −1347.   ■

Example 6

Subtract: 241A − 46C2

$$
\begin{array}{r}
\;1 \\
241A = 241A \\
\underline{-\,46C2 \quad +B93D} \\
DD57
\end{array}
$$

A + D   = 17, carry 1
1 + 1 + 3 = 5
4 + 9   = D
2 + B   = D
Since the number being subtracted is larger, the answer is negative, so we use the complement of each digit, −22A8.   ■

Example 7 Subtract: 5CC1 − F59

$$
\begin{array}{rcl}
 & & 1 \\
5CC1 = & 5CC1 \\
-0F59 & F0A6 \\
\hline
 & 14D67 \\
 & \underline{1} \\
 & 4D68
\end{array}
$$

$1 + 6\quad = 7$

$C + A\quad = 16$, carry 1

$1 + C + 0 = D$

$5 + F = 14$

There is an extra digit; therefore, add 1 as shown. ■

Exercises 16.8

Add the following hexadecimal numbers. Check using decimals.

1. 2 + 9	**2.** 4 + 8	**3.** 5 + F
4. E + 3	**5.** A + D	**6.** E + C
7. 78 + 31	**8.** 22 + 78	**9.** 5B + E2
10. C3 + 6A	**11.** AE + BB	**12.** DC + FA
13. 456 + 327	**14.** 288 + 705	**15.** A5D + 7EA
16. C3E + 8AB	**17.** AAA + BCE	**18.** DAF + CBD
19. 4527 + 8713	**20.** 2851 + 3277	
21. 59C2 + 708F	**22.** 3B25 + 52A1	
23. 5D1E + A2F3	**24.** C7B1 + 2B3D	
25. AC2B + 4C3	**26.** 2FCA + C25	
27. ABFE + 3ACF	**28.** FFEA + D4CE	
29. ABFF + CEED	**30.** FABE + ABED	

Subtract the following hexadecimal numbers. Check using decimals.

31. 2 − E	**32.** 5 − F	**33.** A − D
34. 9 − F	**35.** 9 − 1C	**36.** B − 25
37. F − 35	**38.** 7 − 1B	**39.** 1B − 24
40. 2A − 36	**41.** 12 − A3	**42.** 27 − B4
43. 129 − 248	**44.** 259 − 743	**45.** 43B − 7A2
46. 2C5 − B43	**47.** 2DE − 4CC	**48.** 3AF − 7DD
49. AFA − DAC	**50.** BCB − EAF	
51. 1990 − 2418	**52.** 3855 − 2763	
53. 25A7 − 1992	**54.** 5882 − 7B93	
55. 32B2 − 4B5C	**56.** 7D3F − 51C6	
57. D9BB − BE2C	**58.** ACCF − BA3C	
59. CAAB − FEDA	**60.** BCCF − DDBA	

 ## 16.9 Binary to Hexadecimal Conversion

Changing a Number from Binary Form to Hexadecimal Form

If the binary number is larger than any of the numbers in Table 16.4,

1. insert enough zeros at the beginning of the binary number so that you have a total number of digits divisible by 4,

2. mark the binary number in groups of 4, and

3. use Table 16.4 to make the corresponding hexadecimal substitution for each group of 4 binary digits.

Table 16.4

Decimal	Binary	Hexadecimal
0	0	0
1	1	1
2	10	2
3	11	3
4	100	4
5	101	5
6	110	6
7	111	7
8	1000	8
9	1001	9
10	1010	A
11	1011	B
12	1100	C
13	1101	D
14	1110	E
15	1111	F

Example 1

Change binary 1011100 to hexadecimal form.

Since there are 7 digits, we must add one 0 at the beginning, forming 01011100. We group in 4s and get 0101 and 1100. By Table 16.4, 0101 in binary is 5 in hexadecimal and 1100 is C. The answer is 5C.   ■

Example 2

Change binary 11111 to hexadecimal form.

There are 5 digits, so we must add three zeros at the beginning, forming 00011111. We group in 4s and get 0001 and 1111. By Table 16.4, 0001 in binary is 1 in hexadecimal and 1111 is F. The answer is 1F.   ■

Example 3

Change binary 1110010101 to hexadecimal form.

There are 10 digits, so we must add two zeros at the beginning, forming 001110010101. We group in 4s and get 0011 1001 0101. Using Table 16.4, we get 3 9 5. The answer is 395.■

Example 4

Change binary 1110001111011111 to hexadecimal form.

There are 16 digits; therefore, we need not add any zeros. We group in 4s, get 1110 0011 1101 1111, and then get E 3 D F using Table 16.4. The answer is E3DF.   ■

To change hexadecimal numbers to binary numbers, we reverse the process by replacing each hexadecimal digit by its group of 4 binary equivalent. Note that if fewer than 4 digits are given in Table 16.4, you must add enough zeros at the beginning to make a group of 4.

Changing a Number from Hexadecimal Form to Binary Form

Replace each hexadecimal digit by its group of 4 binary equivalent from right to left.

Example 5

Change hexadecimal 7A2 to binary form.

From Table 16.4, 7 is 111 or 0111, A is 1010, and 2 is 10 or 0010. The answer is 011110100010 or 11110100010. ■

Example 6

Change hexadecimal 8FA2 to binary form.

From Table 16.4, 8 is 1000, F is 1111, A is 1010, and 2 is 0010. The answer is 1000111110100010. ■

Exercises 16.9

Change each binary number to hexadecimal form:

1. 110
2. 1010
3. 1011
4. 111
5. 10111
6. 11001
7. 100100
8. 110110
9. 1000100
10. 1101111
11. 11100110
12. 10011011
13. 100111000
14. 110011100
15. 11100111101
16. 1111110000
17. 110011100111
18. 11110111001
19. 110011001100
20. 110100101001
21. 110100010010001
22. 11001011101101
23. 1111000111001010
24. 1101110010010001

Change each hexadecimal number to binary form:

25. 6
26. E
27. 24
28. 79
29. 2A
30. F4
31. 251
32. 628
33. A32
34. C43
35. 7E4
36. 3F7
37. 4DD
38. 5FC
39. ACD
40. FBF
41. 4A3B
42. 2B7E
43. BCAF
44. CACE

16.10 Hexadecimal Code for Colors

One use of the hexadecimal system is for color codes. The basic computer colors are red, green, and blue (referred to as RGB). We can obtain other colors by using various amounts of these colors. The hexadecimal numbers 00 and FF represent the weakest and the strongest colors in a range.

Red is hexadecimal code FF0000, green is 00FF00, and blue is 0000FF. Various combinations of RGB result in other colors (*e.g.,* white is FFFFFF, black is 000000, misty rose is FFE4E1, midnight blue is 191970, and lawn green is ADFF2F). Charts are available that show the colors and their hexadecimal codes. Part of one such chart is shown in Figure 16.2.

FFFFFF	FFCCFF	FF99FF	FF66FF	FF33FF	FF00FF
FFFFCC	FFCCCC	FF99CC	FF66CC	FF33CC	FF00CC
FFFF99	FFCC99	FF9999	FF6699	FF3399	FF0099
FFFF66	FFCC66	FF9966	FF6666	FF3366	FF0066
FFFF33	FFCC33	FF9933	FF6633	FF3333	FF0033
FFFF00	FFCC00	FF9900	FF6600	FF3300	FF0000

FIGURE 16.2

Chapter 16 Group Activities

1. Working in small groups, complete the following table for multiplication of hexadecimal numbers:

×	0	1	2	3	4	5	6	7	8	9	A	B	C	D	E	F
0																
1																
2																
3																
4																
5																
6																
7																
8																
9																
A																
B																
C																
D																
E																
F																

Working in a small group, multiply two of the following hexadecimal numbers using the table for addition of hexadecimal numbers (Table 16.3) and the table for multiplication of hexadecimal numbers above. Then discuss strategies and what you learned with another *group or groups. Then multiply two other hexadecimal numbers.*

2. 27 × 48   3. 36 × 83   4. 3E × A4

5. F6 × 5B   6. FD × BC   7. AF × AF

Chapter 16 Summary

Glossary of Basic Terms

Binary number system. Numbers with only two symbols or numerals: 0 and 1. (p. 566)

Hexadecimal number system. Numbers with 16 numerals: 0, 1, 2, 3, 4, 5, 6, 7, 8, 9, A = 10, B = 11, C = 12, D = 13, E = 14, and F = 15. (p. 575)

16.2 Addition of Binary Numbers

1. **Addition facts for binary addition:**
   a. $0 + 0 = 0$ **b.** $0 + 1 = 1$
   c. $1 + 0 = 1$ **d.** $1 + 1 = 10$ (p. 568)

16.3 Subtraction of Binary Numbers

1. **Subtraction facts for binary subtraction:**
   a. $0 - 0 = 0$ **b.** $1 - 1 = 0$ **c.** $1 - 0 = 1$
   d. $0 - 1$ is found by "borrowing" and having $10 - 1 = 1$. (p. 569)

16.4 Multiplication of Binary Numbers

1. **Multiplication facts for binary numbers:**
   a. $0 \times 0 = 0$ **b.** $1 \times 0 = 0$
   c. $0 \times 1 = 0$ **d.** $1 \times 1 = 1$ (p. 571)

16.5 Conversion from Decimal to Binary System

1. **Changing a number from decimal form to binary form:**
   a. Write the given number in decimal form.
   b. Divide it by 2.
   c. Write the quotient below the decimal and write the remainder at the right, even if it is 0.
   d. Divide the first quotient by 2 and write the remainder at the right; repeat the process until the final quotient is 0.
   e. Obtain the binary form of the decimal number by using the remainders from each step in order from bottom to top. (p. 572)

16.6 Conversion from Binary to Decimal System

1. **Changing a number from binary form to decimal form:** Find the place value of each digit of the given binary number and multiply it by its corresponding power of 2 or $\frac{1}{2}$. (p. 574)

16.7 Hexadecimal System

1. **Changing a number from hexadecimal form to decimal form:** Find the place value of each digit of the given binary number and multiply it by its corresponding power of 16. (p. 575)

2. **Changing a number from decimal form to hexadecimal form:**
   a. Write the given number in decimal form.
   b. Divide it by 16.
   c. Write the quotient below the decimal and write the remainder at the right, even if it is 0.
   d. Divide the first quotient by 16 and write the remainder at the right; repeat the process until the final quotient is 0.
   e. Obtain the hexadecimal form of the decimal number by using the remainders from each step in order from bottom to top. (p. 575)

16.8 Addition and Subtraction of Hexadecimal Numbers

1. **To add two hexadecimal numbers:** Using the hexadecimal table (Table 16.3), find the first digit in the first vertical column, find the second digit in the first horizontal row, and find their sum at the intersection. Continue this process until all pairs of digits have been added. (p. 576)

2. **To subtract two hexadecimal numbers:** This method uses the idea that subtraction can be written as an addition of the complement (negative) of a hexadecimal number. The following are *complementary pairs* of digits: 0 and F, 1 and E, 2 and D, 3 and C, 4 and B, 5 and A, 6 and 9, and 7 and 8. Change each digit in the number being subtracted to its complement and add the resulting hexadecimal numbers. *Note:* If the final result gives you more digits than you had when you started, add the extra digit 1 to the result to get the answer. (p. 578)

16.9 Binary to Hexadecimal Conversion

1. **Changing a number from binary form to hexadecimal form:** If the binary number is larger than any of the numbers in Table 16.4,
   a. insert enough zeros at the beginning of the binary number so that you have a total number of digits divisible by 4,
   b. mark the binary number in groups of 4, and
   c. use Table 16.4 to make the corresponding hexadecimal substitution for each group of 4 binary digits. (p. 579)

2. **Changing a number from hexadecimal form to binary form:** Replace each hexadecimal digit by its group of 4 binary equivalent from right to left. (p. 581)

 ## Chapter 16 Review

Change each binary number to decimal form:

1. 1101   **2.** 11001   **3.** 110100   **4.** 10110.11

Add the following binary numbers:

5. 11001
  1101

6. 1101001
  101101

7. 1001
  110
  101

Subtract the following binary numbers:

8. 10011
  1010

9. 11011
  1110

Multiply the following binary numbers:

10. 110
  11

11. 11011
  1001

Change each decimal number to binary form:

12. 36   **13.** 205   **14.** 1050

Change each hexadecimal number to decimal form:

15. E1   **16.** 2C   **17.** C1E

Change each decimal number to hexadecimal form:

18. 312   **19.** 52   **20.** 4624

Add the following hexadecimal numbers:

21. 4E + 35   **22.** 1A4 + EF   **23.** 6A12 + 7C2B

Change each binary number to hexadecimal form:

24. 10110   **25.** 100110   **26.** 101110011

Change each hexadecimal number to binary form:

27. 4C   **28.** 365   **29.** B2A   **30.** 4AA1

 ## Chapter 16 Test

Change each binary number to decimal form:

1. 10110   **2.** 10111.101

Add the following binary numbers:

3. 1110101
  11011

4. 11011
  10110
  1110

Subtract the following binary numbers:

5. 111011
  10010

6. 100010
  10111

7. Multiply the following binary numbers: 11011
  1101

Change each decimal number to binary form:

8. 407   **9.** 1142

Change each hexadecimal number to decimal form:

10. D3   **11.** 2F

Change each decimal number to hexadecimal form:

12. 628   **13.** 704

Add the following hexadecimal numbers:

14. 7D + 82   **15.** 36E + 15A   **16.** 23AC + 5F7

Change each binary number to hexadecimal form:

17. 101100   **18.** 1100011

Change each hexadecimal number to binary form:

19. 7F   **20.** 3BD

Cumulative Review

1. Round 2927.404 to **a.** the nearest hundredth and **b.** the nearest ten.

2. Perform the indicated operations and write the result in engineering notation rounded to three significant digits:

$$\frac{\left(612 \times 10^{-6}\right)\left(15 \times 10^{-9}\right)\left(2.7 \times 10^{3}\right)}{\left(82 \times 10^{9}\right)\left(8.16 \times 10^{-12}\right)}$$

3. Convert 72°F to °C.

4. Use the rules for measurement to find the sum of the following set of measurements: 6128 km, 1520 km, 16.28 km, 225 km

5. Do as indicated and simplify:

$$(2a^2 - 5a + 6) - (-2a^2 + 6a - 7) + (3a^2 - 7a - 2)$$

6. Solve: $2(5y - 3) + 4(6y - 1) = 17(2y - 3) - 25y$

7. Solve (round the result to three significant digits):

$$\frac{x}{17} = \frac{14.6}{38.5}$$

8. Find the equation of a line having slope $-\frac{2}{3}$ and y-intercept of 4.5.

9. The perimeter of a rectangle is 39.8 m. Its length is 10.1 m longer than its width. Find its length and width.

10. Factor completely: $30x^2 + 117x - 810$

11. Solve: $3x^2 - 5x = 2$

12. Solve: $7x^2 + 2x + 15 = 0$

13. Find the area of the trapezoid in Illustration 1.

ILLUSTRATION 1

14. Find the volume of the sphere with radius 17.3 in.

15. Given $\cos B = 0.9128$, find angle B in degrees to nearest tenth of a degree.

16. Find side a in the right triangle in Illustration 2.

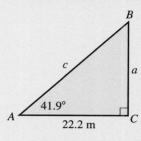

ILLUSTRATION 2

17. Find $\cos 256°$ rounded to four significant digits.

18. Find angle A in the triangle in Illustration 3.

ILLUSTRATION 3

19. Find the mean of the following numbers: 20.2, 27.3, 35.1, 30.6, 29.6, 22.6.

20. Find the median of the data in Exercise 19.

21. Find the standard deviation of the data in Exercise 19.

22. Make a frequency distribution from the following data. The number of defective parts coming off an assembly line per eight-hour shift were as follows:

15, 12, 10, 9, 15, 22, 7, 23, 12, 8, 18, 22, 11, 30, 14, 18, 12, 20, 22, 35, 10, 8, 11, 19, 7, 23, 17, 15, 20, 16, 17, 18, 22, 15, 20, 13.

23. Find the mean using grouped data from Exercise 22.

24. Find the median of the data in Exercise 22.

Do as indicated for the following binary numbers:

25. $1110101 + 10011$ 26. 110001×10111

27. Change the decimal number 612 to binary form.

28. Change the hexadecimal number E5 to decimal form.

29. Change the hexadecimal number 4AB to binary form.

30. Add the following hexadecimal numbers: 2B5 + 1A4D

APPENDIX

A Tables

 ## Table 1 Formulas from Geometry

PLANE FIGURES

In the following tables, a, b, c, and d are lengths of sides, and h is the altitude.

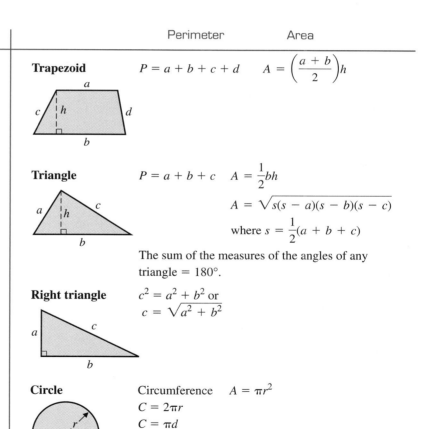

	Perimeter	Area
Rectangle	$P = 2(b + h)$	$A = bh$
Square	$P = 4b$	$A = b^2$
Parallelogram	$P = 2(a + b)$	$A = bh$
Rhombus	$P = 4b$	$A = bh$

	Perimeter	Area
Trapezoid	$P = a + b + c + d$	$A = \left(\dfrac{a + b}{2}\right)h$
Triangle	$P = a + b + c$	$A = \dfrac{1}{2}bh$

$$A = \sqrt{s(s - a)(s - b)(s - c)}$$

where $s = \dfrac{1}{2}(a + b + c)$

The sum of the measures of the angles of any triangle $= 180°$.

Right triangle $\quad c^2 = a^2 + b^2$ or
$\quad c = \sqrt{a^2 + b^2}$

Circle $\quad$ Circumference $\quad A = \pi r^2$
$\quad C = 2\pi r$
$\quad C = \pi d$
$\quad (d = 2r)$
The sum of the measures of the central angles of a circle $= 360°$.

587

GEOMETRIC SOLIDS

In the following table, B is the area of the base, r is the length of the radius, and h is the height.

	Volume	Lateral surface area
Prism	$V = Bh$	
Cylinder	$V = \pi r^2 h$	$A = 2\pi rh$
Pyramid	$V = \dfrac{1}{3}Bh$	
Cone 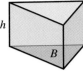	$V = \dfrac{1}{3}\pi r^2 h$	$A = \pi rs$, where s is the slant height.
Sphere	$V = \dfrac{4}{3}\pi r^3$	$A = 4\pi r^2$

Table 2 Electrical Symbols

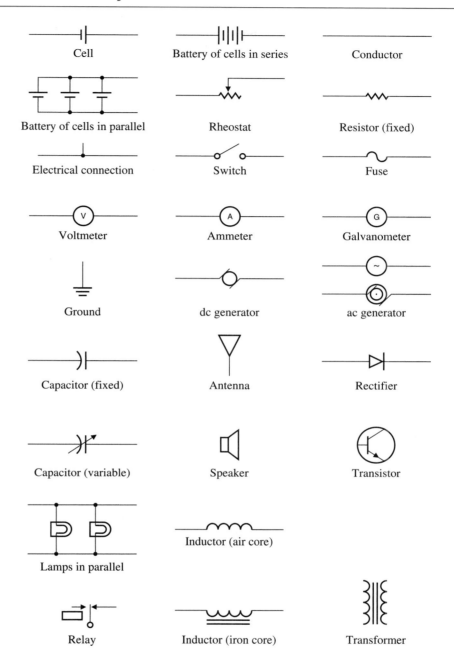

Cell

Battery of cells in series

Conductor

Battery of cells in parallel

Rheostat

Resistor (fixed)

Electrical connection

Switch

Fuse

Voltmeter

Ammeter

Galvanometer

Ground

dc generator

ac generator

Capacitor (fixed)

Antenna

Rectifier

Capacitor (variable)

Speaker

Transistor

Lamps in parallel

Inductor (air core)

Relay

Inductor (iron core)

Transformer

Exponential Equations

Equations in the form $y = x^n$, where the exponent n is a constant and the base is a variable, are called *power equations*. For example, we graphed the simple power equation $y = x^2$ in Chapter 11.

Equations in the form $y = b^x$, where the exponent is a variable and the base b is a constant with $b > 0$ and $b \neq 1$, are called **exponential equations**. Two examples are $y = 2^x$ and $y = 3^{-x}$.

Example 1

Graph the exponential equation $y = 2^x$ by plotting points.

Set up a table, letting x equal some convenient values, and solve for each corresponding value of y as follows:

$$\text{Let } x = -3; \text{ then } y = 2^{-3} = \frac{1}{2^3} = \frac{1}{8}.$$

$$\text{Let } x = -2; \text{ then } y = 2^{-2} = \frac{1}{2^2} = \frac{1}{4}.$$

$$\text{Let } x = -1; \text{ then } y = 2^{-1} = \frac{1}{2^1} = \frac{1}{2}.$$

Let $x = 0$; then $y = 2^0 = 1$.

Let $x = 1$; then $y = 2^1 = 2$.

Let $x = 2$; then $y = 2^2 = 4$.

Let $x = 3$; then $y = 2^3 = 8$.

FIGURE B.1

x	-3	-2	-1	0	1	2	3
y	$\frac{1}{8}$	$\frac{1}{4}$	$\frac{1}{2}$	1	2	4	8

Then plot these points as in Figure B.1. It appears that the points follow the shape of a curve. To confirm, let's try some other points with x values that are not integers. To find the y values, you will need to find these powers with your calculator similar to the steps shown in Section 1.15. The first calculation is illustrated below.

Let $x = 0.5$; then $y = 2^{0.5} = 1.4$.

2 .5

$$\boxed{1.414213562}$$

Let $x = 0.75$; then $y = 2^{0.75} = 1.7$.

Let $x = 1.3$; then $y = 2^{1.3} = 2.5$.

Let $x = 2.7$; then $y = 2^{2.7} = 6.5$.

Let $x = -0.75$; then $y = 2^{-0.75} = 0.59$.

Let $x = -2.7$; then $y = 2^{-2.7} = 0.15$.

x	0.5	0.75	1.3	2.7	-0.75	-2.7
y	1.4	1.7	2.5	6.5	0.59	0.15

FIGURE B.2

Next, plot these points as in Figure B.2. As you can see, these points lie along the same curve as outlined by using a few easy sample ordered pairs.   ■

In general, for $b > 1$, $y = b^x$ is an **increasing function**. That is, as x increases, y also increases.

Example 2

Graph the equation $y = \left(\dfrac{1}{3}\right)^x$.

The laws of exponents for powers of 10 in Section 2.5 also apply to powers of any base. Here, we can write $\dfrac{1}{3} = \dfrac{1}{3^1} = 3^{-1}$ and $\left(\dfrac{1}{3}\right)^x = (3^{-1})^x = 3^{-x}$. We can write this equation as $y = 3^{-x}$.

Set up a table, letting x equal some convenient values, and solve for y as follows:

Let $x = -3$; then $y = 3^{-(-3)} = 3^3 = 27$.

Let $x = -2$; then $y = 3^{-(-2)} = 3^2 = 9$.

Let $x = -1$; then $y = 3^{-(-1)} = 3^1 = 3$.

Let $x = 0$; then $y = 3^{-(0)} = 3^0 = 1$.

Let $x = 1$; then $y = 3^{-1} = \dfrac{1}{3}$.

Let $x = 2$; then $y = 3^{-2} = \dfrac{1}{3^2} = \dfrac{1}{9}$.

Let $x = 3$; then $y = 3^{-3} = \dfrac{1}{3^3} = \dfrac{1}{27}$.

x	-3	-2	-1	0	1	2	3
y	27	9	3	1	$\frac{1}{3}$	$\frac{1}{9}$	$\frac{1}{27}$

Then plot these points as in Figure B.3 and connect with a smooth curve.   ■

In general, for $0 < b < 1$, $y = b^x$ is a **decreasing function**. That is, as x increases, y decreases.

One special exponential equation describes a wide variety of natural growth and decay processes with its base e, which is approximately equal to 2.71828. Examples include population and bacteria growth, radioactive decay, electrical current and voltage analysis, and investment and savings growth. The fundamental equation is $y = e^x$.

FIGURE B.3

Example 3

Graph the equation $y = e^x$.

Again, set up a table, letting x equal some convenient values, and solve for y as follows:

Let $x = -3$; then $y = e^{-3} = 0.050$.

This first calculation with a calculator is illustrated below.

$$0.049787068$$

Note: The right parenthesis may or may not be needed for given calculators.

Let $x = -2$; then $y = e^{-2} = 0.14$.

Let $x = -1$; then $y = e^{-1} = 0.37$.

Let $x = 0$; then $y = e^0 = 1$.

Let $x = 1$; then $y = e^1 = 2.7$.

Let $x = 2$; then $y = e^2 = 7.4$.

Let $x = 3$; then $y = e^3 = 20.1$.

x	-3	-2	-1	0	1	2	3
y	0.050	0.14	0.37	1	2.7	7.4	20.1

Then plot these points as in Figure B.4 and connect with a smooth curve. ■

FIGURE B.4

There are many applications involving exponential equations. When a quantity *increases* so that the amount of increase is proportional to the amount present, we have **exponential growth**. One of the most commonly used exponential equations for exponential or continuous growth is

Exponential Growth

$$y = Ae^{rt}$$

where

y = the amount present at any given time t,

A = the original amount,

e = the natural number,

r = the rate of growth, and

t = the time the growth has occurred.

This exponential growth equation can be used to find interest when compounded continuously, which is very common now. In Example 3 of Section 1.16, we found that $12,682.42 is owed at the end of 3 years if $10,000 is borrowed at 8% per year compounded quarterly and no payments are made on the loan.

Example 4   Find the amount of money owed at the end of 3 years if $10,000 is borrowed at 8% per year compounded continuously and no payments are made on the loan. Here

$$A = \$10,000,$$
$$r = 8\% = 0.08, \text{ and}$$
$$t = 3 \text{ years.}$$
$$y = Ae^{rt},$$
$$y = (\$10,000)\, e^{(0.08)(3)}$$
$$y = \$12,712.49$$

Using a scientific calculator, we have

10000   .08 3

12712.4915

Note: The e^x key usually provides the left parenthesis.

As you can see, interest accumulates more rapidly as the interest is compounded within shorter intervals of time.   ■

When a quantity *decreases* so that the amount of decrease is proportional to the amount present, we have **exponential decay**. One of the most commonly used exponential equations for exponential or continuous decay is

> **Exponential Decay**
> $$y = Ae^{-rt}$$

where

$y =$ the amount present at any given time t,

$A =$ the original amount,

$e =$ the natural number,

$r =$ the rate of decay, and

$t =$ the time the decay has occurred.

Example 5   The discharge current of a capacitor with an initial current of 3.60 A is given by the exponential equation $i = 3.6e^{-55t}$. Find the discharge current after 25 ms.

Here, $t = 25 \text{ ms} = 0.025$ s. (Here, the time must be in seconds.) So, we have

$$i = 3.60e^{-55t}$$
$$i = 3.60e^{(-55)(0.025)}$$
$$i = 0.91 \text{ A}$$

That is, after 25 ms, the discharge current has decreased to 0.91 A.   ■

Exercises B

Graph each equation.

1. $y = 3^x$ **2.** $y = 4^x$

3. $y = 5^x$ **4.** $y = 10^x$

5. $y = \left(\dfrac{1}{2}\right)^x$ **6.** $y = \left(\dfrac{3}{4}\right)^x$

7. $y = \left(\dfrac{2}{3}\right)^x$ **8.** $y = \left(\dfrac{1}{4}\right)^x$

9. $y = 2^{-x}$ **10.** $y = 4^{-x}$

11. $y = 3^{2x}$ **12.** $y = 2^{-2x}$

13. $y = 2^{-4x}$ **14.** $y = 3^{5x}$

15. $y = 2^{2-x}$ **16.** $y = 3^{x+2}$

17. $y = e^{2x}$ **18.** $y = e^{-x}$

19. $y = e^{-3x}$ **20.** $y = e^{-2x}$

21. Mackenzie deposits $5000 in a savings account that pays $3\frac{3}{4}\%$ compounded continuously. Assuming she makes no more deposits, how much money will she have in her account after $2\frac{1}{2}$ years?

22. The amount of bacterial growth in a given culture is given by $N = N_0 e^{0.035t}$, where t is the time in hours, N_0 is the initial amount, and N is the amount after time t. Find how much bacteria we have after 4.00 h if we start with 7500 bacteria.

23. A city with a population of 65,000 has a growth rate of 3.5%. Find its expected population in 5 years.

24. The voltage in a circuit is increasing according to the equation $V = V_0 e^{0.025t}$. Find the voltage in 45.0 ms if the initial voltage is 275 mV. The result will be in mV.

25. Zachary retires with an annual pension of $75,000. With an inflation rate of 5%, what will be his purchasing power after 10 years. *Hint:* $r = -5\%$.

26. A given radioactive element decays according to the equation $y = y_0 e^{-0.0045t}$, where t is the time in seconds, y_0 is the initial amount, and y is the amount remaining after time t. How much of a 75.0-g radioactive element remains after 3.00 minutes?

27. A current flows in a given circuit according to the exponential equation $i = 0.025 e^{-0.175t}$, where t is in ms and i is in mA. Find the amount of current after 4.80 ms.

28. The voltage is discharged in a electric capacitor through a resistor according to the equation $E = E_0 e^{-t/(RC)}$, where E_0 is the original voltage in volts (V), t is the time t for E_0 to decrease to E, R is the resistance in ohms (Ω), and C is the capacitance in farads (F). Find the voltage after 0.500 s if the initial voltage is 25.0 mV, the resistance is 1.85 Ω, and the capacitance is 0.450 F.

APPENDIX **C** **Answers to Odd-Numbered Exercises and All Chapter Review and Cumulative Review Exercises**

Chapter 1

Exercises 1.1 (pages 8–11)
1. 3255 **3.** 1454 **5.** 795,776 **7.** 5164 **9.** 26,008 **11.** 2820 **13.** 4195 Ω **15.** 224 **17.** 39 ft **19.** *I*: 1925 cm^3; *O*: 1425 cm^3
21. 27,216 **23.** 18,172,065 **25.** 35,360,000 **27.** 1809 **29.** 389 **31.** 844 r 40 **33.** 496 mi **35.** 325 cm^3 **37.** 13 km/L
39. $67 **41.** 500 mi **43.** 4820 ft **45.** 9600 **47.** 67 in. from either corner **49.** 44 bu/acre **51. a.** 881 lb **b.** 15 lb
53. 85 bu/acre **55.** $17,200 **57.** 5 A **59.** 24 V **61.** 880 oz **63.** 4 **65.** 9 in. **67.** 19 in. **69.** 213,500 bd ft

Exercises 1.2 (pages 13–14)
1. 2 **3.** 10 **5.** 18 **7.** 131 **9.** 73 **11.** 23 **13.** 16 **15.** 34 **17.** 102 **19.** 137 **21.** 230 **23.** 55 **25.** 0 **27.** 4 **29.** 0
31. 13 **33.** 19 **35.** 26 **37.** 1 **39.** 102 **41.** 85

Exercises 1.3 (pages 17–19)
1. 96 yd^2 **3.** 307,500 ft^2 **5.** 13,943 in^2 **7.** 84 cm^2 **9.** 48 in^2 **11.** 52 in^2 **13.** 108 **15.** 32 gal **17. a.** $118,800 **b.** $257,840
19. 96 m^3 **21.** 720 cm^3 **23.** 3900 in^3 **25.** 600 cm^3 **27.** 11,520 in^3 **29.** 30 ft^3 **31.** 69,480 lb **33.** 14,880 lb **35.** $2750
37. 25 in. $\times$ 37 in. **39.** 128 ft^3

Exercises 1.4 (pages 22–23)
1. 600 **3.** 11,250 **5.** 57,376 **7.** 38,400 **9.** 8 **11.** 5017 **13.** 40 in^2 **15.** 810 ft^2 **17.** 56 m^2 **19.** 648 ft^2 **21.** 32 ft^2
23. 1800 cm^2 **25.** 4500 cm^3 **27.** 172 **29.** 16 **31.** 113.1 **33.** 31.8 tsf

Exercises 1.5 (page 26)
1. a. yes **b.** no **3. a.** yes **b.** yes **5. a.** yes **b.** no **7.** prime **9.** not prime **11.** not prime **13.** not prime **15.** yes **17.** no
19. no **21.** yes **23.** no **25.** yes **27.** yes **29.** no **31.** no **33.** yes **35.** no **37.** no **39.** yes **41.** yes **43.** yes
45. $2 \cdot 2 \cdot 5$ **47.** $2 \cdot 3 \cdot 11$ **49.** $2 \cdot 2 \cdot 3 \cdot 3$ **51.** $3 \cdot 3 \cdot 3$ **53.** $3 \cdot 17$ **55.** $2 \cdot 3 \cdot 7$ **57.** $2 \cdot 2 \cdot 2 \cdot 3 \cdot 5$ **59.** $3 \cdot 3 \cdot 19$
61. $3 \cdot 5 \cdot 7$ **63.** $2 \cdot 2 \cdot 3 \cdot 3 \cdot 7$

Unit 1A Review   (page 27)

1. 241 **2.** 1795 **3.** 2,711,279 **4.** 620 **5.** 262 ft **6.** 254 bu **7.** 42 **8.** 7 **9.** 32 **10.** 499 in^2 **11.** 720 ft^3 **12.** 180 **13.** 10
14. 300 **15.** not prime **16.** prime **17.** yes **18.** no **19.** $2 \cdot 2 \cdot 2 \cdot 5$ **20.** $3 \cdot 3 \cdot 3 \cdot 5$

Exercises 1.6   (pages 32–33)

1. $\frac{3}{7}$ **3.** $\frac{6}{7}$ **5.** $\frac{3}{16}$ **7.** $\frac{1}{3}$ **9.** $\frac{4}{5}$ **11.** 1 **13.** 0 **15.** undefined **17.** $\frac{7}{8}$ **19.** $\frac{3}{4}$ **21.** $\frac{3}{4}$ **23.** $\frac{4}{5}$ **25.** $\frac{3}{10}$ **27.** $\frac{7}{8}$ **29.** $\frac{7}{9}$

31. $15\frac{3}{5}$ **33.** $9\frac{1}{3}$ **35.** $1\frac{1}{4}$ **37.** $9\frac{1}{2}$ **39.** $6\frac{1}{4}$ **41.** $\frac{23}{6}$ **43.** $\frac{17}{8}$ **45.** $\frac{23}{16}$ **47.** $\frac{55}{8}$ **49.** $\frac{53}{5}$

Exercises 1.7   (pages 41–45)

1. 16 **3.** 210 **5.** 48 **7.** $\frac{5}{6}$ **9.** $\frac{5}{32}$ **11.** $\frac{11}{28}$ **13.** $\frac{29}{64}$ **15.** $\frac{7}{20}$ **17.** $1\frac{3}{10}$ **19.** $\frac{37}{48}$ **21.** $\frac{13}{120}$ **23.** $\frac{67}{105}$ **25.** $\frac{1}{8}$ **27.** $\frac{1}{2}$ **29.** $\frac{4}{7}$

31. $\frac{1}{32}$ **33.** $7\frac{1}{4}$ **35.** $2\frac{5}{8}$ **37.** $4\frac{3}{4}$ **39.** $2\frac{5}{16}$ **41.** $13\frac{31}{45}$ **43.** $11\frac{5}{16}$ **45.** $9\frac{13}{24}$ **47.** $2237\frac{1}{4}$ ft **49. a.** $6\frac{1}{4}$ ft **b.** $1\frac{1}{2}$ ft

51. $35\frac{11}{20}$ gal **53.** $1\frac{1}{2}$ gal **55.** $\frac{5}{6}$ h **57.** $1\frac{31}{48}$ tons **59.** $8\frac{1}{4}$ in. **61. a.** $1\frac{1}{2}$ in. **b.** $22\frac{23}{64}$ in. **63. a.** $\frac{1}{4}$ in. **b.** $12\frac{5}{16}$ in.

65. $3\frac{1}{4}$ A **67.** $1\frac{43}{48}$ A **69.** $9\frac{5}{8}$ in. **71. a.** $10\frac{1}{2}$ in. **b.** $\frac{3}{4}$ in. **73. a.** $21\frac{13}{16}$ in. **b.** $\frac{7}{8}$ in. **75.** $14\frac{3}{8}$ in. **77.** $\frac{5}{64}$ in.

79. $1\frac{25}{32}$ in.; $1\frac{19}{32}$ in. **81.** $10\frac{13}{16}$ in.; $3\frac{3}{4}$ in. **83.** $5\frac{11}{16}$ in. **85.** $52\frac{5}{32}$ in. **87.** $2\frac{1}{12}$ cords **89.** $5\frac{1}{2}$ mi

Exercises 1.8   (pages 49–52)

1. 12 **3.** 9 **5.** $\frac{35}{64}$ **7.** $\frac{2}{3}$ **9.** 10 **11.** $\frac{1}{8}$ **13.** $1\frac{1}{4}$ **15.** $\frac{2}{5}$ **17.** $\frac{18}{25}$ **19.** 18 **21.** 40 **23.** $1\frac{2}{33}$ **25.** $1\frac{43}{45}$ **27.** $2\frac{1}{4}$ **29.** $\frac{27}{32}$

31. $\frac{1}{126}$ **33.** $\frac{7}{12}$ **35.** $31\frac{1}{2}$ gal **37.** $45\frac{1}{2}$ in. **39.** $120\frac{3}{4}$ mi/h **41.** 33 ft **43.** 80 bd ft **45.** $1633\frac{1}{3}$ bd ft **47.** $3\frac{27}{32}$ in.

49. $1\frac{7}{8}$ in. **51.** $8\frac{5}{8}$ in. **53.** $8\frac{2}{3}$ min **55.** $2\frac{2}{3}$ ft^3 **57.** $1\frac{1}{4}$ h **59.** 2750 W **61.** 75 W **63.** $2\frac{2}{7}$ A **65.** $7\frac{1}{4}$ ft or 7 ft 3 in.

67. 40 lb **69.** 160 bu/acre **71.** $\frac{1}{2}$ oz **73.** $\frac{1}{2}$ **75.** $1\frac{1}{4}$ gr **77.** 2 **79.** 3 **81. a.** $\frac{3}{4}$ in. **b.** 63 in^3 **83.** 4 Ω **85.** $3\frac{1}{5}$ Ω

87. 75 white; 225 red

Exercises 1.9   (pages 55–56)

1. 43 **3.** 83 **5.** 9 **7.** 10 **9.** 96 **11.** 6 **13.** 8 **15.** 5 **17.** $5\frac{1}{2}$ **19.** $3\frac{1}{2}$ **21.** $3\frac{1}{2}$ **23.** $30\frac{2}{3}$ **25.** 3520 **27.** $15\frac{5}{8}$

29. 6 ft 8 in. **31.** 153 in. **33.** 7 qt **35.** $\frac{66}{125}$ Ω **37.** 2520 in^2 **39. a.** 10,560 ft **b.** 3520 yd **41.** 48 oz **43.** 51 yd

45. $8\frac{1}{2}$ chains **47.** $410\frac{1}{10}$ gr **49.** 75 ft/min **51.** 72 mi/min **53.** $58\frac{2}{3}$ ft/s **55.** 120 ft/min **57.** 16 yd 1 ft 10 in. **59.** 9000 lb

61. 22 paces

Unit 1B Review   (pages 56–57)

1. $\frac{3}{5}$ **2.** $\frac{8}{9}$ **3.** $4\frac{1}{2}$ **4.** $\frac{17}{5}$ **5.** $1\frac{1}{2}$ **6.** $2\frac{23}{24}$ **7.** $\frac{4}{15}$ **8.** $\frac{6}{13}$ **9.** $3\frac{1}{4}$ **10.** 2 **11.** $\frac{7}{8}$ in. **12.** $17\frac{7}{8}$ in. **13.** $17\frac{5}{6}$ in. **14.** $16\frac{2}{3}$ in^2

15. 48 in. **16.** 8 yd **17.** 48 oz **18.** 5 gal **19.** 88 ft/s **20.** 5 ft 8 in.

Exercises 1.10   (pages 63–65)

1. four thousandths **3.** five ten-thousandths **5.** one and four hundred twenty-one hundred-thousandths

7. six and ninety-two thousandths **9.** 5.02; $5\frac{2}{100}$ or $5\frac{1}{50}$ **11.** 71.0021; $71\frac{21}{10,000}$ **13.** 43.0101; $43\frac{101}{10,000}$ **15.** 0.375

17. $0.7\overline{3}$ **19.** 0.34 **21.** $1.\overline{27}$ **23.** $18.\overline{285714}$ **25.** $34.\overline{2}$ **27.** $\frac{7}{10}$ **29.** $\frac{11}{100}$ **31.** $\frac{337}{400}$ **33.** $10\frac{19}{25}$ **35.** 150.888 **37.** 163.204

39. 86.6 **41.** 15.308 **43.** 8.68 **45.** 4.862 **47.** 10.0507 **49.** 1.45 ft × 4.2 ft **51.** 10.8 h **53.** 0.3125 in.
55. $a = 4.56$ cm; $b = 4.87$ cm **57.** 7.94 in. **59.** 4.8125 in. **61.** 2.605 A **63.** 1396.8 Ω **65.** 0.532 in. **67.** 0.22 in. **69.** 1.727 in.
71. 5.09 billion **73.** 1317.5 bbl

Exercises 1.11 (pages 68–69)
1. a. 1700 **b.** 1650 **3. a.** 3100 **b.** 3130 **5. a.** 18,700 **b.** 18,680 **7. a.** 3.1 **b.** 3.142 **9. a.** 0.1 **b.** 0.057
11. a. 0.1 **b.** 0.070

	Hundred	Ten	Unit	Tenth	Hundredth	Thousandth
13.	600	640	636	636.2	636.18	636.183
15.	17,200	17,160	17,159	17,159.2	17,159.17	17,159.167
17.	1,543,700	1,543,680	1,543,679	—	—	—
19.	10,600	10,650	10,650	10,649.8	10,649.83	—
21.	600	650	650	649.9	649.90	649.900

23. 237,000 **25.** 0.0328 **27.** 72 **29.** 1,462,000 **31.** 0.0003376 **33.** 1.01

Exercises 1.12 (pages 72–75)
1. 0.555 **3.** 10.5126 **5.** 9,280,000 **7.** 30 **9.** 15 **11.** 248.23 **13.** 3676.47 **15.** $\frac{7}{12}$ **17.** $\frac{4}{5}$ **19.** 1.2 ft **21.** 119 mi/h
23. 27.7 mi/gal **25.** 9.682 in. **27. a.** 37.76 m **b.** 9.44 m **29.** 80 threads **31.** 2.95 in. **33.** 3000 sheets **35.** $4872
37. 51.20 in. **39.** 337.50 in^3 **41.** 0.5 L **43. a.** 0.056 in. **b.** 4.535 in. **45.** 240 gal **47.** 30.1¢ **49.** 94.2 Ω **51.** 0.288 W
53. 6.20 A **55.** 136.9 Ω **57.** 0.450 gr **59.** 5 **61.** 303.8 nautical mi **63.** 1290 lb **65.** 703,000 tons **67.** 42.7 ft^3

Exercises 1.13 (pages 79–80)
1. 0.27 **3.** 0.06 **5.** 1.56 **7.** 0.292 **9.** 0.087 **11.** 9.478 **13.** 0.0028 **15.** 0.00068 **17.** 0.0425 **19.** 0.00375 **21.** 54%
23. 8% **25.** 62% **27.** 217% **29.** 435% **31.** 18.5% **33.** 29.7% **35.** 519% **37.** 1.87% **39.** 0.29% **41.** 80% **43.** 12.5%
45. $16\frac{2}{3}$% **47.** $44\frac{4}{9}$% **49.** 60% **51.** 32.5% **53.** 43.75% **55.** 240% **57.** 175% **59.** $241\frac{2}{3}$% **61.** $\frac{3}{4}$ **63.** $\frac{4}{25}$ **65.** $\frac{3}{5}$
67. $\frac{93}{100}$ **69.** $2\frac{3}{4}$ **71.** $1\frac{1}{4}$ **73.** $\frac{43}{400}$ **75.** $\frac{107}{1000}$ **77.** $\frac{69}{400}$ **79.** $\frac{97}{600}$

81.

Fraction	Decimal	Percent
$\frac{3}{8}$	0.375	37.5%
$\frac{9}{20}$	0.45	45%
$\frac{9}{50}$	0.18	18%
$1\frac{2}{5}$	1.4	140%
$1\frac{2}{25}$	1.08	108%
$\frac{67}{400}$	0.1675	$16\frac{3}{4}$%

Exercises 1.14 (pages 84–89)
1. P = 60; R = 25%; B = 240 **3.** P = 108; R = 40%; B = 270 **5.** P = unknown; R = 4%; B = 28,000 **7.** P = 21; R = 60%;
B = unknown **9.** P = 2050; R = 6%; B = unknown **11.** 10% **13.** 50% **15.** $1600 **17.** 112.8 **19.** 8.96 V **21.** 74.6%
23. 36.9% **25.** 6.7% **27.** 20% **29.** 766.7 ft **31.** 440 lb total; 352 lb active ingredients, 88 lb inert ingredients
33. 850 gal; 35.7 gal butterfat **35.** 0.3 gr **37.** 25% **39.** 53.3% **41.** 14.2% **43.** $45; $51 **45.** 1558 lb **47.** 74.8%
49. a. 35 deer/mi^2 **b.** 49 deer/mi^2 **51.** $1967.82 **53.** $5625.92

Exercises 1.15 (page 93)
1. 225 **3.** 222 **5.** 0.00000661 **7.** 729 **9.** 562 **11.** 0.00483 **13.** 157 **15.** 2.96 **17.** 68.9 **19.** 42.4 **21.** 0.198

Exercises 1.16 (page 97)
1. a. $300 **b.** $63.89 **3.** $9100.56 **5.** $23,152.64 **7.** $948.10 **9.** $6429.83 **11. a.** $674.40; $24,278.40
b. $710.27; $25,569.71 Thus, choice **a** pays a total of $1291.31 less **13.** $3820.62; $183,389.76

Unit 1C Review   (pages 97–98)

1. 1.625 **2.** $\dfrac{9}{20}$ **3.** 10.129 **4.** 116.935 **5.** 5.854 **6.** 25.6 ft **7.** 160.2 ft **8. a.** 45.1 **b.** 45.06 **9. a.** 45.1 **b.** 45.06

10. 0.11515 **11.** 18.85 **12.** 6 cables; 2 in. left **13.** 0.25 **14.** 72.4% **15.** 69.3 **16.** 2000 **17.** 40% **18.** $17.49 **19.** 2110
20. 9.40

Chapter 1 Review   (pages 102–104)

1. 8243 **2.** 55,197 **3.** 9,178,000 **4.** 226 r 240 **5.** 3 **6.** 43 **7.** 37 **8.** 31 **9.** 340 cm^2 **10.** 30 cm^3 **11.** 10 **12.** 3000

13. no **14.** $2 \cdot 3 \cdot 3 \cdot 3$ **15.** $2 \cdot 3 \cdot 5 \cdot 11$ **16.** $\dfrac{9}{14}$ **17.** $\dfrac{5}{6}$ **18.** $4\dfrac{1}{6}$ **19.** $6\dfrac{3}{5}$ **20.** $\dfrac{21}{8}$ **21.** $\dfrac{55}{16}$ **22.** 2 **23.** $1\dfrac{1}{2}$ **24.** $\dfrac{103}{180}$

25. $14\dfrac{29}{48}$ **26.** $1\dfrac{19}{24}$ **27.** $11\dfrac{3}{5}$ **28.** 5 **29.** $\dfrac{1}{4}$ **30.** 18 **31.** $\dfrac{1}{16}$ **32.** $\dfrac{3}{8}$ **33.** $13\dfrac{11}{25}$ **34.** $A: 3\dfrac{3}{8}$ in. $B: 5\dfrac{7}{16}$ in. **35.** 105

36. 2016 **37.** 24 **38.** 63,360 yd **39.** 0.5625 **40.** $0.41\overline{6}$ **41.** $\dfrac{9}{20}$ **42.** $19\dfrac{5}{8}$ **43.** 168.278 **44.** 17.25 **45.** 68.665

46. 33.72 **47.** 3206.5 **48.** 1.9133 **49.** 3.18 **50.** 20.6 **51. a.** 200 **b.** 248.2 **c.** 250 **52. a.** 5.6 **b.** 5.65 **c.** 5.6491
53. 0.15 **54.** 0.0825 **55.** 6.5% **56.** 120% **57.** $1050

58.

Fraction	Decimal	Percent
$\dfrac{1}{4}$	0.25	25%
$\dfrac{3}{8}$	0.375	37.5%
$\dfrac{5}{6}$	$0.83\dfrac{1}{3}$	$83\dfrac{1}{3}\%$
$8\dfrac{3}{4}$	8.75	875%
$2\dfrac{2}{5}$	2.4	240%
$\dfrac{3}{2000}$	0.0015	0.15%

59. 38.1% **60.** 42.3% **61.** 48 tons **62.** $21\dfrac{7}{8}$ in. **63.** $\dfrac{5}{16}$ in.

64. 30 in. × 41 in. **65.** 4020 **66.** 139

Chapter 2

Exercises 2.1   (pages 111–112)

1. 3 **3.** 6 **5.** 4 **7.** 17 **9.** 15 **11.** 10 **13.** 7 **15.** −2 **17.** −12 **19.** 6 **21.** −9 **23.** 4 **25.** −3 **27.** −1 **29.** −4
31. −6 **33.** −2 **35.** 4 **37.** 9 **39.** 3 **41.** 4 **43.** 7 **45.** −7 **47.** 5 **49.** −20 **51.** −2 **53.** 0 **55.** −5 **57.** −19
59. −12 **61.** 7 **63.** 11 **65.** 4 **67.** 4 **69.** −14

Exercises 2.2   (page 114)

1. −2 **3.** 11 **5.** 12 **7.** 6 **9.** 18 **11.** −6 **13.** −4 **15.** 0 **17.** 1 **19.** −10 **21.** −7 **23.** 15 **25.** −16 **27.** −10 **29.** −2
31. 14 **33.** 23 **35.** −15 **37.** 8 **39.** −4 **41.** 8 **43.** 2 **45.** −15 **47.** −3 **49.** −23 **51.** −4 **53.** −1 **55.** 2 **57.** −8
59. −10

Exercises 2.3   (pages 116–117)

1. 24 **3.** −18 **5.** −35 **7.** 27 **9.** −72 **11.** 27 **13.** 0 **15.** 49 **17.** −300 **19.** −13 **21.** −6 **23.** 48 **25.** 21 **27.** −16
29. 54 **31.** −6 **33.** 24 **35.** 27 **37.** −63 **39.** −9 **41.** −6 **43.** 36 **45.** 30 **47.** −168 **49.** −162 **51.** 5 **53.** −9 **55.** 8
57. 2 **59.** 9 **61.** −4 **63.** 3 **65.** 17 **67.** 40 **69.** 15 **71.** 7 **73.** 4 **75.** −3 **77.** 10 **79.** 8

Exercises 2.4   (page 121)

1. $-\dfrac{3}{16}$ **3.** $\dfrac{1}{16}$ **5.** $-12\dfrac{3}{20}$ **7.** $\dfrac{13}{18}$ **9.** $-\dfrac{1}{20}$ **11.** $1\dfrac{1}{16}$ **13.** $5\dfrac{3}{4}$ **15.** $-1\dfrac{3}{4}$ **17.** $-\dfrac{1}{63}$ **19.** 6 **21.** $-\dfrac{9}{10}$ **23.** $\dfrac{7}{24}$ **25.** $-\dfrac{9}{20}$

27. $\dfrac{3}{8}$ **29.** $1\dfrac{1}{4}$ **31.** $3\dfrac{1}{4}$ **33.** $\dfrac{1}{20}$ **35.** $-3\dfrac{1}{3}$ **37.** -48 **39.** -1 **41.** $-1\dfrac{1}{8}$ **43.** $\dfrac{1}{4}$ **45.** $-\dfrac{1}{4}$

Exercises 2.5   (page 125)

1. 10^{13} **3.** $\dfrac{1}{10^{4}}$ **5.** 10^{3} **7.** 10^{12} **9.** $\dfrac{1}{10^{10}}$ **11.** 10^{3} **13.** $\dfrac{1}{10^{12}}$ **15.** $\dfrac{1}{10^{4}}$ **17.** 10^{2} **19.** $\dfrac{1}{10^{6}}$ **21.** 10^{5} **23.** $\dfrac{1}{10^{4}}$ **25.** 10^{17}
27. 10^{6} **29.** 10^{14}

Exercises 2.6   (page 130)

1. 3.56×10^{2} **3.** 6.348×10^{2} **5.** 8.25×10^{-3} **7.** 7.4×10^{0} **9.** 7.2×10^{-5} **11.** 7.1×10^{5} **13.** 4.5×10^{-6} **15.** 3.4×10^{-8}
17. 6.4×10^{5} **19.** 75,500 **21.** 5310 **23.** 0.078 **25.** 0.000555 **27.** 64 **29.** 960 **31.** 5.76 **33.** 0.0000064
35. 50,000,000,000 **37.** 0.00000062 **39.** 2,500,000,000,000 **41.** 0.000000000033 **43.** 0.0048 **45.** 0.00091 **47.** 0.00037
49. 0.0613 **51.** 0.0009 **53.** 1.0009 **55.** 0.00000000998 **57.** 0.000271 **59.** 2.4×10^{-15} **61.** 3×10^{24} **63.** 1×10^{-6}
65. 1.728×10^{18} **67.** 1.13×10^{-1} **69.** 1.11×10^{-25} **71.** 9×10^{-1} **73.** 6.67×10^{1} **75.** 7.46×10^{5} **77.** 1.17×10^{10}
79. 9.06×10^{-11} **81.** 2.66×10^{24}

Exercises 2.7   (pages 133–134)

1. 28×10^{3} **3.** 3.45×10^{6} **5.** 220×10^{9} **7.** 6.6×10^{-3} **9.** 76.5×10^{-9} **11.** 975×10^{-3} **13.** 57,700
15. 4,940,000,000,000 **17.** 567,000,000 **19.** 0.000026 **21.** 0.000000005945 **23.** 0.00000000001064 **25.** 14.9×10^{18}
27. 19.7×10^{-6} **29.** 588×10^{12} **31.** 15.6×10^{-18} **33.** 339×10^{6} **35.** 123×10^{21} **37.** 8.97×10^{6} **39.** 1.31×10^{12}

Chapter 2 Review   (page 136)

1. 5 **2.** 16 **3.** 13 **4.** 3 **5.** -8 **6.** -3 **7.** -13 **8.** -3 **9.** -11 **10.** 19 **11.** -2 **12.** 0 **13.** -19 **14.** -24 **15.** 36
16. 72 **17.** -84 **18.** 6 **19.** -6 **20.** 5 **21.** $-\dfrac{1}{42}$ **22.** $\dfrac{1}{12}$ **23.** $-3\dfrac{1}{8}$ **24.** $-\dfrac{3}{2}$ **25.** $\dfrac{1}{10^{5}}$ **26.** 10^{9} **27.** $\dfrac{1}{10^{12}}$ **28.** 1
29. 4.76×10^{5} **30.** 1.4×10^{-3} **31.** 0.0000535 **32.** 61,000,000 **33.** 0.00105 **34.** 0.06 **35.** 0.000075 **36.** 0.00183
37. 4.37×10^{-2} **38.** 2.8×10^{14} **39.** 2×10^{20} **40.** 2.025×10^{-15} **41.** 1.6×10^{37} **42.** 6.4×10^{7} **43.** 275×10^{3}
44. 32×10^{6} **45.** 450×10^{-6} **46.** 31,600,000 **47.** 0.746 **48.** 4.73×10^{6} **49.** 24.3×10^{-15} **50.** 46.1×10^{3}

Cumulative Review Chapters 1–2   (page 138)

1. 72 **2.** 51 cm^{2} **3.** 3750 Ω **4.** No **5.** $2 \cdot 3 \cdot 3 \cdot 5 \cdot 7$ **6.** $3\dfrac{5}{9}$ **7.** 4480 ft^{2} **8.** $\dfrac{13}{16}$ **9.** $1\dfrac{1}{16}$ **10.** $1\dfrac{7}{8}$ **11.** $\dfrac{1}{20}$ **12.** 83
13. a. 600 **b.** 615.3 **c.** 620 **d.** 615.288 **14.** 0.074 **15.** $3990 **16.** 662.5 **17.** 9.43% **18.** 6.25% **19.** 10 **20.** -432
21. $-2\dfrac{15}{16}$ **22.** $-\dfrac{25}{64}$ **23.** 3.1818×10^{5} **24.** 0.00213 **25.** $\dfrac{1}{10^{4}}$ **26.** 4.5×10^{3} **27.** 270×10^{-6} **28.** 0.000000281
29. 16,300,000 **30.** 7.02×10^{10} **31.** 4.75×10^{-3} **32.** 3.46×10^{-15} **33.** 2.07×10^{-3}

Chapter 3

Exercises 3.1   (pages 142–143)

1. kilo **3.** centi **5.** milli **7.** mega **9.** h **11.** d **13.** c **15.** μ **17.** 65 mg **19.** 82 cm **21.** 36 μA **23.** 19 hL
25. 18 metres **27.** 36 kilograms **29.** 24 picoseconds **31.** 135 millilitres **33.** 45 milliamperes **35.** metre **37.** ampere
39. litre and cubic metre

Exercises 3.2   (pages 145–146)

1. 1 metre **3.** 1 kilometre **5.** 1 centimetre **7.** 1000 **9.** 0.01 **11.** 0.001 **13.** 0.001 **15.** 10 **17.** 100 **19.** cm **21.** mm
23. m **25.** km **27.** mm **29.** mm **31.** km **33.** m **35.** cm **37.** mm; mm **39.** *A*: 52 mm; 5.2 cm *B*: 11 mm; 1.1 cm
C: 137 mm; 13.7 cm *D*: 95 mm; 9.5 cm *E*: 38 mm; 3.8 cm *F*: 113 mm; 11.3 cm **41.** 52 mm; 5.2 cm **43.** 79 mm; 7.9 cm

45. 65 mm; 6.5 cm **47.** 102 mm; 10.2 cm **49.** 0.675 km **51.** 1.54 m **53.** 0.65 m **55.** 730 cm **57.** 1.25 km **59.** 27.5 cm
61. 12.5 cm **63.** Answers vary.

Exercises 3.3   (pages 147–148)
1. 1 gram **3.** 1 kilogram **5.** 1 milligram **7.** 1000 **9.** 0.01 **11.** 1000 **13.** 1000 **15.** g **17.** kg **19.** g **21.** metric ton
23. mg **25.** g **27.** mg **29.** g **31.** kg **33.** metric ton **35.** 0.875 kg **37.** 85,000 mg **39.** 3600 g **41.** 0.27 g **43.** 0.885 mg
45. 0.375 mg **47.** 2500 kg **49.** 225 metric tons **51.** Answers vary.

Exercises 3.4   (pages 151–152)
1. 1 litre **3.** 1 cubic centimetre **5.** 1 square kilometre **7.** 1000 **9.** 1,000,000 **11.** 100 **13.** 1000 **15.** L **17.** m^2 **19.** cm^2
21. m^3 **23.** mL **25.** ha **27.** m^3 **29.** L **31.** m^2 **33.** ha **35.** cm^2 **37.** cm^2 **39.** 1.5 L **41.** 1,500,000 cm^3 **43.** 85 mL
45. 0.085 km^2 **47.** 8.5 ha **49.** 500 g **51.** 0.675 ha

Exercises 3.5   (page 154)
1. 1 amp **3.** 1 second **5.** 1 megavolt **7.** 43 kW **9.** 17 ps **11.** 3.2 MW **13.** 450 Ω **15.** 1000 **17.** 0.000000001
19. 1,000,000 **21.** 0.000001 **23.** 350 mA **25.** 0.35 s **27.** 3 h 52 min 30 s **29.** 0.175 mF **31.** 1.5 MHz

Exercises 3.6   (page 156)
1. b **3.** c **5.** b **7.** c **9.** d **11.** 25 **13.** 617 **15.** 3.2 **17.** −26.7 **19.** −108.4

Exercises 3.7   (pages 159–160)
1. 3.63 **3.** 15.0 **5.** 366 **7.** 66.0 **9.** 81.3 **11.** 46.6 mi **13.** 10.8 mm **15.** 30.3 L **17. a.** 38 oz **b.** 1.08 kg **19. a.** 300 ft
b. 91.4 m **21. a.** 12.7 cm **b.** 127 mm **23.** 2.51 m^2 **25.** 1260 ft^2 **27.** 116 cm^2 **29.** 1170 ft^2 **31.** 11.5 m^3 **33.** 279,000 mm^3
35. 2,380,000 cm^3 **37.** $2.25/$ft^2$; $405/frontage ft **39.** 34.4 acres **41.** 13.9 ha **43.** 0.619 acre **45.** 80 acres
47. 9397 lb/acre; 168 bu/acre **49.** 0.606 acre **51.** 363 lb/in^2 **53.** 29.1 m/s

Chapter 3 Review   (page 162)
1. milli **2.** kilo **3.** M **4.** μ **5.** 42 mL **6.** 8.3 ns **7.** 18 kilometres **8.** 350 milliamperes **9.** 50 microseconds **10.** 1 L
11. 1 MW **12.** 1 km^2 **13.** 1 m^3 **14.** 0.65 **15.** 0.75 **16.** 6100 **17.** 4.2×10^6 **18.** 1.8×10^7 **19.** 25,000 **20.** 25,000
21. 2.5 **22.** 6×10^5 **23.** 250 **24.** 22.2 **25.** −13 **26.** 0 **27.** 100 **28.** 81.7 **29.** 38.4 **30.** 142 **31.** 1770 **32.** 162
33. 176 **34.** 6.07 **35.** c **36.** a **37.** d **38.** d **39.** b **40.** b **41.** a
42.

Prefix	Symbol	Power of 10	Sample unit	How many?	How many?
tera	T	10^{12}	m	10^{12} m = 1 Tm	1 m = 10^{-12} Tm
giga	G	10^9	W	10^9 W = 1 GW	1 W = 10^{-9} GW
mega	M	10^6	Hz	10^6 Hz = 1 MHz	1 Hz = 10^{-6} MHz
kilo	k	10^3	g	10^3 g = 1 kg	1 g = 10^{-3} kg
hecto	h	10^2	Ω	10^2 Ω = 1 hΩ	1 Ω = 10^{-2} hΩ
deka	da	10^1	L	10^1 L = 1 daL	1 L = 10^{-1} daL
deci	d	10^{-1}	g	10^{-1} g = 1 dg	1 g = 10^1 dg
centi	c	10^{-2}	m	10^{-2} m = 1 cm	1 m = 10^2 cm
milli	m	10^{-3}	A	10^{-3} A = 1 mA	1 A = 10^3 mA
micro	μ	10^{-6}	W	10^{-6} W = 1 μW	1 W = 10^6 μW
nano	n	10^{-9}	s	10^{-9} s = 1 ns	1 s = 10^9 ns
pico	p	10^{-12}	s	10^{-12} s = 1 ps	1 s = 10^{12} ps

Chapter 4

Exercises 4.1   (page 169)
1. 3 **3.** 4 **5.** 4 **7.** 4 **9.** 3 **11.** 4 **13.** 3 **15.** 3 **17.** 3 **19.** 4 **21.** 4 **23.** 5 **25.** 2 **27.** 6 **29.** 4 **31.** 6 **33.** 4 **35.** 2

Exercises 4.2   (page 173)
1. a. 0.01 A **b.** 0.005A **3. a.** 0.01 cm **b.** 0.005 cm **5. a.** 1 km **b.** 0.5 km **7. a.** 0.01 mi **b.** 0.005 mi **9. a.** 0.001 A
b. 0.0005 A **11. a.** 0.0001 W **b.** 0.00005 W **13. a.** 10 Ω **b.** 5 Ω **15. a.** 1000 L **b.** 500 L **17. a.** 0.1 cm **b.** 0.05 cm

19. a. 10 V **b.** 5 V **21. a.** 0.001 m **b.** 0.0005 m **23. a.** $\frac{1}{3}$ yd **b.** $\frac{1}{6}$ yd **25. a.** $\frac{1}{32}$ in. **b.** $\frac{1}{64}$ in. **27. a.** $\frac{1}{16}$ mi **b.** $\frac{1}{32}$ mi **29. a.** $\frac{1}{9}$ in^2 **b.** $\frac{1}{18}$ in^2

Exercises 4.3A   (pages 176–177)

1. 27.20 mm **3.** 63.55 mm **5.** 8.00 mm **7.** 115.90 mm **9.** 71.45 mm **11.** 10.25 mm **13.** 34.60 mm **15.** 68.45 mm **17.** 5.90 mm **19.** 43.55 mm **21.** 76.10 mm **23.** 12.30 mm

Exercises 4.3B   (page 180)

1. 1.362 in. **3.** 2.695 in. **5.** 0.234 in. **7.** 1.715 in. **9.** 2.997 in. **11.** 0.483 in. **13.** 1.071 in. **15.** 2.503 in. **17.** 0.316 in. **19.** 4.563 in. **21.** 2.813 in. **23.** 0.402 in.

Exercises 4.4A   (pages 183–184)

1. 4.25 mm **3.** 3.90 mm **5.** 1.75 mm **7.** 7.77 mm **9.** 5.81 mm **11.** 10.28 mm **13.** 7.17 mm **15.** 8.75 mm **17.** 6.23 mm **19.** 5.42 mm

Exercises 4.4B   (pages 186–187)

1. 0.237 in. **3.** 0.314 in. **5.** 0.147 in. **7.** 0.820 in. **9.** 0.502 in. **11.** 0.200 in. **13.** 0.321 in. **15.** 0.170 in. **17.** 0.658 in. **19.** 0.245 in.

Exercises 4.5   (pages 192–193)

1. a. 14.7 in. **b.** 0.017 in. **3. a.** 16.01 mm **b.** 0.737 mm **5. a.** 0.0350 A **b.** 0.00050 A **7. a.** All have the same accuracy. **b.** 0.391 cm **9. a.** 205,000 Ω **b.** 205,000 Ω and 45,000 Ω **11. a.** 0.04 in. **b.** 15.5 in. **13. a.** 0.48 cm **b.** 43.4 cm **15. a.** 0.00008 A **b.** 0.91 A **17. a.** 0.6 m **b.** All have the same precision. **19. a.** 500,000 Ω **b.** 500,000 Ω **21.** 18.1 m **23.** 94.8 cm **25.** 97,000 W **27.** 840,000 V **29.** 19 V **31.** 459 mm or 45.9 cm **33.** 126.4 cm **35.** 8600 mi **37.** 35 mm or 3.5 cm **39.** 65.4 g **41.** 0.330 in. **43.** 26.0 mm **45.** 12.7 ft **47.** 67 lb **49.** 1.3 gal **51.** 124 gal **53.** 10.66666 **55.** 9 lb

Exercises 4.6   (pages 195–196)

1. 4400 m^2 **3.** 1,230,000 cm^2 **5.** 901 m^2 **7.** 0.13 ΩA **9.** 7360 cm^3 **11.** 4.7×10^9 m^3 **13.** 35 A^2 Ω **15.** 2500 in^2 **17.** 40 m **19.** 340 V/A **21.** 2.1 km/s **23.** $3\bar{0}0$ V^2/Ω **25.** 4.0 g/cm^3 **27.** $9\bar{0}0$ ft^3 **29.** 28 m **31.** 28.8 hp **33.** $1\bar{0}00$ m^3 **35.** 270 ft^3/min **37.** 4700 in^3 **39.** 9.6 gal/h **41.** 1560 in^2 **43.** 19 ft^3 **45.** 9 **47.** 580 ppm **49.** 2.25 mi^3

Exercises 4.7   (pages 198–199)

1. 100 lb; 50 lb; 0.0357; 3.57% **3.** 1 rpm; 0.5 rpm; 0.000571; 0.06% **5.** 0.001 g; 0.0005 g; 0.00588; 0.59% **7.** 1 g; 0.5 g; 0.25; 25% **9.** 0.01 g; 0.005 g; 0.00225; 0.23% **11.** 0.01 kg; 0.005 kg; 0.005; 0.5% **13.** 0.001 A; 0.0005 A; 0.0122; 1.22%

15. $\frac{1}{8}$ in.; $\frac{1}{16}$ in.; 0.00526; 0.53% **17.** 1 in.; 0.5 in.; 0.00329; 0.33% **19.** 13.5 cm **21.** 19.7 g **23.** Answer in text.

	Upper limit	Lower limit	Tolerance interval
25.	$6\frac{21}{32}$ in.	$6\frac{19}{32}$ in.	$\frac{1}{16}$ in.
27.	$3\frac{29}{64}$ in.	$3\frac{27}{64}$ in.	$\frac{1}{32}$ in.
29.	$3\frac{25}{128}$ in.	$3\frac{23}{128}$ in.	$\frac{1}{64}$ in.
31.	1.24 cm	1.14 cm	0.10 cm
33.	0.0185 A	0.0175 A	0.0010 A
35.	26,000 V	22,000 V	4000 V
37.	10.36 km	10.26 km	0.10 km

39. $53,075 **41.** $1,567,020.60

Exercises 4.8   (pages 203–204)

1. 360 Ω; ±10%   **3.** 830,000 Ω; ±20%   **5.** 1,400,000 Ω, ±20%   **7.** 70 Ω; ±5%   **9.** 500,000 Ω; ±20%   **11.** 10,000,000 Ω, ±20%
13. yellow, gray, red   **15.** violet, red, orange   **17.** blue, green, yellow   **19.** red, green, silver   **21.** yellow, green, green
23. violet, blue, gold   **25. a.** 36 Ω   **b.** 396 Ω   **c.** 324 Ω   **d.** 72 Ω   **27. a.** 166,000 Ω   **b.** 996,000 Ω   **c.** 664,000 Ω
d. 332,000 Ω   **29. a.** 3.5 Ω   **b.** 73.5 Ω   **c.** 66.5 Ω   **d.** 7 Ω

Exercises 4.9   (pages 209–212)

1. +1.16 mm   **3.** −0.67 mm   **5.** −3.40 mm   **7.** +1.74 mm   **9.** −4.08 mm   **11.** −0.056 in.   **13.** +0.231 in.   **15.** −0.188 in.
17. +0.437 in.   **19.** 6 V   **21.** 6.4 V   **23.** 0.4 V   **25.** 1.4 V   **27.** 40 V   **29.** 230 V   **31.** 7 Ω   **33.** 12 Ω   **35.** 11 Ω   **37.** 35 Ω
39. 85 Ω   **41.** 300 Ω

Chapter 4 Review   (pages 214–216)

1. 3   **2.** 2   **3.** 3   **4.** 3   **5.** 2   **6.** 3   **7.** 4   **8.** 5   **9. a.** 0.01 m   **b.** 0.005 m   **10. a.** 0.1 mi   **b.** 0.05 mi   **11. a.** 100 L   **b.** 50 L
12. a. 100 V   **b.** 50 V   **13. a.** 0.01 cm   **b.** 0.005 cm   **14. a.** 10,000 V   **b.** 5000 V   **15. a.** $\frac{1}{8}$ in.   **b.** $\frac{1}{16}$ in.   **16. a.** $\frac{1}{16}$ mi
b. $\frac{1}{32}$ mi   **17.** 42.35 mm   **18.** 1.673 in.   **19.** 11.84 mm   **20.** 0.438 in.   **21. a.** 36,500 V   **b.** 9.6 V   **22. a.** 0.0005 A   **b.** 0.425 A
23. 720,000 W   **24.** 4$\overline{0}$0 m   **25.** 400,000 V   **26.** 1900 cm³   **27.** 5.88 m²   **28.** 1.4 N/m²   **29.** 130 V²/Ω   **30.** 0.0057; 0.57%
31. 0.00032; 0.03%   **32.** 2200 Ω; 1800 Ω   **33.** 120,000 Ω; ±20%   **34.** 0.85 Ω; ±5%   **35.** −0.563 in.   **36.** 8.4 V   **37.** 2.6 Ω

Cumulative Review Chapters 1–4   (pages 217–218)

1. 27   **2. a.** 32,520   **b.** 32,518.61   **3.** 60   **4.** 102   **5.** $\frac{1}{10^5}$ **6.** 8.70 × 10⁵   **7.** m   **8.** 25 kg   **9.** 250 microseconds
10. 1 mega amp   **11.** 120,000 m   **12.** 2.5 m   **13.** 0.05 kg   **14.** 4.06 metric tons   **15.** 186.8°F   **16.** 10°C   **17.** 1,050 cm²
18. 0.12 km   **19.** 10,000 mL   **20.** 2   **21.** 4   **22. a.** 0.01 mm   **b.** 0.005 mm   **23. a.** 0.1 lb   **b.** 0.05 lb   **24.** 77.75 mm
25. 3.060 in.   **26.** 7.53 mm   **27.** 0.537 in.   **28.** 34,900 km   **29.** 46.0 L   **30.** 42,000 cm²   **31.** 33.7 ft   **32. a.** 0.001 cm
b. 0.0005 cm   **c.** 0.000234   **d.** 0.02%   **33.** 165 V   **34.** 110 Ω

Chapter 5

Exercises 5.1   (page 222)

1. 83   **3.** 8   **5.** 42   **7.** −6   **9.** −128   **11.** 1   **13.** −5   **15.** $\frac{3}{2}$ **17.** −24   **19.** $-9\frac{3}{5}$ **21.** −13   **23.** $-1\frac{3}{5}$ **25.** $\frac{1}{3}$
27. 331,776   **29.** −72   **31.** 50   **33.** 8   **35.** −78   **37.** −1   **39.** 25   **41.** 2

Exercises 5.2   (pages 225–226)

1. $a + b + c$ **3.** $a + b + c$ **5.** $a − b − c$ **7.** $x + y − z + 3$ **9.** $x − y − z − 3$ **11.** $2x + 4 + 3y + 4r$
13. $3x − 5y + 6z − 2w + 11$ **15.** $−5x − 3y − 6z + 3w + 3$ **17.** $2x + 3y − z − w + 3r − 2s − 10$
19. $−2x + 3y − z − 4w − 4r + s$ **21.** $2b$ **23.** $3x^2 + 10x$ **25.** $3m$ **27.** $a + 12b$ **29.** $6a^2 − a + 1$ **31.** $3x^2 + 3x$
33. $−1.8x$ **35.** $\frac{8}{9}x − \frac{1}{8}y$ **37.** $2x^2y − 2xy + 2y^2 − 3x^2$ **39.** $5x^3 + 3x^2y − 5y^3 + y$ **41.** 1   **43.** $3x + 4$
45. $5 − x$ **47.** $3y − 7$ **49.** $4y + 5$ **51.** 5   **53.** 6   **55.** 28   **57.** $\frac{5}{4}x − \frac{8}{3}$ **59.** $12x + 36y$ **61.** $−36x^2 + 48y^2$ **63.** $2x + 19$
65. $−8.5y − 4$ **67.** $−42$ **69.** $4n + 8$ **71.** $1.8x − 7$ **73.** $−6n − 2$ **75.** $−5x + 6$ **77.** $−1.05x − 8.4$

Exercises 5.3   (page 229)

1. binomial   **3.** binomial   **5.** monomial   **7.** trinomial   **9.** binomial   **11.** $x^2 − x + 1$; 2nd   **13.** $7x^2 + 4x − 1$; 2nd
15. $5x^3 − 4x^2 − 2$; 3rd   **17.** $4y^3 − 6y^2 − 3y + 7$; 3rd   **19.** $−7x^5 − 4x^4 + x^3 + 2x^2 + 5x − 3$; 5th   **21.** $7a^2 − 10a + 1$
23. $9x^2 − 5x$ **25.** $9a^3 + 4a^2 + 5a − 5$ **27.** $4x + 4$ **29.** $5x^2 + 18x − 22$ **31.** $5x^3 + 13x^2 − 8x + 7$ **33.** $8y^2 − 5y + 6$
35. $4a^3 − 3a^2 − 5a$ **37.** $2x^2 + 6x + 2$ **39.** $−3x^2 + 2x + 2$ **41.** $a + 3b$ **43.** $7a − 4b − 3x + 4y$ **45.** $x^2 − 3x + 5$
47. $7w^2 − 24w − 6$ **49.** $x^2 − 4x − 10$ **51.** $4x^2 − 4x + 12$ **53.** $−6x^2 + 2x − 10$ **55.** $x^3 + 3x − 3$ **57.** $8x^5 − 18x^4 + 5x^2 + 1$

Exercises 5.4   (page 232)

1. $-15a$ **3.** $28a^3$ **5.** $54m^4$ **7.** $32a^8$ **9.** $-26p^2q$ **11.** $30n^3m$ **13.** $21a^4b$ **15.** $\dfrac{3}{8}x^3y^4$ **17.** $24a^3b^4c^3$ **19.** $\dfrac{3}{16}x^3y^5z^3$

21. $-371.64m^3n^2p^2$ **23.** $-255a^6b$ **25.** x^6 **27.** x^{24} **29.** $9x^8$ **31.** $-x^9$ **33.** x^{10} **35.** x^{30} **37.** $25x^6y^4$ **39.** $225m^4$

41. $15{,}625n^{12}$ **43.** $9x^{12}$ **45.** $8x^9y^{12}z^3$ **47.** $-32h^{15}k^{30}m^{10}$ **49.** -408 **51.** -144 **53.** $17{,}712$ **55.** 16 **57.** -1728

59. $291{,}600$ **61.** -1080 **63.** 324 **65.** -324 **67.** 16

Exercises 5.5   (page 234)

1. $4a + 24$ **3.** $-18x^2 - 12y$ **5.** $4ax^2 - 6ay + a$ **7.** $3x^3 - 2x^2 + 5x$ **9.** $6a^3 + 12a^2 - 20a$ **11.** $-12x^3 + 21x^2 + 6x$

13. $-28x^3 - 12xy + 8x^2y$ **15.** $3x^3y^2 - 3x^2y^3 + 12x^2y^2$ **17.** $-6x^3 + 36x^5 - 54x^7$ **19.** $5a^4b^2 - 5ab^5 - 5a^2b^3$ **21.** $\dfrac{28}{3}mn - 8m^2$

23. $16y^2z^3 - \dfrac{8}{35}yz^4$ **25.** $-5.2a^6 - 10a^3 - 4a$ **27.** $1334.4a^3 + 1668a^2$ **29.** $24x^4y - 16x^3y^2 + 20x^2y^3$

31. $\dfrac{1}{2}a^3b^3 - \dfrac{1}{3}a^2b^5 + \dfrac{5}{9}ab^6$ **33.** $-19x^2 + 8x$ **35.** $3x^2y - 2x^2y^2 + 2x^2y^3 - 7xy^3$ **37.** $x^2 + 7x + 6$ **39.** $x^2 + 5x - 14$

41. $x^2 - 13x + 40$ **43.** $3a^2 - 17a + 20$ **45.** $12a^2 - 10a - 12$ **47.** $24a^2 + 84a + 72$ **49.** $15x^2 - 4xy - 4y^2$ **51.** $4x^2 - 12x + 9$

53. $4c^2 - 25d^2$ **55.** $91m^2 + 32m - 3$ **57.** $x^8 - 2x^5 + x^2$ **59.** $10y^3 - 24y^2 - 32y + 16$ **61.** $24x^2 - 78x - 6y^2 - 39y$

63. $g^2 - 3g - h^2 + 9h - 18$ **65.** $10x^7 - 3x^6 - x^5 + 44x^4 + x^3 - x^2 + 16x - 2$

Exercises 5.6   (page 236)

1. $3x^2$ **3.** $\dfrac{3x^8}{2}$ **5.** $\dfrac{6}{x^2}$ **7.** $\dfrac{2x}{3}$ **9.** x **11.** $\dfrac{5}{2}$ **13.** $\dfrac{5a^2}{b}$ **15.** $\dfrac{8}{mn}$ **17.** 0 **19.** $23p^2$ **21.** -2 **23.** $\dfrac{36}{r^2}$ **25.** $-\dfrac{23x^2}{7y^2}$ **27.** $\dfrac{8}{7a^3b^2}$

29. $\dfrac{4x^2z^4}{9y}$ **31.** $2x^2 - 4x + 3$ **33.** $x^2 + x + 1$ **35.** $x - y - z$ **37.** $3a^4 - 2a^2 - a$ **39.** $b^9 - b^6 - b^3$ **41.** $-x^3 + x^2 - x + 4$

43. $12x + 6x^2y - 3$ **45.** $8x^3z - 6x^2yz^2 - 4y^2$ **47.** $4y^3 - 3y - \dfrac{2}{y}$ **49.** $\dfrac{1}{2x^2} - 3 - 2x^2$

Exercises 5.7   (pages 238–239)

1. $x + 2$ **3.** $3a + 3$ r 11 **5.** $4x - 3$ r -3 **7.** $y + 2$ r -3 **9.** $3b - 4$ **11.** $6x^2 + x - 1$ **13.** $4x^2 + 7x - 15$ r 5

15. $2x^2 - 2x - 12$ **17.** $2x^2 - 16x + 32$ **19.** $3x^2 + 10x + 20$ r 34 **21.** $2x^2 + 6x + 30$ r 170 **23.** $2x^3 + 4x + 6$

25. $4x^2 - 2x + 1$ r -2 **27.** $x^3 - 2x^2 + 4x - 8$ **29.** $3x^2 - 4x + 1$

Chapter 5 Review   (page 241)

1. a **2.** 0 **3.** 1 **4.** -2 **5.** 50 **6.** 2 **7.** 1 **8.** 9 **9.** -30 **10.** $-\dfrac{9}{2}$ **11.** -12 **12.** 5 **13.** $6y - 5$ **14.** $6 - 8x$

15. $10x + 27$ **16.** $3x^3 + x^2y - 3y^3 - y$ **17.** binomial **18.** 4 **19.** $8a^2 + 5a + 2$ **20.** $9x^3 + 4x^2 + x + 2$ **21.** $10x^2 - 7x + 4$

22. $24x^5$ **23.** $-56x^5y^3$ **24.** $27x^6$ **25.** $15a^2 + 20ab$ **26.** $-32x^2 + 8x^3 - 12x^4$ **27.** $15x^2 - 11x - 12$ **28.** $6x^3 - 24x^2 + 26x - 4$

29. $\dfrac{7}{x}$ **30.** $5x^2$ **31.** $4a^2 - 3a + 1$ **32.** $3x - 4$ **33.** $3x^2 - 4x + 2$

Chapter 6

Exercises 6.1   (page 249)

1. 6 **3.** 17 **5.** $10\dfrac{1}{2}$ **7.** 19.5 **9.** 5.2 **11.** 301 **13.** 7 **15.** 20 **17.** $4\dfrac{1}{14}$ **19.** 0 **21.** 0 **23.** 16 **25.** 392 **27.** -4 **29.** -2

31. 2 **33.** -2 **35.** 32 **37.** -4 **39.** 18 **41.** $5\dfrac{2}{3}$ **43.** 42 **45.** $10\dfrac{1}{2}$ **47.** $-\dfrac{1}{3}$ **49.** -1 **51.** $\dfrac{77}{135}$ or 0.570 **53.** $1\dfrac{5}{7}$ **55.** $2\dfrac{12}{29}$

57. $-3\dfrac{1}{3}$ **59.** 25

Exercises 6.2   (pages 251)

1. 8 **3.** 4 **5.** 4 **7.** 3 **9.** 6 **11.** 6 **13.** -5 **15.** 7 **17.** $2\dfrac{1}{5}$ **19.** -9 **21.** 3 **23.** 13 **25.** $\dfrac{3}{4}$ **27.** -1 **29.** 4

Exercises 6.3   (pages 253–254)

1. 5 **3.** $\frac{2}{5}$ **5.** $-1\frac{1}{3}$ **7.** 6 **9.** -5 **11.** $7\frac{1}{5}$ **13.** -5 **15.** 0 **17.** $1\frac{1}{7}$ **19.** 9 **21.** 14 **23.** 45 **25.** 4 **27.** 22 **29.** -1

31. 10 **33.** 18 **35.** 2 **37.** -8 **39.** -2 **41.** -1 **43.** -7 **45.** $8\frac{1}{4}$ **47.** $-1\frac{1}{3}$ **49.** -3 **51.** 6 **53.** -2 **55.** 0 **57.** $\frac{19}{24}$

59. $6\frac{9}{44}$

Exercises 6.4   (page 258)

1. 8 **3.** 5 **5.** 3 **7.** 16 **9.** 1 **11.** 50 **13.** 6 **15.** 1 **17.** 24 **19.** 4 **21.** -56 **23.** 6 **25.** -3 **27.** $\frac{2}{3}$ **29.** $-\frac{1}{2}$

31. 1 **33.** $\frac{1}{5}$ **35.** 3 **37.** $\frac{2}{3}$ **39.** 5 **41.** $\frac{1}{7}$

Exercises 6.5   (page 260)

1. $x - 20$ **3.** $\frac{x}{6}$ **5.** $x + 16$ **7.** $26 - x$ **9.** $2x$ **11.** $6x + 28 = 40$ **13.** $\frac{x}{6} = 5$ **15.** $5(x + 28) = 150$ **17.** $\frac{x}{6} - 7 = 2$

19. $30 - 2x = 4$ **21.** $(x - 7)(x + 5) = 13$ **23.** $6x - 17 = 7$ **25.** $4x - 17 = 63$

Exercises 6.6   (pages 264–265)

1. 10 in. **3.** 92 incandescent and 164 fluorescent bulbs **5.** \$825 to John; \$1650 to Maria; \$2475 to Betsy **7.** 10 cm by 20 cm

9. 105 ft by 120 ft **11.** 42 ft; 42 ft; 38 ft **13.** $5\frac{1}{4}$ ft; $6\frac{3}{4}$ ft **15.** 8 @ \$6.50; 12 @ \$9.50 **17.** \$4500 @ 2.5%, \$3000 @ 4%

19. 20 L **21.** 320 mL of 30%; 480 mL of 80% **23.** 4 qt **25.** 558 ft^3

Exercises 6.7   (pages 267–268)

1. $\frac{E}{I}$ **3.** $\frac{F}{m}$ **5.** $\frac{C}{\pi}$ **7.** $\frac{V}{lh}$ **9.** $\frac{A}{2\pi r}$ **11.** $\frac{v^2}{2g}$ **13.** $\frac{Q}{I}$ **15.** vt **17.** $\frac{V}{I}$ **19.** $4\pi Er^2$ **21.** $\frac{1}{2\pi CX_C}$ **23.** $\frac{2A}{h}$ **25.** $\frac{QJ}{I^2t}$

27. $\frac{5}{9}(F - 32)$ or $\frac{5F - 160}{9}$ **29.** $C_T - C_1 - C_3 - C_4$ **31.** $\frac{-By - C}{A}$ **33.** $\frac{Q_1 + PQ_1}{P}$ or $\frac{Q_1}{P} + Q_1$ **35.** $\frac{2A}{a + b}$ **37.** $\frac{l - a}{n - 1}$

39. $\frac{Ft}{V_2 - V_1}$ **41.** $\frac{Q}{w(T_1 - T_2)}$ **43.** $\frac{PV}{2\pi} - 3960$

Exercises 6.8   (pages 270–271)

1. a. $l = \frac{A}{w}$ **b.** 23.0 **3. a.** $h = \frac{3V}{\pi r^2}$ **b.** 20.0 **5. a.** $m = \frac{2E}{v^2}$ **b.** 2000 **7. a.** $t = \frac{v_f - v_i}{a}$ **b.** 9.90 **9. a.** $h = \frac{v_f^2 - v_i^2}{2g}$

b. 576 **11. a.** $r_1 = \frac{L - 2d}{\pi} - r_2$ **b.** 3.00 **13. a.** $r = \frac{Wv^2}{Fg}$ **b.** 1900 **15. a.** $b = \frac{2A}{h} - a$ **b.** 49.0 **17. a.** $d = \frac{2V}{lw} - D$

b. 3.00 **19. a.** $a = \frac{2S}{n} - l$ **b.** 16.6 **21.** 324 W **23.** 6.72 ft **25.** 12.1 m **27.** 15.0 Ω **29.** 0.43 in.

Exercises 6.9   (pages 274–275)

1. 4.80 Ω **3.** 18.0 Ω **5.** 40.0 Ω **7.** 2.50 cm **9.** 44.5 cm **11.** 9.00 Ω **13.** 318 Ω **15.** 3240 Ω **17.** 6.00 μF
19. 1.91×10^{-6} F **21.** 1.74×10^{-8} F **23.** 219 Ω

Chapter 6 Review   (pages 276–277)

1. $1\frac{1}{2}$ **2.** -4 **3.** 57 **4.** 24 **5.** -7 **6.** 9 **7.** 8 **8.** 3 **9.** 6 **10.** $\frac{3}{5}$ **11.** 1 **12.** 6 **13.** 4 **14.** -2 **15.** 6 **16.** $\frac{1}{2}$ **17.** $2\frac{2}{3}$

18. $-7\frac{1}{2}$ **19.** 26 **20.** 2 **21.** 5 **22.** 6 in. by 18 in. **23.** 4.5 L of 100%; 7.5 L of 60% **24.** $g = \frac{F}{W}$ **25.** $A = \frac{W}{P}$

26. $t = 2L - 2A - 2B$ **27.** $m = \frac{2k}{v^2}$ **28.** $T_1 = \frac{P_1 T_2}{P_2}$ **29.** $v_0 = 2v - v_f$ **30.** 347 **31.** 19.5 **32.** 37 **33.** 30.0 Ω **34.** 70.6 μF

Cumulative Review Chapters 1–6   (page 278)

1. $2 \cdot 2 \cdot 2 \cdot 3 \cdot 29$ **2.** 8.1% **3.** 0.0003015 **4.** 2.85×10^4 **5.** 50,000 **6.** 38.3° **7.** 43.4 **8. a.** 2 **b.** 1 **c.** 5
9. a. 55.60 mm **b.** 2.189 in. **10.** 0.428 in. **11.** 494,000 W **12.** $-8x + 8y$ **13.** $2y^3 + 4y^2 + 5y - 11$ **14.** $27y^9$

15. $-2x^3 + 6x^2 - 8x$ **16.** $12y^4 - 16y^3 + 3y^2 + 5y - 2$ **17.** $20x^2 - 7xy - 6y^2$ **18.** $\frac{43}{9xy^2}$ **19.** $-80x^6y^8$

20. $x^2 - 3x + 4 - \frac{40}{x + 5}$ **21.** $6x^2 - 11xy + 15y^2$ **22.** $\frac{7}{2}$ **23.** 56 **24.** -6 **25.** $2\frac{2}{5}$ **26.** $\frac{11}{2}$ **27.** $a = 2C - b - c$

28. 11.1 m **29.** $7x = 250$ **30.** 5 ft × 10 ft

Chapter 7

Exercises 7.1 (pages 282–283)

1. $\frac{1}{5}$ **3.** $\frac{1}{3}$ **5.** $\frac{5}{3}$ **7.** $\frac{1}{5}$ **9.** $\frac{2}{1}$ **11.** $\frac{1}{32}$ **13.** $\frac{9}{14}$ **15.** $\frac{11}{16}$ **17.** $\frac{2}{1}$ **19.** $\frac{4}{1}$ **21.** $\frac{7}{5}$ **23.** $\frac{1}{4}$ **25.** 30 mi/gal **27.** 46 gal/h

29. 50 mi/h **31.** $\frac{3}{8}$ lb/gal **33.** $\frac{32}{5}$ or 6.4 to 1 **35.** 33 gal/min **37.** $\frac{1}{275}$ **39.** $\frac{12}{1}$ **41.** 36 lb/bu **43.** 25 gal/acre **45.** $1.69/ft

47. $105/ft^2 **49.** $\frac{3}{4}$ **51.** $\frac{4}{25}$ **53.** 50 mg/cm^3 **55.** 4 mg/cm^3 **57.** 1 male cougar/150 mi^2 **59.** 50 drops/min **61.** 45 drops/min

63. $3\frac{1}{3}$ h **65.** $8\frac{1}{3}$ h

Exercises 7.2 (pages 288–289)

1. a. 2, 3 **b.** 1, 6 **c.** 6 **d.** 6 **3. a.** 9, 28 **b.** 7, 36 **c.** 252 **d.** 252 **5. a.** 7, w **b.** x, z **c.** $7w$ **d.** xz **7.** yes; 30 = 30

9. no; 60 ≠ 90 **11.** yes; 12 = 12 **13.** 3 **15.** $5\frac{3}{5}$ **17.** 14 **19.** $3\frac{1}{3}$ **21.** 35 **23.** −7.5 **25.** 3.5 **27.** 0.5 **29.** 126 **31.** 20.6

33. $37\frac{1}{3}$ **35.** 38.2 **37.** 818 **39.** 9050 **41.** 21.3 **43.** 44 ft^3 **45.** $228,000 **47.** $144 **49.** 10 gal **51.** 52.5 lb

53. 108 lb **55.** 1250 ft **57.** 595 turns **59.** 177 hp **61.** 26.4 gal **63.** 14 cm^3 **65.** 6 cm^3 **67.** 23.3% **69.** 5.9% **71. a.** 18%

b. 117 lb **73.** 19 hL **75. a.** 13.3% **b.** 33.3% **c.** 53.3% **77.** 83 ft 4 in. **79.** 70 lb **81.** 35 turns **83.** 4.5 mL **85.** 5 mL

87. 1.5 mL **89.** 30 mL

Exercises 7.3 (pages 294–295)

For all answers the allowable error is ±8 mi or ±8 km.

1. 48 mi **3.** 60 mi **5.** 86 mi **7.** 104 mi **9.** 96 mi **11.** 520 km **13.** 298 km **15.** 323 km **17.** 12.5 cm **19.** 100 cm^2

21. 1 cm × 1 cm **23.** 6.5 cm **25.** 6.5 cm

27.

Side of square = 1 ft

29.

Side of a square = 3 mm

31. a. yes **b.** no **33.** 20 : 1 **35.** 9 : 1 **37.** 68 lb
39. 238,500 lb **41.** 81 : 1 **43.** 25 : 1 **45.** 400 lb **47.** 400 lb
49. 640 lb **51.** 225 mi **53.** 141

Exercises 7.4 (pages 297–299)

1. 36 rpm **3.** 80 rpm **5.** 22 cm **7.** 31 in. **9.** 520 rpm **11.** 50 in. **13.** 11 in. **15.** 10 cm; 17 cm **17.** 160 teeth

19. 1008 rpm **21.** 576 teeth **23.** 96 rpm **25.** 144 rpm **27.** 40 teeth **29.** 10 in. **31.** 120 lb **33.** 135 lb **35.** $\frac{1}{2}$ ton or 1000 lb

37. 64 cm **39.** $25\frac{1}{3}$ in. **41.** 27 cm **43.** 3 A

Chapter 7 Review (pages 300–301)

1. $\frac{1}{4}$ **2.** $\frac{3}{2}$ **3.** $\frac{2}{1}$ **4.** $\frac{11}{18}$ **5.** yes **6.** no **7.** 1 **8.** 30 **9.** 24 **10.** 40 **11.** 106 **12.** 41.3 **13.** 788 **14.** 529 **15.** $187.50

16. 1250 ft **17.** 216 h **18.** 300 lb **19.** Jones 58%; Hernandez 42% **20.** 8.8% **21. a.** direct **b.** inverse **22.** direct

23. 362.5 mi **24.** 17.6 in. **25.** 562.5 rpm **26.** 75 rpm **27.** 50 kg **28.** 180 lb **29.** 2 A **30.** 108 h

Chapter 8

Exercises 8.1 (page 309)

1. (3, 2) (8, −3)(−2, 7) **3.** (2, −1) (0, 5) (−2, 11) **5.** (0, −2) $\left(2, -\frac{1}{2}\right)$ (−4, −5) **7.** (5, 4) (0, 2) $\left(-3, \frac{4}{5}\right)$

9. (2, 4) (0, −5) (−4, −23) **11.** (2, 10) (0, 4) (−3, −5) **13.** (2, −3) (0, 7) (−4, 27) **15.** (3, 10) (0, 4) (−1, 2)

17. (4, 14) (0, 4) (−2, −1) **19.** (1, 2) $\left(0, -\frac{5}{2}\right)$ (−3, −16) **21.** (2, 3) (0, 3) (−4, 3) **23.** (5, 4) (5, 0) (5, −2) **25.** $y = \frac{6 - 2x}{3}$

27. $y = \dfrac{7 - x}{2}$ **29.** $y = \dfrac{x - 6}{2}$ **31.** $y = \dfrac{2x - 9}{3}$ **33.** $y = \dfrac{2x + 6}{3}$ **35.** $y = \dfrac{-2x + 15}{3}$ **37.** $A(-2, 2)$ **39.** $C(5, -1)$

41. $E(-4, -5)$ **43.** $G(0, 5)$ **45.** $I(4, 2)$

47–65.

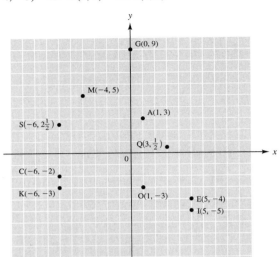

Exercises 8.2   (page 316)

1.

3.

5.

7.

9.

11.

13.

15.

17.

19.

21.

23.

25.

27.

29.

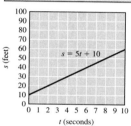

	Independent	Dependent
31.	t	s
33.	V	R
35.	t	i
37.	t	v
39.	t	s

41.

43.

45.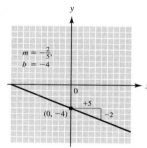

Exercises 8.3   (pages 322–323)

1. 1   **3.** 6   **5.** $-\dfrac{5}{4}$ **7.** 0   **9.** undefined   **11.** $\dfrac{3}{5}$ **13.** -2 **15.** $\dfrac{3}{7}$ **17.** 0   **19.** 6   **21.** -5 **23.** $-\dfrac{3}{5}$ **25.** $\dfrac{1}{4}$ **27.** $\dfrac{5}{2}$
29. undefined   **31.** parallel   **33.** perpendicular   **35.** perpendicular   **37.** neither   **39.** parallel

Exercises 8.4   (pages 327–328)

1.

3.

5.

7.

9.

11.

13.

15.

17.

19.

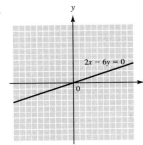

21. $y = 2x + 5$ **23.** $y = -5x + 4$

25. $y = \frac{2}{3}x - 4$ or $2x - 3y = 12$

27. $y = -\frac{6}{5}x + 3$ or $6x + 5y = 15$

29. $y = -\frac{3}{5}x$ or $3x + 5y = 0$

31.

33.

35.

37.

39.

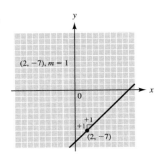

41. $2x - y = 1$ **43.** $3x - 4y = -15$ **45.** $3x - 2y = 22$ **47.** $10x + 3y = -33$
49. $x + y = 2$ **51.** $2x + 3y = 13$ **53.** $2x + y = 2$ **55.** $x - 3y = -3$
57. $x - 2y = -1$ **59.** $x + y = 6$

Chapter 8 Review   (pages 330–331)

1. $\left(3, \frac{5}{2}\right)(0, 4)(-4, 6)$ **2.** $(3, -2)(0, -4)(-3, -6)$ **3.** $y = -6x + 15$ **4.** $y = \frac{3x + 10}{5}$ or $y = \frac{3}{5}x + 2$ **5.** $A\ (3, 5)$

6. $B\ (-2, -6)$ **7.** $C\ (2, -1)$ **8.** $D\ (4, 0)$ **9–12.**

13.

14.

15.

16.

17.

18.

19.

20.

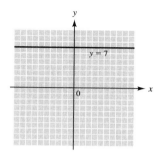

21. $\dfrac{9}{7}$ **22.** 1 **23.** 4 **24.** $-\dfrac{2}{5}$ **25.** $\dfrac{5}{9}$ **26.** perpendicular **27.** neither
28. parallel **29.** perpendicular **30.**

31.

32.

33.

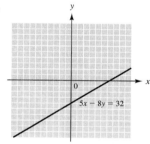

34. $y = -\dfrac{1}{2}x + 3$ or $x + 2y = 6$ **35.** $y = \dfrac{8}{3}x$ or $8x - 3y = 0$ **36.** $y = 0$

37.

38.

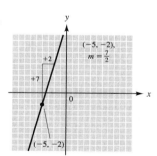

39. $x + y = 6$ **40.** $x + 4y = -20$
41. $x - y = 5$ **42.** $x - 2y = 12$

Cumulative Review Chapters 1–8 (page 332)

1. 5 **2.** 15 **3. a.** $4\dfrac{3}{16}$ in. **b.** $2\dfrac{7}{8}$ in. **4.** 98.5 **5.** 1.116×10^3 **6.** 0.061 m **7.** 5 **8.** 7.82 mm **9.** 350 m^3 **10.** $7x - 11$

11. $-40x^4y^4$ **12.** $8x^2 - 6xy$ **13.** $-\dfrac{3}{5}$ **14.** $-\dfrac{52}{25}$ **15.** $V = \dfrac{3s - t}{2}$ **16.** $\dfrac{1}{13}$ **17.** $\dfrac{8}{27}$ **18.** 60 **19.** 41.0 **20.** 1450 **21.** $102.60

22. 170 mi **23.** 10 teeth **24.** (3, 2) (0, 4) (−3, 6) **25.** $y = \dfrac{7 - 4x}{2}$

26.

27.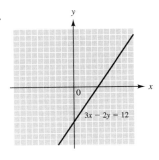

28. 1   **29.** $x - 2y = 10$ **30.** neither

Chapter 9

Exercises 9.1   (pages 339–340)

1.

3.

5.

7.

9.

11.

13.

15.

17.

19.

21.

23.

25.

27.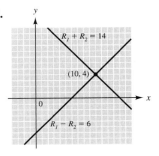

29. 4 ft³ of concrete; 16 ft³ of gravel

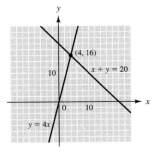

Exercises 9.2 (page 345)

1. $(2, 1)$ **3.** $(4, 2)$ **5.** $(-2, 7)$ **7.** $(3, 1)$ **9.** $(-5, 5)$ **11.** $(5, 1)$ **13.** $(1, -2)$ **15.** $(-1, 4)$ **17.** $(7, -5)$ **19.** $(4, -1)$

21. $(-2, 5)$ **23.** $\left(2, 1\frac{1}{4}\right)$ **25.** $(-2, 4)$ **27.** coincide   **29.** $(2, -1)$ **31.** parallel   **33.** $\left(\frac{1}{2}, 5\right)$ **35.** coincide   **37.** parallel

39. $(72, 30)$ **41.** $\left(\frac{1}{4}, \frac{2}{5}\right)$ **43.** $(9, -2)$

Exercises 9.3 (page 347)

1. $\left(\frac{12}{5}, \frac{36}{5}\right)$ or $(2.4, 7.2)$ **3.** $(10, 2)$ **5.** $(6, 6)$ **7.** $(4, 2)$ **9.** $(-3, -1)$ **11.** $(1, 4)$ **13.** $(-2, 2)$ **15.** $\left(2\frac{1}{2}, -12\frac{1}{2}\right)$ **17.** $(1, 3)$
19. $(9, -2)$ **21.** $(3, 4)$ **23.** $(6, 6)$ **25.** $(2, 4)$

Exercises 9.4 (pages 351–353)

1. 42 cm, 54 cm   **3.** $2\frac{1}{2}$ h @ 180 gal/h; $3\frac{1}{2}$ h @ 250 gal/h   **5.** 26 h @ \$32; 22 h @ \$41   **7.** 30 lb @ 5%; 70 lb @ 15%
9. 2700 bu corn; 450 bu beans   **11.** 200 gal of 6%; 100 gal of 12%   **13.** 5 @ 3 V; 4 @ 4.5 V   **15.** 105 mL @ 8%; 35 mL @ 12%
17. 5 min @ 850 rpm; 9 min @ 1250 rpm   **19.** 2 h @ setting 1; 3 h @ setting 2   **21.** 160 L of 3%; 40 L of 8%   **23. a.** 6.8 A
b. 1.2 A   **25.** 5 h @ 140 cm^3/h; 3 h @ 100 cm^3/h   **27.** 31 of 2 cm^3; 11 of 5 cm^3 **29.** 4 one-bedroom and 9 two-bedroom
31. 100 Ω, 450 Ω   **33.** 30 cm, 90 cm   **35.** 40 cm × 80 cm   **37.** 150 mA   **39.** 680 ft × 720 ft   **41. a.** 6 ft × 14 ft
b. 8 ft × 12 ft = 96 ft^2 The second room is 12 ft^2 larger.   **43.** 30 ft × 75 ft   **45.** 76 in.; 68 in.   **47.** 85 lb corn; 15 lb soybean meal

Chapter 9 Review (page 355)

1.

2.

3.

4.

5.

6.
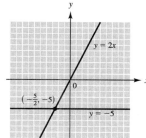

7. $(3, 4)$ **8.** $(3, 1)$ **9.** $(4, 3)$ **10.** no common solution—lines parallel   **11.** $(6, -2)$ **12.** $(3, 1)$

13. infinitely many solutions—lines coincide   **14.** $(4, 8)$ **15.** $(-1, -2)$ **16.** $\left(1\frac{9}{25}, 10\frac{11}{25}\right)$

17. 20 amp @ \$3.50; 15 amp @ \$11.50   **18.** 132.5 ft by 57.5 ft   **19.** 20 mH; 70 mH   **20.** 35 ft; 55 ft

Chapter 10

Exercises 10.1 (pages 359–360)

1. $4(a + 1)$ **3.** $b(x + y)$ **5.** $5(3b - 4)$ **7.** $x(x - 7)$ **9.** $a(a - 4)$ **11.** $4n(n - 2)$ **13.** $5x(2x + 5)$ **15.** $3r(r - 2)$
17. $4x^2(x^2 + 2x + 3)$ **19.** $9a(a - x^2)$ **21.** $10(x + y - z)$ **23.** $3(y - 2)$ **25.** $7xy(2 - xy)$ **27.** $m(12x^2 - 7)$ **29.** $12a(5x - 1)$
31. $13mn(4mn - 1)$ **33.** $2(26m^2 - 7m + 1)$ **35.** $18y^2(2 - y + 3y^2)$ **37.** $3m(2m^3 - 4m + 1)$ **39.** $-2x^2y^3(2 + 3y + 5y^2)$
41. $3abc(abc + 9a^2b^2c - 27)$ **43.** $4xz^2(x^2z^2 - 2xy^2z + 3y)$

Exercises 10.2　(page 362)

1. $x^2 + 7x + 10$　**3.** $6x^2 + 17x + 12$　**5.** $x^2 - 11x + 30$　**7.** $x^2 - 14x + 24$　**9.** $2x^2 + 19x + 24$　**11.** $x^2 + 4x - 12$
13. $x^2 - 19x + 90$　**15.** $x^2 - 6x - 72$　**17.** $8x^2 - 18x - 35$　**19.** $8x^2 + 6x - 35$　**21.** $14x^2 + 41x + 15$　**23.** $3x^2 - 19x - 72$
25. $6x^2 + 47x + 35$　**27.** $169x^2 - 104x + 16$　**29.** $120x^2 + 54x - 21$　**31.** $100x^2 - 100x + 21$　**33.** $4x^2 - 16x + 15$
35. $4x^2 + 4x - 15$　**37.** $6x^2 + 5x - 56$　**39.** $6x^2 + 37x + 56$　**41.** $16x^2 + 14x - 15$　**43.** $2y^2 - 11y - 21$　**45.** $6n^2 + 3ny - 30y^2$
47. $8x^2 + 26xy - 7y^2$　**49.** $\frac{1}{8}x^2 - 5x + 48$

Exercises 10.3　(page 365)

1. $(x + 2)(x + 4)$　**3.** $(y + 4)(y + 5)$　**5.** $3(r + 5)^2$　**7.** $(b + 5)(b + 6)$　**9.** $(x + 8)(x + 9)$　**11.** $5(a + 4)(a + 3)$
13. $(x - 4)(x - 3)$　**15.** $2(a - 7)(a - 2)$　**17.** $3(x - 7)(x - 3)$　**19.** $(w - 6)(w - 7)$　**21.** $(x - 9)(x - 10)$　**23.** $(t - 10)(t - 2)$
25. $(x + 4)(x - 2)$　**27.** $(y + 5)(y - 4)$　**29.** $(a + 8)(a - 3)$　**31.** $(c - 18)(c + 3)$　**33.** $3(x - 4)(x + 3)$　**35.** $(c + 6)(c - 3)$
37. $(y + 14)(y + 3)$　**39.** $(r - 7)(r + 5)$　**41.** $(m - 20)(m - 2)$　**43.** $(x - 15)(x + 6)$　**45.** $(a + 23)(a + 4)$　**47.** $2(a - 11)(a + 5)$
49. $(a + 25)(a + 4)$　**51.** $(y - 19)(y + 5)$　**53.** $(y - 16)(y - 2)$　**55.** $7(x + 2)(x - 1)$　**57.** $6(x^2 + 2x - 1)$　**59.** $(y - 7)(y - 5)$
61. $(a + 9)(a - 7)$　**63.** $(x + 4)(x + 14)$　**65.** $2(y - 15)(y - 3)$　**67.** $3x(y - 3)(y - 3)$　**69.** $(x + 15)^2$　**71.** $(x - 9)(x - 17)$
73. $(x + 12)(x + 16)$　**75.** $(x + 22)(x - 8)$　**77.** $2b(a + 6)(a - 4)$　**79.** $(y - 9)(y + 8)$

Exercises 10.4　(pages 366–367)

1. $x^2 - 9$　**3.** $a^2 - 25$　**5.** $4b^2 - 121$　**7.** 9991 or $(10,000 - 9)$　**9.** $9y^4 - 196$　**11.** $r^2 - 24r + 144$　**13.** $16y^2 - 25$
15. $x^2y^2 - 8xy + 16$　**17.** $a^2b^2 + 2abd + d^2$　**19.** $z^2 - 22z + 121$　**21.** $s^2t^2 - 14st + 49$　**23.** $x^2 - y^4$　**25.** $x^2 + 10x + 25$
27. $x^2 - 49$　**29.** $x^2 - 6x + 9$　**31.** $a^2b^2 - 4$　**33.** $x^4 - 4$　**35.** $r^2 - 30r + 225$　**37.** $y^6 - 10y^3 + 25$　**39.** $100 - x^2$

Exercises 10.5　(page 369)

1. $(a + 4)^2$　**3.** $(b + c)(b - c)$　**5.** $(x - 2)^2$　**7.** $(2 + x)(2 - x)$　**9.** $(y + 6)(y - 6)$　**11.** $5(a + 1)^2$　**13.** $(1 + 9y)(1 - 9y)$
15. $(7 + a^2)(7 - a^2)$　**17.** $(7x + 8y)(7x - 8y)$　**19.** $(1 + xy)(1 - xy)$　**21.** $(2x - 3)^2$　**23.** $(R + r)(R - r)$　**25.** $(7x + 5)(7x - 5)$
27. $(y - 5)^2$　**29.** $(b + 3)(b - 3)$　**31.** $(m + 11)^2$　**33.** $(2m + 3)(2m - 3)$　**35.** $4(x + 3)^2$　**37.** $3(3x + 1)(3x - 1)$　**39.** $a(m - 7)^2$

Exercises 10.6　(pages 371–372)

1. $(5x + 2)(x - 6)$　**3.** $(2x - 3)(5x - 7)$　**5.** $(6x - 5)(2x - 3)$　**7.** $(2x + 9)(4x - 5)$　**9.** $(16x + 5)(x - 1)$　**11.** $4(3x + 2)(x - 2)$
13. $(5y + 3)(3y - 2)$　**15.** $(4m + 1)(2m - 3)$　**17.** $(7a + 1)(5a - 1)$　**19.** $(4y - 1)^2$　**21.** $(3x - 7)(x + 9)$　**23.** $(4b - 1)(3b + 2)$
25. $(5y + 2)(3y - 4)$　**27.** $(10 + 3c)(9 - c)$　**29.** $(3x - 5)(2x - 1)$　**31.** $(2y^2 - 5)(y^2 + 7)$　**33.** $(2b + 13)^2$　**35.** $(7x - 8)(2x - 5)$
37. $7x(2x + 5)^2$　**39.** $5a(2b + 7)(b - 5)$

Chapter 10 Review　(page 373)

1. $c^2 - d^2$　**2.** $x^2 - 36$　**3.** $y^2 + 3y - 28$　**4.** $4x^2 - 8x - 45$　**5.** $x^2 + 5x - 24$　**6.** $x^2 - 13x + 36$　**7.** $x^2 - 6x + 9$
8. $4x^2 - 24x + 36$　**9.** $1 - 10x^2 + 25x^4$　**10.** $6(a + 1)$　**11.** $5(x - 3)$　**12.** $x(y + 2z)$　**13.** $y^2(y + 18)(y - 1)$　**14.** $(y - 7)(y + 1)$
15. $(z + 9)^2$　**16.** $(x + 8)(x + 2)$　**17.** $4(a^2 + x^2)$　**18.** $(x - 9)(x - 8)$　**19.** $(x - 9)^2$　**20.** $(x + 15)(x + 4)$　**21.** $(y - 1)^2$
22. $(x - 7)(x + 4)$　**23.** $(x - 12)(x + 8)$　**24.** $(x + 11)(x - 10)$　**25.** $(x + 7)(x - 7)$　**26.** $(4y + 3x)(4y - 3x)$　**27.** $(x + 12)(x - 12)$
28. $(5x + 9y)(5x - 9y)$　**29.** $4(x - 13)(x + 7)$　**30.** $5(x + 12)(x - 13)$　**31.** $(2x + 7)(x + 2)$　**32.** $(4x - 1)(3x - 4)$
33. $(6x + 5)(5x - 3)$　**34.** $(12x - 1)(x + 12)$　**35.** $2(2x - 1)(x - 1)$　**36.** $(6x + 7y)(6x - 7y)$　**37.** $2(7x + 3)(2x + 5)$
38. $3(5x - 7)(2x + 1)$　**39.** $4x(x + 1)(x - 1)$　**40.** $25(y + 2)(y - 2)$

Cumulative Review Chapters 1–10　(page 374)

1. 6　**2. a.** 746.8　**b.** 750　**3.** $-\dfrac{8}{3}$　**4. a.** 3.18×10^{-4}　**b.** 318×10^{-6}　**5.** 0.625 kg　**6.** 64.6 ft^2　**7.** 70 V　**8.** 95 cm^3
9. $15x - 14$　**10.** $8a + c$　**11.** -60　**12.** 17 m; 12 m　**13.** 19.1　**14.** 10 in.　**15.** $(3, 2), (0, 4), (-3, 6)$　**16.** $y = 3x - 5$

17.

18.

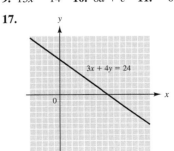

19. many solutions; lines coincide　**20.** $\left(\dfrac{23}{11}, \dfrac{14}{11}\right)$

21. $(2, -4)$　**22.** no solution; lines parallel

23. $\left(\dfrac{1}{3}, -\dfrac{2}{5}\right)$　**24.** 6 days @ $53.95; 10 days @ $42.95

25. $6x^2 + x - 40$　**26.** $25x^2 - 70xy + 49y^2$

27. $15x^2 - 46x + 35$　**28.** $7x(x + 3)(x - 3)$

29. $4x^2(x + 3)$　**30.** $(2x + 1)(x - 4)$

Chapter 11

Exercises 11.1 (page 378)

1. $-4, 3$ **3.** $-5, 4$ **5.** $2, -1$ **7.** $1, -1$ **9.** $7, -7$ **11.** $-2, -3$ **13.** $7, -3$ **15.** $10, -4$ **17.** $0, 9$ **19.** $12, -9$ **21.** $-8, 2$

23. $-\dfrac{5}{2}, -\dfrac{2}{5}$ **25.** $\dfrac{5}{2}, -\dfrac{5}{2}$ **27.** $\dfrac{4}{3}$ **29.** $0, -3$ **31.** Base: 12 m; Height: 11 m **33.** 5 in. $\times$ 8 in.

Exercises 11.2 (page 381)

1. $a = 1, b = -7, c = 4$ **3.** $a = 3, b = 4, c = 9$ **5.** $a = -3, b = 4, c = 7$ **7.** $a = 3, b = 0, c = -14$ **9.** $-3, 2$ **11.** $-9, 1$

13. $0, -\dfrac{2}{5}$ **15.** $\dfrac{5}{4}, -\dfrac{7}{12}$ **17.** $1.35, -1.85$ **19.** $0, \dfrac{5}{3}$ **21.** $-1, \dfrac{3}{2}$ **23.** $-1.38, -0.121$ **25.** $-\dfrac{1}{4}, -1$ **27.** $2.38, -2.38$

29. $15.5, -0.453$ **31.** $4.15, -2.49$ **33.** $0.206, -2.40$

Exercises 11.3 (pages 384–385)

1. a. 4 s; 8 s **b.** 1.42 s; 10.6 s **c.** 16 s **3.** 3 ft $\times$ 7 ft **5.** 5 ft $\times$ 35 ft **7.** 20 m $\times$ 60 m **9.** 4.5 ft $\times$ 9 ft **11. a.** 5 cm $\times$ 5 cm
b. 7500 cm^3 **13.** 2 ft **15. a.** 1400 ft^2 **b.** 2600 ft^2 **c.** $A = l(150 - l) = 150\, l - l^2$
d.

Length (ft)	30	40	50	60	70	80	90	100	110	120	130	140
Area (ft^2)	3600	4400	5000	5400	5600	5600	5400	5000	4400	3600	2600	1400

e. The maximum area given in the table is 5600 ft^2 when the length is 70 ft or 80 ft.
f.

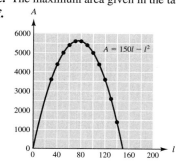

g. From the graph: When the length $=$ the width $= 75$ ft, the maximum area is 5625 ft^2.
17. 65 ft $\times$ 141 ft

Exercises 11.4 (page 389)

1.

3.

5.

7.

9.

11.

13.

15.

17.

19.

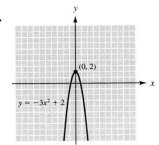

Exercises 11.5   (page 393)

1. $7j$ **3.** $3.74j$ **5.** $1.41j$ **7.** $7.48j$ **9.** $13j$ **11.** $5.20j$ **13.** $-j$ **15.** j **17.** $-j$ **19.** 1   **21.** -1 **23.** $-j$ **25.** two rational roots
27. two imaginary roots   **29.** two irrational roots   **31.** two imaginary roots   **33.** two imaginary roots   **35.** $3 + j; 3 - j$
37. $7 + 2j; 7 - 2j$ **39.** $-4 + 5j; -4 - 5j$ **41.** $-0.417 + 1.08j; -0.417 - 1.08j$ **43.** $1 + 1.15j; 1 - 1.15j$
45. $-0.8 + 0.4j; -0.8 - 0.4j$ **47.** $0.2, -3$ **49.** $-0.5 + 0.866j; -0.5 - 0.866j$

Chapter 11 Review   (pages 394–395)

1. $a = 0$ or $b = 0$ **2.** $0, 2$ **3.** $2, -2$ **4.** $3, -2$ **5.** $0, \dfrac{6}{5}$ **6.** $7, -4$ **7.** $9, 5$ **8.** $6, -3$ **9.** $-\dfrac{8}{3}, -4$ **10.** $6, -\dfrac{2}{3}$

11. $0.653, -7.65$ **12.** $\dfrac{5}{2}, -3$ **13.** $4.45, -0.449$ **14.** $2.12, -0.786$ **15.** length 9 ft, width 4 ft   **16. a.** $4\ \mu s, 8\ \mu s$

b. $6\ \mu s$ **c.** $2.84\ \mu s, 9.16\ \mu s$

17.

18.

19. $6j$ **20.** $8.54j$ **21.** 1   **22.** $-j$
23. one rational root   **24.** two imaginary roots
25. $2 + j, 2 - j$ **26.** $0.6 + 0.663j, 0.6 - 0.663j$
27. 2.5 ft by 8.5 ft   **28.** 3 in.

Chapter 12

Exercises 12.1   (pages 403–405)

1. acute   **3.** right   **5.** acute   **7.** obtuse   **9.** right, perpendicular   **11. a.** 1, 2; 2, 4; 3, 4; 1, 3; 5, 6; 6, 8; 7, 8; 5, 7
b. 1, 4; 2, 3; 5, 8; 6, 7   **13.** $\angle 2 = 57°; \angle 3 = 57°; \angle 4 = 123°$ **15.** $61°$ **17.** $52°$ **19. a.** Yes   **b.** Yes   **21.** $90°$ **23.** $15°$
25. $148°$ **27.** $152°$ **29.** triangle   **31.** hexagon   **33.** quadrilateral   **35.** heptagon

Exercises 12.2   (pages 408–410)

1. 60.0 cm; 225 cm^2 **3.** 32.0 m; 48 m^2 **5.** 45.0 m; 78 m^2 **7.** 89.4 in.; 411 in^2 **9.** 36.8 cm; 85 cm^2 **11.** 24.0 cm   **13.** 13.4 ft
15. 352 in^2 **17. a.** 59.4 mi   **b.** 197 mi^2 **19.** 16 in.; 48 in.   **21.** 9   **23.** 36 ft   **25.** 6 pieces   **27.** 28 in^2 **29. a.** 208 in^2
b. $59\overline{0}$ in^2 **31.** $40\overline{0},000$ acres   **33.** 7 squares   **35.** $\$927.35$ **37.** 250 ft^2 **39.** 568.6 ft   **41.** 704 ft^2 **43.** $20\overline{0}$ ft^2 **45.** 105 ft

Exercises 12.3   (pages 415–419)

1. 10.0 m   **3.** 25.0 m   **5.** 17.0 m   **7.** 13.5 cm   **9.** 1460 km   **11.** 350 ft   **13.** 59.9 in.   **15.** 33.4 in.   **17.** 2.83 in.   **19.** 16.6 in.
21. 116 V   **23.** 21 A   **25.** $127\ \Omega$ **27.** $2.6\ \Omega$ **29.** 158 m^2; 60.8 m   **31.** 15.6 cm^2; 18.00 cm   **33.** 395 m^2; 108.7 m
35. 29.9 m^2; 36.0 m   **37.** 96.0 cm^2; 48.0 cm   **39.** 10.8 ft   **41.** 74.3 mi   **43.** 145 in^2 **45.** $81°$ **47.** $60°$ **49.** 5 bags
51. a. 9% **b.** 50.2 ft   **53.** 71.1 acres

Exercises 12.4   (pages 421–423)

1. 5   **3.** Yes, all corresponding angles of △ABO and △DCO are equal.   **5.** $5\frac{1}{3}$ **7.** 120 ft   **9. a.** 12.0 m   **b.** 17.0 m   **11.** 10.6 in.
13. $55\overline{0}$ ft   **15.** 20.0 in.   **17.** $6\overline{0}$ in.   **19.** $A = 1.33$ ft; $B = 2.67$ ft; $C = 4.00$ ft; $D = 5.33$ ft; $E = 6.67$ ft; cross-piece is 10.0 ft
21. a. 3 in. $= \frac{3}{4}$ in.; 6 in. $= 1\frac{1}{2}$ in.; 2 in. $= \frac{1}{2}$ in.; 1 in. $= \frac{1}{4}$ in.; 2.5 in. $= \frac{5}{8}$ in.   **b.** 2.33 oz
23. lengths: 36 in., 39 in., 53.1 in.; a 12-ft board must be purchased

Exercises 12.5   (pages 426–430)

1. a. 31.4 in.   **b.** 78.5 in²   **3. a.** 28.9 mm   **b.** 66.6 mm²   **5. a.** 352 mi   **b.** 9890 mi²   **7.** 171°   **9.** 43.4°   **11.** 3.00 cm
13. 10.0 m   **15.** 90°   **17.** 12.6 ft²   **19.** 3.98 in²; 7.07 in.   **21.** 50.3 in.   **23.** 2670 ft   **25.** 13.4 in.   **27.** 1.25 in.   **29.** 1332 cm²
31. 24.42 in.   **33. a.** 2.50 ft   **b.** 3.54 ft   **35.** 72°   **37.** 30°   **39.** 18,600 mi   **41.** 171°   **43.** 14°   **45. a.** 120°   **b.** 60°
c. 120°   **47. a.** 60°   **b.** 120°   **c.** 60°   **49. a.** 5.6 in²   **b.** 4.13 in.   **c.** 2.26 in.   **d.** 8.4 oz   **51. a.** 817 ft   **b.** 1.22 acres

Exercises 12.6   (pages 434–435)

1. 180°   **3.** 0.367 rad   **5.** 60°   **7.** $\frac{3\pi}{4}$ rad; 2.36 rad   **9.** $\frac{4\pi}{3}$ rad; 4.19 rad   **11.** $\frac{4\pi}{5}$ rad; 2.51 rad   **13.** 31.4 cm   **15.** 4.71 cm
17. 104 cm   **19.** 2.80 rad   **21.** 98.5°   **23.** 8.83 cm   **25.** $\frac{5}{\pi}$ rps or 1.59 rps   **27.** 92.2 cm   **29.** 25.8 cm   **31. a.** 26.2 m
b. 327 m²   **c.** 56 m²

Exercises 12.7   (pages 439–441)

1. a. $44\overline{0}$ in²   **b.** 622 in²   **c.** 912 in³   **3.** 84 in²; 36 in³   **5.** $6\overline{0}$ ft³   **7. a.** 2490 m²   **b.** 1230 m²   **c.** 188 m³   **d.** $49\overline{0}0$ m²
9. 324 ft³   **11.** $3\overline{0}$ in.   **13.** 12 ft 1 in. by 9 ft 11 in.   **15.** $9\overline{0}$ in. by 113 in.   **17.** 8   **19.** $36\overline{0}0$ ft³

Exercises 12.8   (pages 444–446)

1. 13,600 mm³   **3.** 8,280,000 L   **5.** 42.5 ft   **7.** 56.5 in³   **9.** 1140 ft³   **11.** 1.79 in.   **13.** 28.3 in³   **15. a.** 2050 mm²   **b.** 2470 mm²
17. 101,000 cm²   **19.** 18.0 in³   **21.** 1.38 in²   **23.** 6.1 gal   **25.** 25 kg   **27. a.** 68.0 in.   **b.** 13,800 lb   **c.** 2.08 gal   **d.** 88,400 lb
29. 15.5 ft   **31.** 10,100 bu

Exercises 12.9   (pages 450–453)

1. 147 in³   **3.** $11\overline{0}0$ m³   **5.** 69,700 mm³   **7.** 1440 mm³   **9.** 312 ft³   **11. a.** 1010 cm³   **b.** 427 cm²   **13.** 1520 bu   **15.** 225 in²
17. 37,000 ft³, $1\overline{0}0$ truckloads   **19.** 15.0 m³   **21.** 108,000 ft³; 9420 ft²   **23.** 28.0 in.   **25.** 33.9 ft³   **27.** yes, 13.2%   **29.** 125 gal

Exercises 12.10   (pages 454–455)

1. a. 804 m²   **b.** 2140 m³   **3. a.** 4120 in²   **b.** 24,800 in³   **5.** 114,000 m³   **7.** 1530 in³   **9.** 7.56 in.   **11.** 933,000 gal
13. a. 0.375 or $\frac{3}{8}$ **b.** 0.25 or $\frac{1}{4}$ **c.** $\frac{3}{r}$ **15. a.** 2140 ft²   **b.** 16,800 ft³   **c.** 301,000 lb   **d.** 39 gal   **17.** 380 ft³

Chapter 12 Review   (pages 460–462)

1. acute   **2.** obtuse   **3.** $\angle 1 = 59°$; $\angle 2 = 121°$; $\angle 3 = 59°$; $\angle 4 = 59°$ **4.** adjacent or supplementary   **5.** 20°   **6. a.** quadrilateral
b. pentagon   **c.** hexagon   **d.** triangle   **e.** octagon   **7.** 36.00 cm; 60.0 cm²   **8.** 360.4 m; $80\overline{0}0$ m²   **9.** 43.34 cm; 102 cm²
10. 7.73 m   **11.** 1.44 cm   **12.** 159 m²; 58.7 m   **13.** 796 m²; 159.5 m   **14.** 4.73 cm²; 10.93 cm   **15.** 41.9 m   **16.** 3.85 cm   **17.** 86°
18. 8.0 m   **19.** 423 cm²; 72.9 cm   **20.** 12.1 cm   **21.** 216°   **22.** $\frac{2\pi}{15}$ rad or 0.419 rad   **23.** 10°   **24.** 42.2 cm   **25.** 1.55 rad
26. 291 cm   **27.** 1790 m²; 2150 m²   **28.** 5760 m³   **29.** 1260 cm³   **30.** 70.4 cm³   **31. a.** 70.4 cm²   **b.** 95.5 cm²   **32.** 131 m³
33. a. 57,300 m³   **b.** 6380 m²   **34. a.** 869 m³   **b.** $44\overline{0}$ m²   **35.** 65,900 in³; 6150 in²   **36. a.** 13 gal; $520   **b.** $27\overline{0}0$ lb

Cumulative Review Chapters 1–12   (pages 463–464)

1. -4 **2.** $13\frac{3}{4}$ in.   **3.** b   **4. a.** 100 L   **b.** 50 L   **5.** $2x^2 - 8x - 4$ **6.** $-8x^3 + 18x^2 + 23x - 15$ **7.** -7 **8.** 19.4

9. 34 lb/bu   **10.** 56 lb   **11.** $y = \dfrac{5x - 10}{8}$ **12.** $-\dfrac{7}{3}$ **13.** $(4, 8)$ **14.** 325 Ω, 975 Ω **15.** $6x^2 + 11x - 35$

16. $16x^2 - 24x + 9$ **17.** $5x(x^2 - 3)$ **18.** $(x - 7)(x + 4)$ **19.** $\dfrac{7}{2}, -3$ **20.** $\dfrac{3}{2}, -1$ **21.** $0.4, -3$ **22.** $-1.72, 4.22$

23.

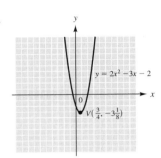

24. $\angle 1 = \angle 2 = 82°$; $\angle 3 = \angle 5 = 98°$ **25.** $59°$ **26.** 76.3 cm, 265 cm^2 **27.** 6.59 m, 1.86 m^2
28. 191 cm^2, 49.0 cm   **29.** 7.31 cm   **30.** 987 cm^3; 568 cm^2

Chapter 13

Exercises 13.1   (pages 469–470)
1. a **3.** c **5.** a **7.** B **9.** B **11.** 60.0 m   **13.** 55.2 mi   **15.** 17.7 cm   **17.** 265 ft   **19.** $35\overline{0}$ m   **21.** 1530 km   **23.** 59,900 m
25. 0.5000   **27.** 0.5741   **29.** 0.5000   **31.** 0.7615   **33.** 2.174   **35.** 0.8686   **37.** 0.5246   **39.** 0.2536   **41.** 0.05328   **43.** 0.6104
45. 0.3929   **47.** 0.6800   **49.** 52.6°   **51.** 62.6°   **53.** 54.2°   **55.** 9.2°   **57.** 30.8°   **59.** 40.8°   **61.** 10.98°   **63.** 69.07°   **65.** 8.66°
67. 75.28°   **69.** 43.02°   **71.** 20.13°   **73.** sin A and cos B; cos A and sin B

Exercises 13.2   (page 472)
1. $A = 35.3°$; $B = 54.7°$ **3.** $A = 24.4°$; $B = 65.6°$ **5.** $A = 43.3°$; $B = 46.7°$ **7.** $A = 45.0°$; $B = 45.0°$ **9.** $A = 64.05°$; $B = 25.95°$
11. $A = 52.60°$; $B = 37.40°$ **13.** $A = 39.84°$; $B = 50.16°$ **15.** $A = 41.85°$; $B = 48.15°$ **17.** $A = 11.8°$; $B = 78.2°$
19. $A = 31.11°$; $B = 58.89°$ **21.** $A = 15.9°$; $B = 74.1°$ **23.** $A = 64.65°$; $B = 25.35°$

Exercises 13.3   (page 474)
1. $b = 40.6$ m; $c = 54.7$ m   **3.** $b = 278$ km; $c = 365$ km   **5.** $a = 29.3$ cm; $b = 39.4$ cm   **7.** $a = 10.4$ cm; $c = 25.9$ cm
9. $a = 6690$ km; $c = 30,\overline{0}00$ km   **11.** $b = 54.92$ m; $c = 58.36$ m   **13.** $a = 161.7$ mi; $b = 197.9$ mi   **15.** $a = 4564$ m; $c = 13,170$ m
17. $b = 5950$ m; $c = 6270$ m   **19.** $a = 288.3$ km; $c = 889.6$ km   **21.** $a = 88.70$ m; $b = 163.0$ m   **23.** $a = 111$ cm; $b = 231$ cm

Exercises 13.4   (page 476)
1. $B = 39.4°$; $a = 37.9$ m; $b = 31.1$ m   **3.** $A = 48.8°$; $b = 234$ ft; $c = 355$ ft   **5.** $A = 22.6°$; $B = 67.4°$; $a = 30.0$ mi
7. $B = 37.9°$; $b = 56.1$ mm; $c = 91.2$ mm   **9.** $B = 21.2°$; $a = 36.7$ m; $b = 14.2$ m   **11.** $A = 26.05°$; $B = 63.95°$; $c = 27.33$ m
13. $B = 60.81°$; $a = 1451$ ft; $b = 2597$ ft   **15.** $A = 67.60°$; $B = 22.40°$; $c = 50.53$ m   **17.** $a = 319$ m; $b = 365$ m; $B = 48.9°$
19. $b = 263.8$ ft; $c = 1607$ ft; $A = 80.55°$ **21.** $a = 268.6$ m; $A = 44.90°$; $B = 45.10°$ **23.** $a = 14\overline{0}0$ ft; $c = 9470$ ft; $A = 8.5°$

Exercises 13.5   (pages 479–484)
1. 18.8 m   **3.** 5.2 m   **5.** 58.2°; 32.3 ft   **7.** 5°   **9.** 92.2 ft   **11.** 28.1°   **13.** 5.70 cm   **15.** 95.3 m   **17. a.** 430 Ω **b.** 38°
c. 37°;   **19. a.** 53.7 V   **b.** 267 V   **c.** 236 V   **21.** 3.7529 in.   **23.** 9.5°   **25. a.** 3.46 in.   **b.** 2.04 in.
27. $A = 55.2°$; $x = 22.8$ ft   **29.** $x = 4.50$ cm; $y = 3.90$ cm   **31.** $a = 12.4$ ft; $b = 30.6$ ft   **33. a.** 7.4 ft   **b.** 3.1 ft

Chapter 13 Review   (pages 486–487)
1. 29.7 m   **2.** B **3.** Hypotenuse   **4.** 43.8 m   **5.** sin A **6.** Length of side adjacent to $\angle A$
7. $\dfrac{\text{length of side opposite } \angle B}{\text{length of side adjacent to } \angle B} = \dfrac{32.2 \text{ m}}{29.7 \text{ m}}$ **8.** 0.8070   **9.** 1.138   **10.** 0.4023   **11.** 45.5°   **12.** 10.4°   **13.** 65.8°   **14.** 39.2°
15. 50.8°   **16.** 7.17 m   **17.** 8.25 m   **18.** $b = 21.9$ m; $a = 18.6$ m; $A = 40.4°$ **19.** $b = 102$ m; $c = 119$ m; $B = 58.8°$
20. $A = 30.0°$; $B = 60.0°$; $b = 118$ mi   **21.** 4950 m   **22.** 5250 ft   **23.** 14.0°

Chapter 14

Exercises 14.1 (page 496)

1. 0.6820   3. −0.4067   5. −0.4352   7. −0.9724   9. 0.7701   11. −0.4390

13.

15.

17.

19.

21.

23.

25. 25 V; −33 V   27. 3.5 A; −3.5 A   29. 2.0×10^{-4} s
31. 1.8 Hz   33. 8.8×10^9 Hz or 8.8 GHz   35. 2.25 m/s
37. 1.53 m

Exercises 14.2 (page 500)

1. 120°; 3

3. 60°; 8

5. 40°; 10

7. 720°; 6

9. 540°; 3.5

11. 144°; 4

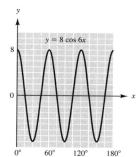

13. 360°; 1; 30° left

15. 360°; 2; 60° right

17. 120°; 4; 60° left

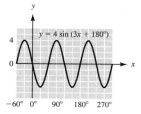

19. 90°; 10; 30° right

21. 720°; 5; 180° left

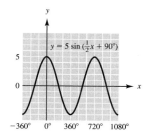

23. 1440°; 10; 720° left

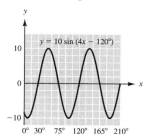

Exercises 14.3   (pages 502–503)

1. $c = 24.9$ m; $B = 41.5°$; $C = 70.5°$ **3.** $a = 231$ ft; $c = 179$ ft; $C = 42.9°$ **5.** $a = 14.4$ cm; $b = 32.9$ cm; $C = 130.3°$
7. $b = 32.5$ m; $c = 22.4$ m; $B = 82.9°$ **9.** $b = 79.3$ mi; $B = 145.0°$; $C = 14.9°$ **11.** $a = 6910$ m; $c = 5380$ m; $A = 74.9°$
13. $a = 26.4$ km; $A = 54.2°$; $C = 20.3°$ **15.** $c = 38\overline{0}$ ft; $B = 15.1°$; $C = 148.4°$ **17.** 63.5 m   **19.** 7.29 km   **21.** 214 ft

Exercises 14.4   (page 508)

1. a. one solution   **b.** $B = 28.2°$; $C = 113.8°$; $c = 62.9$ m   **3. a.** two solutions   **b.** $B = 28.7°$; $C = 125.7°$; $c = 517$ m
and $B = 151.3°$; $C = 3.1°$; $c = 34.4$ m   **5.** no solution   **7. a.** two solutions   **b.** $A = 77.0°$; $B = 31.8°$; $b = 132$ cm
and $A = 103.0°$; $B = 5.8°$; $b = 25.4$ cm   **9. a.** one solution   **b.** $A = 44.6°$; $C = 30.4°$; $c = 17.3$ mi
11. a. two solutions   **b.** $C = 34.3°$; $B = 114.2°$; $b = 656$ m and $C = 145.7°$; $B = 2.8°$; $b = 35.2$ m   **13.** no solution
15. a. two solutions   **b.** $C = 15.3°$; $A = 156.7°$; $a = 1280$ m and $C = 164.7°$; $A = 7.3°$; $a = 412$ m   **17.** 496 ft or 78.5 ft
19. 9.99 ft, 28.3°

Exercises 14.5   (pages 512–514)

1. $a = 21.3$ m; $C = 74.0°$; $B = 51.0°$ **3.** $c = 476$ ft; $B = 36.9°$; $A = 28.1°$ **5.** $A = 33.7°$; $B = 103.5°$; $C = 42.8°$
7. $A = 148.7°$; $C = 12.0°$; $b = 3070$ ft   **9.** $A = 30.7°$; $B = 107.3°$; $C = 42.0°$ **11.** 65.3 m   **13.** $C = 30.5°$; $A = 40.9°$
15. 61.0°; 30.0°; 89.0°   **17.** 132 mi   **19.** 147 in.   **21.** 43.6 ft or 19.7 ft   **23. a.** 59.5°   **b.** 70.5°   **c.** 4.92 m   **d.** 6.33 m
25. 114.3°   **27.** 36 yd   **29.** 31.5 ft   **31.** 134 m

Chapter 14 Review   (page 516)

1. −0.7536   **2.** −0.4970   **3.** 0.7361

4.

5.

6. 120°; 5

7. 90°; 3

8. 360°; 4; 60° left
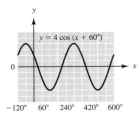

9. 180°; 6; 90° right
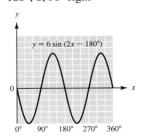

10. $c = 258$ m; $A = 42.5°$; $C = 84.8°$ **11.** $B = 43.2°$;
$b = 62.2$ cm; $a = 79.6$ cm   **12.** $b = 129$ m; $C = 109.0°$; $A = 53.5°$
13. $A = 66.8°$; $B = 36.2°$; $C = 77.0°$ **14.** $c = 33.4$ m; $B = 33.1°$;
$C = 117.4°$ and $B = 146.9°$; $C = 3.6°$; $c = 2.36$ m
15. $c = 3010$ m; $A = 20.5°$; $C = 141.0°$ and $c = 167$ m;
$A = 159.5°$; $C = 2.0°$ **16.** $A = 27.3°$; $B = 59.7°$; $C = 93.0°$
17. no solution   **18. a.** 36.6°   **b.** 52.8 m   **19.** 9.40 in.
20. a. 23.5 m   **b.** 16.7 m   **c.** 15.1 m   **d.** 7.04 m

Cumulative Review Chapters 1–14   (pages 517–518)

1. 8   **2.** $-\dfrac{7}{30}$ **3.** 4.18×10^4 **4.** 198   **5.** 13 Ω   **6. a.** 0.0001 mm   **b.** 0.00005 mm   **c.** 0.02778   **d.** 2.78%   **7.** $6a^2 - 2a + 2$

8. 8   **9.** $\dfrac{26}{21}$ **10.** 102 m²   **11.** $y = \dfrac{8 - 5x}{3}$ **12.** Graphs coincide; many solutions.   **13.** $(x - 14)(x + 12)$ **14.** $3(x + 7)(x - 9)$

15. $5, -\dfrac{2}{3}$ **16.** 2.27; −1.77   **17.** 515 ft²; 126.3 ft   **18.** 13.7 ft   **19.** 6.38 m³   **20.** 2.379   **21.** 51.6°   **22.** 55.8°   **23.** 10.7 cm

24. 19.1 cm   **25.** pitch $\dfrac{2}{3}$; 173 in.   **26.** −0.9812   **27.**   **28.** $P = 720°$, $A = 2$ **29.** 65.2°   **30.** 25.8°

Chapter 15

Exercises 15.1 (pages 521–522)

1. $4020 **3.** $1990 **5.** $1560 **7.** $3600 **9.** $3080 **11.** 2.0 million **13.** United States **15.** 2.5 million **17.** 21.0 million **19.** 2.6 million

21.

23.

25.

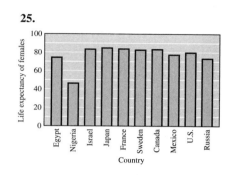

Exercises 15.2 (pages 525–526)

1. 94° **3.** 55° **5.** 270° **7.** 219° **9.** 15° **11.** 328°

13.

15.

17.

19.

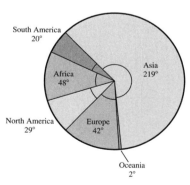

Exercises 15.3 (pages 526–528)

1.

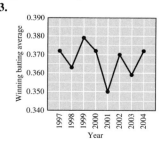

3.

5. 29.52 in. Fri. near midnight **7.** 29.34 in. **9.** 98%
11. 90° **13.** 95% **15. a.** 61 ppmv **b.** 1995

Exercises 15.4   (page 530)

1. 900 mW; 14 dB   **3.** 85 mW   **5.** 30 dB; 32 dB   **7.** 16 dB   **9.** 14 dB; 16 dB

Exercises 15.5   (pages 531–532)

1. 6911   **3.** 2020̄   **5.** 1.096   **7.** 51   **9.** 4.58   **11.** 47.89 cm   **13.** 0.2619 in.   **15.** 30,292 mi   **17.** 73°   **19.** 161 kW
21. 5248 million tons

Exercises 15.6   (pages 534–535)

1. 6185.5   **3.** 2023   **5.** 1.102   **7.** 53   **9.** 4.17   **11.** 47.87 cm   **13.** 0.2618 in.   **15.** 28,985 mi   **17.** 72.5°   **19.** 169 kW   **21.** 163
23. 107   **25.** 23   **27.** 2.85 mm   **29.** 2 modes: 7291, 7285   **31.** 28   **33.** 6.2 in.   **35.** 0.024 in.   **37. a.** 9   **b.** 9   **c.** 9

Exercises 15.7   (page 537)

1. 6060   **3.** 447   **5.** 0.208   **7.** 73   **9.** 8.54   **11.** 0.62 cm   **13.** 0.0010 in.   **15.** 19,393 mi   **17.** 20°F   **19.** 101 kW   **21.** 1930̄
23. 150̄   **25.** 0.064   **27.** 25   **29.** 2.45   **31.** 0.26   **33.** 0.0003 in.   **35.** 7221 mi   **37.** 7°F   **39.** 39 kW

Exercises 15.8   (pages 541–544)

1. 59   **3.** 7.7 months   **5.** 36 h   **7.** 42 lb   **9.** 11   **11.** 4.2 days   **13.** 99 strokes   **15.** 58 players

Exercises 15.9   (page 546)

1. 9   **3.** 4 months   **5.** 5.0 h   **7.** 15 lb   **9.** 4   **11.** 3 days   **13.** 12 strokes   **15.** 38 players   **17. a.** 4 mm   **b.** 1.3 mm

Exercises 15.10   (pages 549–550)

1. a.

b. Out of control; in 16, the value was outside of limits.

3. a.

b. process in control

5. a.

b. process in control

Exercises 15.11   (pages 552–553)

1.

3.

5.

7.

9. a.

b. positive correlation

11. a.

b. no linear correlation

Exercises 15.12   (pages 555–556)

1. Approximately 95%. **3. a.** 63 **b.** 6.6 **c.** No **5. a.** 3544 **b.** 1034 **c.** Yes   **7.** 68,000 blouses of sizes 10, 12, and 14

27,000 blouses of sizes 8 and 16

2700 blouses of sizes 4, 6, and 18

300 blouses of size 2

Exercises 15.13   (pages 557–558)

1. [A, 2, 3, 4, 5, 6, 7, 8, 9, 10, J, Q, K] **3.** [J diamonds, Q diamonds, K diamonds, J hearts, Q hearts, K hearts]

5. [{red, red}, {red, white}, {red, white}] **7.** $\frac{1}{13}$ **9.** $\frac{1}{6}$ **11.** 0 **13.** $\frac{1}{7}$ **15.** $\frac{1}{7}$ **17.** $\frac{13}{2500}$

Exercises 15.14   (page 559)

1. $\frac{1}{16}$ **3.** $\frac{5}{32}$ **5.** $\frac{9}{25}$ **7.** $\frac{1}{12}$ **9.** $\frac{1}{28,561}$ **11.** $\frac{1}{216}$ **13.** $\frac{1}{49}$ **15.** $\frac{84}{6859}$

Chapter 15 Review   (page 562)

1. 126° **2.** 202°   **3.**

Students in 2000

College students 75°

Primary students 193°

Grades 9–12 students 92°

4.

Millions of students

Primary Grades 9–12 College
Students in 2000

5. 74°F **6.** 7.0036 mm
7. 7.0036 mm **8.** 0.0001 mm
9. a. 49.9 **b.** 18.3 **10.** 84.4
11. 83 **12.** 7.4 **13.** 19.2
14. 5.2 **15. a.** [1, 2, 3, 4, 5]
b. $\frac{3}{5}$ **16.** $\frac{5}{324}$

Chapter 16

Exercises 16.1   (page 567)

1. 3 **3.** 6 **5.** 9 **7.** 51 **9.** 47 **11.** 78 **13.** 59 **15.** 156 **17.** 140 **19.** 63

Exercises 16.2   (page 569)

1. 1000 **3.** 1011 **5.** 1100 **7.** 10000 **9.** 100001 **11.** 100100 **13.** 1000111 **15.** 1000101 **17.** 1110101 **19.** 100100111
21. 100110 **23.** 1010001 **25.** 1000000 **27.** 1000011 **29.** 111111

Exercises 16.3   (pages 570–571)

1. 100 **3.** 110 **5.** 10 **7.** 1 **9.** 1010 **11.** 11100 **13.** 10011 **15.** 10 **17.** 1001 **19.** 1000111 **21.** 10 **23.** −10 **25.** 10
27. −11 **29.** −11101 **31.** 10111010 **33.** −101110 **35.** 110000010

Exercises 16.4   (page 571)

1. 1111 **3.** 1100 **5.** 11001 **7.** 100100 **9.** 100001 **11.** 1001000 **13.** 1010100 **15.** 1001110011 **17.** 1111110
19. 100001001

Exercises 16.5   (page 573)

1. 1110 **3.** 111111 **5.** 1001000 **7.** 10100 **9.** 11000 **11.** 100000 **13.** 100101 **15.** 1100100 **17.** 1101111 **19.** 1110001

Exercises 16.6   (page 574)

1. 2.5 **3.** 2.25 **5.** 5.75 **7.** 4.125 **9.** 4.625 **11.** 12.75 **13.** 7.875 **15.** 26.5625 **17.** 4.8125 **19.** 3.8125

Exercises 16.7   (page 576)

1. 37 **3.** 293 **5.** 30 **7.** 197 **9.** 1031 **11.** 2594 **13.** 3010 **15.** 2859 **17.** 3294 **19.** 10,843 **21.** EB **23.** 3A **25.** D18
27. 15DA0 **29.** 8316

Exercises 16.8   (page 579)

1. B **3.** 14 **5.** 17 **7.** A9 **9.** 13D **11.** 169 **13.** 77D **15.** 1247 **17.** 1678 **19.** CC3A **21.** CA51 **23.** 10011 **25.** B0EE
27. E6CD **29.** 17AEC **31.** −C **33.** −3 **35.** −13 **37.** −26 **39.** −9 **41.** −91 **43.** −11F **45.** −367 **47.** −1EE
49. −2B2 **51.** −A88 **53.** C15 **55.** −18AA **57.** 1B8F **59.** −342F

Exercises 16.9   (page 581)

1. 6 **3.** B **5.** 17 **7.** 24 **9.** 44 **11.** E6 **13.** 138 **15.** 33D **17.** 4E7 **19.** CCC **21.** 2891 **23.** F1CA **25.** 110
27. 100100 **29.** 101010 **31.** 1001010001 **33.** 101000110010 **35.** 11111100100 **37.** 10011011101 **39.** 101011001101
41. 100101000111011 **43.** 1011110010101111

Chapter 16 Review   (page 584)

1. 13 **2.** 25 **3.** 52 **4.** 22.75 **5.** 100110 **6.** 10010110 **7.** 10100 **8.** 1001 **9.** 1101 **10.** 10010 **11.** 11110011
12. 100100 **13.** 11001101 **14.** 10000011010 **15.** 225 **16.** 44 **17.** 3102 **18.** 138 **19.** 34 **20.** 1210 **21.** 83 **22.** 293
23. E63D **24.** 16 **25.** 26 **26.** 173 **27.** 1001100 **28.** 1101100101 **29.** 101100101010 **30.** 100101010100001

Cumulative Review Chapters 1–16   (page 585)

1. a. 2927.40 **b.** 2930 **2.** 37.0×10^{-9} **3.** 22.2°C **4.** 7890 km **5.** $7a^2 - 18a + 11$ **6.** $-\dfrac{41}{25}$ **7.** 6.45 **8.** $4x + 6y = 27$

9. 4.9 m × 15.0 m **10.** $3(5x - 18)(2x + 15)$ **11.** $-\dfrac{1}{3}, 2$ **12.** $0.143 \pm 1.46j$ **13.** 306 cm^2 **14.** 21,700 in^3

15. 24.1° **16.** 19.9 m **17.** −0.2419 **18.** 44.0° **19.** 27.6 **20.** 28.5 **21.** 5.5

22.

Interval	Midpoint	Frequency
5.5–10.5	8	7
10.5–15.5	13	11
15.5–20.5	18	10
20.5–25.5	23	6
25.5–30.5	28	1
30.5–35.5	33	1

23. 16 **24.** 16 **25.** 10001000 **26.** 10001100111
27. 1001100100 **28.** 229 **29.** 10010101011 **30.** 1D02

Appendix B   (page 595)

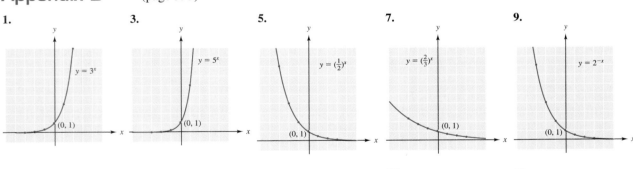

1. $y = 3^x$ (0, 1)   **3.** $y = 5^x$ (0, 1)   **5.** $y = (\tfrac{1}{2})^x$ (0, 1)   **7.** $y = (\tfrac{2}{3})^x$ (0, 1)   **9.** $y = 2^{-x}$ (0, 1)

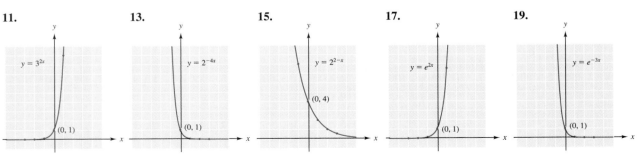

11. $y = 3^{2x}$ (0, 1)   **13.** $y = 2^{-4x}$ (0, 1)   **15.** $y = 2^{2-x}$ (0, 4)   **17.** $y = e^{2x}$ (0, 1)   **19.** $y = e^{-3x}$ (0, 1)

21. $5491.43 **23.** 77,400 **25.** $45,490 **27.** 0.0108 mA

Index